# 21st Chemnitz Seminar on Materials Engineering



## 21. Werkstofftechnisches Kolloquim

IOP Conference Series: Materials Science and Engineering
Volume 480

Chemnitz, Germany
6 – 7 March 2019

ISBN: 978-1-5108-8272-0
ISSN: 1757-8981

**Printed from e-media with permission by:**

Curran Associates, Inc.
57 Morehouse Lane
Red Hook, NY 12571

**Some format issues inherent in the e-media version may also appear in this print version.**



Printed by Curran Associates, Inc. (2019)

For permission requests, please contact the Institute of Physics at the address below.

Institute of Physics
Dirac House, Temple Back
Bristol BS1 6BE UK

Phone: 44 1 17 929 7481
Fax: 44 1 17 920 0979

techtracking@iop.org

**Additional copies of this publication are available from:**

Curran Associates, Inc.
57 Morehouse Lane
Red Hook, NY 12571 USA
Phone: 845-758-0400
Fax: 845-758-2633
Email: curran@proceedings.com
Web: www.proceedings.com

# Table of contents

## Volume 480

**21st Chemnitz Seminar on Materials Engineering – 21. Werkstofftechnisches Kolloquium**

**6–7 March 2019, Chemnitz, Germany**

**Accepted papers received: 4 February 2019**
**Published online: 5 March 2019**

### Preface



### Papers

#### Invited Talks



#### Thermal Coatings

IOP Conf. Series: Materials Science and Engineering **480** (2019) 011001 doi:10.1088/1757-899X/480/1/011001

# Preface

The Chemnitz Seminar on Materials Engineering – Werkstofftechnisches Kolloquium – is a well-established scientific meeting that is hosted, this year already for the 21st time, by the Institute of Materials Science and Engineering (Institut für Werkstoffwissenschaft und Werkstofftechnik, IWW) of Chemnitz University of Technology.

Scientists and engineers both from academia and industry attend this annual meeting to present and discuss most recent research results and current issues. The topics of the scientific sessions range from general challenges in materials engineering and testing to specific applications such as long-term stability of aluminium alloys for electrical connections. Further key topics are composite materials, surface engineering, joining technologies and additive manufacturing.

Following the encouraging trend of the last years, most of our contributors have again decided to publish their work in this international, peer-reviewed journal. Consequently, this issue of *IOP Conference Series: Materials Science and Engineering* provides an excellent overview of the main topics of the Seminar.

Chemnitz, March 2019

Prof. Thomas Lampke
Prof. Guntram Wagner
Prof. Martin F.-X. Wagner

IOP Conf. Series: Materials Science and Engineering **480** (2019) 011002 doi:10.1088/1757-899X/480/1/011002

# Open challenges in additive manufacturing – a materials engineering perspective

**T Niendorf***

Institute of Materials Engineering – Metallic Materials, University of Kassel, Mönchebergstraße 3, 34125 Kassel, Germany

* e-mail: niendorf@uni-kassel.de

**Abstract**. Additive manufacturing (AM), also termed 3D-printing, receives considerable attention since years. Due to the layer-wise nature of manufacturing using techniques such as selective laser melting/ laser beam melting (SLM/LBM), components of unprecedented complexity can be obtained. Numerous studies in the field focused on general processability of alloys. Besides relative density and tensile properties, however, data crucially needed for robust application of AM processed alloys are rare. Even if commercially available alloys such as Ti-6Al-4V, stainless steel 316L and Ni-base superalloy IN718 are employed most oftenly, processing conditions prevailing in AM lead to the evolution of unique microstructures. Thereby, it is important to note that geometry and microstructure are established simultaneously, strongly affecting each other. Based on in-depth knowledge of processing-microstructure-property-damage relationships superior performance can be realized. However, inadequate processing conditions and post-treatments can severely deteriorate properties. Future potentials and drawbacks of additive manufacturing from a materials engineering point of view will be presented and discussed. Measures to be accomplished, deduced from state-of-the-art, will be highlighted.

IOP Conf. Series: Materials Science and Engineering **480** (2019) 011002 doi:10.1088/1757-899X/480/1/011002

# Microstructural and mechanical characterization of ultrasonic metal welded NiTi shape memory alloy wires

**A Gester[*] and G Wagner**

TU Chemnitz, Professur Verbundwerkstoffe und Werkstoffverbunde, Erfenschlager Str. 73, 09125 Chemnitz

[*] e-mail: andreas.gester@mb.tu-chemnitz.de

**Abstract.** Due to their shape memory effect and pseudoelasticity shape memory alloys (SMA) find a large variety of applications as sensors as well as actors in automotive, aeronautics or medical sectors, among many others. For example, they could be used as actors for controlling the opening and closing of ventilation lids without additional energy input. An application as sensors could be a mesh of SMA wires embedded in hybrid materials for analyzing the exact position and amount of potentially occurring stresses during operation. For realizing such applications, it is necessary to interconnect these SMAs to control units, measurement devices or other SMAs whereas particularly the joining of thin wires can be a challenging task. Especially the commercially available NiTi SMAs are prone to negative changes of the transformation behavior due to an increased heat input in joining processes, e.g. soldering or fusion welding. This is also the cause for the formation of brittle intermetallics in the joint zone, thus decreasing the overall joint strength. Additionally the contact resistance at the joint zone increases due to the aforementioned intermetallic phases, which influence the behavior of SMAs with electrical activation in particular. Ultrasonic metal welding is a pressure welding technology, which allows welding materials without melting them. The joint formation occurs at low temperatures and in solid state. As a result of this, ultrasonic metal welding poses a viable solution for joining SMAs while avoiding negative impacts on the transformation properties of the joining partners. Recent investigations focus on the joining of NiTi wires with diameters of 25 µm - 250 µm to NiTi and copper sheets. For validating proper function of the wire and sheet clamping system vibrometric measurements of the fixture have been performed. The microstructure of the joints has been characterized via light as well as scanning electron microscopy. Additionally, the mechanical properties of the joints have been analyzed via micro tensile tests.

IOP Conf. Series: Materials Science and Engineering **480** (2019) 011002 doi:10.1088/1757-899X/480/1/011002

# Influence of sputtering temperature and layer thickness on the electrical performance of thin-film strain sensors consisting of nickel-carbon composite

**C Karapepas[1,*], D Nestler[2] and G Wagner[1]**

[1] Chemnitz University of Technology, Faculty of Mechanical Engineering, Institute of Materials Science and Engineering, Chair of Composites and Material Compounds, Chemnitz, Germany
[2] Chemnitz University of Technology, Faculty of Mechanical Engineering, Institute of Lightweight Structures, Endowed Chair Textile Plastic Composites and Hybrid Compounds, Chemnitz, Germany

[*] e-mail: christos.karapepas@mb.tu-chemnitz.de

**Abstract.** Hybrid laminates consisting of fiber-reinforced thermoplastic films and metallic thin sheets are successively replacing thermoset based systems due to their obvious advantages of higher formability and aptitude for mass production. In order to monitor the material under operating condition, hybrid laminates need to be equipped with smart sensor units. Artifact-free integration of commercial strain gauges into hybrid laminates is almost impossible. Therefor a new thin film strain sensor must be developed. The aim of this work is to evaluate the influence of the layer thickness as well as the elevated temperature during the sputtering process on the electrical performance of Ni-C strain sensors. The analyses of the deposited Ni-C films manufactured by means of a magnetron sputtering process were investigated the sheet resistance and the change of temperature coefficients of resistance of different layer thickness and after different sputtering temperatures. In addition, Raman spectroscopy was utilized to investigate the phase development after different sputtering temperatures. It can be seen that the gauge factor gets doubled while optimizing the layer thickness. When the sputtering temperature is increased, the graphitic phase formation is preferred and the impurities are reduced. These results will be discussed in the paper. Solution concepts will be shown.

IOP Conf. Series: Materials Science and Engineering **480** (2019) 011002 doi:10.1088/1757-899X/480/1/011002

# Development of tailored hybrid laminates: Manufacturing of basalt fiber reinforced thermoplastic orthoses with aluminum thin sheets

**C Karapepas[1,*], M Trautmann[1], A Todt[1], A Al-Obaidi[2], S Nendel[3], V Kräusel[2] and G Wagner[1]**

[1] Chemnitz University of Technology, Faculty of Mechanical Engineering, Institute of Materials Science and Engineering, Chair of Composites and Material Compounds, Chemnitz, Germany
[2] Chemnitz University of Technology, Faculty of Mechanical Engineering, Institute for Machine Tools and Production Processes, Chair for Forming and Joining Lightweight Structures and Polymer Technology, Chemnitz, Germany
[3] Cetex Institute for Textile and Processing Machines non-profit GmbH, Chemnitz, Germany

[*] e-mail: christos.karapepas@mb.tu-chemnitz.de

**Abstract.** Nowadays, orthoses are made from fiber-reinforced thermoset based composites with a high manual labor input. These thermoset based orthoses are no longer formable, which brings forth a significant disadvantage. Hence, hybrid laminates consisting of fiber-reinforced thermoplastic films and thin metal sheets can replace successive thermoset based systems due to their advantages of higher formability and the suitability for mass production. In the present work, various surface treatment methods like pickling or mechanical blasting have been used on thin metal sheets to increase the adhesive and shear strength of the produced thermoplastic based hybrid laminates. The modified metal sheets were further combined with basalt fiber reinforced interlayers to manufacture overlap samples which were used to determine the tensile shear strength. In addition, the roughness of the modified metal sheets has also been investigated. Moreover, the consolidation parameters such as pressure, holding time and temperature have been varied for the production of hybrid laminates using hot-pressing process and then the microstructural images have been recorded. Finally, the mechanical properties of the produced hybrid laminates have been tested by means of three-point bending test and the interlaminar shear strength has been analyzed.

IOP Conf. Series: Materials Science and Engineering **480** (2019) 011002 doi:10.1088/1757-899X/480/1/011002

# MOCVD of $Al_2O_3$ based electrical insulation coatings on SiC fiber bundles for sensor applications

**S Palaniyappan*, N Rudolph, H Ahmad, K Roder and G Wagner**

Institute of Materials Science and Engineering, Chemnitz University of Technology, Chemnitz, Germany

* e-mail: saravanan.palaniyappan@mb.tu-chemnitz.de

**Abstract**. For an artefact free functionalization of any fiber reinforced polymer matrix composites (PMCs), the integration of commercial strain gauges need to be replaced with highly miniaturized sensors that are dimensionally similar to the dimensions of the reinforcements. By this way, the effect of sensor integration in altering the original properties of the PMCs could be avoided to a greater extent. But whenever a single fiber is to be employed as a strain sensor for the structural health monitoring (SHM) of PMCs, an electrical insulation around the sensor fiber becomes evident, if the surrounding fibers are electrically conductive, as in the case of carbon fiber reinforced polymers (CFRPs). This electrical insulation assists the sensor fiber in delivering an accurate sensory data to the data acquisition system without any electrical conflicts. For this purpose, an aluminium oxide (Al2O3) based electrical insulation coating has been fabricated on the silicon carbide (SiC) fiber bundles using metal-organic chemical vapor deposition (MOCVD) technique. The coated bundles have been analyzed using Raman spectroscopy and X-ray diffraction (XRD) methods to characterize the phase compositions of the coatings. Also microstructural analyses have been carried out using scanning electron microscopy (SEM) to determine the coating thickness and morphological characteristics of the coated fiber bundles.

IOP Conf. Series: Materials Science and Engineering **480** (2019) 011002 doi:10.1088/1757-899X/480/1/011002

# Galvanic deposition of Iridium as catalyst for PEM electrolyzers

**J Näther[1,*], F Köster[1], T Hülser[2], U Rost[3], M Brodmann[3], D Pascal[3] and L Holtkotte[3]**

[1] Hochschule Mittweida, Fakultät INW, Fertigungs- und Werkstofftechnik, Technikumplatz 17, 09648 Mittweida
[2] Institut für Energie- und Umwelttechnik e.V., Bliersheimer Straße 58-60, 47229 Duisburg
[3] Westfälisches Energieinstitut, Westfälische Hochschule, Gelsenkirchen Bocholt Recklinghausen, University of Applied Sciences, Neidenburger Str. 43, 45897 Gelsenkirchen

[*] e-mail: naether@hs-mittweida.de

**Abstract**. The storage of energy generated from renewable sources is the key technology on the way to sustainable power industry. PEM electrolyzers could provide some advantages compared to state of the art alkaline electrolyzers due to the high gas purity and the discontinuous operation mode. The high costs for PEM electrolyzers inhibited the industrial use, but during the last years different research teams proposed some encouraging approaches to enhance these systems. Among other ideas the reduced use of expensive precious metal catalyst, like iridium or platinum, is an appropriate path to reduce the production costs. The general objective of this work is the development of cost effective electrode systems based on optimized iridium/titanium oxide layers for anodic use in PEM water electrolysis. In that topic, recent studies focus on novel supports for Ir nanoparticles with enhanced stability, to increase the electrochemical active surface area of the applied catalyst material and, thus, increasing precious metal utilization. Accordingly, Ir electrocatalyst nanoparticles were galvanostatically applied onto a sintered Ti electrode, which was coated with $TiO_2$-x nanoparticles, previously. Preliminary results indicate an improvement for activity and stability of PEM electrolyzers.

IOP Conf. Series: Materials Science and Engineering **480** (2019) 011002 doi:10.1088/1757-899X/480/1/011002

# On the performance of additively manufactured, non-stochastic, cellular lattice structures

**T Arold, F Brenne and T Niendorf**

University of Kassel, Institute of Materials Engineering – Metallic Materials, Mönchebergstraße 3, 34125 Kassel, Germany

**Abstract.** Additive manufacturing (AM) methods such as laser beam melting (LBM) and selective electron beam melting (SEBM) allow for realization of completely new designs, e.g. resource-efficient light-weight structures, which can be tailored to perfectly meet force flow.One kind of such light-weight components are non-stochastic cellular lattices. These cellular structures are characterized by a distinct deformation behaviour, i.e. stretch- and bending-dominated behaviour, depending on the cell type employed. However, in addition to the elementary nature of deformation, process-induces defects such as pores, anisotropy and the general deformation characteristics of the bulk material employed are of highest relevance. The current study focuses on the influence of the deformation mode and material conditions on the mechanical performance under static and cyclic load of non-stochastic cellular structures. Stainless steel 316L and the well-known titanium alloy Ti-6Al4V are considered. The obtained results show a clear correlation between deformation mode, local strain distribution, ductility and fatigue life.

IOP Conf. Series: Materials Science and Engineering **480** (2019) 011002 doi:10.1088/1757-899X/480/1/011002

# Mo-Si-B alloys produced by additive manufacturing

**M Krüger***

Forschungszentrum Jülich

* e-mail: m.krueger@fz-juelich.de

**Abstract**. Current research on high temperature metallic materials focuses on Mo-Si-B alloys, which are potential candidates for novel turbine materials or other high temperature applications. For structural applications in a wide temperature range, the most interesting alloy compositions in the ternary Mo-Si-B system are located within the three-phase field between the Mo solid solution phase (Moss) and the silicides Mo5SiB2 (T2) and Mo3Si (A15), known as the so-called "Berczik-triangle". Various alloy compositions within the "Berzcik-triangle" were investigated in the past, and it could be shown that such alloys have a good combination of creep strength, oxidation resistance and fracture toughness. The (near-)eutectic ternary alloys with a fine distribution of Moss,Mo5SiB2 and Mo3Si phases, displayed the best overall oxidation resistance for temperatures between 600 °C and 1300 °C. Furthermore, outstanding creep properties for near-eutectic Mo-Si-B alloys, e.g. Mo-17.5Si-8B, even above temperatures of 1100 °C, indicate the enormous potential of this type of microstructure. However, ingot processing of this class of materials is challenging due to the high melting point of Mo-Si-B materials being typically >2000 °C (eutectic Moss-Mo3Si-Mo5SiB2 alloy ~ 2000 °C). Due to this reason, different multi-step powder metallurgical processes were typically used in the past to produce dense Mo-Si-B samples under laboratory conditions. The introduction of a one-step process for this type of material via additive manufacturing (AM) or 3D printing represents an important innovation that will allow the production of complex bulk materials with various geometries. Compared to more conventional alloys like steels or titanium alloys that were shown to be satisfactorily processed by AM production routes, the specific challenge in terms of Mo-Si-B alloys is the ultra-high melting point and the accompanied problems of interlayer bonding and defects that may occur during cooling. In our approach we present an industrial gas atomization process leading to nearly spherical Mo-Si-B powder particles that were tailored for processing via laser metal deposition as a method for AM. Here we show the very first study on the printability of these near-eutectic Mo-Si-B powders and the resulting bulk microstructures after AM processing.

IOP Conf. Series: Materials Science and Engineering **480** (2019) 011002 doi:10.1088/1757-899X/480/1/011002

# Potentials and challenges of additive manufacturing – From prototyping to direct manufacturing

**T Niendorf**

University of Kassel, Institute of Materials Engineering – Metallic Materials, Mönchebergstraße 3, 34125 Kassel, Germany

**Abstract.** 3D printing, also known as additive manufacturing, offers unprecedented freedom of design already being applied in industry, e.g. cooling channels for tools, light-weight structures or functionally-integrated parts and components. In the automotive sector, metallic additive manufacturing is seen as a supplement to conventional manufacturing processes. Aim is the enhancement and use of already available technologies. Highlighting three project examples in different areas of application, different additive manufacturing processes and their process chain will be demonstrated. Additionally, the resulting challenges are presented and an insight into current and future research activities as well as potentials of additive manufacturing in the automotive environment is given.

IOP Conf. Series: Materials Science and Engineering **480** (2019) 011002 doi:10.1088/1757-899X/480/1/011002

# Crack detection in fatigued specimens with a center hole made of a Ni-base alloy

**A Butz***

BAM, Unter den Eichen 87, 12205, Berlin, DE

* e-mail: adam.butz@bam.de

**Abstract.** Measuring cracks in flat specimens with holes is a difficult task. This is especially true when no marking of the crack surface is performed and no assumption on the cracks' shape is made. Current work on the detection and characterization of cracks in such specimens is presented that is based on the use of thermographic imaging for surface crack detection and -measuring as well as multisensor data-fusion for crack shape classification.

IOP Conf. Series: Materials Science and Engineering **480** (2019) 011002 doi:10.1088/1757-899X/480/1/011002

# Fatigue behaviour of an ultra-high strength austenitic-martensitic steel treated by novel Q&P processing

**M Droste[1,*], M Wendler[2], O Volkova[1] and H Biermann[1]**

[1] Institute of Materials Engineering, Technische Universität Bergakademie Freiberg
[2] Institute of Iron and Steel Technology, Technische Universität Bergakademie Freiberg

[*] e-mail: matthias.droste@iwt.tu-freiberg.de

**Abstract.** The fatigue properties of three variants of an ultra-high strength cast stainless steel (Fe-14.2Cr-3.3Ni-3Mn-0.4Si-0.11N-0.15C, wt.%) obtained by novel quenching and partitioning (Q&P) processing were investigated. After solution annealing at 1150 °C and subsequent water quenching the steel possessed an almost fully austenitic microstructure. By cooling to different cryogenic temperatures (-20 °C, -60 °C, -120 °C), different fractions of alpha'-martensite (45 vol.%, 39 vol.%, 22 vol.%) were generated in the steel matrix. Afterwards, the retained austenite was stabilized by an annealing step at 450 °C for 3 minutes provoking the diffusion of interstitials carbon and nitrogen from the supersaturated alpha'-martensite to austenite. The three steel variants were cyclically deformed in total strain controlled tests with strain amplitudes in the range of 0.25 % to 1.2 %. As expected, the -120 °C variant exhibited the highest stress amplitudes in the beginning of the tests whereas the highest fractions of fatigue-induced alpha'-martensite were formed in the -20 °C variant due to the lowest austenite stability. Overall, the -120 °C variant exhibited the best mechanical properties and the best fatigue life. Furthermore, the formation of stacking faults and deformation bands was observed, with the latter acting as nucleation sites for alpha'-martensite.

IOP Conf. Series: Materials Science and Engineering **480** (2019) 011002 doi:10.1088/1757-899X/480/1/011002

# Precipitation behavior and corrosion resistance of nickel-free, high-nitrogen austenitic stainless steels

**P Rosemann[1,*], C Müller[2], S Roßberg[2], P Pensel[2], T Halle[2] and A Burkert[1]**

[1] Bundesanstalt für Materialforschung und -prüfung (BAM), Unter den Eichen 87, 12205, Berlin, Germany
[2] Institut für Werkstoff- und Fügetechnik, Otto-von-Guericke Universität Magdeburg, Universitätsplatz 2, 39106, Magdeburg, Germany

[*] e-mail: paul.rosemann@bam.de

**Abstract**. Nickel-free, nitrogen alloyed austenitic stainless-steels, with about 19 wt.-% Mn and 0,8 wt. % N, are an interesting alternative to classic CrNi austenitic stainless steels due to their superior mechanical properties (Rm > 900 MPa, A5 > 50 %, Av > 350 J) in the solution annealed condition. The formation of chromium-rich nitrides during suboptimal heat treatment, processing or application leads to an inhomogeneous distribution of alloying elements in the microstructure, which may reduce the corrosion resistance. Consequently, an accurate knowledge of the sensitization behavior is indispensable for the use of nickel-free, high-nitrogen austenitic stainless steels. The relationship between artificial aging, phase formation and corrosion resistance was investigated on the alloys X8CrMnN18-19 (1.3815) and X8CrMnMoN18-19-2 (1.4456), both alloyed with 0.8 wt.-% Nitrogen, in the present work. The microstructural evolution was studied by SEM and the corrosion resistance was characterized with the electrochemical potentiodynamic reactivation (EPR) and the KorroPad-test. Both alloys showed increased susceptibility to pitting corrosion within critical aging parameters. Finally, a sensitization diagram was described successfully for both alloys showing the positive effect of molybdenum on the precipitation and sensitization behavior.

IOP Conf. Series: Materials Science and Engineering **480** (2019) 011003 doi:10.1088/1757-899X/480/1/011003

# Peer review statement

All papers published in this volume of *IOP Conference Series: Materials Science and Engineering* have been peer reviewed through processes administered by the proceedings Editors. Reviews were conducted by expert referees to the professional and scientific standards expected of a proceedings journal published by IOP Publishing.

IOP Conf. Series: Materials Science and Engineering **480** (2019) 012001 doi:10.1088/1757-899X/480/1/012001

# A new generation of highly flexible and soft tools for high precision surface finishing with embedded micro-diamonds

**M Dvorak[1,*], F Mielemeier[1] and H Adams[2]**

[1] P&S Powder and Surface GmbH, Salzkotten, Germany
[2] Horst Adams, ADAMCO Inc., Houston, TX, USA

[*] e-mail: m.dvorak@powderandsurface.de

**Abstract.** Fully automated and process-integrated, high-quality surface polishing is still an extremely challenging task. In particular, the final polishing of complex surfaces without affecting the part geometry often remains a problem. A newly developed technology allows the direct embedding of diamonds into the surfaces of polymer and elastomer polishing tissues and sponges without the use of any adhesives. The same technology can be applied for making micro-diamond enhanced brushes with very fine and highly flexible filaments for polishing surfaces of complex geometries as well as customized polishing tools for direct integration in processing machines. This article highlights some of the possibilities as well as limitations of a new generation of flexible, diamond-coated polishing tools.

## 1. Introduction

The constant request for miniaturization of mechanical components, higher working pressures at injectors and valves, as well as the demand for extended life time, require ever better surface finishings. Typical examples are valve surfaces, gear edges, cutting edges, chip flutes of tools and pressing surfaces of molds. In order to increase the efficiency of series production, it is often desirable to integrate the polishing step directly into the production process. In order to provide solutions to this demand, we have developed a new process for the direct embedding of micro- and submicron diamonds into polymer surfaces, allowing the making of novel technical brushes with highly flexible filaments as well as nanoparticle enhanced elastic precision finishing tools made of foams.

## 2. State of the art

### *2.1. Technical brushes*

Brushes are particularly suitable for polishing complex shaped surfaces. Due to the flexibility of the filaments the effects of varying contact conditions can be much better compensated for as compared to other finishing methods. After getting into contact with the workpiece surface, the filaments bend toward the plane of rotation of the polishing tool, following the principle of the least resistance and then perform a grazing motion over the workpiece surface [1]. Brush filaments usually consist of plastics materials like PA, polyester PBT and LDPE. At elevated working temperatures of up to 200 °C the more temperature-resistant though more expensive high-performance polymers PPS and PEEK are preferred. Abrasive brushes made from compound materials containing abrasive particles have been used for many years as a highly flexible machining method in industrial production. Typical application areas are surface cleaning, reprocessing components, deburring, edge rounding of cutting edges, surface pre-treatment, surface activation or for polishing complex surfaces made of metals, hard metals and ceramics [2-4].

**Figure 1.** Typical brush shapes: roller, brush and cup brushes.

**Figure 2.** Mini brushes. From left: brush, pot- and pipe brush.

Abrasive plastic brush filaments are usually made by extrusion. The plastic granules fed to the extruder usually consist of thermoplastics like PET, PA or the more temperature-resistant but considerably more expensive PPS and PEEK. Hard and abrasive materials like $Al_2O_3$, SiC, or diamonds are homogeneously co-extruded and embedded in the polymer matrix at a typical volume ratio of 20 to 30 %. Higher volume ratios will result in too brittle filaments. Figures 3 and 4 show a scanning electron micrograph of a typical filled filament. It can be seen, that a substantial portion of the abrasive particles is completely enclosed by the polymer matrix.

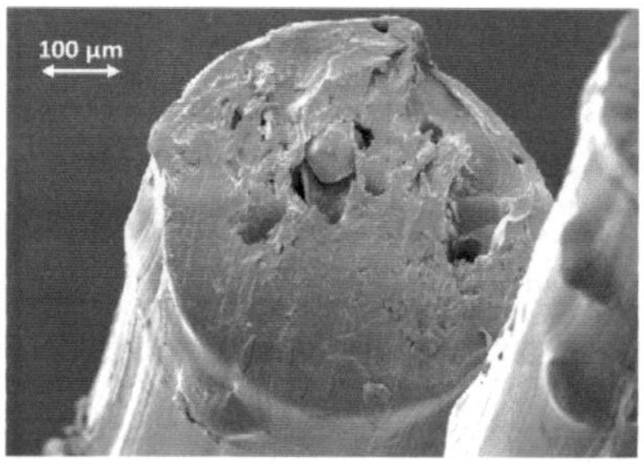


**Figure 3.** SEM image of a particle-filled brush filament. Head view.

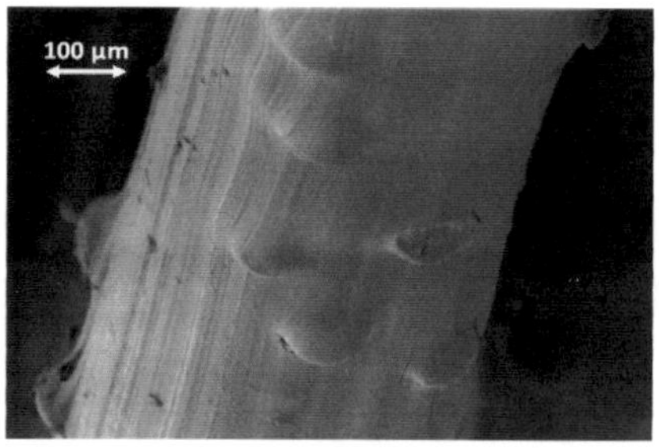


**Figure 4.** SEM image of a particle-filled brush filament. Side view.

Filled filaments are more brittle and stiffer than unfilled filaments. In most cases, brushes are made up of tufts of filaments rather than single filaments. For tuft stiffness [5]:

- The stiffness (resistance to bending) of a tuft of round filament is proportional to the 2nd power of the filament diameter.
- The stiffness of a tuft of round filament is inversely proportional to the 3rd power of the length of that tuft.

Therefore, the advantage of easy contour following of highly flexible filaments is consequently limited when using brushes with filled filaments. Furthermore, wear of the brush causes the change of tuft length and, as a consequence, dramatic change of the brush stiffness. Latter makes automated application of brushes made of filled filaments for high end finishing operations nearly impossible. A possible solution would be to make filled bristles of smaller diameter filaments. However, the practical feasibility of brushes made of very small diameter filled filaments is limited by their commercial availability. The minimum diameter is 0.3 mm. Another disadvantage of filled filaments is that the abrasive particles must get laid open from the polymer matrix by the polishing process itself, requiring

IOP Conf. Series: Materials Science and Engineering **480** (2019) 012001 doi:10.1088/1757-899X/480/1/012001

a high relative speed of 20 – 30 m/s between brush and workpiece. The stiffness of a brush increases with rotation speed. Although, the high cutting speeds are needed to partially melt the surface of the filaments. This ensures a continuous release of the abrasive particles. Polishing of the surface and grinding of the tool (brush filaments) take place simultaneously. The resulting thin layer of melted polymer in the contact zone reduces the efficiency of the abrasive particles, especially for particle sizes below 10 microns. The melting of the matrix can also lead to a contamination of the workpiece surface with polymer material, which must be cleaned off subsequently.

### *2.2. Foams*

For manual and mechanical polishing of larger surfaces mainly soft and highly elastic foams made of ethylene vinyl acetate (EVA) or polyethylene (PE) are used as the polishing tool substrate because their high elasticity and porosity provide for an excellent distribution of the contact pressure. More expensive polishing pads are made of PU (polyurethane) and are preferably used in the optical industry for fine polishing of optical glasses [6].

If foams are to be outfitted with one or more abrasive surfaces it can be done by coating the foam with phenolic or epoxy resins and subsequently electrostatically depositing abrasive particles onto the surface before the resins are cured. Another method is to just paste a previously hard particle coated fleece or other type of PA fabric to the foam surface. The most critical part of this setup is the interface between the foam and the abrasive layer since in general it tends to reduce the flexibility of the tool which in turn reduces its contour following capability.

As it is the case with abrasive brushes made of filled filaments, also in coated foams the abrasive particles are wrapped-in or encapsulated by the adhesives and become exposed to the open only by friction during the polishing process. Compared to brushes, the variety of commercially available foam polishing tools is relatively small and limited mainly to pads and sheet products of different thicknesses causing that elastic foams coated with hard materials are currently being used less frequently in industrial applications.

## 3. Adhesive-free diamond coated brushes and elastic foams

In this program we have developed a new and patented method for the direct, adhesive and binder-free implantation of diamonds onto the surface of polishing tools. The main focus of the development was to combine the advantages of flexible and elastic tools with the higher removal rate of tools with bonded / guided particles. In the here presented new and revolutionary setup micro-diamonds are directly implanted into the surface and the near-surface areas of brush filaments and elastic foams thereby firmly attaching the diamonds to the carrier material surface. The inner composition of the carrier material remains totally unaffected in its mechanical and technological properties since there are no macroscopic or microscopic interface layers necessary in this setup.

### *3.1. Construction and properties of diamonds implanted brushes*

A plastic fiber or a brush filament having diamonds implanted only on the surface has the same high flexibility as an unoccupied filament. Thus, very soft brushes can be designed that can easily follow the changing contours of complex surfaces during automated brushing without damaging them. So far, fibers with a minimum diameter of 15 µm have been successfully surface-implanted with diamonds. Typical fiber materials used are thermoplastics like PET, PA, PPS and PEEK. However, the particle size should not exceed 25% of the filament diameter. With decreasing filament diameter, the particle load becomes even higher compared to filled filaments. For a particle diameter of 25 % of the filament diameter the particle load can reach nearly 55 vol%. Additionally, the diamonds are free standing and ready for immediate use.

Figure 5 illustrates the good surface coverage and excellent adhesion of diamonds deposited using the newly developed and patented implantation method. It can also be seen that the diamonds are mostly exposed to the open and that the smoothly rounded filament tips show a good uniformity. The exposed abrasive particles allow for dry polishing of surfaces with the brushes moving at relatively low relative speeds while at the same time ensuring high abrasion rates. The relative speed is in a very beneficial range of only 0.2 to 5 m/s as compared to 20 to 30 m/s when using brushes with filled filaments. At these low speeds only an extremely low level of unwanted friction abrasion and wear takes place, rendering any subsequent cleaning work virtually unnecessary.

**Figure 5.** PA fiber surfaces implanted with monocrystalline diamonds.

Depending on the filament length and diameter, polishing work can be carried out even in deep and recessed areas down to the centimeter range on surfaces with complex shapes. Typical penetration depths are in the range of 0.3 to 5 mm. A further advantage of the high flexibility of the filaments is that smaller fluctuations in the tool positioning do not have a decisive influence on the quality of the surface treatment. Positioning errors in the millimeter range merely lead to a slight bending of the filaments or the filaments just yield to the incoming object (for surfaces with fine topological structures in the millimeter range). Only in the effective range (0.2 – 3 mm penetration depth) does the relative position of the filaments towards the surface enable active engagement of the diamonds. This property allows even the processing of the interior of comparatively deep and narrow grooves and also undercuts without affecting the geometry of the component surface.

*3.2. Structure and properties of diamond implanted foams*

So far, two different foam types have been investigated for their suitability for adhesive-free occupation with diamonds. On the one hand mixed-cell polyurethane (PUR) foam elastomers and on the other hand open-cell, medium-hard foam made of low-density polyethylene (PE-LD), which is often used for final polishing of paint coated surfaces. The PE-LD foam shows a net-like structure while the PU foams are more similar to a bubble structure (see figure 6). Tools made of PUR foams could be coated with diamonds with grain sizes starting from 0.1 µm up to 200 µm, while the fine mesh structure of the PE-LD foam prevented the deposition of diamonds larger than 60 µm.

The open-cell structure of the PE-LD foam was found to be beneficial for the dry polishing of softer materials like paints, polycarbonate, Al, Cu, stainless steel as well as Titanium and its alloys. Due to the smaller contact area and because of the lower coefficient of friction (µ against steel is between 0.25 up to 0.3) compared to the polyurethane foam (µ against steel = 0.5), the surface undergoing polishing is heating up to a much lesser extent. Also, this structure favors the uptake of a liquid cooling medium. The only disadvantage is the early softening point of about 80 °C. However, surface temperature measurements during dry polishing of polymer-based gel coats on GFK-structures never exceeded temperatures of 50 °C (polishing was done at RT).

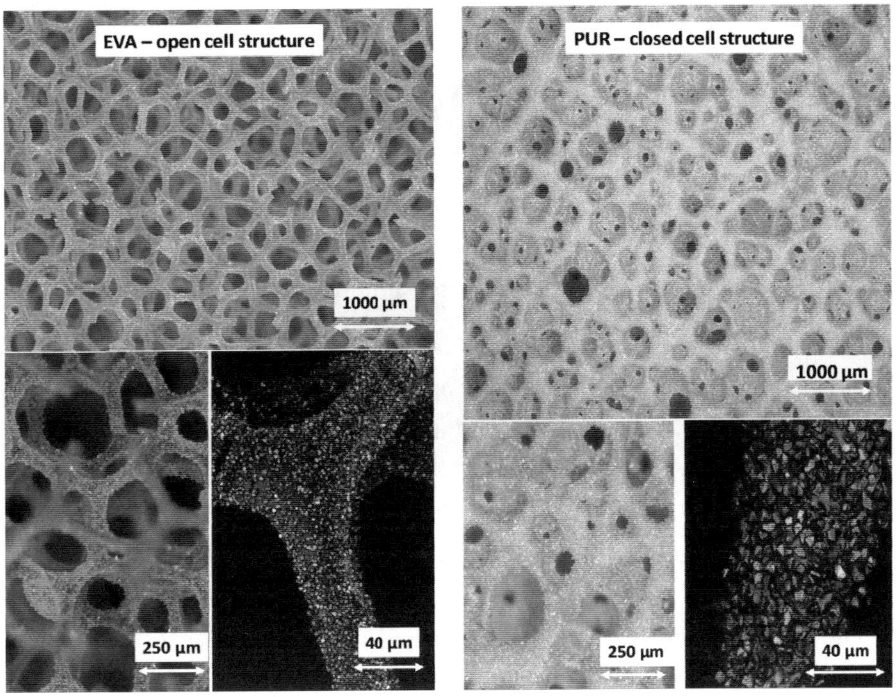


**Figure 6.** Surface of monocrystalline diamond implanted foams; grain size 6 - 12 µm; left hand – open cell structured PE-LD foam; right hand – closed cell structured elastomer PUR foam.

The mixed cell structure and the rubber like behavior of the tested PUR foams showed a higher dimensional stability and abrasion resistance than the PE-LD foams. When finishing hard materials (hardened steels, hard metals and ceramics) and for the finest finishing quality, the better results were obtained with PUR foams and diamond grain sizes between 3 – 30 µm. However, experimental work is still in progress. So far, we have confirmed that both foam types can be used for dry as well as wet polishing, however quantitative comparative investigations are still in progress.

First attempts to produce defined geometric polishing shapes according to FEPA standard "Shapes and dimensions for superabrasive wheels" were carried out [7]. In particular, the elastomer PUR foam allowed the economic production of very small tools with a minimum diameter down to 1 mm, made from large samples of plate and block material. Using shaped samples cut out of large plates increases the reproducibility and homogeneity of the foam properties, which is a crucial requirement for automated industrial applications of polishing tools.

## 4. Applications

The following paragraph gives some application examples for the use of highly flexible and elastic polishing tools manufactured following the above described methods. Figure 7 shows the final polishing result of a watch case made of TiC-Ni carbide. The flexible filaments provide for a non-destructive yet effective polishing of the complex shaped surface. Table 1 shows the roughness parameters $R_a$ and $R_z$ of a TiC-Ni Cermet watch case before and after a two-minute brushing treatment with our technology in comparison to an eight hours conventional frictional grinding. The filament length is 25 mm with a filament diameter of 250 µm and a brush diameter (roll brush) of 120 mm. Synthetic, mono-crystalline diamonds with a grain size of 3 – 6 µm were implanted on the surface of the PA 6.6 filaments. The relative brushing speed was 1.3 m/s. The penetration depth was in the range of several millimeters. Results with a finishing surface quality comparable to a two minutes brushing using our technology were achieved by conventional frictional grinding only after a processing time of 8 hours in an aqueous SiC suspension (1 – 3 µm) with cylindrical ceramic grinding wheels.

**Figure 7.** Soft brush finishing of a watch case (grey subject left hand); rotating brush (right hand).

**Table 1.** Comparison of surface roughness of a TiC-Ni cermet watch case, finished by different techniques.

| | $R_z$ in µm | $R_a$ in µm |
|---|---|---|
| Initial state | 0.65 | 0.09 |
| Brushed using new technology, processing time 2 min | 0.33 | 0.04 |
| Conventional friction grinding, processing time 8 h | 0.45 | 0.04 |

*4.1. Removal of droplets*

PVD hard coatings, in particular applied by ARC-PVD or magnetron sputtering, often have process related microscopic irregularities on the component surface, so-called droplets. To ensure the full functionality of the coating, these droplets must be removed. In case of hard coatings (e.g. TiC, TiN, DLC) this might be a challenging task. Figure 8 shows the AFM surface profile of a DLC hard coating before and after brushing using our technology. Brush type: roller brush, OD 220 mm, B 80 mm, filament length 250 µm, filament diameter 200 µm, filament type PA 6.6, diamond grain size 8 – 12 µm. The size of the surface area to be treated was 5 x 20 $mm^2$ with a positive crown radius of about 100 mm. The decisive factor in this application is that the geometry of the component and the functional thickness of the coating have to remain unchanged. Our brushing process solved the problem. The soft filaments guarantee for protecting the surface from any damaging while the bonded diamond grain has enough energy to remove even hard microscopic droplets.

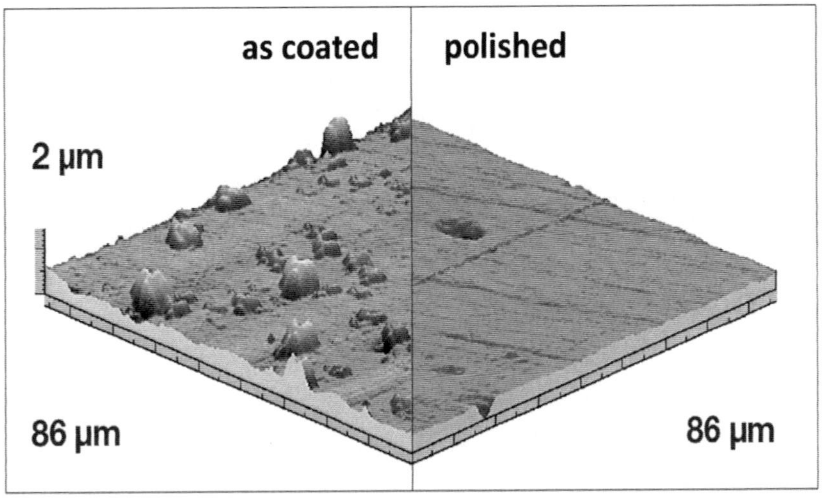


**Figure 8.** AFM measurement of a DLC coating before and after brushing. Brush advance 2 mm; Relative speed 2.3 m/s; Processing time 20 s.

The problem of increasing the load bearing of smaller components of complex geometry is a constantly recurring challenge. Figure 9 shows the photograph of a PUR foam tool with a head diameter of 2 mm. This polishing pen is intended for integration into a machining center. The Pt value of 100Cr6 valve seats could be reduced from 0.65 µm to 0.16 µm within 5 seconds while the geometry of the valve seat remained unchanged within the required tolerance range. The same tool coated with 10 µm diameter diamonds was used to decrease the surface roughness of a finely structured injection mold made of tool steel. The original surface roughness of 150 nm $R_a$ was reduced to 5 nm $R_a$.

**Figure 9.** Polishing pen with head made of diamond-coated PUR foam; monocrystalline diamonds, grain size 8-12 µm.

*4.2. Cutting edge rounding*
Other common problem areas are defined edge rounding, polishing of the chip flute, as well as the reduction of the chipping of tools. All these parameters, when optimized through proper polishing, can contribute to a significant increase in lifetime of the workpiece. The flexible brush filaments of the brush technology presented here in combination with the further advantage of the freestanding diamonds allows for damage free high processing speeds while simultaneously enabling machining of complex workpiece geometries like for instance HM miniature tools. For example, a radius change of 1 µm/s could be achieved on HM cutting edges with a radius of 5 µm. The roller brushes used were of the same construction as described in section 4.1.

Figure 10 illustrates the possibilities for edge processing of an HSS tool. Also for this task a roller brush was used as described in section 4.1. The positioning was done in the millimeter range in order to ensure optimum polishing of the flutes.

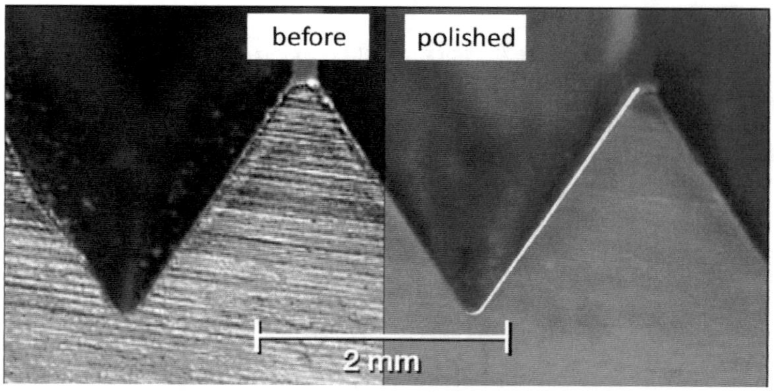


**Figure 10.** Edges of an HSS tool before and after brushing.

### *4.3. Polishing gel coats*

While the previous examples were predominantly hard surface treatments, the next examples also demonstrate the benefit of using these new tools for the treatment of soft materials.

For instance, polishing weathered gel coats on GFRP surfaces with diamond-implanted highly elastic foams enabled an increase in processing speed by a factor of 2 to 4, compared to the usual processing with uncoated polishing sponges in combination with suspensions. The pads were made both of soft polyurethane foam as well as PE-LD foam, implanted with diamond grain sizes of 0.5 µm, 1 – 3 µm and 8 – 12 µm. Even heavy matting and gray haze could be removed in a single processing step. A comparable surface quality can be achieved using the conventional procedure only with at least two sequential operations, i.e. pre-grinding (use of abrasive suspensions with P1500 and P2000 grits) and polishing (use of abrasive suspensions with P3000). We demonstrated that for achieving the same surface quality using our new technology a sponge covered with diamonds in the grain size range 8-12 µm is sufficient. The abrasion rate could be influenced by the applied pressure. Relatively low speeds in the range of 300-500 rpm were recommended for polishing pads with a diameter of 150 mm. Although a subsequent use of foam pads with even finer grains is no longer necessary it can lead to an even higher gloss. In general, the polishing was carried out dry. Paint abrasion and debris trapped in the sponge pores could be easily removed from the sponge in a water bath without any detergent necessary. A decrease in the abrasion rate over the duration of the test series was not observed. It should be noted that until now surfaces up to a size of 50 $m^2$ have been polished successfully with a single pad. Only extremely weathered surfaces or paint surfaces with deep scratches required an additional polishing step. A PUR sponge proved to be particularly suitable with diamonds of grain size D63. Another advantage of dry polishing is that the surface stays free of liquid chemicals and other contaminants. Immediately after the polishing the surface is highly activated, so that subsequently applied protective layers show a much better surface bonding. Previous tests reveal an up to a factor 3 longer lifetime of the protective coatings (sealings). Also, the surface treatment can be easily done in direct sunlight and both at elevated temperatures as well as below freezing. The polishing can be done even in saltwater conditions due to the good dust trapping properties of the pads in combination with the dry processing (no liquid chemicals).

## 5. Conclusions

Direct and adhesive-free embedding of diamonds into the surface of soft, flexible and highly elastic polishing tools such as brushes and moldings made of foams opens up new ways for the efficient finishing of complex and finely structured surfaces. Current applications are only the first steps into a very wide range of possible future applications.

**Acknowledgments**

Part of this work was done in close cooperation with Reishauer AG, Switzerland.

**References**

[1] Przyklenk J 1985 *Bestimmen des Bürstverhaltens anhand einer Einzelborste* IPA-IAO Forschung und Praxis vol 87, ed, Przyklenk H J Warnecke und H J Bullinger (Berlin, Heidelberg: Springer) pp 105 - 110 DOI 10.1007/978-3-642-82628-3
[2] Beier H M 1990 *Industrielles Entgraten; Theorie, Praxis, Probleme, Lösungen* (Berlin: Verlag Technik) ISBN 3341008152, 9783341008157
[3] Landenberger D 2008 Feinbearbeitung automatisiert: Bürstspanen im Werkzeug- und Formenbau *Form + Werkzeug* **2**
[4] Tikal F, Bienemann R und Heckmann L 2009 *Schneidkantenpreparation* Berichte aus Industrie und Forschung (Kassel: kassel university press GmbH) ISBN-10: 3899584945
[5] DuPont 2014 *Filament Performance in Brushes* [DuPont Online brochure] p 6
[6] Wächter D 2017 *Poliermittelträger für das Polieren optischer Gläser* Ergebnisse aus der Produktionstechnik D82 Diss RWTH Aachen University (Aachen: Apprimus Verlag) ISBN 972-3-86359-582-1
[7] FEPA 2010 Standard: Shapes and dimensions for precision superabrasive wheels (Paris: Federation of European Producers of Abrasives) 2010
[8] Fischer K Sbicego S David S G 2011 Wissenschaftliche Dokumentation Proxyt® (Ivoclar Vivadent AG, FL - 9494 Schaan)

IOP Conf. Series: Materials Science and Engineering **480** (2019) 012002 doi:10.1088/1757-899X/480/1/012002

# Temperature distribution on thermally sprayed heating conductor coatings

**K Bobzin, M Öte, M A Knoch and I Alkhasli***

RWTH Aachen University/ Surface Engineering Institute

* e-mail: alkhasli@iot.rwth-aachen.de

**Abstract.** Cavities in injection moulding can be coated to increase the corrosion and wear resistance or to adapt the heat flux through the cavity wall. The latter is to be achieved by thermal spraying of $TiO_2/Cr_2O_3$ on the surface and its electrical heating. A thermally sprayed $Al_2O_3$ coating serves as an electrical insulator between the steel substrate and the $TiO_2/Cr_2O_3$ coating. In this study, the feasibility of a homogeneous surface heating is investigated. The heating behaviour was analysed using a thermographic camera. Depending on the process parameters during the coating process and used electrical current, inhomogeneity in the temperature distribution was detected. The observed inhomogeneity was distributed in linear patterns in form of "hot lines" perpendicular to the electrical current. To identify the root cause of the observed inhomogeneity, numerical modelling was utilized. The results of the simulations were validated with the experimental measurements. It was found that the cracks in the coating microstructure were the prime cause of the temperature increase that was distributed in linear patterns. Moreover, even though the crack distribution does not have a directional preference, the pronounced heating of the cracks aligned in the perpendicular direction to the electrical current created the temperature pattern.

## 1. Introduction

Thermal spraying has been used to apply coatings that protect parts from corrosion, wear and thermal loads for a long time. In recent years, novel applications of thermally sprayed coatings such as sensors, electronic components and thermoelectric devices have been developed [1, 2]. This work focuses on thermally sprayed coatings as an electric heater for the application in the injection moulding tools. **Figure 1** shows a coating system concept, consisting of a heating element coating surrounded by two $Al_2O_3$ coatings for electrical insulation. The application of thermal spraying to produce electrical heaters on part surfaces with vacuum plasma spraying, air plasma spraying as well as high-velocity oxygen fuel spraying has been analysed in the literature [3–6]. The feasibility of using metallic materials, such as Ni, NiCr and NiAl, as a heating element up to 600 °C has been established. In this work, a semiconductor material was chosen for the heating element coating for its advantages over metallic materials. Due to higher electrical resistance of semiconductors, the same amount of heating power can be achieved with less electrical current flowing through them. This would allow for thinner electrical cabling that connects the heating element with the power source, which could simplify the integration of the heating element in the injection moulding tool. Moreover, the affinity of semiconductors with ceramics leads to a better bonding between the two and their similar thermal expansion coefficients ensures lower residual stress at the interface of the insulator and the heating element coating.

IOP Conf. Series: Materials Science and Engineering **480** (2019) 012002 doi:10.1088/1757-899X/480/1/012002

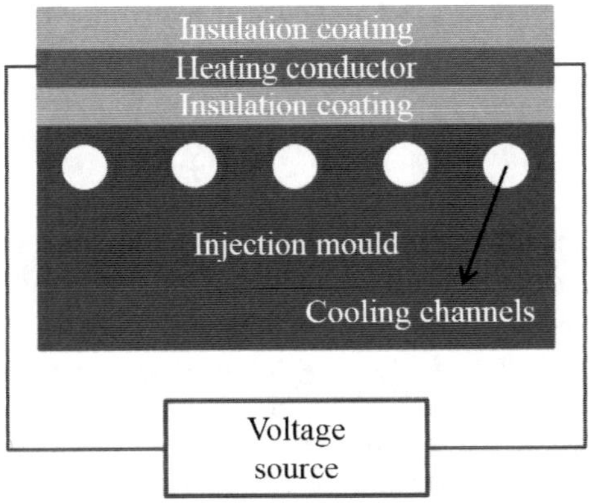


**Figure 1.** Schematic description of electrical heater coating concept for the application in injection moulding.

Prototypes were manufactured on steel substrates and electrically heated while their heating behaviour has been investigated using a thermal camera. The material used as an electric heater is $TiO_2/Cr_2O_3$. An $Al_2O_3$ coating was applied for insulation on the surface of a steel substrate. Analysis of the surface temperature of the coating by thermal imaging during its electrical heating has shown varying level of homogeneity. The factors influencing the temperature distribution on the coating surface are the process parameters used during the application of the coating, the coating thickness and the rate at which the electrical heater coating has been heated. Using appropriate spraying parameters and heating the coating at moderate to high rates, it is possible to achieve a homogeneous temperature distribution, which is a desirable outcome for the application in injection moulding [7–9].

Certain coating and operating parameters resulted in appearance of spots and lines with significantly higher temperatures compared to their immediate vicinity. This study aims to identify the root cause for this inhomogeneity so that they can be prevented. The temperature inhomogeneity exhibited a directional preference depending on the direction of the electrical current. Numerical modelling of the resistive coating heating was used to investigate the underlying physical phenomena that causes the directional preference of the temperature inhomogeneity. The numerical model is validated experimentally, for the case of the coating heating up with a homogeneous temperature distribution, by comparing the peak temperatures. In the case of the coating system which shows hot lines in its temperature distribution, only a qualitative comparison between experimental and numerical results was feasible. In the next sections, the heating conductors and the thermal spraying process will be briefly introduced. Afterwards, the experimental setup for the coating application and analysis is presented. The obtained results will be presented and discussed and finally, the conclusion is drawn.

### *1.1. Electrical heating of heating conductors*

In an electrical heater, the electrical energy ($E_{el}$) is converted into the heat energy ($Q_W$). The correlation between power and heat energy is defined by the following equation:

$$Q_W = E_{el} = P\ t \tag{1}$$

where P denotes the power of the electrical heater and t the duration of the heating period. The power of the electrical heater is determined by the following equation:

$$P = U\ I = \frac{U^2}{R} = R\ I^2 \tag{2}$$

where U corresponds to the effective electrical voltage, I to the electrical current and R to the electrical resistance. For a heating element with a known electrical resistance, the electrical power can be calculated based either on the electrical voltage or the electrical current. The electrical resistance of a heating element depends on material properties and its shape:

$$R = \rho(T)\,\frac{l}{A} \tag{3}$$

where ρ(T) represents temperature dependent resistivity, l the length of the element in the direction of the electrical current and A the cross-sectional area of the element perpendicular to the electrical current. The ratio of the element's length to its cross-sectional area is referred to as the shape factor.

The material used in this work, $TiO_2/Cr_2O_3$, is a semiconductor whose electric and thermoelectric properties can be influenced by controlling the process parameters during the spraying process [10, 11]. The electrical conductivity of a semiconductor generally increases with increasing temperature. This is explained by the increase of intrinsic movement of electrons at higher temperatures, which increases the probability of an electron leaving its orbit. With the help of the heat energy introduced into the semiconductor, an electron is moved to a higher energy level, which is called the conduction band, leaving an electron hole in the so called valance band. The energy difference between the energy levels defines the band gap. The electrons from the neighbouring atoms tend to fill the electron hole, leaving another electron hole in their place. The movement of the electron holes creates a movement of positive charge across the material. Materials where the carrier of the electrical charge is the electron holes are called the p-type semiconductors. It is possible to improve the conductivity of a semiconductor by doping its crystal lattice with impurities. In the case of $TiO_2$, the electrical conductivity is induced by oxygen vacancies. $Cr_2O_3$ is used to increase the stability of the sub-stoichiometric oxides [12].

### *1.2. Thermal spraying*

The heating element as well as the insulator coatings in this work were manufactured using air plasma spraying. It is a process variant of thermal spraying, where molten particles impact on the part surface and undergo rapid solidification. During this rapid solidification, intrinsic residual stresses build up in the coating. The resulting coating has a microstructure with varying porosity, depending on the impact characteristics of the particle and therefore, on the process parameters. Due to different thermal expansion coefficients of the coating and substrate materials, extrinsic residual stresses build up as the coating and the substrate cool down. The intrinsic and extrinsic residual stresses superpose and can have a negative impact on the mechanical characteristics of a coating system. A higher coating thickness leads to a residual stress increase and can cause cracks in the coating microstructure [13].

## 2. Experimental setup

### *2.1. Thermal spraying of the coatings*

The coating concept developed for the injection moulding application, as shown in **Figure 1**, consists of a heating element coating which is encased by two insulation coatings. The coating system that was investigated in this work, however, consists of a single electrical insulator coating between the substrate and the heating element coating. The insulation coating on top of the heating element coating is necessary to electrically protect the molten plastic material from the heating element material and to chemically protect the heating element from the molten plastic. Since the temperature distribution of the heating element is investigated in this work, the top layer of coating system (**Figure 1**) of has not been considered. This enables a closer investigation of the temperature distribution and the failure mechanisms of the heating element coating. The process parameters that were used to manufacture the coating system are listed in **Table 1**.

**Table 1**. Spraying parameters of the insulation and the heating element coating.

| Parameter | Insulator/Heating element |
|---|---|
| Plasma generator | F4MB-XL |
| Argon [SLPM] | 42 |
| Hydrogen [SLPM] | 9 |
| Electrical current [A] | 600 |
| Net power [kW] | 36 |
| Spraying distance [mm] | 100 |
| Conveyor disc rotation speed [%] | 27/20 |
| Injector Argon [SLPM] | 6,5 |
| Meander velocity [mm/s] | 875/1,000 |
| Meander step [mm] | 5 |

The tooling steel substrates used in this work have the size of 102x61x9 $mm^3$. First, the substrate surfaces have been prepared by grit blasting and coated with an $Al_2O_3$ powder (GTV 40.05.0 Aluminium oxide 99.9%) with particle size fraction of D = -25+5 μm. The resulting insulator coating had the thickness of t = 250 μm and covered the entire surface of the substrate. On top of the $Al_2O_3$ coating, a $TiO_2/Cr_2O_3$ powder (Ceram 80/20 molten) with particle size fraction of D = -25 + 5 μm was used to apply the second layer. The resulting heating element coating had the thickness of t = 110 μm and covered an area smaller than the surface of the substrate with the size of 80x50 $mm^2$. This was necessary to eliminate a possible contact of the heating element coating with the substrate along the substrate edges and achieved by masking the edges on the insulator surface prior to spraying. The microstructure of both coatings can be seen in the cross-section image, which is shown in **Figure 2**. The final layout of the coating system can be seen in **Figure 3**. Another coating system has been manufactured with the same set of process parameters, except with the thickness of the heating element coating of t = 400 μm. Higher thickness of the second coating system was necessary to ensure residual stresses large enough to initiate crack formation. In contrast to the t = 110 μm thick coating, the coating with the thickness of t = 400 μm exhibited temperature inhomogeneity in form of the hot lines in preliminary tests (**Figure 6**). This is an expected behaviour due to the forced crack formation.

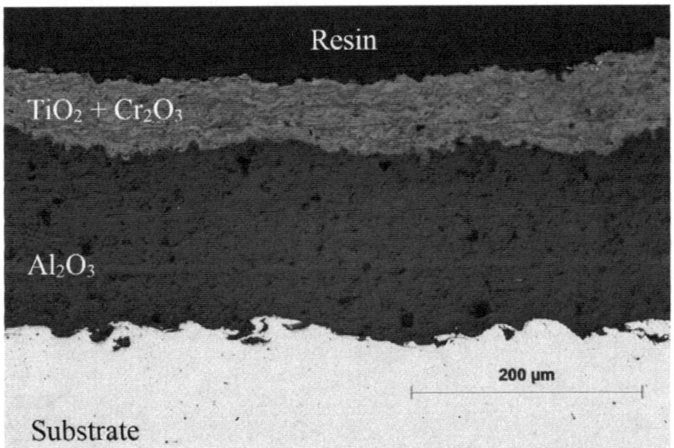


**Figure 2.** Cross-sectional picture of the microstructure of the insulation and heating element coatings.

IOP Conf. Series: Materials Science and Engineering **480** (2019) 012002 doi:10.1088/1757-899X/480/1/012002

*2.2. Electrical heating and thermal imaging*

The experimental setup for the electrical heating of the coatings is schematically shown in **Figure 3**. The contact plates have been clammed on the sides of the heating element coating and connected to the voltage source. The distance between the contact plates is L = 70 mm, which is shorter than the full length of the heating element coating L = 80 mm. The power supply unit CHROMA 61604 (dataTec) was used as the voltage source, which can supply both alternating and direct currents.

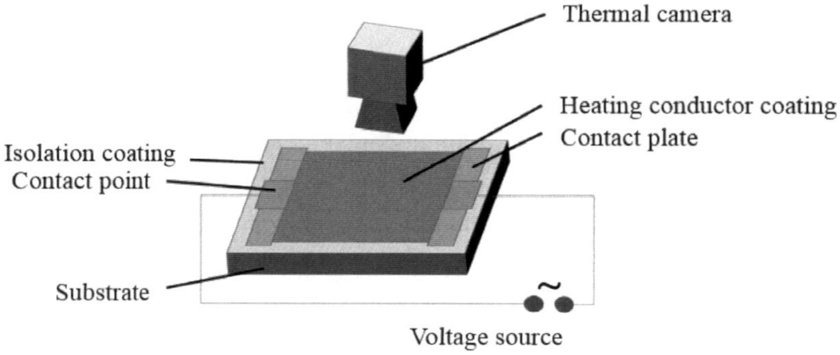


**Figure 3.** Schematic description of the experimental setup for the heating and thermal imaging measurements.

In this work the alternating current was used, which can supply a maximum current magnitude of I = 8 A due to the device limitations. The electrical voltage is given to the power supply as an input, whereas the resultant electrical current flowing through the heating element coating recorded on-line. Since the semiconductor coating ceramic is a better electrical conductor at higher temperatures, as measured and shown in **Figure 4**, the electrical current increases as the coating system, the substrate and the mounting system which is in contact with the substrate, gradually warmed up until reaching the equilibrium state. Lower values of electrical voltage were applied to avoid hitting the device limitation of I = 8 A as the system reached the equilibrium. The temperature dependent electrical resistance values for the heating conductor coating, which are shown in **Figure 4**, were obtained with the help of the temperature measurements on the coating surface by thermal imaging, the temperature dependent electrical current value and the electrical voltage input. Unlike the metallic heating element coatings, where the resistance of the heating element increases with temperature [5, 6], the resistance of the semi-conductor ceramic coatings decrease with increasing temperature as it is confirmed in this study as well.

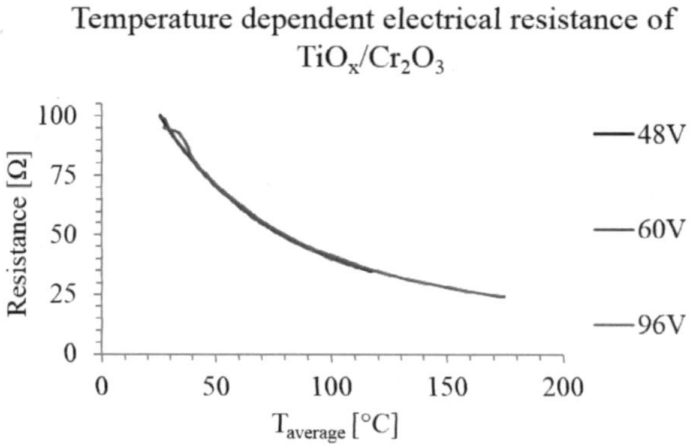


**Figure 4.** Measured dependence of the electrical resistance on temperature.

An infrared camera ThermaCam SC500 (Flir Systems) was used for the thermal imaging. The camera has an image resolution of IR = 320x240 pixels, recording frequency of f = 50 Hz and a temperature measurement accuracy of ΔT = ± 2 °C. A macro lens was mounted on the thermal camera, which allowed to zoom in on the coating surface and record the surface temperatures of the coating with a pixel resolution of PR = 8.4 pixels/mm. It should be noted that due to the recording with the macro lens, the margins of the coating were not captured on the thermal images. The emissivity value of the coating surface has a major effect on the precision of the temperature measurement with an infrared camera. The emissivity value of the heating element coating was calibrated based to the boiling temperature of water and determined to be ε = 0.48. **Figure 5** shows a heating element coating with the thickness of t = 110 μm being heated with the electrical voltages of U = 15 V and U = 20 V.

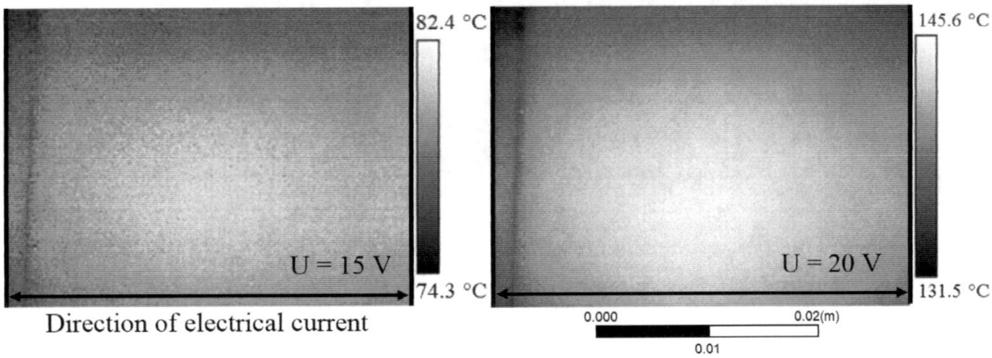


**Figure 5.** Thermal images of the coating system with the heating element coating thickness of t = 110 μm at electrical voltage of U = 15 V (left) and U = 20 V (right).

Upon turning on the voltage source, the temperature of the coating rose gradually and reached quasi-equilibrium state. In accordance with the preliminary tests, hot lines were not detected neither during the heating up, nor at the equilibrium stage. This corresponds to expectations, since low residual stresses within the coating could not have caused crack formation. The thermal images shown correspond to the steady state temperatures of the coating surface. After the coating temperatures have reached their peak value, the temperature of the substrate side near the contact plate were measured to be T = 70 °C for U = 15 V and T = 112 °C for U = 20 V with an infrared temperature sensor, since the sides of the substrate are not visible on the thermal images. These values will be used as a boundary condition in the numerical modelling to take into account the cooling effect of the entire experimental setup on the substrate, and therefore, on the coating. **Figure 6** shows four different directions of electrical current flow rotated around the spot in the middle of the sample. Each temperature pattern shows a directional alignment perpendicular to the electrical current.

IOP Conf. Series: Materials Science and Engineering **480** (2019) 012002 doi:10.1088/1757-899X/480/1/012002

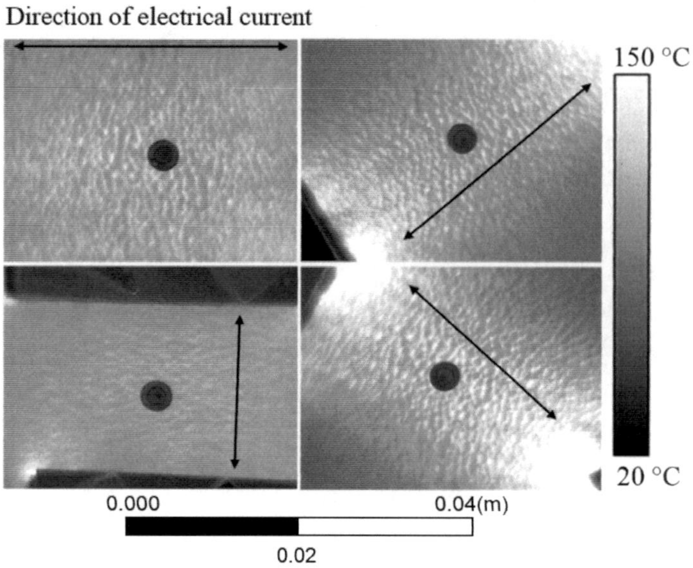


**Figure 6.** Temperature inhomogeneity aligned perpendicular to the electrical current direction seen in thermal imaging for the heating element coating with the thickness of t = 400 μm.

## 3. Numerical modelling

For the numerical modelling of the resistive heating of the coating system, the Thermal-Electric module of the simulation package Ansys® has been used. The substrate and the coating geometries were modelled according to the dimensions given in section *2.1*, except for the length of the heating element coating. It was taken as the distance between the contact plates L = 70 mm to ensure correct application of the electrical voltage boundary condition. The material properties used in the simulations were taken from the material database "Sample materials data from Granta Design" in Ansys Workbench and are given in **Table 2**. The material properties in the range of T = 15 – 150 °C were assumed to be constant due to negligible variation in the given temperature range. An exception is the temperature dependence of electrical resistivity of $TiO_2/Cr_2O_3$. It was taken into account as the resistivity of $TiO_2/Cr_2O_3$ decreases significantly at high temperature and this phenomenon has a major effect on the heating behaviour of the coating system.

The boundary conditions used in the model are the electrical voltage on the sides of the $TiO_2/Cr_2O_3$ coating, cooling due to the radiation and convection as well as the temperature boundary condition on the sides of the substrate. The simulation results are shown in **Figure 7** for the electrical voltage boundary condition of U = 15 V and U = 20 V. Both results show the same temperature distribution, with their temperature peaks centred in the middle of the heating element coating. The maximum temperatures of T = 83.3 °C and 145.6 °C respectively show a very good correspondence to the experimental measurements of the maximum coating temperature of T = 82.4 and 145.6 °C respectively. This shows the suitability of the developed numerical model for the electrical heating of heating element coating systems.

IOP Conf. Series: Materials Science and Engineering **480** (2019) 012002 doi:10.1088/1757-899X/480/1/012002

**Table 2.** Material properties used in the simulations.

| Property/Material | Value |
| --- | --- |
| Density/Steel substrate | 7,860 kg/m$^3$ |
| Thermal conductivity/Steel substrate | 37.5 W/m K |
| Specific heat/Steel substrate | 473 J/kg K |
| Density/Aluminum oxide | 3,700 kg/m$^3$ |
| Thermal conductivity/Aluminium oxide | 24 W/m K |
| Specific heat/Aluminium oxide | 880 J/kg K |
| Density/$TiO_2/Cr_2O_3$ | 4,000 kg/m$^3$ |
| Thermal conductivity/$TiO_2/Cr_2O_3$ | 3 W/m K |
| Specific heat/$TiO_2/Cr_2O_3$ | 690 J/kg K |
| Resistivity/$TiO_2/Cr_2O_3$ | 0.0003-0.0015* ohm m |

*Temperature dependent values based on the plot in **Figure 4**.

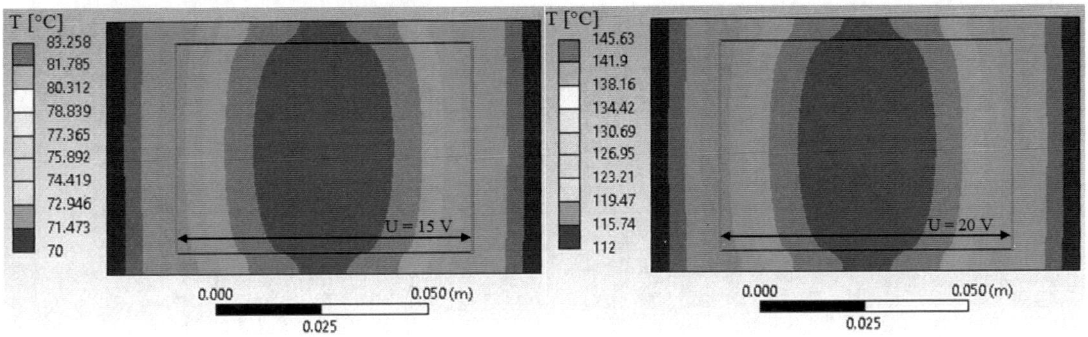


**Figure 7.** Temperature distribution on the coating systems obtained by numerical modelling. The applied electrical voltages are U = 15 V (left) and U = 20 V (right).

The hypothesis of this work is that the hot lines seen in **Figure 6** are caused by the cracks in the heating element coating, which emerge at higher coating thicknesses due to residual stress. In order to test this hypothesis, random artificial cracks with the depth of d = 380 μm are introduced into the heating element coating with the thickness of t = 400 μm (**Figure 8**). As can be seen in the figure, the artificial cracks are exaggerated in size and do not represent a realistic distribution. It can be also seen that the distribution of the cracks follow no directional pattern.

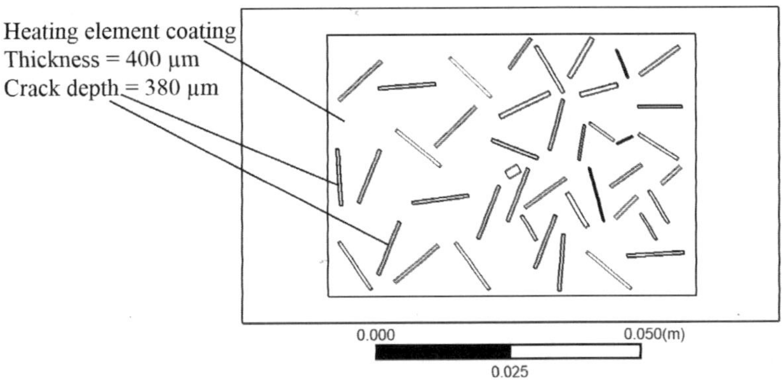


**Figure 8.** The pattern of artificial cracks in the heating element coating.

IOP Conf. Series: Materials Science and Engineering **480** (2019) 012002 doi:10.1088/1757-899X/480/1/012002

A voltage of U = 60 V was applied horizontally and vertically on the sides of the heating element coating. The resulting patterns of the temperature (T), current density (I/A) and Joule heating (JH) are illustrated in **Figure 9**. The directional preference of the temperature inhomogeneity is evident. Moreover, it can be seen that the current density and therefore the Joule heating are significantly higher in the introduced cracks and show a directional preference corresponding to the direction of the applied voltage. This supports the hypothesis that the inhomogeneity in the temperature distribution in form of hot lines can be caused by the cracks in the coating microstructure which have no directional preference. It should be noted that the coating temperature was significantly lower than it would be expected from a uniform coating with the same coating thickness and applied electrical voltage. This can be explained by the fact that the introduced cracks significantly increased the electrical resistance of the coating, thus leading to a lesser amount of electrical heating.

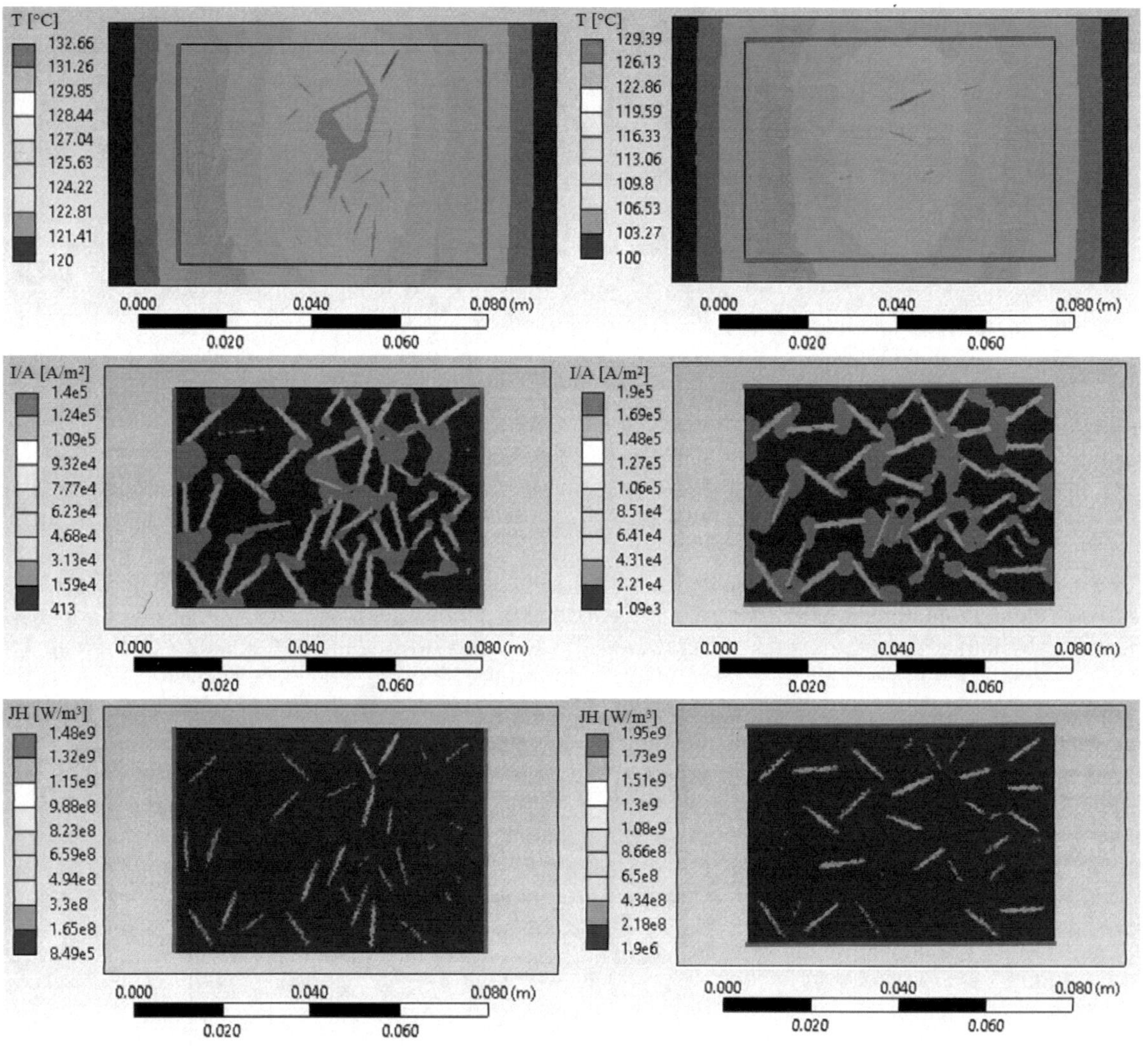


**Figure 9.** Results of the electrical heating simulation of a heating element coating with artificial cracks. The electrical voltage was applied horizontally (left column) and vertically (right column).

## 4. Summary and outlook

An electrical heater coating system with the heating element coating thickness of t = 110 µm has been thermally sprayed and analysed with thermal imaging under application of electrical voltages of U = 15 V and U = 20 V. The results of the numerical modelling showed good correspondence with the experimental temperature measurements. Based on this model, the resistive heating of a coating system with a random pattern of artificially introduced cracks was simulated. Based on the direction of the applied electrical voltage, directional preference in the temperature, electrical current density and Joule heating distributions could be detected. It was seen that the introduction of the artificial cracks had a negative impact on the resistive heating characteristics of the coating system. While a quantitative experimental verification was feasible for the simulation of the homogeneously heated coatings, only a qualitative comparison was made for the case with the temperature inhomogeneity in form of hot lines. In future works, the cracks in the coating microstructure should be experimentally analysed with respect to their size and depth distribution as well as to their orientations. By introducing more realistic artificial cracks in the numerical model, a quantitative comparison will become feasible for this case as well.

**Acknowledgments**
All presented investigations were conducted in the context of the Collaborative Research Centre SFB1120 "Precision Melt Engineering" at RWTH Aachen University and funded by the German Research Foundation (DFG). For the sponsorship and the support we wish to express our sincere gratitude.

**References**

[1] Longtin J P, Zuo L, Hwang D, Fu G, Tewolde M, Chen Y and Sampath S 2013 Fabrication of Thermoelectric Devices Using Thermal Spray: Application to Vehicle Exhaust Systems *J Therm Spray Tech* **22** 577–87
[2] Bobzin K, Öte M, Knoch M A, Liao X, Hopmann C and Ochotta P 2016 Investigation of Influencing Factors on the Transfer of Wire Arc Sprayed Zn Coatings for the Metallization of Plastic Parts: In: International Thermal Spray Conference 24-29
[3] Prudenziati M, Cirri G and Bo P D 2006 Novel High-Temperature Reliable Heaters in Plasma Spray Technology *J Therm Spray Tech* **15** 329–31
[4] Michels D, Hadeler J and Lienhard V J H 1998 High-Heat-Flux Resistance Heaters from VPS and HVOF Thermal Spraying *Experimental Heat Transfer* **11** 341–59
[5] Prudenziati M 2008 Development and the Implementation of High-Temperature Reliable Heaters in Plasma Spray Technology *J Therm Spray Tech* **17** 234–43
[6] Prudenziati M and Gualtieri M L 2008 Electrical Properties of Thermally Sprayed Ni- and Ni20Cr-Based Resistors *J Therm Spray Tech* **17** 385–94
[7] Hopmann C, Filz P, Nikoleizig P and Schmitz M 2016 Precision melt engineering- Controlling the solidification process in injection moulding: In: 28. Internationales Kolloquium Kunststofftechnik 306-350
[8] Bobzin K, Hopmann C, Öte M, Knoch M A, Alkhasli I, Dornebusch H and Schmitz M 2017 Tailoring the heat transfer on the injection moulding cavity by plasma sprayed ceramic coatings *IOP Conf. Ser.: Mater. Sci. Eng.* **181** 12013
[9] Liu J T, Zhao G Q, Guan Y J and Wang G L 2010 Coupled Simulation of Plastic Part and Mold Temperatures in the Filling Process of Rapid Heat Cycle Injection Molding *AMR* **97-101** 3179–82
[10] Lee H, Han S J, Chidambaram Seshadri R and Sampath S 2016 Thermoelectric properties of in-situ plasma spray synthesized sub-stoichiometry $TiO_2$-x *Scientific reports* **6** 36581
[11] Skopp A, Kelling N, Woydt M and Berger L-M 2007 Thermally sprayed titanium suboxide coatings for piston ring/cylinder liners under mixed lubrication and dry-running conditions *Wear* **262** 1061–70

IOP Conf. Series: Materials Science and Engineering **480** (2019) 012002 doi:10.1088/1757-899X/480/1/012002

[12] Sōmiya S, Hirano S and Kamiya S 1978 Phase relations of the $Cr_2O_3/TiO_2$ system *Journal of Solid State Chemistry* **25** 273–84
[13] Tucker R C JR 2013 *ASM Handbook: Volume 5A:* (ASM International)

IOP Conf. Series: Materials Science and Engineering **480** (2019) 012003 doi:10.1088/1757-899X/480/1/012003

# Modelling of particle impact using modified momentum source method in thermal spraying

**K Bobzin, M Öte, M A Knoch, I Alkhasli and S R Dokhanchi***

Surface Engineering Institute, RWTH Aachen University, Germany

* e-mail: dokhanchi@iot.rwth-aachen.de

**Abstract.** Thermal spraying is a coating process in which the feedstock material is accelerated and impacts on a substrate in form of molten or semi-molten particles. Particle impact simulation is helpful for understanding the coating build-up during thermal spraying. In this study, a computational fluid dynamics (CFD) model, based on the volume of fluid (VOF) approach, is used to model the impact and solidification of nickel particles on a flat substrate in 2D and 3D. Temperature dependent viscosity and momentum source are commonly used for solidification modelling. The former is accurate, but computationally too expensive for multiple particle impact simulations. In the momentum source method, the momentum equation of the particle is manipulated in order to reduce its velocity to zero as it solidifies. ANSYS Fluent employs this method for solidification. However, this method is proven to be inadequate for the simulation of multiple-particle solidification. In the context of this study, a modification to this method has been introduced. Temperature dependent viscosity and validated numerical studies from literature are used to validate the modified method. The developed method is proven to be capable of simulating the deposition of a 60 μm thick coating in a more feasible computational time, in comparison to temperature dependent viscosity method.

## 1. Introduction

Thermal spraying, as a coating technology, is categorized into three major process variants: flame spraying, electric arc spraying, and plasma spraying. In thermal spraying, the feedstock material in form of powder, wire or rod is heated to molten or semi-molten state. The resultant heated particles are accelerated in a high-temperature free jet and impact on a prepared substrate. Through the impact of subsequently approaching particles, a coating is built up [1].

Plasma spraying is characterized by particle velocities of up to v = 800 m/s [2] and high plasma temperatures in the range of T = 6,000 to T = 15,000 °C, considerably above the melting temperature of any known material [1]. Injected particles are usually in the size range of 20 – 90 μm [2]. Immediately after impact, the heated particles spread and deform on the substrate. During the spreading of particles, rapid solidification with cooling rates in the range of $\dot{q} = 10^7 - 10^8$ K/s occurs due to the heat transfer from the liquid material to the underlying material and to the ambient atmosphere [3]. Therefore, particle deformation on the substrate, cooling down and solidification occur simultaneously.

The coating microstructure is directly related to the particle impact process. Therefore, it is essential to have a detailed understanding of the dynamics of particle impact on the substrate for a better control of the coating build-up. The deposition of the particles in plasma spraying can be poorly observed experimentally, since the splat formation and solidification occur in a few microseconds [4].

Hence, many studies have been dedicated to numerical and analytical investigation of particle impact, splat formation and solidification [5–9].

The volume of fluid (VOF) method provides a region-following scheme with minimum storage requirements [10]. This method is utilized for modelling two or more immiscible fluids by tracking the volume fraction of each of the fluids in two or three-dimensional meshes. Simulation time is a big challenge in numerical problems based on VOF method, since tracking the fluid phases is computationally expensive. Most of the computational cost of VOF algorithms are ascribed to the cells that form interfaces between different fluids [11]. In case of the presented study, the interfaces are the cells in which coating material and gas coexist.

The solidification and melting problems in thermal spraying are associated with many complex physical phenomena. Among others, heat transfer is an important phenomenon, which can affect the evolution of particle impact process. A comprehensive review on major methods of mathematical modelling of solidification and melting problems was done by Hu et al. [12]. They have reviewed the heat transfer analysis of major solidification methods commonly used in literature, such as the enthalpy method and source term method. The main feature of the enthalpy method is that the evolution of the latent heat is described by the enthalpy. In this method, the energy equation can be defined in terms of the enthalpy as a function of temperature. In the source term method, a heat source or a heat sink is introduced into the energy equation as an additional term. This is a popular numerical method for modelling solidification problems due to its easy implementation and low computational cost. [12]

Pasandideh-Fard et al. [13] developed a 3D model to simulate the impact and solidification of a molten particle on a flat substrate. They used the fixed velocity approach for solidification, in which the solid is defined as liquid with infinite density and zero velocity. Yet, it is not a physical assumption to consider the material with infinite density. In previous works at the Surface Engineering Institute, the impact and solidification of partially yttria-stabilized zirconia (PYSZ) particles was modelled, both on a flat and a rough substrate, using the temperature dependent dynamic viscosity approach for solidification [14]. In another work at the Surface Engineering Institute, the deposition of multiple particles with diameters between 20 and 60 μm was modelled in 3D, resulting in a 70 μm thick coating using temperature dependent viscosity approach [15]. Here it must be noticed that, this solidification method led to high computational cost in the range of several weeks of computation time. Zheng et al. [16] developed a three-dimensional model to simulate the impact of a single molten droplet on a solid substrate during plasma spraying. They have utilized the momentum source method of ANSYS Fluent, which is based on an enthalpy-porosity technique, for modelling the solidification of a single droplet. While these studies helped to increase the understanding of the particle impact and solidification in thermal spraying, they either only considered the impact of single particles or high computational times were required in case of subsequent impact of a large number of particles.

The goal of this study is to develop a numerical model for simulating the deposition of a Ni-coating in a feasible computation time. First, the temperature dependent viscosity approach was utilized to simulate the particle impact and solidification. Due to its high computational expense, this solidification method is identified to be impractical to simulate an entire coating. However, since the temperature dependent viscosity method is based on a fundamental physical phenomenon, it is used as reference to validate the developed model. For further validation, the results in [8] were reproduced to compare the developed model with a simulation, which was experimentally validated. Afterwards, the solidification model in ANSYS Fluent, which is based on the momentum source method, was used. However, an unphysical phenomenon occurred during the interaction of liquid and solid phases, which made this method inadequate for the simulation of multiple-particle solidification. In this study, a modified momentum source term is introduced which enables the simulation of multiple particle solidification. Chandra and Mostaghimi et al. [8], at the Centre for Advanced Coating Technologies, used nickel as the feedstock material to study the deposition of particles in plasma spraying on a stainless steel surface using both numerical simulations and experiments. In this study, nickel is used as feedstock material to compare the results of a three-dimensional particle impact simulation, using modified momentum source method, to the results of Chandra and Mostaghimi et al. [8]. Finally, the validated model is employed to simulate the deposition of 100 molten nickel particles on a flat substrate in a 2D domain. Furthermore, the porosity of the simulated coating microstructure is calculated by image processing.

## 2. Numerical Method

In order to model the particle impact process, Computational Fluid Dynamics (CFD) approach is used. ANSYS Fluent is employed as the simulation tool to implement this method. Feedstock material and gas are the two phases to be considered in the simulation. The Volume of Fluid (VOF) method is employed to study this multiphase problem by tracking the particle and gas phases throughout the domain. Figure 1 shows the boundary and initial conditions of the simulation. In a first run, particle in-flight properties like their temperatures and velocities are taken over from [8], as this work is used as a reference for the validation. Furthermore, the simulations are conducted with different particle velocities as well as with different particle and substrate temperatures in order to analyse the functionality of the model for different processing conditions.

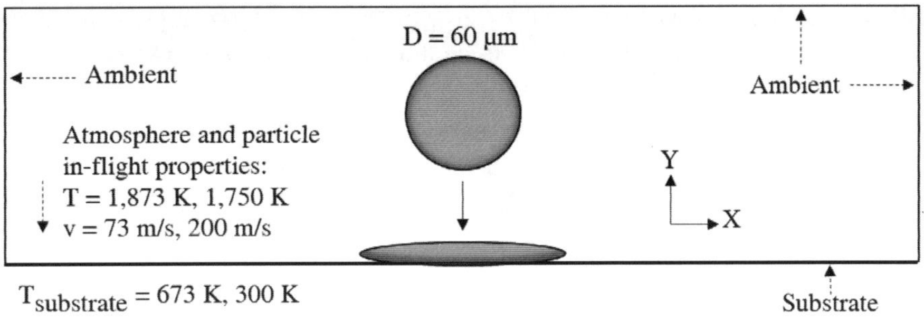


**Figure 1.** Schematic diagram of the simulation domain for the particle impact.

The substrate is modelled as a rigid wall with a constant temperature. The material properties of nickel used in the simulation are listed in Table 1. The surface tension coefficient of molten nickel is set to 1.77 N/m as reported in [17].

**Table 1.** Material properties of nickel [18].

| Property | Unit | Value |
|---|---|---|
| Density | kg m$^{-3}$ | 7,850 |
| Specific heat | J kg$^{-1}$ K$^{-1}$ | 735 |
| Thermal conductivity | W m$^{-1}$ K$^{-1}$ | 69.2 |
| Viscosity of molten nickel | kg m$^{-1}$ s$^{-1}$ | 0.0055 |
| Latent heat of fusion | J kg$^{-1}$ | 292,000 |
| Melting temperature | K | 1,728 |

### *2.1. Solidification Model*

Solidification depends on splat thickness, thermal diffusivities of both sprayed feedstock material and underlying solid material, and the thermal contact resistance between the flattening particle and substrate. It directly affects the deformation behavior, the splat shape and the coating microstructure [19]. Thus, solidification is an important physical phenomenon during particle impact. Reliable modelling of this complex problem represents the key challenge of particle impact simulation in thermal spraying. In this study, temperature dependent viscosity and momentum source are considered as the two commonly used solidification modelling methods. Based on the latter, a modified solidification method is developed to simulate the deposition of an entire coating.

*2.1.1. Temperature dependent viscosity.* In this approach, the solid material is modelled as liquid with high viscosity. It assumes that as the particle cools down, it rapidly becomes more viscous and as a result, the particle velocity decreases. A user defined function (UDF) is used to implement this method. The temperature is calculated in every iteration and based on this, the appropriate value of viscosity is set. Temperature dependent viscosity is an accurate method for solidification modelling, since it is based on an underlying physical phenomenon. However, this method is extremely computationally expensive. In this method, the time step size of the simulation calculations decreases exponentially as the viscosity of the particle increases, respectively as the particle solidifies. This increases the total simulation time dramatically. Therefore, the method is impractical for simulating the deposition of an entire coating.

*2.1.2. Momentum source method of ANSYS Fluent.* ANSYS Fluent employs a momentum source which is based on an enthalpy-porosity technique [20]. In this method, the appropriate momentum sink terms are added to the conservation of momentum equations and hereby, the velocity of the particle is reduced to zero as it solidifies. The momentum source (S) is defined as:

$$S = \frac{(1-\beta)^2}{(\beta^3+\varepsilon)} A_{mush} v \tag{1}$$

In equation (1), β is the liquid volume fraction which is defined here as the volumetric proportion of the fluid phase consisting of the liquid particle phase together with the gas phase in a cell, ε is a small number (0.001) to prevent division by zero, v is the fluid velocity and $A_{mush}$ is the mushy zone parameter which is a constant number. β is computed based on an enthalpy balance at each iteration. In the mushy zone, the liquid volume fraction (β) lies between 0 and 1. $A_{mush}$ determines the rate at which the source term reduces the velocity. The higher $A_{mush}$, the more the source influences the momentum equation. However, very large values may cause the solution to diverge.

In case of a fully molten material, the liquid fraction is one. Thus, according to equation (1), the source term would be zero. In fully solidified regions, β is zero. Thus, the source stops the particle motion.

When using the momentum source method in ANSYS Fluent, an unphysical interaction between solidified particles and molten material of another particle was observed. Upon contact between two such particles, the motionless solid particle is accelerated and thrown out of the simulation domain. This behavior is only observed when simulating the subsequent impact of several particles. The unphysical behavior is assumedly caused by an unknown numerical error in implementation of the method. This study is focused on solving this problem by modifying the momentum source term as explained in the next section.

*2.1.3. Modified momentum source method.* To circumvent the problem described above, a modification to the momentum source method for solidification modelling is introduced. Equation (2) is the developed function, which replaces equation (1) to manipulate the conservation of momentum equations of the particles.

$$S = -\,C\, VF_P\, K(\rho)\, v \tag{2}$$

IOP Conf. Series: Materials Science and Engineering **480** (2019) 012003 doi:10.1088/1757-899X/480/1/012003

In equation (2), C is a numerical constant, $VF_P$ is particle volume fraction which is the volumetric proportion of particle phase under or above the melting temperature, K(ρ) is a function of mixture density and v is the velocity vector.

The numerical constant used in equation (2) is adjusted iteratively based on the viscosity method to assure a physically valid implementation. The momentum source is proportional to the velocity and is directed in the opposite direction. Therefore, when the velocity of the particle is positive, a negative momentum source is added to the momentum equation as well.

In the momentum source term of the Fluent, the volume fraction of the fluid phase (β), consisting of the liquid particle phase and the gas phase, is considered. For the particular application of particle impact in thermal spraying, it would be seen favorable to distinguish between the liquid particle phase and the gas phase. For this purpose, the parameter $VF_P$ has been introduced instead of β. The particle volume fraction in equation (2) could be liquid or solid. Therefore, a temperature condition is implemented to apply the momentum source only on the solid phase. If the temperature is below or equal to the melting temperature, the source is computed according to equation (2) to stop the particle deformation. If the temperature is above the melting point, the source is set to zero. In equation (1), there is no temperature condition and the momentum source depends on the liquid volume fraction (β).

The parameter K(ρ), according to equation (3), controls the rate of increase in the momentum source with respect to the inertia of the existing materials in a single cell. In equation (3), mixture density is the average density of a single cell in which both coating material and gas coexist. The mixture density is equal to the density of feedstock material in cells in which the particle volume fraction corresponds to one. In mushy zone, the amount of mixture density is proportional to the volume fraction of the coexisting phases.

$$K(\rho) = \frac{(\text{mixture density - gas density})^{n}}{(\text{feedstock density - gas density})^{n-1}} \tag{3}$$

Figure 2 shows the plot of K(ρ) versus mixture density for different n values. The value of feedstock density is taken from Table 1 and gas density is set to 1.225 kg $m^{-3}$. In mushy zone, the cells with lower mixture density have less inertia. The amount of momentum source for these cells should be lower in comparison to the cells with higher mixture density. The parameter K(ρ) adapts the amount of momentum source to the inertia of the cells exponentially. For higher n values, the effect of the momentum source is reduced for the cells containing lower mixture density. However, by choosing a very high n value, the rate of stopping deformation decreases because the source will act only on highly pure particle phase. After running several simulations with increasing n in the momentum source, n = 5 was selected to add an appropriate momentum sink term to the momentum equation of the particle.

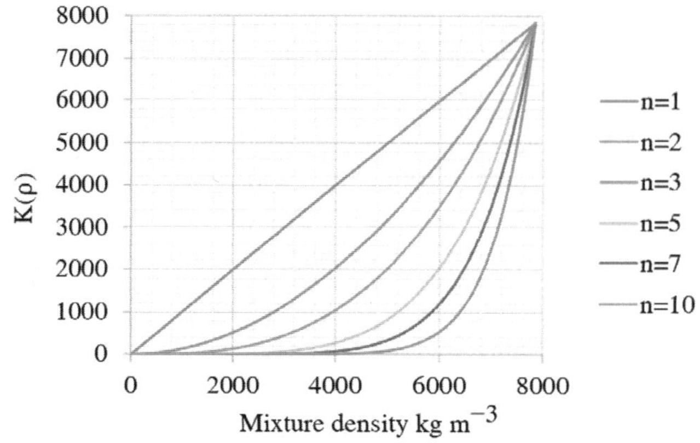


**Figure 2.** Plot K(ρ) versus mixture density for different values of n.

Furthermore, both the particle volume fraction ($VF_P$) and the parameter K(ρ) in equation (2) contribute to eliminate the effect of momentum source on the gas phase. Without considering these parameters, the particle does not touch the substrate and deforms in the air as shown in Figure 3-a. The lack of contact between the substrate and particle influences the cooling down of the particle and consequently the spreading and flattening behavior cannot be modelled properly. This problem is due to the significant influence of the momentum source on the gas phase. To simulate the particle solidification, the momentum source should be applied only on the particle phase. However, Fluent applies the momentum source to the mixture of phases. Together with high values of $A_{mush}$, this leads to an unphysical outcome of simulation as shown in Figure 3-a. Hence, in the presence of $VF_P$ and K(ρ), as introduced in Figure 2, this undesired affect could be eliminated (see Figure 3-b).

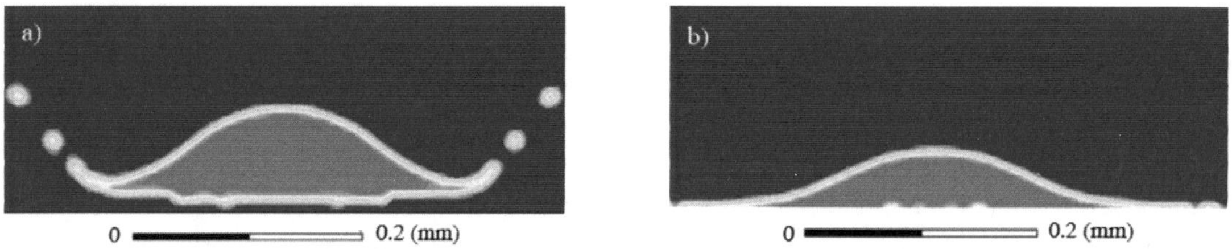


**Figure 3.** a) Problem in case of acting momentum source on the mixture phase. b) Minimizing the effect of momentum source on the mixture phase.

It should be noticed that the modified momentum source term presented in equation (2) is one of the alternative approaches to adjust the source term in equation (1) for particle impact in thermal spraying. The possibility of using other parameters to modify the momentum source term to be well-functional in thermal spraying applications is not excluded.

## 3. Results and discussion

In order to validate the developed model, the results of a 3D particle impact simulation using the modified momentum source are compared to the results of Chandra and Mostaghimi et al. [8] (see Figure 4). The model presented by Chandra and Mostaghimi et al. was also experimentally validated [8]. Therefore, it is a suitable validation case for this study. In both simulations, nickel is used as the feedstock material. Small variations might be attributed to the nickel material properties, since this information is not provided in [8]. The particle diameter, impact velocity, particle and substrate temperature are set to D = 60 μm, v = 73 m/s, $T_p$ = 1,873 K and $T_s$ = 673 K respectively. In Figure 4,

IOP Conf. Series: Materials Science and Engineering **480** (2019) 012003 doi:10.1088/1757-899X/480/1/012003

the splats exhibit a similar disk-shaped morphology, which shows a good agreement of both splat length and the development of splat morphology with the investigations in literature [3, 8].

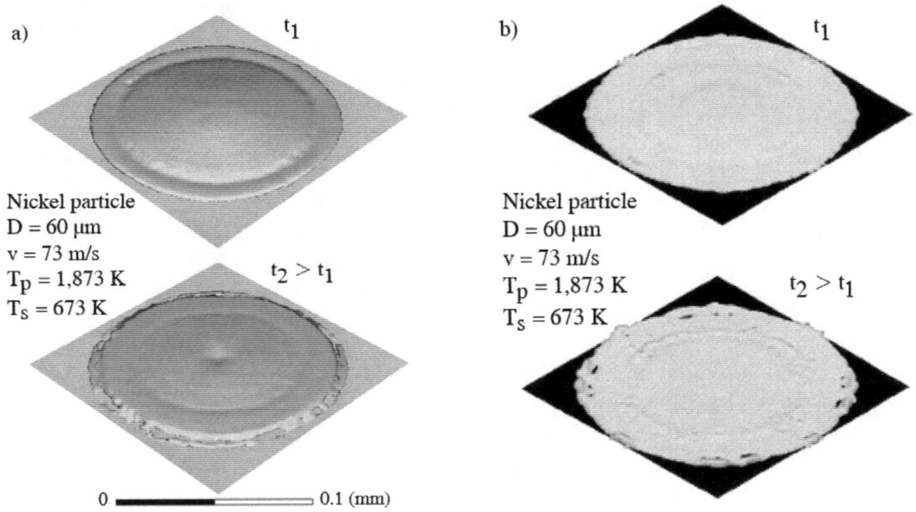


**Figure 4.** Particle impact model in 3D using a) modified momentum source and b) fixed velocity method reported in [8].

As a further validation, the results of the simulation, which follows the temperature dependent viscosity approach, is used as a reference. Figure 5 shows the results of the impact of a 60 μm diameter molten nickel particle using the modified momentum source (a) and the temperature dependent viscosity approach (b). In this case, a higher particle velocity of v = 200 m/s and a slightly lower particle temperature $T_p$ = 1,750 K are considered in order to analyse the functionality of the method for different particle in-flight properties. In this simulation, the substrate temperature is set to $T_s$ = 300 K. As the simulation time increases, the particles spread and deform on the substrate. In Figure 5, t = 0 corresponds to the time point of the first contact of the particle to the substrate. The particles still deform at the time shown in the figure. The splat deformation in both cases are compared visually by making an animation sequence with the volume fraction contours. The splat length of both cases are in the same order of magnitude. Therefore, the results of temperature dependent viscosity can be replicated by the modified momentum source method. The main motivation of using momentum source, instead of temperature dependent viscosity for solidification modelling, is that the momentum source allows the use of significant larger time steps. In this case, the time step size in the momentum source simulation was ca. $10^4$ times larger than the simulation using the temperature dependent viscosity approach. Consequently, the computation time was reduced approximately by the factor of 100.

IOP Conf. Series: Materials Science and Engineering **480** (2019) 012003 doi:10.1088/1757-899X/480/1/012003

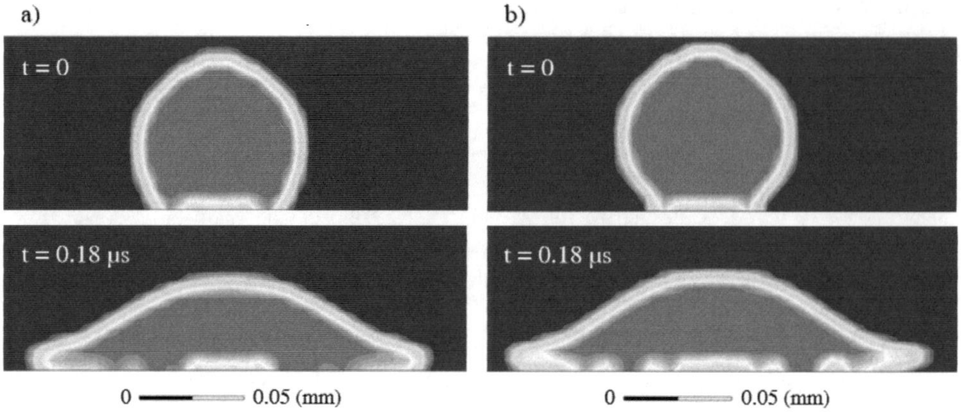


**Figure 5.** Particle impact model in 2D using a) modified momentum source and b) temperature dependent viscosity.

The modified momentum source method yields comparable results as the temperature dependent viscosity method, as well as the numerical and experimental studies in literature [8]. Hence, the developed model is suitable to simulate multiple particles in 3D and also the deposition of an entire coating.

The substrate temperature is vital for the spreading and final shape of the impacted particles [8]. The effect of substrate temperature on the spreading and final shape of the impacted particles was investigated. Typically, during a plasma spray process, the splats result in disk-like shape when the substrate temperature is above 300–400°C [3]. Figure 4 shows the simulation results of disk-shaped morphology of splats, where the impact occurs on a hot substrate. In a further simulation, successive impact of two particles is simulated in order to analyse the functionality of the developed model in a slightly more complicated model setup. In this case, the substrate temperature is reduced to $T_s = 300$ K. Figure 6 shows the successive states of two nickel particles during the impact on a non-heated substrate in a three-dimensional simulation. The initial temperature of the particles is set to T = 1,750 K. Thus, the particles are fully molten prior to the impact. The particles have a diameter of D = 60 μm and an impact velocity of v = 200 m/s. The parameters were changed in comparison to Figure 4 to investigate the possible splash development for faster particles impacting on a cold substrate. Although, an apparent splashing behavior cannot be triggered for the chosen particle in-flight properties and a disk-shaped flattening of the particles is evident, the model is obviously capable of resolving the interaction between two successively impacting particles. Nevertheless, a partial splashing of the second particle in the region of the interaction with the underlying particle can be resolved using the developed model. It is assumed that, a radial jetting takes place at the regions depicted as splashing area in Figure 6, as a result of the altered spreading velocity and solidification rate of the second particle due to the interaction with the underlying one.

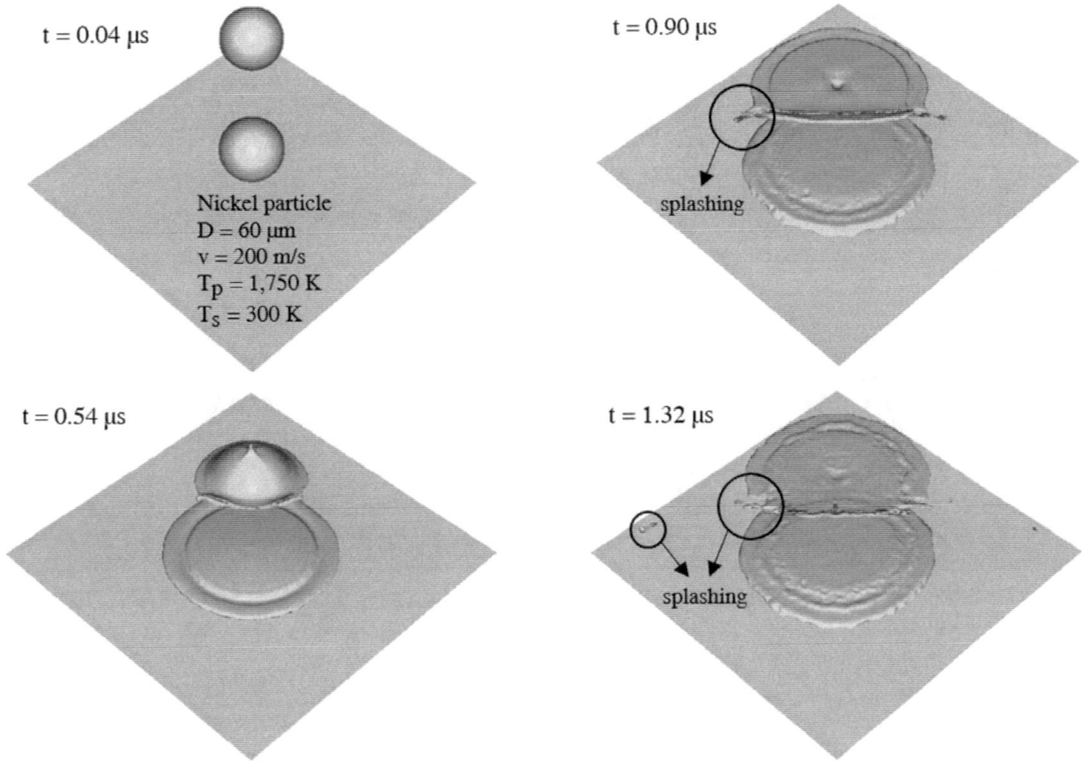


**Figure 6.** 3D model of particle impact using modified momentum source method for solidification.

In a further step, the developed model is utilized to simulate the deposition of a higher number of particles. Figure 7 shows a coating microstructure resulting from the impact of 100 nickel particles with the diameter, temperature and velocity of D = 60 μm, T = 1,750 K and v = 200 m/s respectively. In this simulation, the substrate temperature is set to $T_s$ = 300 K. In a real APS process, the particles have different velocities, temperatures and sizes. The aim of this study was to develop a numerical model for simulating the entire coating in a feasible computation time. To demonstrate the feasibility of the developed model, the result of a coating simulation with constant particle parameters is presented. Here, an average time interval of 5 microseconds is chosen between two consecutive particles. The particles are introduced to random locations in the simulation domain. The porosity (ø) of the simulated coating microstructure is calculated to be ø = 3.52 % by processing of the dark and light grey pixels of the coating image in MATLAB. The thickness of the coating is 63 μm. The developed model can be further used to understand the development of the particles deposition and also predicting the porosity of a coating microstructure in thermal spraying. In order to predict the porosity with a reasonable accuracy, different velocities, temperatures and dimensions of successively impacting particles need to be considered in the simulations. The developed model is capable of conducting simulations with different particle in-flight properties as well. The results of these simulations will be addressed in the further publications.

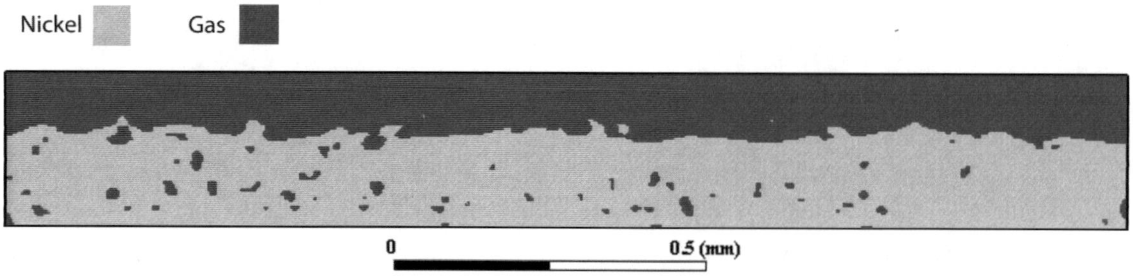


**Figure 7.** 2D model of a coating microstructure.

## 4. Summary and conclusions

Since the splat formation and solidification in thermal spraying occur in a few microseconds, it is extremely difficult to observe the dynamics of particle impact experimentally. Hence, modelling of particle impact process is of great interest.

The aim of this study was modelling and simulation of the deformation, cooling down and solidification of multiple nickel particles on a solid substrate during thermal spraying. A Computational Fluid Dynamics (CFD) model, based on the volume of fluid (VOF) method, was utilized in ANSYS Fluent as the simulation tool to model the particle impact both in 2D and 3D. In this respect, two common solidification models, temperature dependent viscosity and momentum source method, were presented. The advantages and limitations of each solidification model were further studied. Finally, a modified method for particle solidification was introduced to eliminate the implementing limitations of previous methods. The developed model has been validated with the results of the temperature dependent viscosity method as well as other numerical and experimental studies in literature. The modified method is able to simulate a coating microstructure in a feasible computation time.

## Acknowledgments

All presented investigations were conducted in the context of the Collaborative Research Centre SFB1120 "Precision Melt Engineering" at RWTH Aachen University and funded by the German Research Foundation (DFG). For the sponsorship and the support, we wish to express our sincere gratitude.

## References

[1] Davis JR 2004 Introduction to thermal spray processing *Handbook of thermal spray technology editor, Asm International and the Thermal Spray Society, America* 3–13

[2] Pawłowski L 2008 *The Science and engineering of thermal spray coatings* 2nd edn (Chichester, Hoboken, NJ: John Wiley & Sons)

[3] Vardelle A, Moreau C, Themelis N J and Chazelas C 2015 A Perspective on Plasma Spray Technology *Plasma Chem Plasma Process* **35** 491–509

[4] Goutier S, Vardelle M and Fauchais P 2011 Last developments in diagnostics to follow splats formation during plasma spraying *J. Phys.: Conf. Ser.* **275** 12003

[5] Vardelle M, Vardelle A, Leger A C, Fauchais P and Gobin D 1995 Influence of particle parameters at impact on splat formation and solidification in plasma spraying processes *Journal of Thermal Spray Technology* **4** 50–8

[6] Ghafouri-Azar R, Mostaghimi J, Chandra S and Charmchi M 2003 A stochastic model to simulate the formation of a thermal spray coating *Journal of Thermal Spray Technology* **12** 53–69

[7] Chandra S and Fauchais P 2009 Formation of solid splats during thermal spray deposition *Journal of Thermal Spray Technology* **18** 148–80

[8] Pasandideh-Fard M, Pershin V, Chandra S and Mostaghimi J 2002 Splat shapes in a thermal spray coating process: simulations and experiments *Journal of Thermal Spray Technology* **11** 206–17
[9] Yang K, Liu M, Zhou K and Deng C 2012 Recent developments in the research of splat formation process in thermal spraying *Journal of Materials* **2013**
[10] Hirt C W and Nichols B D 1981 Volume of fluid (VOF) method for the dynamics of free boundaries *Journal of computational physics* **39** 201–25
[11] Borrell R, Jofre L, Lehmkuhl O and Castro J 2013 Parallelization strategy for the Volume-of-Fluid method on unstructured meshes *Procedia Engineering* **61** 198–203
[12] Hu H and Argyropoulos S A 1996 Mathematical modelling of solidification and melting: a review *Modelling and Simulation in Materials Science and Engineering* **4** 371
[13] Pasandideh-Fard M, Chandra S and Mostaghimi J 2002 A three-dimensional model of droplet impact and solidification *International Journal of Heat and Mass Transfer* **45** 2229–42
[14] Bobzin K, Bagcivan N, Parkot D and Petković I 2010 Simulation of PYSZ particle impact and solidification in atmospheric plasma spraying coating process *Surface and Coatings Technology* **204** 1211–5
[15] Beck T *et al* 2010 Modeling of Coating Process, Phase Changes, and Damage of Plasma Sprayed Thermal Barrier Coatings on Ni-Base Superalloys *Adv. Eng. Mater.* **12** 110–26
[16] Zheng Y Z, Li Q, Zheng Z H, Zhu J F and Cao P L 2014 Modeling the impact, flattening and solidification of a molten droplet on a solid substrate during plasma spraying *Applied Surface Science* **317** 526–33
[17] Brillo J and Egry I 2005 Surface tension of nickel, copper, iron and their binary alloys *Journal of materials science* **40** 2213–6
[18] Dhiman R, McDonald A G and Chandra S 2007 Predicting splat morphology in a thermal spray process *Surface and Coatings Technology* **201** 7789–801
[19] Fauchais P, Fukumoto M, Vardelle A and Vardelle M 2004 Knowledge Concerning Splat Formation: An Invited Review *Journal of Thermal Spray Technology* **13** 337–60
[20] Voller V R and Prakash C 1987 A fixed grid numerical modelling methodology for convection-diffusion mushy region phase-change problems *International Journal of Heat and Mass Transfer* **30** 1709–19

IOP Conf. Series: Materials Science and Engineering **480** (2019) 012004 doi:10.1088/1757-899X/480/1/012004

# Influence of external magnetic fields on the coatings of a cascaded plasma generator

**K Bobzin[1], M Öte[1], M A Knoch[1], H Heinemann[1,*], S Zimmermann[2] and J Schein[2]**

[1] RWTH Aachen University, Aachen / Surface Engineering Institute
[2] University of Federal Forces Germany, Munich / Lab for Plasma Technology and Mathematics

* e-mail: heinemann@iot.rwth-aachen.de

**Abstract.** Cascaded single cathode single anode plasma generators represent a novel design in plasma spraying. They promise to combine the advantages of cascaded plasma generators, e.g. a higher plasma enthalpy and stability, with the simple design of the well-established non-cascaded single cathode single anode generators. In this paper, the electric arc of such a cascaded plasma torch (SinplexPro™-90, Oerlikon Metco) is manipulated by magnets mounted on the torch casing with the aim of moving the anodic attachment point. To analyze the effect on the coating properties, alumina (Amdry™ 6062) coatings are deposited using two operating parameters and two magnets of different remanences. The resulting coatings and their porosity, microhardness as well as phase composition are presented. Furthermore, measurements of the height profile of the spray spot, exhibiting an asymmetric shape, are shown. Consequently, the influence of the different process parameters and applied external magnetic fields on the coating properties is discussed.

## 1. Introduction

The development and application of plasma torches has gained importance in recent years. In atmospheric plasma spraying, plasma torches are used to generate thermal plasmas. The high temperature of the resulting plasma jet allows to process nearly any congruently melting material [1]. The most commonly used plasma torches have a single cathode and a single anode [2]. This design leads to a rather unstable behavior of the electric arc [3] and thus, the plasma jet [4], which, in turn, results in varying particle in-flight properties [5, 6]. Novel plasma torch designs utilize a cascaded neutrode between the cathode and anode to increase the enthalpy of the plasma jet and to reduce its fluctuations [7, 8]. Examples of such torches are the TriplexPro™ (Oerlikon Metco), Delta (GTV), KK (AMT) and the SinplexPro™ (Oerlikon Metco). In contrast to the first two, the SinplexPro™ and the KK have only one anode and one cathode, thereby combining the simplicity of a single cathode single anode plasma torch with the expected benefits of a longer and more stable electric arc due to the cascaded design.

It is not conclusively investigated whether the position of the anodic attachment point of the arc stays constant during the process or a movement of the attachment point applies for different process parameters. If the arc attachment point is fixed, it can be expected, that the increased length of the arc in case of the cascaded design leads to a higher intensity of the thermal loads on the anode and consequently to an increased thermal degradation. If this assumption applies, one possible method to reduce the thermal load on the anode might be to force the attachment point to change its position by applying a magnetic field, which induces a Lorentz force on the electric arc [9–12]. With the utilization of a permanent magnet to apply magnetic fields, as proposed in this study, it is expected that another position on the anode will be physically more favorable for the electric arc to attach on. Such a manipulation could affect the plasma jet and thus, the particle injection, the deposition rate and the coating properties. In the current state of the study, the investigations focus merely on the feasibility of such a manipulation and its possible effects on the coating properties. To the knowledge of the authors, there has not been any study regarding the influence such a manipulation on the coating properties in case of cascaded single anode single cathode plasma generators, so far. Here it must be noted that, Guggenheim et al. have already reported a possible movement of the arc attachment point by means of permanent magnets in [13].

Therefore, the effects of an external magnetic field on the coating properties applied using a cascaded single anode single cathode plasma torch were investigated with the aim of proving the feasibility of moving the electric arc and its attachment point in such a plasma torch. The possible influence of a manipulation of anodic arc-attachment point on the resulting coatings properties, i.e. porosity, microhardness, thickness and phase composition, is the main focus of this study.

## 2. Experimental and materials

To manipulate the electric arc and anodic arc root, permanent magnets were mounted on the plasma torch (SinplexPro™-90; Oerlikon Metco, Pfäffikon, Switzerland) as shown in **Figure 1**. The magnets were placed outside of the plasma torch on top of its anode. The setup results in a magnetic field which is aligned perpendicular to the current flow. This leads to a Lorentz force ($F_L$) acting on the electric arc, which directs it towards the permanent magnet and should move the attachment point in the same direction. Permanent magnets with different remanences of B = 680 mT and 1,200 mT were used for the manipulation. As the aim of these investigations was proving the feasibility of this setup, the chosen remanences were not adapted to the process parameters.

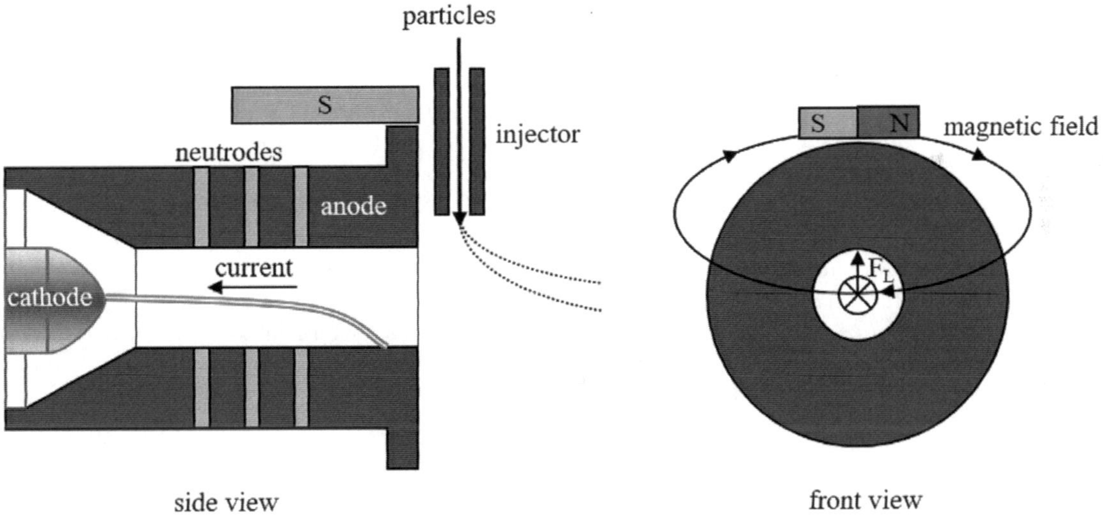


**Figure 1.** Setup of the plasma torch and the permanent magnet for the manipulation of the electric arc (conventional current direction)

As a first step, the attachment point of the electric arc was determined by head-on images of the plasma torch. This procedure is comparable to investigations conducted in [4, 14]. The setup used in this work is shown in **Figure 2**. To capture the highly dynamic events in the plasma torch, a high-speed camera (dimax HD; PCO AG, Kelheim, Germany) with a framerate of f = 34,473 Hz (Δt =29 μs) was used. Thermal plasmas exhibit strong emissions. To attenuate the brightness of the

IOP Conf. Series: Materials Science and Engineering **480** (2019) 012004 doi:10.1088/1757-899X/480/1/012004

resulting images, a stack of neutral density filters (OD0.9+OD0.5+OD0.4) were mounted on the lens, which were combined with a bandpass filter for a wavelength of 694 nm (FWHM: 10 nm). Prior to the recordings, an alignment image was taken as shown in the upper right part of **Figure 2**. This image was used to determine the position of the nozzle in the following head-on recordings of the plasma column.

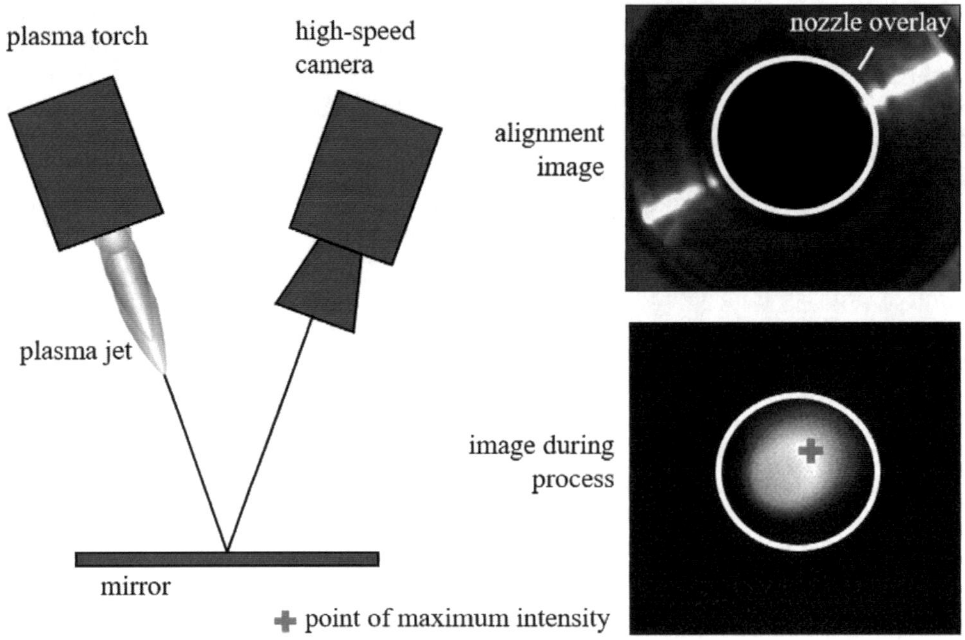


**Figure 2.** Determination of the azimuthal position of the electric arc by means of high-speed videography and corresponding image analysis.

In the lower right part of **Figure 2**, an image recorded during the process is shown. The point with the maximum intensity is assumed to represent the center of the electric arc, as the light emissions are higher in this region due to the elevated temperatures. Additionally, this point indicates the direction of the arc attachment point. To establish a statistical basis for the determination of the arc attachment point, 256 consecutive images were recorded and evaluated. Subsequently, the points of maximum intensity were determined for all images using a custom image processing tool.

The influence of the external magnetic field was investigated using the parameters given in **Table 1**. With each parameter set, alumina coatings were applied on two specimens to observe the effect of the external magnetic fields on the coating properties. The feedstock powder (Amdry™-6062; Oerlikon Metco, Pfäffikon, Switzerland) consists of $Al_2O_3$ particles with a nominal size distribution of -45+22 µm. Round specimen with a diameter of 25 mm and a thickness of 7 mm made of S235JR structural steel were used as substrate material. Prior to the coating deposition, the substrates were grit blasted to a roughness of $R_a \approx 8.5$ µm and a preparation grade of Sa 3. During the deposition, the plasma torch was moved in a meander shape at a standoff distance of $d_s = 100$ mm over the substrate material with a transversal speed of $v = 1$ m/s, a gap of $d_g = 5$ mm between the meander for $n = 20$ passes in total. The powder was injected with a carrier gas flow of $\dot{V}_g = 5$ SLPM at a powder feeding rate of $\dot{m}_p = 17$ g/min through an injector head with an internal diameter of 2 mm.

To observe the influence of the magnetic field manipulation on the coating thickness, it was measured at ten positions at two specimens using the eddy current method (DUALSCOPE®MP40 EAW3.3; Helmut Fischer GmbH, Sindelfingen, Germany). The hardness of the coatings was determined by microindentation (MICROMET™ 1; Bühler Ltd. Illinois, USA) in the cross section with a load of $m_i = 100$ g, which was held for a period of $t = 15$ s. For each parameter set, $n_i = 20$

IOP Conf. Series: Materials Science and Engineering **480** (2019) 012004 doi:10.1088/1757-899X/480/1/012004

points on two specimens were investigated to achieve a statistically relevant database. Furthermore, the porosity of the coatings was measured at these cross sections by optical analysis of ten images with the software ImageJ. Phase identification of the specimen were conducted using an X-ray diffractometer (XRD 3000; Seifert, Ahrensberg, Germany), which is equipped with a Cu-Kα X-ray source generating a wavelength of 1.540598 Å. The incidence angle was kept constant at $\omega = 10°$ while $2\theta$ was measured between 20° and 80° in steps of 0.05° holding each position for 5 s.

**Table 1.** Designation of the investigated process parameters and the applied external magnetic flux density.

| Designation | Current [A] | Gas flow rate Ar/$H_2$ [SLPM] | Magnetic flux density B [mT] |
|---|---|---|---|
| $A_{0mT}$ | 540 | 52 / 8 | 0 |
| $A_{680mT}$ | 540 | 52 / 8 | 680 |
| $A_{1,200mT}$ | 540 | 52 / 8 | 1,200 |
| $B_{0mT}$ | 360 | 52 / 8 | 0 |
| $B_{680mT}$ | 360 | 52 / 8 | 680 |
| $B_{1,200mT}$ | 360 | 52 / 8 | 1,200 |

To gain a further insight in the influence of the manipulation on the footprint, spray spots were applied on S235JR substrates, with the dimensions of 50 x 50 x 5 mm³, by keeping the plasma torch in front of these specimen at a stand-off distance of 100 mm for a duration of 3 s. The height profile of these spray spots was measured by confocal laser microscopy (VKX 210; Keyence Deutschland GmbH, Neu-Isenburg, Germany).

## 3. Results and discussion

In **Figure 3**, the results of the head-on recordings are displayed for the parameter set A. The accumulated red dots represent the points of maximum intensity for 256 consecutive head on images. The dashed lines are the center lines of the anode nozzle, the dotted line is an auxiliary line used to follow the movement of the arc attachment point. Without an external magnet field, the 256 points of maximum intensity are located in a restricted area at the top right of the center of the nozzle. This indicates a preferential azimuthal attachment point of the electric arc for this parameter even without any manipulation by an external magnetic field. It must be note that further investigations are necessary to conclude, whether this preferential azimuthal attachment point is valid for a longer time and if they apply for other process parameters as well.

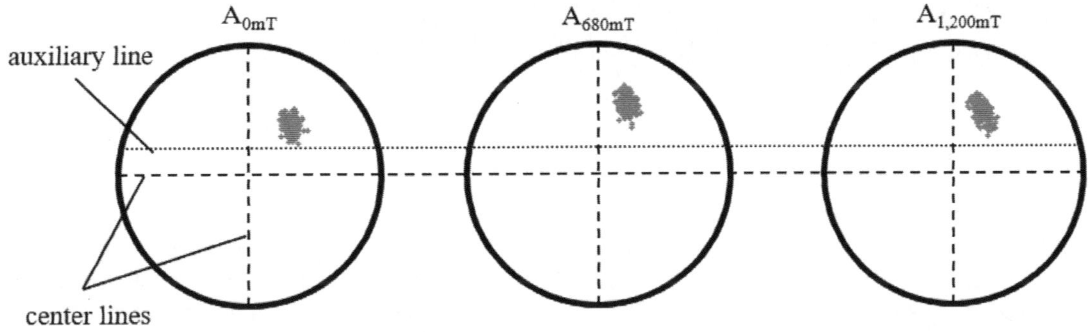


**Figure 3.** Arc attachment points, determined by high-speed videography for the parameter A with no external magnetic field, $A_{0mT}$, and the same parameter manipulated by two magnets with different magnetic flux densities, $A_{680mT}$ and $A_{1,200mT}$.

The manipulation by an external magnetic field of 680 mT led to an upward movement of measured maximum intensity, implying a position change of the arc attachment point. While moving upwards, the measured maximum intensity was shifted towards the vertical center line, which is ascribed to the circular form of the anode. A similar movement of the attachment point was observed for the process parameter $A_{1,200mT}$, even though this shift seems to be slightly smaller. The observed movement of maximum intensity corresponds to the expected direction for a movement of the attachment point. The external magnetic field induces a Lorentz force, which affects the electric arc and consequently moves the arc attachment point to a different position. For the parameter set B, which is not shown here, a similar but somewhat smaller movement of the arc attachment point was observed. These first results prove the feasibility of the proposed manipulation of an electric arc in a cascaded one cathode one anode plasma torch by an external magnetic field. However, these results need to be validated as the overall timespan of the consecutive 256 images was limited to $t_{rec} \approx 7.4$ ms.

**Figure 4** shows micrographs for the non-manipulated samples $A_{0mT}$ and $B_{0mT}$. The parameter $A_{0mT}$ led to a coating thickness of about $t = 236$ µm. Parameter $B_{0mT}$ exhibits a coating thickness of $t = 131$ µm. Considering the relatively low powder feeding rate of $\dot{m}_p = 17$ g/min, these coating thicknesses are quite remarkable indicating a high deposition efficiency for the parameter set A. The porosities of these coatings were determined to be approx. $P = 7.4$ % for parameter $A_{0mT}$ and $P = 10.9$ % for parameter $B_{0mT}$.

**Figure 4.** Polished cross sections of the coatings $A_{0mT}$ and $B_{0mT}$.

**Table 2** shows the measured thickness and its standard deviation as well as the determined porosity of the coatings applied with the parameter sets A and B. While the thickness of the coating $A_{0mT}$ was about 236 µm, the manipulation of the plasma torch by the external magnetic fields led to an increased coating thickness of 285 µm and 284 µm. This indicates that there is an influence of the manipulation on the coating properties. However, this effect was only observed for the parameter set A. Parameter set B shows a rather constant coating thickness in the range of 124 µm to 131 µm. The same can be stated for the porosity of the coatings. In case of parameter set A, the porosity ranges between 7.2 % and 8.3 %. In case of the parameter B set, it varies between 10.2 % and 11.4 %. These values show no significant influence of the arc manipulation on the fraction of the pores. Yet, as this visual determination of the porosity is an imprecise procedure, capturing rather drastic effects, there could be an effect on the porosity which might not be detected by this method.

**Table 2.** Thickness and porosity of the alumina coatings.

| designation | thickness [µm] | standard deviation [µm] | porosity [%] |
|---|---|---|---|
| $A_{0mT}$ | 236 | 23.1 | 7.4 |
| $A_{680mT}$ | 285 | 13.4 | 7.2 |
| $A_{1,200mT}$ | 284 | 13.6 | 8.3 |
| $B_{0mT}$ | 131 | 11.2 | 10.9 |
| $B_{680mT}$ | 127 | 13.3 | 11.4 |
| $B_{1,200mT}$ | 124 | 8.8 | 10.2 |

In order to further investigate the influence of external magnetic fields, the microhardness values ($HV_{0.1}$) of the coatings were determined. The results are displayed in **Figure 5**. The mean value of the hardness of the coatings applied with parameter set A ranges between 1,158 $HV_{0,1}$ and 1,229 $HV_{0,1}$, compared to the hardness of the coatings applied with parameter set B, which is in a range of 993 $HV_{0,1}$ to 1,064 $HV_{0,1}$. The difference in hardness between parameter set A and B is in accordance with the porosities of the coatings given in **Table 2**. The higher hardness of parameter A seems to be based on the less porous structure of these coatings. Yet again, there is no discernible difference between the microhardness of the coatings due the manipulation by different external magnetic fields detectable. The depicted differences are too small compared to the statistical variance to identify any trend.

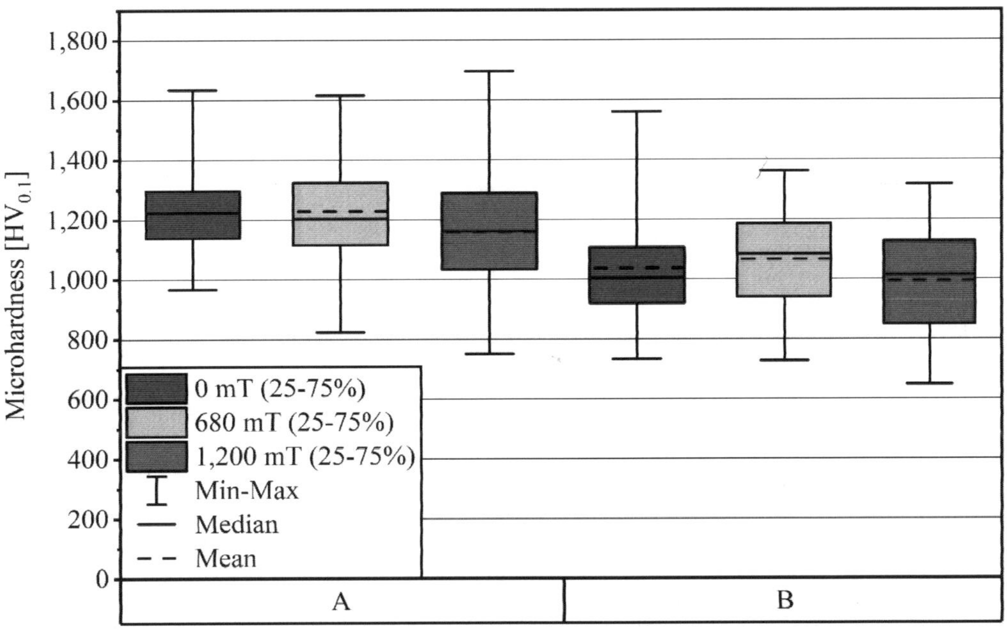


**Figure 5.** Box-and-whisker plot of the microhardness ($HV_{0.1}$) of the alumina coatings for the parameters A and B affected by the external magnetic fields.

**Figure 6** shows the measured XRD spectra of the alumina feedstock powder (black) and the resulting coatings as well as the positions of the peaks for $\alpha$-$Al_2O_3$ (JCPDS 01-088-0826) and $\gamma$-$Al_2O_3$ (JCPDS

00-056-0457). The spectrum of the analyzed powder exhibits the peaks of $\alpha$-$Al_2O_3$, demonstrating that the feedstock material consists of the $\alpha$-phase only. In the spectra of the resulting coatings, the peaks of $\alpha$-$Al_2O_3$ are comparably small in case of the parameter set B and are barely visible for the coatings applied by the parameter set A. The dominant peaks of all coatings correspond to $\gamma$-$Al_2O_3$, which is the typically prevailing phase in thermally sprayed alumina coatings [15, 16]. The rapid solidification of the molten alumina particles during the impact on the substrate leads to the formation of the $\gamma$-$Al_2O_3$-phase, while the formation of $\alpha$-$Al_2O_3$ is negligible for comparably low substrate temperatures [17, 18].

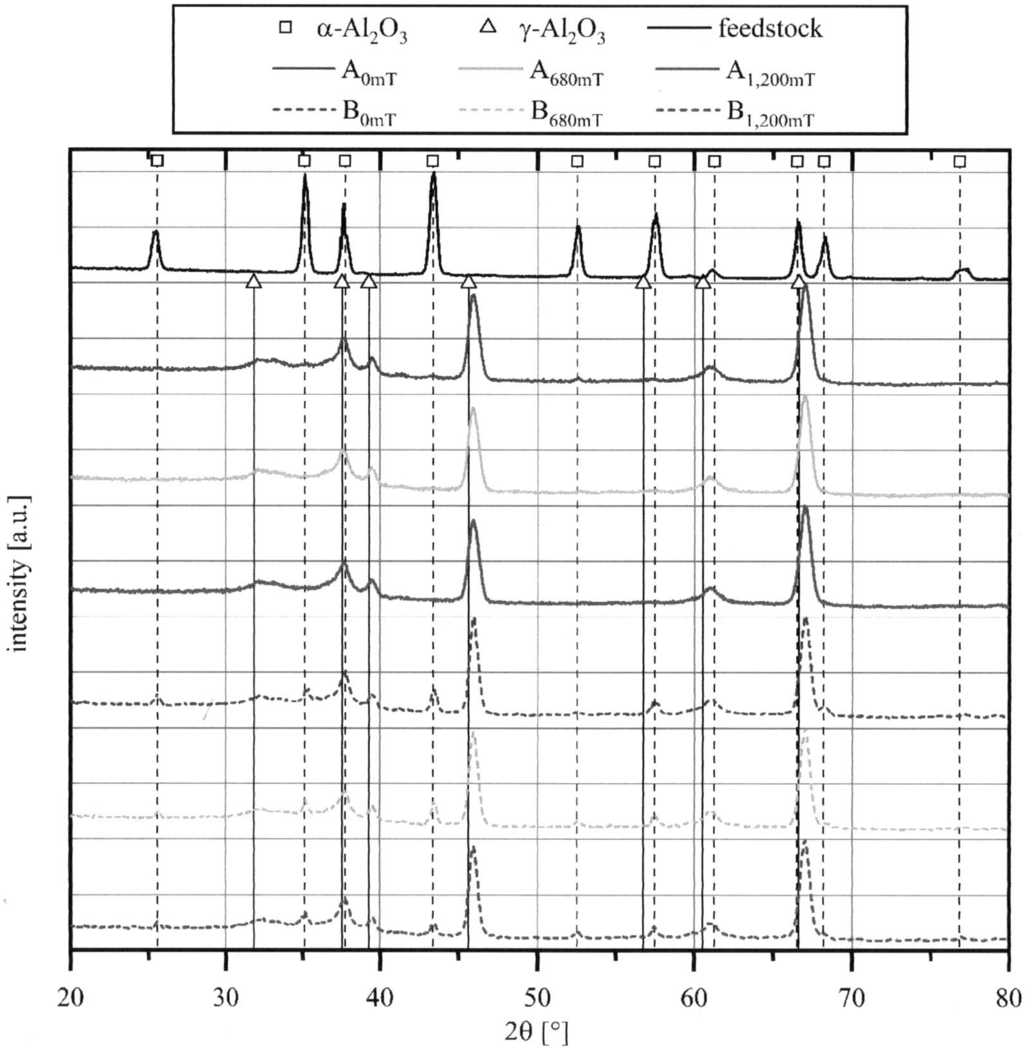


**Figure 6.** XRD spectra of the alumina feedstock powder (Amdry™ 6062) and the applied coatings as well as the positions of the peaks for $\alpha$-$Al_2O_3$ (JCPDS 01-088-0826) and $\gamma$-$Al_2O_3$ (JCPDS 00-056-0457). The spectra were normalized to their respective highest peak.

The high contents of $\gamma$-$Al_2O_3$ for parameter $A_{0mT}$ indicate, that the alumina particles were nearly fully molten during the flight. As a result, it can be stated that the cascaded single cathode single anode plasma torch operating with parameter set A is capable of fully melting und thus effectively processing alumina. The spectra of the coatings $A_{680mT}$ and $A_{1,200mT}$ exhibited little difference to the

IOP Conf. Series: Materials Science and Engineering **480** (2019) 012004 doi:10.1088/1757-899X/480/1/012004

spectrum of the $A_{0mT}$ coating. Similarly, for parameter set B there is no significant difference in the XRD spectra caused by the influence of external magnetic fields.

To visualize the influence of the magnetic manipulation on the deposition of the particles on the substrate, spray spots were applied for parameter set A. Spray spots are a simple and comparatively fast way to detect the local distribution of molten particles in the free jet. The upper part of **Figure 7** shows the measured height profiles of the spray spots for the parameter $A_{0mT}$ as well those of the manipulated parameters $A_{680mT}$ and $A_{1,200mT}$. The direction of the particle injection is also illustrated. The particles are injected in top-down direction. For all three cases, the spray spot of the SinplexPro™-90 exhibits an asymmetric shape. Two effects are most likely responsible for this shape. The first might be the swirl of the plasma flow which deflects the particles injected from above and thus, prevents the formation of a circular spray spot. The second reason might be the fixed position of the arc attachment point leading to an asymmetric distribution of temperatures and velocities in the cross section of the plasma jet. It is known that a fixed anodic arc attachment greatly affects the plasma gas properties in the region of particle injection thus having a huge impact on the particle in-flight properties [7]. Consequently, a fixed and decentered attachment point could promote the asymmetric distribution of the injected powder in the plasma jet resulting in an asymmetric spray spot.

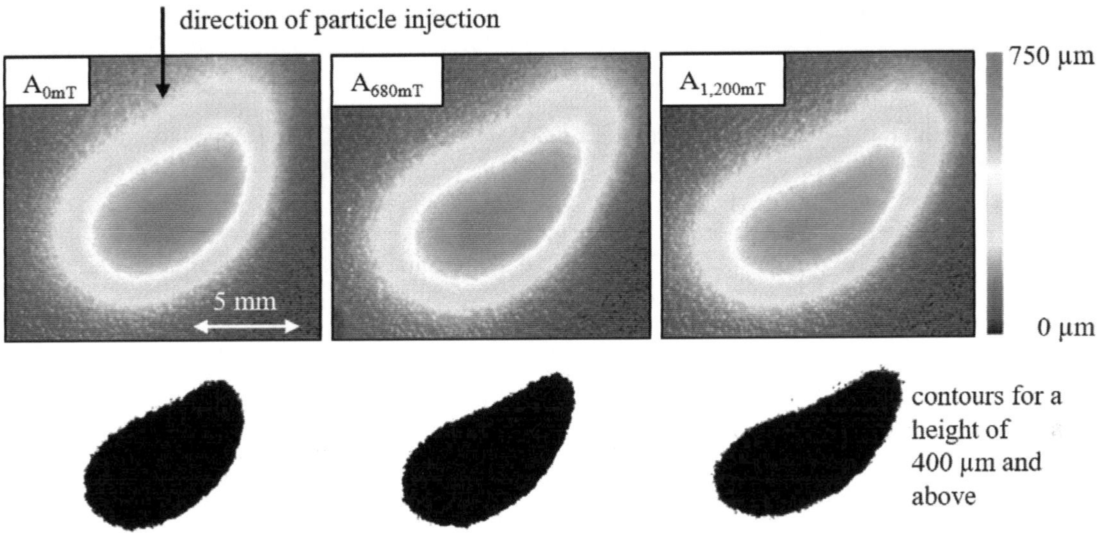


**Figure 7.** Influence of the arc manipulation on the shape of the spray spot. Top: Height profile of the spray spots for the process parameter A under influence of the external magnetic fields. Bottom: Corresponding contours for a height of 400 µm and above.

If the preferential attachment point of the electric arc is manipulated by external magnetic fields, it should have an influence on the shape of the spray spot as well. This effect can be observed by comparing the height profiles shown in **Figure 7**. To facilitate a comparison, the contours for a height of 400 µm and above were drawn. They visualize the effect of the shift in the arc attachment position due to the magnetic manipulation on the shape of the spray spot. The shape of the spray spot becomes slightly broader with increasing magnetic remanence. Although this qualitative comparison shows only slight differences, it is in accordance with the measured maximum intensities shown in **Figure 3**. It can be expected to observe a more pronounced effect by using magnets with higher magnetic remanences or at different positions.

## 4. Conclusions

This study investigated the possible effects of external magnetic fields on a cascaded plasma torch. The position of the arc attachment point was determined and its change due to the manipulation was observed. Subsequently, the effect of this position change on the spray spot was visualized. The influence of anodic attachment manipulation on the applied coatings and their properties was examined. The results can be summarized as follows:

- Cascaded single cathode single anode plasma torches, like the SinplexPro™-90, can exhibit a preferential arc attachment point for certain parameter sets
- It is possible to manipulate the anodic arc root in such plasma torches by external magnetic fields
- This manipulation has an influence on the particle deposition leading to different shapes of the spray spot
- The effect on deposition showed, so far, no pronounced influence on the resulting coatings' porosity, phase composition and hardness

As the presented results are first investigations, further research is needed especially using magnets with adjusted magnetic flux densities and at different positions to confirm these results and to evaluate the effects of the manipulation by magnetic fields in plasma spraying in detail. Supplementary measurements of the coating hardness and its porosity are necessary to improve the statistical basis and to detect a possible effect of the manipulation on the coating properties. The authors aim to investigate the effect of the arc attachment position on the temperature and velocity distribution of the plasma jet. For this purpose, simulations of the plasma torch as well as the resulting plasma jet will be conducted and supplemented by computer tomographic measurements of the plasma jet temperatures.

**Acknowledgments**
The authors gratefully acknowledge the financial support of the German Research Foundation (DFG) within the Project ''Beeinflussung des Lichtbogenverhaltens im kaskadierten DC-Ein-Kathoden-Ein-Anoden-Plasma-Generator (DC-EKEAPG) zur Verbesserung der Prozessstabilität und der Schichteigenschaften'' (GZ: BO 1979/50-1 and ZI 1311/4-1).

## References

[1] Heimann R B 2008 *Plasma spray coating: Principles and applications* 2nd edn (Weinheim: Wiley-VCH)
[2] Fauchais P L 2004 Understanding plasma spraying *J. Phys. D: Appl. Phys.* **37** R86-R108
[3] Eckert E R G, Pfender E and Wutzke S A 1967 Study of electric arc behavior with superimposed flow *AIAA Journal* **5** 707–13
[4] Duan Z and Heberlein J V 2002 Arc Instabilities in a Plasma Spray Torch *J. Therm. Spray Technol.* **11** 44–51
[5] Bisson J F, Gauthier B and Moreau C 2003 Effect of Plasma Fluctuations on In-Flight Particle Parameters *J. Therm. Spray Technol.* **12** 38–43
[6] Bisson J F and Moreau C 2003 Effect of Direct-Current Plasma Fluctuations on In-Flight Particle Parameters: Part II *J. Therm. Spray Technol.* **12** 258–64
[7] L. Marqués J, Forster G and Schein J 2009 Multi-Electrode Plasma Torches: Motivation for Development and Current State-of-the-Art *TOPPJ* **2** 89–98
[8] Vardelle A, Moreau C, Themelis N J and Chazelas C 2015 A Perspective on Plasma Spray Technology *Plasma Chem Plasma Process* **35** 491–509
[9] Kotalik P and Nishiyama H 2002 An effect of magnetic field on arc plasma flow *IEEE Trans. Plasma Sci.* **30** 160–1
[10] Minoo H, Arsaoui A and Bouvier A 1995 An analysis of the cathode region of a vortex-stabilized arc plasma generator *J. Phys. D: Appl. Phys.* **28** 1630–48

[11] Park J M, Kim K S, Hwang T H and Hong S H 2004 Three-Dimensional Modeling of Arc Root Rotation by External Magnetic Field in Nontransferred Thermal Plasma Torches *IEEE Trans. Plasma Sci.* **32** 479–87

[12] Szente R N, Munz R J and Drouet M G 1988 Arc velocity and cathode erosion rate in a magnetically driven arc burning in nitrogen *J. Phys. D: Appl. Phys.* **21** 909–13

[13] Guggenheim L, Schwenk A, Zimmermann S, Schein J and Landes K 2018 Untersuchungen zum Einfluss von Permanentmagneten auf das physikalische Verhalten von kaskadierten, wandstabilisierten Lichtbögen und keramisch gespritzten Schichten *7. GTV Kolloquium Thermisches Spritzen & Laser Cladding (Luckenbach, 7.09.2018)* ed K Nassenstein and K von Niessen pp 95–101

[14] Dorier J-L, Gindrat M, Hollenstein C, Salito A, Loch M and Barbezat G 2001 Time-resolved imaging of anodic arc root behavior during fluctuations of a DC plasma spraying torch *IEEE Trans. Plasma Sci.* **29** 494–501

[15] Ault N N 1957 Characteristics of Refractory Oxide Coatings Produced by Flame-Spraying *J American Ceramic Society* **40** 69–74

[16] McPherson R 1989 A review of microstructure and properties of plasma sprayed ceramic coatings *Surface and Coatings Technology* **39-40** 173–81

[17] McPherson R 1980 On the formation of thermally sprayed alumina coatings *J Mater Sci* **15** 3141–9

[18] McPherson R 1973 Formation of metastable phases in flame- and plasma-prepared alumina *J Mater Sci* **8** 851–8

IOP Conf. Series: Materials Science and Engineering **480** (2019) 012005 doi:10.1088/1757-899X/480/1/012005

# Novel Fe-based and HVAF-sprayed coating systems for large area applications

**K Bobzin, M Öte, M A Knoch and J Sommer***

Institut für Oberflächentechnik, RWTH Aachen University, Aachen

* e-mail: sommer@iot.rwth-aachen.de

**Abstract.** In paper-machines, components such as dryer cylinders are often exposed to a combined load of wear and corrosion and, therefore, require an adequate surface protection. A high thermal conductivity of these coatings is desired, to increase the overall thermal efficiency. In this study, a novel high velocity air-fuel- (HVAF-) sprayed FeCrB/WC-Co coating is compared to an industrially established sealed FeCrBMn coating applied by wire-arc-spraying (WAS) in terms of wear behaviour of aged samples and thermal conductivity. The chosen feedstock material had powder fractions of -32 +11 µm and -20 +3 and were applied with a powder feed rate of $\dot{m}$ = 200 g/min by means of HVAF. A high powder feed rate and fine powder fractions were considered to reduce the required time to coat large areas and to reduce the post-processing effort. The samples were aged in saturated NaCl-solution steam at a temperature of T = 105 °C for t = 100 h. For the aged HVAF samples, a low influence of the aging on the wear rate was observed for both powder fractions. In contrast, for the aged WAS reference coating, a noticeable higher wear rate was measured. The thermal conductivity of the HVAF-sprayed coatings is significantly influenced by the employed powder fraction.

## 1. Introduction

In order to dry raw material during paper production, dryer cylinders are used. These cylinders are heated by hot steam, which flows through the cylinders [1] and are typically made of grey cast iron or mild steel such as 1.0038. In many applications, both materials have a poor wear and corrosion resistance and need to be protected against corresponding loading conditions, e.g. with thermally sprayed coatings. In case of dryer cylinders, it is state of the art to apply Fe-based coatings by WAS or cemented carbides by high velocity oxygen-fuel (HVOF) spraying [2]. Both coating systems have several advantages and disadvantages.

Fe-based coatings applied by WAS exhibit a good cost-efficiency, due to the inexpensive raw materials and the high achievable wire feed rates of $\dot{m} \approx$ 100 g/min [3]. A high corrosion and wear resistance of Fe-based materials can be achieved by including a high amount of Cr and B in the alloy respectively [4, 5]. The microstructure of Fe-based WAS coatings is porous and the coating surface is rough in the as-sprayed condition [6]. Therefore, a time consuming grinding process is necessary to achieve the required surface roughness. Experience shows that high coating thicknesses of $s_{ct}$ = 700 – 1.200 µm [7, 8] are necessary to achieve the required corrosion protection. The corrosion resistance can be further increased by sealing the pores using an epoxy resin, which requires an additional process step. When using epoxy resin-based sealers, the long-term temperature resistance of the sealer must be considered as well. The high coating thicknesses and a low thermal conductivity of these coatings result in a decreased heat flow from the inner part of the cylinder to the coated surface. For this reason, the coating insulates the cylinder thermally and the thermal efficiency of the whole system can decrease.

Thinner, more near net shape and denser coatings can be produced by applying cemented carbides with HVOF-spraying [1]. For these coatings, a lower powder feed rate, higher hardness and higher material prices have to be considered. Fe-based feedstock materials can be a cost efficient alternative to cemented carbide coating systems. It has already been shown that novel Fe-based and HVOF-sprayed coating systems can protect mild steel substrates as good as industrially established cemented carbides, depending on the load collective [9, 10]. A further trend in thermal spraying is the use of finer feedstock materials [11], due to the achievable increased powder deposition efficiency, decreased porosity and surface roughness. However, experience shows that processing of finer powder feedstock materials by means of HVOF spraying can result in nozzle clogging.

The HVAF process is a modification of the HVOF process [12, 13]. It uses air instead of $O_2$ as oxidant [12, 13]. Due to the $N_2$ in the air, high gas flow rates and a lower flame temperature, compared to HVOF, are achievable. The high gas flow rates enable high powder feed rates [14] and high particle velocities [13]. The lower flame temperature can result in a lower particle temperature, which might be helpful to process finer feedstock materials without nozzle clogging [12]. According to a study of Verstak et al., the lower particle temperature and the high particle velocities can result in dense and smooth coatings with a low oxide content [14]. These results were also confirmed in [12, 13, 15–17].

In the studies [18–20], a novel HVAF sprayed FeCrB/WC-Co coating system with high wear and corrosion resistance was developed. FeCrB/WC-Co was chosen as a cost efficient feedstock material in the powder fractions of -32 +11 µm and -20 +3 µm [19, 20]. Using the HVAF-process, it was possible to produce dense, near net shape coatings with high powder feed rates and to process the finer feedstock materials. These investigations showed that the new coating system provides comparable or even better corrosion and wear resistance than the sealed WAS FeCrBMn reference coating system, depending on the powder feed rate and powder fraction. In the current study, the novel HVAF-sprayed FeCrB/WC-Co coating system is analysed with regard to the wear behaviour in aged condition. Moreover, the thermal conductivity and the thermal resistance of the final coating system have been analysed. The aging of the samples was performed in a corrosive environment at increased temperature. These conditions are similar to those in paper machines.

## 2. Experimental setup and materials

The mild steel 1.0038 (S235JR) is widely used for dryer cylinders and was, therefore, chosen as substrate material. The substrates with the dimensions of 40 x 50 x 8 mm³ were cleaned and roughened using grit blasting. The coatings were applied onto the prepared surfaces via HVAF and WAS. The chemical compositions of the used feedstock materials are given in Table 1.

**Table 1:** Chemical compositions of FeCrB/WC-Co and FeCrBMn (SP112) feedstock material, as given by the producer companies [wt.%].

| Alloy | Fe | Cr | B | C | W | Co | Si | Mn |
|---|---|---|---|---|---|---|---|---|
| FeCrBMn (Sp112) | Bal. | 27.5 - 29 | 3.8 | 0.1 | - | - | 1.5 | 1.5 |
| FeCrB/WC-Co | Bal. | 20 - 23 | 3.5 - 4 | 1 - 2 | 10 - 12 | 1 – 1-5 | | |

The wire FeCrBMn (Corodur Fülldraht GmbH, Willich, Germany) is often used for protecting components against corrosion at increased temperature and, chosen therefore as reference. For the HVAF process, the powder FeCrB/WC-Co (Above Material Technology Co., Ltd., Beijing, China) was chosen as feedstock material. Both materials FeCrB/WC-Co and FeCrBMn have high amounts of Cr and B to achieve a high corrosion and wear resistance. WC-Co was added to the FeCrB-powder to further increase the wear resistance.

The process parameters used in order to apply HVAF and WAS coatings are given in [18–20]. An overview of all samples and the varied process parameters in this study are given in Table 2. The FeCrBMn coating applied by means of WAS was used as reference and is designated hereafter as WAS-Ref. The WAS-system G30/4SF-Push-LD/U2 of Oerlikon Metco AG (Winterthur, Switzerland) was used to apply the reference coatings. To increase the corrosion resistance of the sample WAS-Ref, it was sealed with the epoxy based industrially established sealer #HM2407 (Diamant Metallplastic GmbH, Mönchengladbach, Germany). The HVAF-spraying system AK-07 (Kermetico Inc., Benicia, USA) was used to apply the FeCrB/WC-Co coatings.

**Table 2**: Overview of the samples investigated in this study.

| Sample name | Process | Powder fraction [μm] | Powder feed rate ṁ [g/min] |
|---|---|---|---|
| WAS-Ref* | WAS | Cored wire | 110 |
| Fe/W_D-32 | HVAF | -32 +11 | 200 |
| Fe/W_D-20 | HVAF | -20 +3 | 200 |

*sealed

The experimental setup developed in [19] was used in this study to apply HVAF coatings, see Figure 1. With this experimental setup a thin layer thickness of approximately 13 μm can be achieved, even for high powder feed rates of ṁ = 200 g/min. Using this setup, crack-free coating could be produced with high powder feed rates [19, 20].

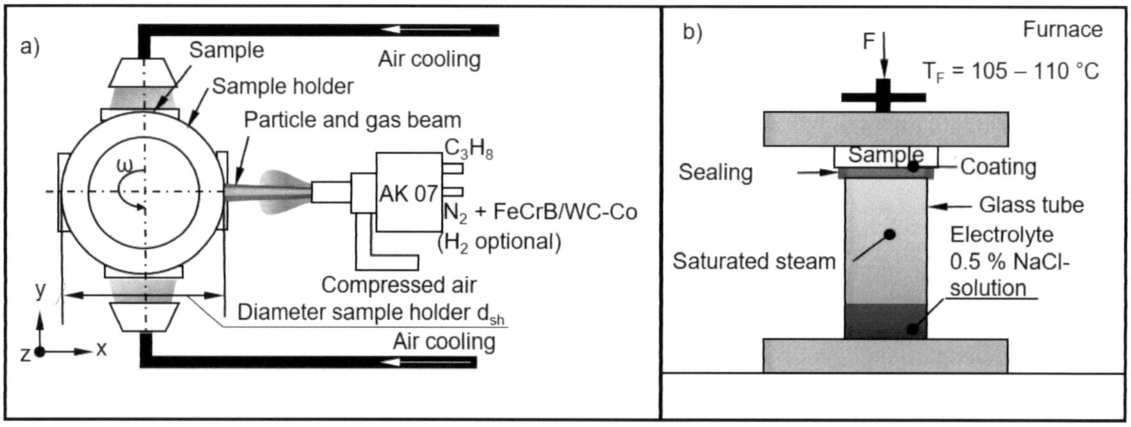


**Figure 1**: Schematic drawing of the setup a) for the application of the HVAF-sprayed coatings and b) the aging of the coatings at higher temperatures.

*2.1 Aging test and wear behaviour of aged samples*

Dryer cylinders are exposed to a combined load of wear and corrosion. Corrosion reactions can change the chemical bonds, increase the volume by including additional atoms or lead to material loss by dissolution. Any of these mechanisms can reduce the cohesion between two splats which in turn, can reduce the wear resistance. To investigate the influence of corrosion on the wear behaviour, coated samples were aged in application-oriented corrosive conditions and then investigated regarding their wear behaviour using the Pin-on-Disc (PoD) -test to simulate the combined load. Before the aging, all samples were ground and polished to Ra ≈ 0.5 μm using a 10 μm diamond pad to ensure all sample have comparable initial conditions. The polished samples were then aged using the experimental setup illustrated in Figure 1 b). The coatings were exposed to saturated NaCl steam at a temperature of T = 105 – 110 °C for a time period of t = 100 h. Subsequently, the wear resistance of the aged samples was investigated using a PoD tribometer from CSM instruments (Freiburg, Germany). $Al_2O_3$-balls with a diameter of $Ø_{cb}$ = 6 mm were used as counterbody, in order to force an abrasive type of wear

IOP Conf. Series: Materials Science and Engineering **480** (2019) 012005 doi:10.1088/1757-899X/480/1/012005

and to avoid adhesion. The further PoD-test parameters are listed in Table 3. The wear tracks were analysed using the confocal laser scanning microscope (CLSM) VKX-210 (Keyence Corp, Osaka, Japan). Furthermore, the cross-section of the wear tracks was determined using CLSM and the wear rate was calculated using equation (1). The wear rates of the aged samples were compared to the wear rates of not-aged samples, determined in studies [19, 20]. In a last step, cross-sections of the aged and worn samples were prepared and investigated using the light microscope Zeiss Axiophot (Carl Zeiss AG, Oberkochen, Germany).

**Table 3:** PoD test parameters.

| Parameter | Value |
|---|---|
| Wear track diameter d | 10 mm |
| Wear track length s | 1,000 m |
| Normal load F | 10 N |
| Rotation speed v | 100 mm/s |
| Test temperature | Room temperature (RT) |

$$K\left[\frac{\mathrm{mm}^3}{\mathrm{Nm}}\right] = V \cdot \left[F\,(s \cdot \frac{d}{d_t})\right]^{-1} \tag{1}$$

*2.2 Determination of the thermal conductivity and thermal resistance*

The thermal properties of the wear and protection coatings influence the energy efficiency of the dryer cylinder and therefore the overall operating costs. A higher thermal conductivity and a thinner coating, compared to the reference, can decrease the thermal resistance of the system consisting of coating and cylinder. The thermal conductivity λ depends on the thermal diffusivity α, the density ρ and the thermal capacity $c_p$ of the coating. All values were measured using freestanding coatings. These coatings were produced on a special substrate, which enables peeling of the coatings from the substrate. In contrast to the samples used for aging and wear investigation, the samples used for determining the thermal properties of coatings were produced with a powder feed rate of $\dot{m}$ = 100 g/min, due to the speed limit of the robot. The further process parameters were kept constant. The thermal conductivity λ of coatings was calculated using equation (2) [21]. The density ρ was measured using the gas pycnometer AccuPyc 1330 (Micromeritics GmbH, Aachen, Germany) at T = RT. It was assumed that the density stays constant in the considered temperature range. The thermal diffusivity α was determined using the laserflash system Flashline 4010 (Anter Corp., Pittsburgh, USA) and the heat capacity $c_p$ was determined using the thermo analysis system Setsys Evolution (Setaram Instrumentation, Caluire, France). Both parameters were determined at a temperature of T = 200 °C. In the literature, steam temperatures between 100 °C < T < 300 °C are mentioned for dryer cylinders, depending on the type of paper [1, 22].

$$\lambda\left[\frac{W}{\mathrm{mK}}\right] = \alpha \cdot \rho \cdot c_p \tag{2}$$

The thermal resistance of the different coatings was calculated according to equation (3) [21]. The following assumptions and approximations were made;

- Length of rotary dryer (L) = 3 m
- Inner radius of rotary dryer ($r_{r1}$) = 0.9 m
- Outer radius of rotary dryer ($r_{r2}$) = 0.91 m
- Thermal conductivity of rotary dryer ($\lambda_r$) = 48 W/mK
- Coating thickness WAS-Ref ($s_{ct}$) = 500 µm
- Coating thickness of HVAF-sprayed coatings ($s_{ct}$) = 250 µm

$$R = \frac{1}{2\pi L}\left[\frac{1}{\lambda_c}\ln\left(\frac{r_{c2}}{r_{c1}}\right) + \frac{1}{\lambda_r}\ln\left(\frac{r_{r2}}{r_{r1}}\right)\right] \tag{3}$$

## 3. Results and discussion

*3.1 Aging test and wear behaviour of aged samples*

The corrosive atmosphere of dryer cylinders was simulated by a saturated NaCl-steam at a temperature of T = 105 -110 °C. In initial conditions, the surface of all coatings and the substrate were comparable and shined metallic bright, see Figure 2. After the aging, the substrate surface was heavily corroded. This result show that the mild steel 1.0038 has no protection against corrosion under investigated conditions. For WAS-Ref, Fe/W_D-32 and Fe/W_D-20, selective corrosion and crevice corrosion were observed at the surface. Crevice corrosion could only be observed at the edge where the sample is in contact with the sealing.

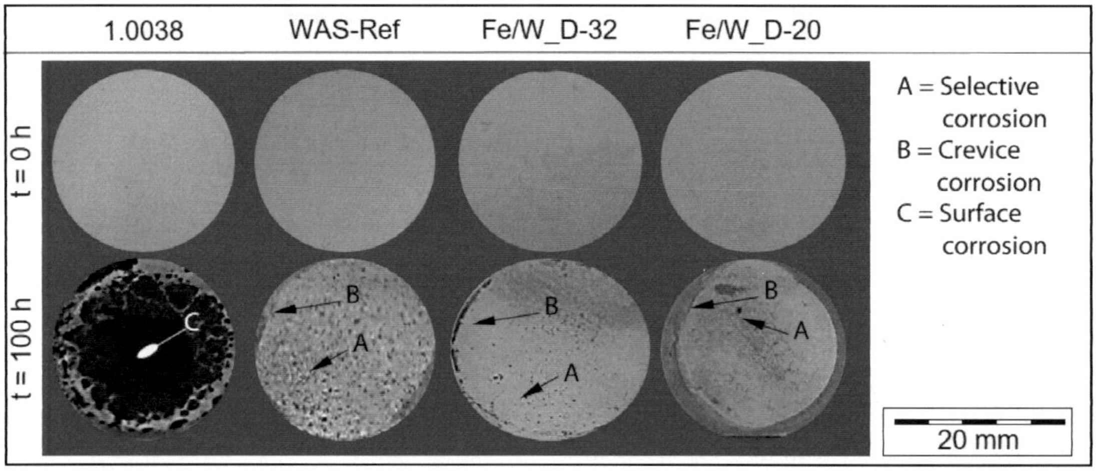


**Figure 2:** Images of the aged coating surfaces at initial conditions and after t = 100 h in saturated steam of 0.5 % NaCl-solution at T = 105 °C

Selective corrosion can be observed distributed evenly over the entire surface of WAS-Ref. In contrast the samples Fe/W_D-32 and Fe/W_D-20 exhibit selective corrosion only in a few areas. Qualitatively, Fe/W_D-20 shows less selective corrosion than Fe/W_D-32. This result confirms the good corrosion resistance of the coatings presented in [20]. It is assumed that selective corrosion is a result of a local Cr segregation. In study [23], the local reduction of Cr content, due to the formation of Borides (Fe, Cr), was observed and proven for HVOF-sprayed FeCrNiBC and FeCrNiMoBC coatings. Selective corrosion can be traced back to the local differences in the amount of free Cr in the coatings, as a result of boride formation and related Cr segregation.

The microstructure of the aged samples is presented in Figure 3 a). No under corrosion was observed in the interface between the coating and the substrate, for all coatings. WAS-Ref exhibits oxidized particles in deeper layers, see Figure 3 a). These particles could have formed during the spraying process or during the aging. In contrast to WAS-Ref, oxidized areas could not be observed in deeper layers for the HVAF-sprayed samples Fe/W_D-32 and Fe/W_D-20. Both samples show a very dense microstructure after the aging, which is comparable to that in as-sprayed condition [19, 20].

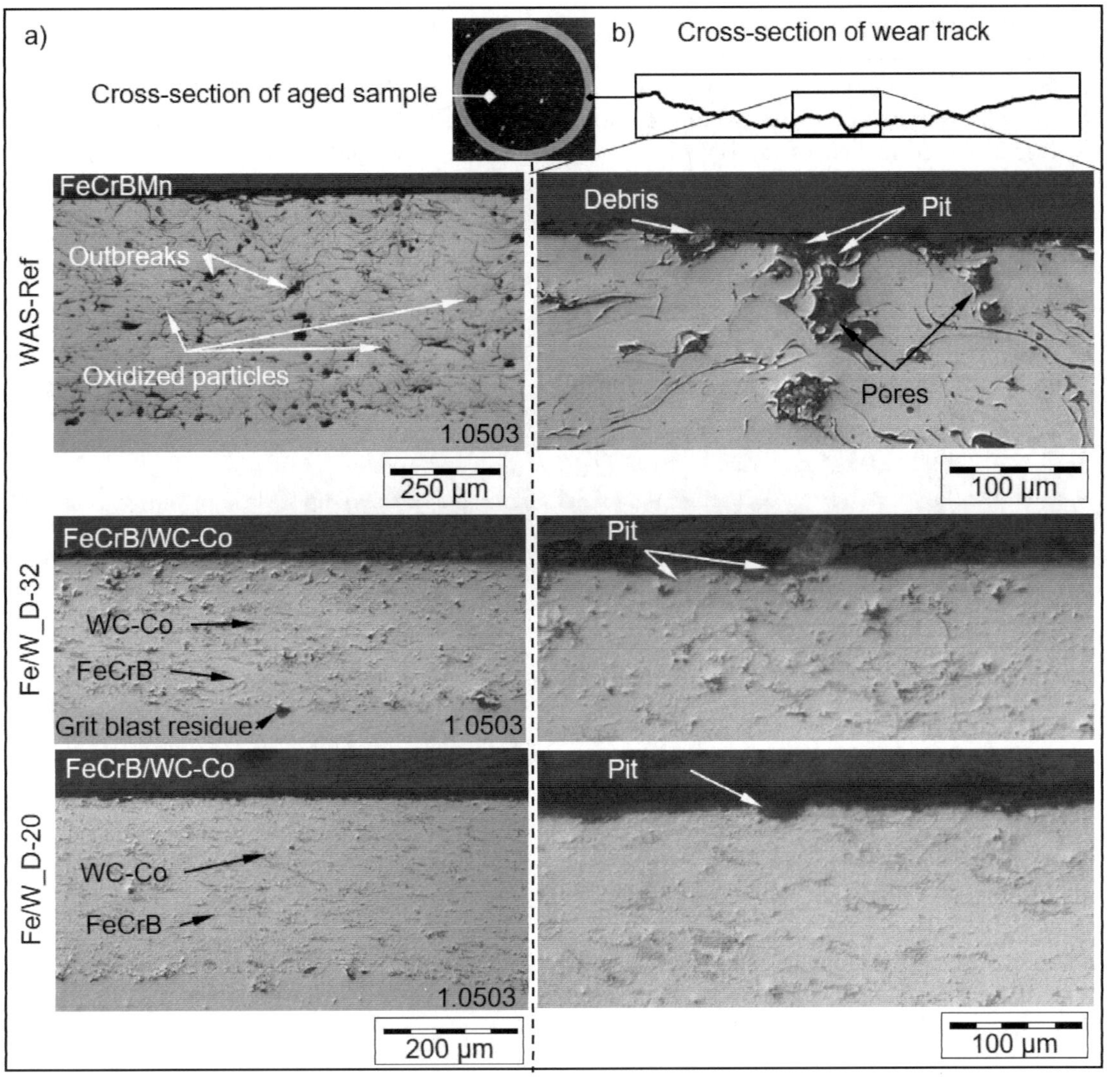


**Figure 3:** Cross-section of the a) aged sample and of b) the wear track.

In Figure 3 b), cross sections of the wear tracks are presented. Pits in the surface of the wear tracks can be observed for all samples. Compared to the HVAF-sprayed coatings, significantly bigger pits in the wear track can be observed for WAS-Ref. Furthermore, bigger pores can be observed in the vicinity of the wear track for WAS-Ref. It is assumed that these pores are a result of the metallographic preparation and caused by a weakened cohesion in the coating after the aging and wear tests. Using light microscopy and a magnification of 500x, no cracks or a delamination could be observed, for all coatings.

The cross sections and surface images of the wear tracks captured by CLSM are presented in Figure 4. The CLSM-images were taken with the integrated light microscope and coloured in a post processing step to visualize the surface topography of the wear track. The widths of the cross-sections of the wear tracks in not-aged and aged conditions of the HVAF-sprayed conditions looks significantly smaller compared to those of WAS-Ref. A further interesting aspect, which can be observed in the cross-section areas is that the width and height of the wear tracks are comparable for the samples Fe/W_D-32 and Fe/W_D-20 in aged and not-aged conditions. In contrast, the height and width of the wear track of WAS-Ref increases in the aged condition. This aspect is a first indication that the

IOP Conf. Series: Materials Science and Engineering **480** (2019) 012005 doi:10.1088/1757-899X/480/1/012005

cohesion in the coating WAS-Ref has been reduced after the aging process due to the formed corrosion products between the splats. Furthermore, non-homogeneous wear tracks, pits and debris were observed for all samples, see Figure 4. Compared to the HVAF-sprayed coatings, the pits and debris of WAS-Ref are significantly more pronounced. It is assumed, that the significantly smaller pits and debris observed in case of the HVAF-sprayed coatings in the aged condition can be attributed to the dense microstructure, the high cohesion of these coatings and the smaller particle size during spraying.

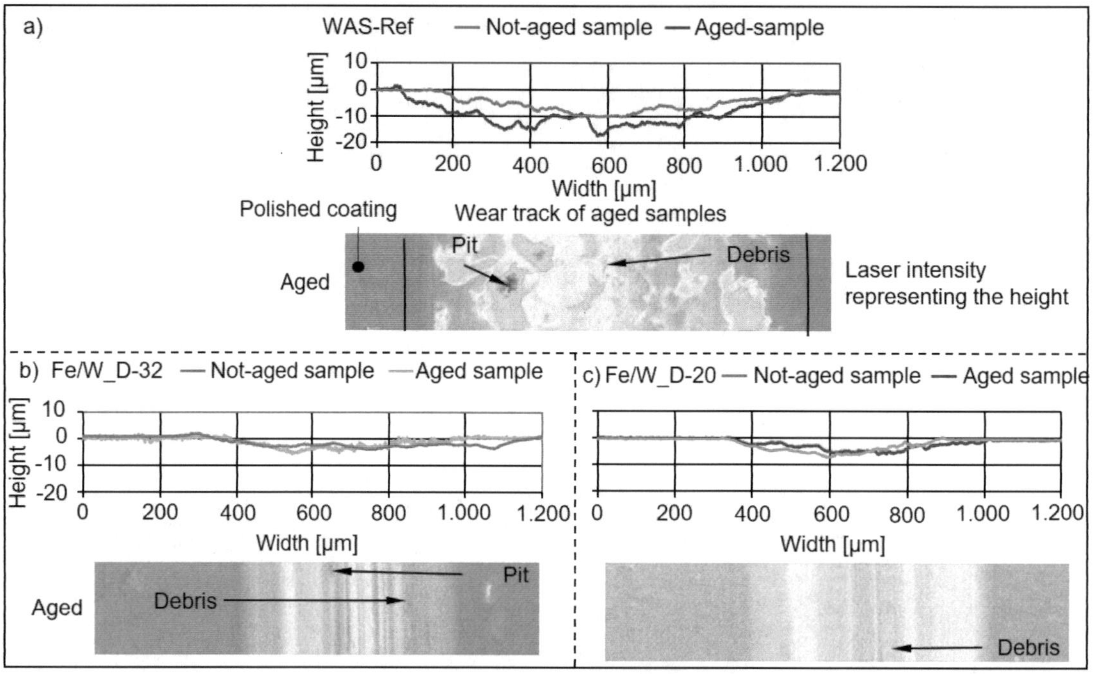


**Figure 4:** Wear tracks of the aged and not-aged samples and CLSM surface images of the wear tracks of the aged samples for sample a) WAS-Ref, b) Fe/W_D-32 and Fe/W_D-20.

The wear tracks were analysed quantitatively regarding the wear rates of aged and not-aged samples, see Figure 5. Since the substrate severely corroded, the aged samples is not measured for substrate. The wear rates of Fe/W_D-32 and Fe/W_D-30 for aged and not-aged samples are in a similar range between $K = 5.0 \cdot 10^{-6}$ and $K = 6.1 \cdot 10^{-6}$ mm³/Nm. In contrast, the wear rate of the sample WAS-Ref increases significantly after aging. Another interesting aspect can be observed for the deviations in measured wear rates. In case of aged WAS-Ref, the deviation from mean value of wear rate increases significantly after PoD-test. The deviations in case of HVAF-sprayed samples are barely affected by the aging. The higher wear rate and increase in the deviation of the measured values in case of the aged WAS-Ref sample are can be ascribed to a decreased cohesion of the coating, which was caused by corrosion. Although Figure 2 shows selective corrosion of the HVAF-sprayed coatings in a few areas, it seems that the cohesion of Fe/W_D-32 and Fe/W_D-20 remains at a sufficient level under the used test conditions.

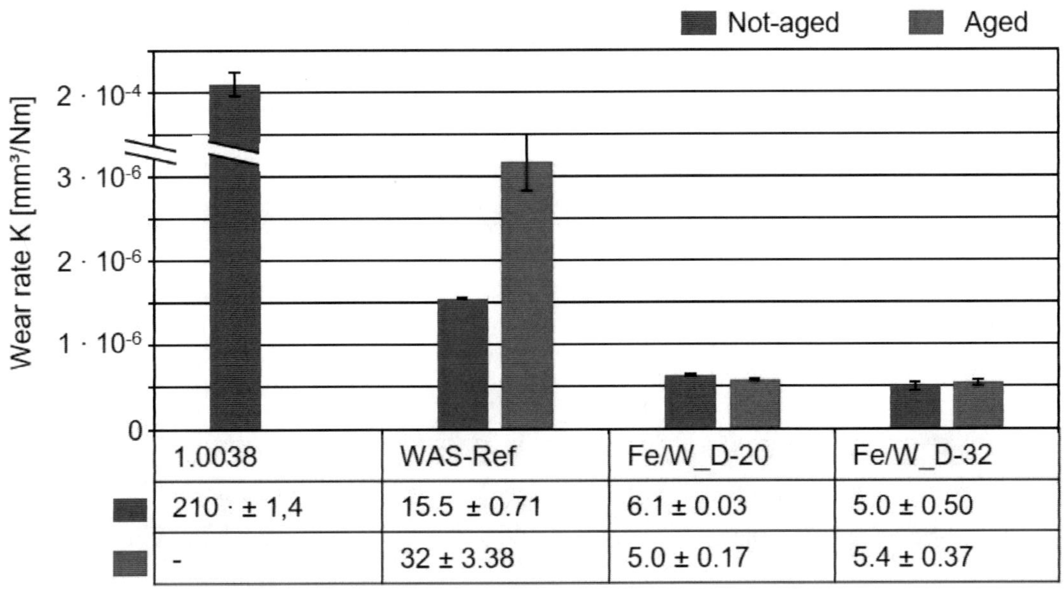


**Figure 5:** Wear rates of the substrate, the WAS-Ref, Fe/W_D-20 and Fe/W_D-32 samples in aged and not-aged conditions; not-aged values taken from [19, 20]. The error bars represent the standard deviation.

*3.2 Determination of the thermal conductivity and thermal resistance*

Thermal properties are important to describe the thermal insulation of thermally sprayed coatings and thus, are measured in the following. In Table 4 the measured values for the density, specific heat capacity and thermal diffusivity of the coatings are given. The deviations from measured values of thermal diffusivity are not presented, since this results in zero after rounding to third decimal places.

**Table 4:** Determined density, specific heat capacity and thermal diffusivity of the coatings.

| | Density coating [g/cm³] | Specific heat capacity [J/gK] | Thermal diffusivity [cm²/s] |
|---|---|---|---|
| WAS-Ref | 6.936 ± 0.049 | 0.519 ± 0.003 | 0.023 |
| Fe/W_D-32 | 7.332 ± 0.095 | 0.772 ± 0.004 | 0.018 |
| Fe/W_D-20 | 7.186 ± 0.029 | 0.676 ± 0.002 | 0.010 |

The calculated thermal conductivities and thermal resistances are shown in Figure 6. The highest thermal conductivity was determined for Fe/W_D-32 with a value of $\lambda = 10.20$ W/mK. Compared to WAS-Ref, the thermal conductivity of Fe/W_D-32 is higher, which reduces the thermal insulation of the dryer cylinder. The lowest thermal conductivity was measured for Fe/W_D-20 with a value of $\lambda = 4.94$ W/mK. Although the samples Fe/W_D-32 and Fe/W_D-20 were produced with a feedstock material of the same chemical composition and both coatings show a very dense microstructure [19, 20], the thermal conductivity of both coatings is significantly different. Two reasons are assumed to be responsible for this phenomenon. A finer powder fraction can result in a higher amount of splat interfaces. These splat interfaces are working as a barrier, which decreases the thermal conductivity. A partly amorphous microstructure was assumed for the fine powder coatings in [20]. Amorphous phases might influence the thermal conductivity, as well. In general, it is assumed that metallic amorphous coatings exhibit a lower thermal conductivity compared to crystalline coatings [24–26]. The thermal conductivity depends among others on the mean free path of electrons and phonons [27]. To decrease the thermal conductivity the mean free path can be decreased by scattering defects in the coating, to hinder the mobility of electrons and/or phonons. In [26] it was reported that in case of coatings with a mount of amorphous phases, the phonon contribution to thermal conductivity is suppressed and therefore the thermal conductivity decreases.

Both theories could explain a lower thermal conductivity of the fine powder coating. Which of these assumptions applies here, must be evaluated in more detail in further investigations, e.g. using transmission electron microscope (TEM).

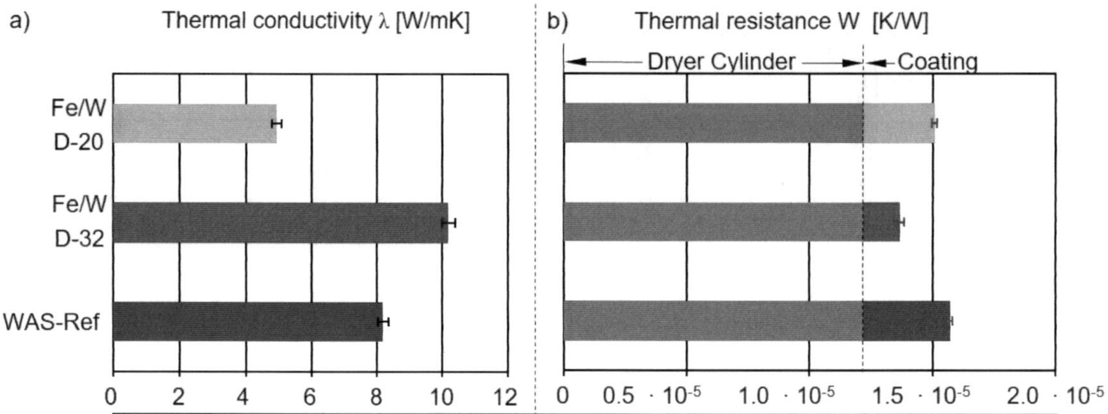


| | | Dryer Cylinder | Coating |
|---|---|---|---|
| | 4.94 ± 0.16 | 1.22 | 0.29 ± 0.02 |
| | 10.20 ± 0.18 | 1.22 | 0.15 ± 0.02 |
| | 8.19 ± 0.17 | 1.22 | 0.35 ± 0.02 |
| | | All values $10^{-5}$ | |

**Figure 6:** Determination of the a) thermal conductivity and the b) thermal resistance at a temperature of T = 200 °C

Besides the thermal conductivity, the coating thickness influences the thermal efficiency of dryer cylinders considerably. The thermal resistance of a dryer cylinder with different coatings was calculated. The bar chart presented in Figure 6 b) shows that Fe/W_D-32 has the lowest thermal resistance, which is the result of the low coating thickness and high thermal conductivity. Although the lowest thermal conductivity was measured for Fe/W_D-20, a theoretical cylinder coated would this coating would still exhibit a lower thermal resistance compared to the same cylinder with the reference coating. This is a result of the higher thickness of WAS-Ref, which need to be assured in order to ensure sufficient corrosion protection. A higher thermal resistance can insulate the dryer cylinder and therefore results in a higher energy consumption. For this reason, the new developed HVAF-sprayed coating system shows a high potential with regard to increasing the energy efficiency in paper machines.

## 4. Conclusions

The goal of this study was to investigate the thermal properties and the wear behaviour of aged samples of the novel HVAF-sprayed FeCrB/WC-Co coating system developed in [18–20]. Therefore, the HVAF-sprayed samples were produced with a powder feed rate of $\dot{m}$ = 200 g/min and a coating thickness of $s_{ct}$ ≈ 250 µm. Subsequently, the samples were aged for t = 100 h at T = 100 - 105 °C in a saturated steam atmosphere of a 0.5 % NaCl-solution. Afterwards, PoD-tests were performed on the aged samples. The wear tracks were analysed and compared to not-aged samples. Furthermore, the thermal conductivity and thermal resistance of the coatings were determined. The novel coating system developed by means of HVAF-spraying were compared to an industrial established sealed FeCrBMn coating system, which was produced with a wire feed rate of $\dot{m}$ = 110 g/min and a coating thickness of $s_{ct}$ = 750 µm by means of WAS. The results can be summarized as follows:

- WAS-Ref showed selective corrosion evenly distributed over the whole surface after aging. In contrast, only slight selective corrosion was observed for the HVAF sprayed FeCrB/WC-Co coatings.
- No under corrosion was observed for any samples.
- The wear rate of the HVAF-sprayed samples was nearly constant for aged and not-aged samples.
- The wear rate of WAS-Ref increased significantly after aging.
- The thermal conductivity of HVAF-sprayed coatings depend strongly on used powder fraction.
- A finer powder fraction leads to a reduced thermal conductivity.
- Since HVAF coatings allow a thinner deposition, the thermal resistance of the HVAF-sprayed coatings was lower compared to the WAS-Ref sample independent of used powder fraction.

In the used laboratory tests, the HVAF-sprayed coatings had a superior wear behaviour in aged conditions compared to the sealed WAS FeCrBMn reference coating. The superior corrosion and wear resistance in not-aged condition was already proven in [19, 20]. The reduced thermal insulation promises a higher efficiency of the paper dryers. In conclusion, the novel HVAF-sprayed FeCrB/WC-Co coatings system has a high potential for large area applications such as for dryer cylinders. Nevertheless, more application oriented investigations and field tests are needed to verify the promising results shown in this study.

**Acknowledgments**
This work is based on the DFG project BO 1979/45-1. The authors gratefully acknowledge the financial support of the German Research Foundation (DFG).

**References**

[1] Blechschmidt J 2013 *Taschenbuch der Papiertechnik* (München: CARL HANSER VERLAG GMBH)
[2] Vuoristo P and Nylén P 2007 Industrial and Research Activities in Thermal Spray Technology in the Nordic Region of Europe *J Therm Spray Tech* **16** 466–71
[3] Salimi Jazi H (ed) 2012 *Advanced Plasma Spray Applications* (InTech)
[4] Osmanda A, Battenbough A and Staines A 2012 Fe-Cr and Fe-Cr-Ni based HVOF Coating Systems Customised for Wear and Corrosion Protection *Thermal Spray Bulletin* **64** 117–22
[5] Schwenk A HVOF-K2-Spritzen von Eisenbasislegierungen für Anwendungen in der Druckindustrie *Proceedings 8. HVOF Kolloquium 2009*
[6] Espallargas N 2015 *Future Development of Thermal Spray Coatings: Types, design, manufacture and applications (Woodhead publishing series in metals and surface engineering* vol 65*)* (Amsterdam: Woodhead)
[7] Tenkula J, Hellmann B and Majava J Yankee cylinder with a plasma-sprayed carbide coating *US5123152A*
[8] Perdikaris C Methods of coating yankee dryer drums *EP1158066A1*
[9] Bobzin K, Öte M, Linke T F and Malik K M 2016 Wear and Corrosion Resistance of Fe- Based Coatings Reinforced by TiC Particles for Application in Hydraulic Systems *J Therm Spray Tech* **25** 365–74
[10] Bobzin K, Zhao L, Öte M and Königstein T 2017 Novel Fe-based Wear and Corrosion Resistant Coatings by Three-Cathode Plasma Technology *Surface and Coatings Technology* **318** 288–92

[11] Tillmann W, Vogli E, Baumann I, Matthaeus G and Ostrowski T 2008 Influence of the HVOF Gas Composition on the Thermal Spraying of WC-Co Submicron Powders (−8 + 1 μm) to Produce Superfine Structured Cermet Coatings *J Therm Spray Tech* **17** 924–32
[12] Wang Q, Zhang S, Cheng Y, Xiang J, Zhao X and Yang G 2013 Wear and corrosion performance of WC-10Co4Cr coatings deposited by different HVOF and HVAF spraying processes *Surface and Coatings Technology* **218** 127–36
[13] Guo R Q, Zhang C, Chen Q, Yang Y, Li N and Liu L 2011 Study of Structure and Corrosion Resistance of Fe-Based Amorphous Coatings Prepared by HVAF and HVOF *Corrosion Science* **53** 2351–6
[14] Verstak A and Baranowksi V AC-HVAF sprayed tungsten carbide: properties and applications *Proceedings ITSC 2006* pp 643–8
[15] Bolelli G *et al* 2015 Tribology of HVOF- and HVAF-Sprayed WC–10Co4Cr Hardmetal Coatings: A Comparative Assessment *Surface and Coatings Technology* **265** 125–44
[16] Hulka I, Şerban V A, Niemi K, Vuoristo P and Wolf J 2012 Comparison of Structure and Wear Properties of Fine-Structured WC-CoCr Coatings Deposited by HVOF and HVAF Spraying Processes *SSP* **188** 422–7
[17] Bolelli G, Hulka I, Koivuluoto H, Lusvarghi L, Milanti A, Niemi K and Vuoristo P 2014 Properties of WC–FeCrAl Coatings Manufactured by Different High Velocity Thermal Spray Processes *Surface and Coatings Technology* **247** 74–89
[18] Bobzin K, Öte M, Knoch M A and Sommer J Development of HVAF-sprayed Novel Fe-Based Coatings for Large Area Applications *Proceedings ITSC 2017* pp 339–45
[19] Bobzin K, Öte M, Knoch M A and Sommer J 2018 Influence of Powder Feed Rate on Corrosion and Wear Properties of Fe-based HVAF Coatings *IOP Conf. Ser.: Mater. Sci. Eng.* **373** 12012
[20] Bobzin K, Öte M, Knoch M A and Sommer J Influence of Grain Size on the Corrosion and Wear Behviour of HVAF-Sprayed Fe-Based Coatings *Proceedings ITSC 2018*
[21] 2013 *VDI-Wärmeatlas (VDI-Buch)* 11th edn (Berlin: Springer Berlin)
[22] Karsten B 2009 *Handbuch für Dampf- und Kondensatanlagen* (Karlstein)
[23] Milanti A, Matikainen V, Bolelli G, Koivuluoto H, Lusvarghi L and Vuoristo P 2016 Microstructure and Sliding Wear Behavior of Fe-Based Coatings Manufactured with HVOF and HVAF Thermal Spray Processes *J Therm Spray Tech* **25** 1040–55
[24] Bobzin K, Öte M and Königstein T 2017 Investigation of Amorphous/Nanocrystalline Iron-Based Thermal Barrier Coatings *J Therm Spray Tech* **26** 388–97
[25] Shin D-i, Gitzhofer F and Moreau C 2007 Properties of Induction Plasma Sprayed Iron Based Nanostructured Alloy Coatings for Metal Based Thermal Barrier Coatings *J Therm Spray Tech* **16** 118–27
[26] Shin D-i, Gitzhofer F and Moreau C 2007 Thermal property evolution of metal based thermal barrier coatings with heat treatments *J Mater Sci* **42** 5915–23
[27] Kingery W D, Bowen H K and Uhlmann D R 1976 *Introduction to ceramics (Wiley series on the science and technology of materials)* 2nd edn (New York: Wiley)

IOP Conf. Series: Materials Science and Engineering **480** (2019) 012006 doi:10.1088/1757-899X/480/1/012006

# Surface inspection of joint areas by means of laser-induced breakdown spectroscopy

**P Doraciak[1,*], A Bünting[1] and T Lampke[2]**

[1] Process Development, Daimler AG, 70546 Stuttgart, Germany
[2] Institute of Materials Science and Engineering, Chemnitz University of Technology, 09125 Chemnitz, Germany

[*] e-mail: patrik.doraciak@daimler.com

**Abstract.** The quality of laser welded joints is dependent on different factors like the process-parameters, weld configuration and the cleanliness of the joint area. Filmic contaminations on the joint area can be oil, grease and other lubricants, which are used for machining, corrosion protection or component assembly. Residuals of lubricants can vaporize during the laser welding process and may lead to weld smoke as well as spatters and results in welding defects like cracks and pores in the weld seam. Laser-induced breakdown spectroscopy (LIBS) was investigated successfully and applied to detect contaminants and finally avoid these welding defects. 1.7131 steel was used as model-substrate. The ground sample surfaces were characterized by means of surface energy as well as fluorescence-intensity and LIBS spectra of the cleaned surfaces were recorded. The effect of the laser pulse energy and the delay time between laser pulse and the spectra recording was investigated. Furthermore, a forming lubricant was applied on the sample surfaces and reflectometry was successfully used for the film thickness measurements. LIBS spectra of the contaminated surfaces are presented and the impacts of the oil film thickness on the appearance of the spectra are outlined. Additionally, the examination presents the evaluation of the spectra with the linear correlation and the peak intensity ratio.

## 1. Introduction

The condition of joint areas in terms of filmic contaminations has a major impact on the quality of laser welded joints [1-3]. Technical products, which are processed by laser welding, are usually treated beforehand by various machining processes. Residuals of oil, grease and other lubricants from metal forming, milling, drilling and other machine processing can be found on the surfaces. Additionally, residuals from chemical cleaning, corrosion protection and the assembly remain on the surfaces. These filmic contaminations can lead to weld smoke, spatters and the ejection of melted material during the welding process as well as defects in the welded joint like cracks and pores [4-6]. Current strategies to ensure an adequate high weld seam quality are adjusted on the application and reach from prior chemical cleaning as well as laser cleaning up to 100 % end of line ultrasonic tests of the weld seams.

Technical regulations demand cleaned surfaces without filmic contaminations such as oil, grease and other lubricants in general to avoid these welding defects [1, 2, 7, 8]. However, a limited selection of measurement methods is available for the detection of organic or carbon-based contaminations on surfaces. Most of them are intended for laboratory use or the composition of the contaminations have to be known sufficiently [9, 10]. Usually, these requirements aren't given. For this reason, the dependable usage isn't possible or the measurement methods aren't suitable for the industrial inline usage [9-11].

Laser-induced breakdown spectroscopy (LIBS) is a well-known atom emission spectroscopy based measurement technique. The measurement method is used in industrial applications from production control in the steel industry through to the fast classification of metals in the scrap metal sorting by now [12, 13]. LIBS is also used as a tool for surface analysis [14-16] in diverse applications. Especially, [11] used LIBS with a high-resolution echelle spectrometer for the detection of surface contaminations on CFK surfaces and [17] used LIBS with a low resolution spectrometer for the process control of a laser ablation process. Summarized, the LIBS measurement technique is a promising approach for the detection of filmic contaminations on metallic surfaces, suitable for industrial inline use.

The presented study examines the possibility to detect contaminations on surfaces with the laser-induced breakdown spectroscopy. Therefore, a processing routine for samples with a constant and reproducible cleanliness condition as well as cleanliness condition characterization methods are investigated. Additionally, LIBS spectra of cleaned and coated samples are recorded and the possibilities for the contamination detection are examined.

## 2. Experimental and materials

In context of this study, the case hardening steel 16MnCr5 (1.7131) was used as a model substrate. 16MnCr5 offers good mechanical properties and is a typically used material in the automotive field for powertrain components. The composition of 16MnCr5 is shown in table 1.

**Table 1.** Composition of 16MnCr5 (1.7131) in wt.-%.

| | C | Mn | Cr | Si | S | P |
|---|---|---|---|---|---|---|
| 16MnCr5 | 0.14 to 0.19 | 1.0 to 1.3 | 0.8 to 1.1 | $\leq 0.4$ | $\leq 0.035$ | $\leq 0.035$ |

Every single sample was cut from metal sheets and had the dimensions of $V = (70 \times 30 \times 5)\ mm^3$. The samples were ground with a grinding machine Saphir 350 / Rubin 520 from ATM GmbH (Mammelzen, Germany) and a special sample holder. In order to obtain a polished and reflective surface, SiC papers with grit sizes up to 2500 were used. For the surface roughness measurements, the profile method according to DIN EN ISO 4287 was used on three samples with eight single measurements on each sample to get average $R_z$ and $R_a$ roughness values. Figure 1b) depicts sketches of the samples with the measurement positions and directions of the surface roughness measurements (see black arrows).

Every sample was cleaned in a multi-stage cleaning procedure to get a consistent surface cleanliness condition. The entire cleaning routine is shown in figure 2. The samples were hand cleaned with demineralized water after the grinding process to remove abrasive particles. Afterwards, the samples were cleaned in an ultrasonic cleaning process for $t = 15$ min and $T = 30$ °C in a mixture of Green Star BMP from Koch-Chemie GmbH (Unna, Germany) and demineralized water at a ratio of 1:20. To remove surfactants from the ultrasonic cleaning, the samples were rinsed with demineralized water and dried with compressed air subsequently. Furthermore, the samples were cleaned with ethyl alcohol, 2-propanol and petroleum benzine for $t = 5$ min in each case. The used solvents are suitable for analytical use. After each solvent bath the samples were dried with compressed air.

Two different methods were applied for the cleanliness condition characterization of the samples. In the first step, the average surface energies were determined for six samples. For this purpose an OCA 25 contact angle measurement device of DataPhysics Instrument GmbH (Filderstadt, Germany) was used. It is offering an automatically drop application, a visualization of the drops and a calculation of the contact angles. Contact angles were recorded at three different positions on each sample as shown in figure 1c) and the surface energy $\sigma$ was calculated. As test liquids demineralized water, diiodomethane, ethylene glycol and thiodiglycol were used because of their different polar and disperse shares. The calculation of the surface energy was performed with the method from Owens, Wendt, Rabel & Kälble (OWRK) [18-20].

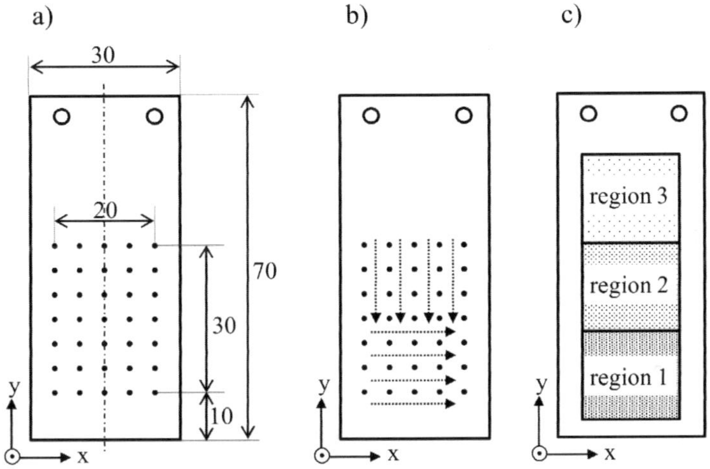


**Figure 1.** Sketches of the samples. a) Dimension of the samples with the position and the size of the measurement field for fluorescence measurements, LIBS measurements and film thickness measurements. b) Positions and directions of the surface roughness measurements. c) Positions of the three regions for the surface energy measurements.

The fluorescence intensity of the sample surfaces were measured as second method for the cleanliness condition characterization. For this purpose, the F-Scanner from Fraunhofer ipm (Freiburg, Germany) was used. The F-Scanner is a device to detect laser-induced fluorescence emitted from organic or carbon-based contaminations on surfaces. Therefore, the sample surface is inspected with a semiconductor laser operating at a wavelength of $\lambda = 405$ nm. Thus, contaminations on the surface are excited to fluorescence. The emitted fluorescence is detected in a wavelength range between $\lambda = 420$ nm and $\lambda = 520$ nm. The voltage at the detector $U_F$ or the percentage value of the fluorescence intensity related to the USFS 461 standard is given for every pixel as measurement result [21]. The measurement range is adjustable so that impacts from surface roughness or contaminations, which exhibit different fluorescence intensities, could be compensated. The detector voltage $U_F$ was recorded at 35 positions of every sample for an average fluorescence value as shown in figure 1a).

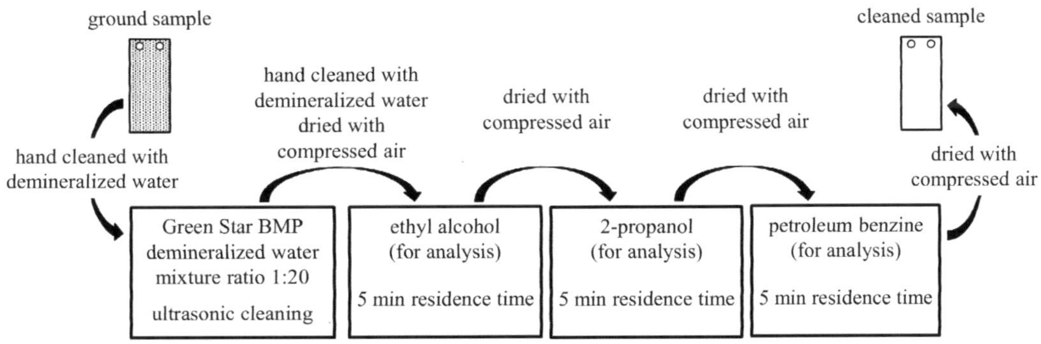


**Figure 2.** Multi-stage cleaning routine for the ground samples.

For the examination of contaminated surfaces, the forming lubricant Wisura Akalit CF10 from Fuchs Wisura GmbH (Bremen, Germany) was applied on the samples. This forming lubricant is used in some other investigations as a reference contamination [9, 22]. Varying film thicknesses were achieved by dip-coating with different coating velocities in the range from $v = 100$ mm·min$^{-1}$ to $v = 530$ mm·min$^{-1}$ and variable lubricant solvents mixtures [23, 24]. The film thickness was measured with a reflectometry measurement system. Reflectometry is normally used for layer thickness measurements in the semiconductor industry but can also be used for the measurement of transparent oil films. Polychromatic light is directed perpendicularly to the surface and is partially reflected at the interface between the ambient air and the oil film. The other part of the light is transmitted and finally reflected at the interface between the oil film and the substrate. The two reflected light beams interfere due to the optical path difference. The interfered light beam is coupled into an optical fiber and the interference pattern, which is mainly depending on the film thickness, is recorded by a fiber-coupled spectrometer. With the knowledge of the optical properties of the substrate and the oil, the film thickness can be calculated with a computer software. The oil film thickness was measured according to figure 1a) at 35 positions (see black dots) on the sample to achieve an average value of the film thickness.

Laser-induced breakdown spectroscopy is a method for the detection of chemical elements and belongs to the group of atom emission spectroscopy. The method is widely described in [25, 26] as well as [27] and used for research and industrial applications. For this study a standard laboratory setup was used as illustrated in figure 3. The pulsed laser radiation for the plasma ignition is provided by a flash lamp pumped Q-switched Nd:YAG laser (Q-Smart 450, Quantel Laser, Les Ulis Cedex, France) operating at a wavelength of $\lambda = 1064$ nm with a pulse repetition rate of $f_{rep} = 10$ Hz. The pulse duration is fixed at $t_p = 6$ ns (FWHM). The laser beam was focused onto the surface with a plano-convex lens with a focal length of $f = 200$ mm and an anti-reflection coating for $\lambda = 532$ nm and $\lambda = 1064$ nm. The plasma emission was coupled out by a dichroic mirror with an average reflection $R > 95$ % in a wavelength range of 420 nm $\leq \lambda \leq$ 900 nm. The plasma emission was observed by the fiber coupled czerny-turner spectrometer HR2000+ from OceanOptics Inc. with a 25 µm entrance slit and a grit with 600 grooves mm$^{-1}$, centered on $\lambda = 500$ nm. The CCD detector offers 2048 pixels and a resulting spectral recording range of 393 nm $\leq \lambda \leq$ 845 nm at a resolution of $\Delta\lambda = 0.22$ nm. The shortest integration time of the CCD is $t_I = 1$ ms.

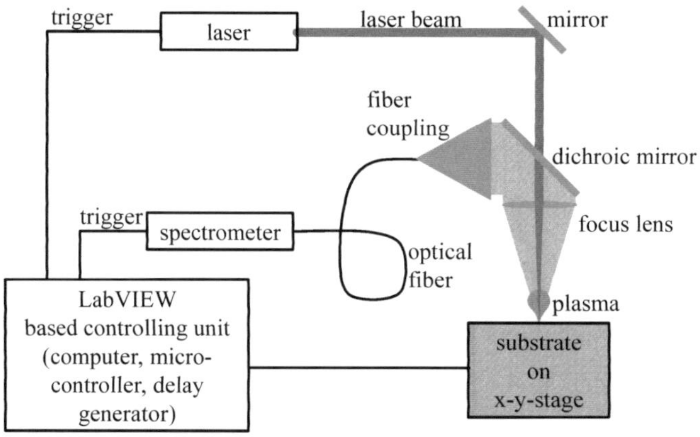


**Figure 3.** Standard LIBS measurement setup for laboratory use [25].

The measurement system is equipped with two linear stages from Physik Instrumente (PI) (Karlsruhe, Germany) offering an x-y-measurement range of $A = (180 \times 180)$ mm². The stages as well as the laser and spectrometer are controlled by a self-developed software tool based on LabVIEW and a microcontroller type chipKIT Max32 from Digilent Inc (Pullman, USA). The LabVIEW software tool sets the program for the linear stages. The microcontroller triggers two delay generators (AFG3051C and AFG3022B, Tektronix GmbH, Köln, Germany), which supply trigger signals for the laser flash lamp, laser Q-switch and the spectrometer. The duration between the firing of the laser pulse and the start of the integration time is defined as delay time $t_d$. The recorded spectra are saved on the computer for the evaluation process.

Two different ways for the evaluation of the spectral data are investigated. The linear correlation coefficient $r$, with

$$r = \frac{\sum_{i=1}^{m}(x_i - \bar{x})(y_i - \bar{y})}{\left[\sum_{i=1}^{m}(x_i - \bar{x})^2\right]^{1/2}\left[\sum_{i=1}^{m}(y_i - \bar{y})^2\right]^{1/2}} \quad (1),$$

is a calculated value for the relationship of two data sets. It ranges from -1 to 1 at which $r = 1$ means a 100 % positive correlation, $r = -1$ means a 100 % negative correlation and $r = 0$ means no correlation. The linear correlation coefficient $r$ is invariant against linear variations. It means that an overall offset or a factorization of the data set doesn't influence the correlation coefficient $r$ [28].

Hence, it's used for the evaluation of spectral data in many other LIBS applications like the sample identification or the recognition of solid layers [17, 28]. Usually, a set of measured spectra are linearly correlated with a reference spectrum. The calculation of the reference spectrum is described in equation (2). A set of $S_i$ to $S_n$ spectra is given in which $n$ is the number of recorded spectra, $i \in (1,...n)$ Every single spectrum $S_i$ has a set of 2048 intensity values $I_{\mathrm{i,\,P1}}$ to $I_{\mathrm{i,\,P2048}}$. With this data, the reference spectrum can be calculated by

$$S_{\mathrm{Ref}} = \begin{pmatrix} \frac{1}{n}\sum_{i=1}^{n} I_{i,\mathrm{P1}} \\ \vdots \\ \frac{1}{n}\sum_{i=1}^{n} I_{i,\mathrm{P2048}} \end{pmatrix}. \quad (2)$$

A further method for the investigation of LIBS spectra is the calculation of the peak intensity ratio $k$ [25]. Two peak intensities from different elements are used. This technique is often applied with calibration curves for the determination of element concentrations in materials. For the investigation of coating thicknesses, two different element lines are necessary. The first element line occurs typically in the spectrum of the substrate (without coating) and is used as a reference line. The second element line is characteristic for the coating and occurs in the spectrum of the coating. If the thickness of the coating increases, the emission intensity of the second line increases, too, because of the higher element occurrence in the ablated material. The emission intensity of the first line decreases concurrently and the intensity ratio of the two element lines increases [11, 12, 29]. The calculation of the peak intensity ratio $k$ is given by the equation

$$k = \frac{I_{\lambda_2}}{I_{\lambda_1}}. \quad (3)$$

## 3. Results and discussion

First of all, the surface roughness of the ground samples was measured. The cleanliness condition of the cleaned surfaces was investigated with the measurement of the surface energy and the laser-induced fluorescence. LIBS spectra of the cleaned surface were recorded and the impact of laser pulse energy and delay time determined. Additionally, occurred element lines were determined. In the

IOP Conf. Series: Materials Science and Engineering **480** (2019) 012006 doi:10.1088/1757-899X/480/1/012006

second part of the study, samples with applied oil films were investigated. The film thickness was measured with reflectometry and the LIBS measurement of the equal samples were carried out.

### *3.1. Characterization of cleaned sample surfaces*

The surface roughness measurement was performed using the profile method for x- and y-direction. Three samples with four single measurements for both directions on each sample were executed and the average $R_z$ and $R_a$ values were calculated. The results of the measurements are shown in table 2. The average $R_z$ values amount between $0.12 \pm 0.01$ µm and $0.13 \pm 0.02$ µm in x- and y-direction. The average $R_a$ values amount between $0.016 \pm 0.001$ µm and $0.018 \pm 0.002$ µm, as well in both directions. All single average $R_z$ and $R_a$ values exhibit only marginal standard deviations in the range of $0.01\ \mu m \leq s_{Rz} \leq 0.02\ \mu m$ and $0.001\ \mu m \leq s_{Ra} \leq 0.003\ \mu m$. As expected, the samples exhibit a nearly constant and reproducible surface roughness without significant differences in x- and y-direction. This can be explained by the well-defined and automated grinding process.

**Table 2.** Surface roughness $R_z$ and $R_a$ of the samples in x- and y-direction.

| | x-direction | | y-direction | |
|---|---|---|---|---|
| Sample No. | $R_z$ / µm | $R_a$ / µm | $R_z$ / µm | $R_a$ / µm |
| 1 | 0.12 ± 0.01 | 0.016 ± 0.001 | 0.12 ± 0.01 | 0.015 ± 0.001 |
| 2 | 0.13 ± 0.01 | 0.016 ± 0.001 | 0.13 ± 0.02 | 0.017 ± 0.003 |
| 3 | 0.12 ± 0.01 | 0.016 ± 0.002 | 0.13 ± 0.01 | 0.018 ± 0.002 |

The surface energy was determined at three sites on six samples to get average values of the surface energy as well as disperse and polar shares. Figure 4 shows the results. The average surface energy for all six samples amount $\bar{\sigma} = 46.2 \pm 0.5$ mN/m with $\bar{\sigma}_{Disperse} = 4.9 \pm 0.5$ mN/m and $\bar{\sigma}_{Polar} = 41.3 \pm 0.4$ mN/m. The average surface energy values of the single samples are very consistent and in a range of $\sigma_{min} = 45.7 \pm 0.7$ mN/m and $\sigma_{max} = 47.0 \pm 0.9$ mN/m. The disperse share is between $\sigma_{disperse,\ min} = 4.42 \pm 0.43$ mN/m and $\sigma_{disperse,\ max} = 5.82 \pm 1.35$ mN/m, the polar share is between $\sigma_{polar,\ min} = 41.01 \pm 1.02$ mN/m and $\sigma_{polar,\ max} = 42.12 \pm 0.12$ mN/m.

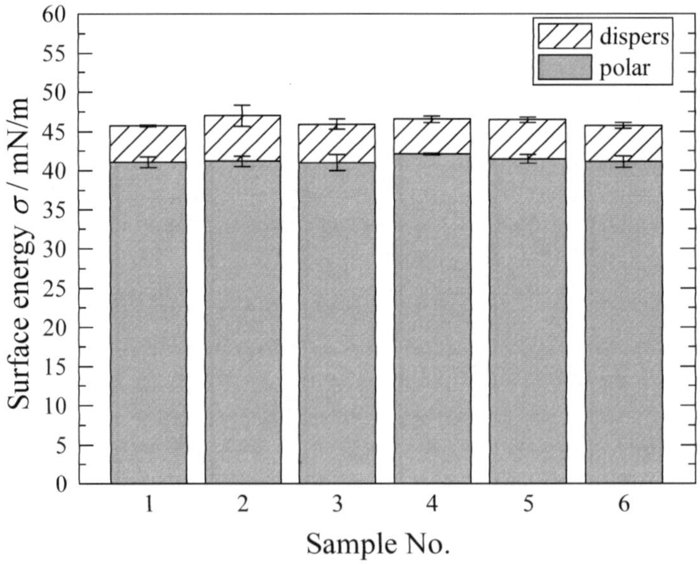


**Figure 4.** Surface energy of the six measured samples. The disperse and polar shares are illustrated.

The results of the fluorescence measurements for 115 samples are shown in figure 5. Every data point represents the average detector voltage $U_F$ of 35 measurements on the sample. As well as the standard deviation, the 95 % confidence interval is shown for every sample. The average detector voltage for all 115 samples is very low with $\overline{U}_F = 3.21 \cdot 10^{-3} \pm 0.31 \cdot 10^{-3}$ V. Differences in the values appear in the third digit after the decimal point. The minimal und maximum values amount $U_{F,\,min} = 2.15 \cdot 10^{-3} \pm 0.29 \cdot 10^{-3}$ V and $U_{F,\,max} = 4.05 \cdot 10^{-3} \pm 0.83 \cdot 10^{-3}$ V. For comparison, a film thickness $d_s$ of $d_s = 125 \pm 6$ nm of Wisura Akalit CF 10 lead to an average detector voltage of $\overline{U}_F = 266.2 \cdot 10^{-3} \pm 13.8 \cdot 10^{-3}$ V. The results of the fluorescence measurements demonstrate that a constant and reproducible cleanliness condition can be produced.

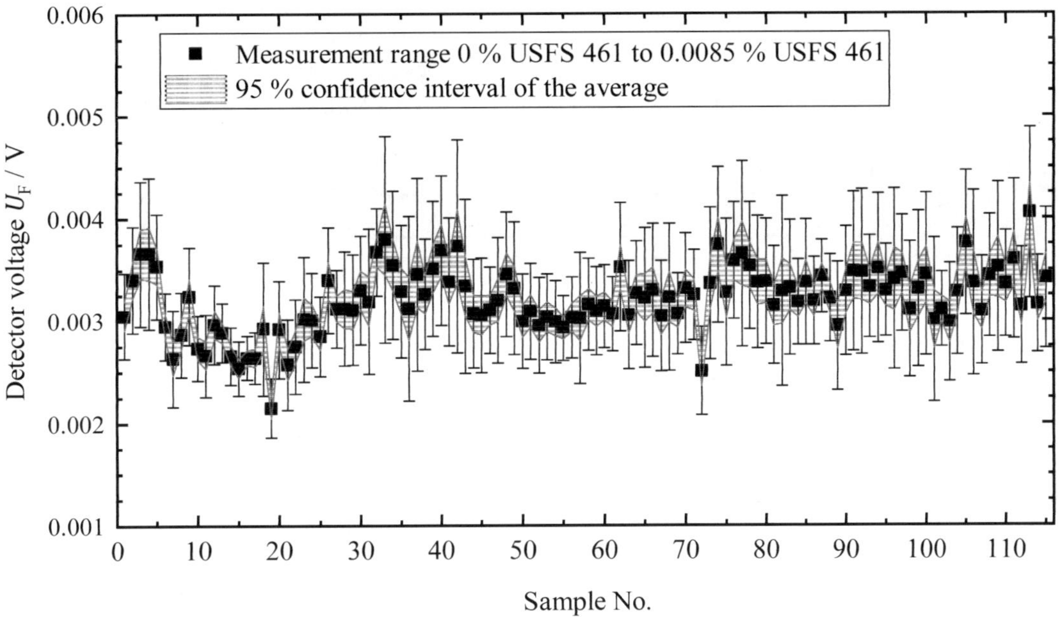


**Figure 5.** Fluorescence intensity, measured as detector voltage $U_F$ of 115 cleaned samples. The standard deviations as well as the 95 % confidence intervals are shown.

*3.2. LIBS measurements of cleaned sample surfaces*

The LIBS laboratory setup as shown in figure 3 was used for the LIBS spectra recording of the cleaned samples. The cleaned samples were overall placed at a constant distance to the lens. Different power densities $\Psi$ in the range of 2.07 GW·cm$^{-2}$ to 3.21 GW·cm$^{-2}$ were used to analyze the impact of the laser pulse energy. The delay time between the firing of the laser pulse and the recording of the spectrum was varied from $t_d = 500$ ns to $t_d = 6000$ ns. 35 spectra were recorded for every combination of power density and delay time and the average spectra were calculated.

Figure 6 shows the change of the spectra with increasing delay time for the power density of $\Psi = 2.83$ GW·cm$^{-2}$. In the time range till $t = 1$ µs after the laser pulse, the strong continuum emission is visible [26]. In the time range of $t = 1$ µs to $t = 3$ µs, spectra with decreasing continuum emission and increasing emission intensity of the discrete element lines can be observed. In the time range $t > 3$ µs, the emission intensity of the element lines decreases fast. This behaviour is also described in basic literature of laser-induced breakdown spectroscopy [26, 27, 30].

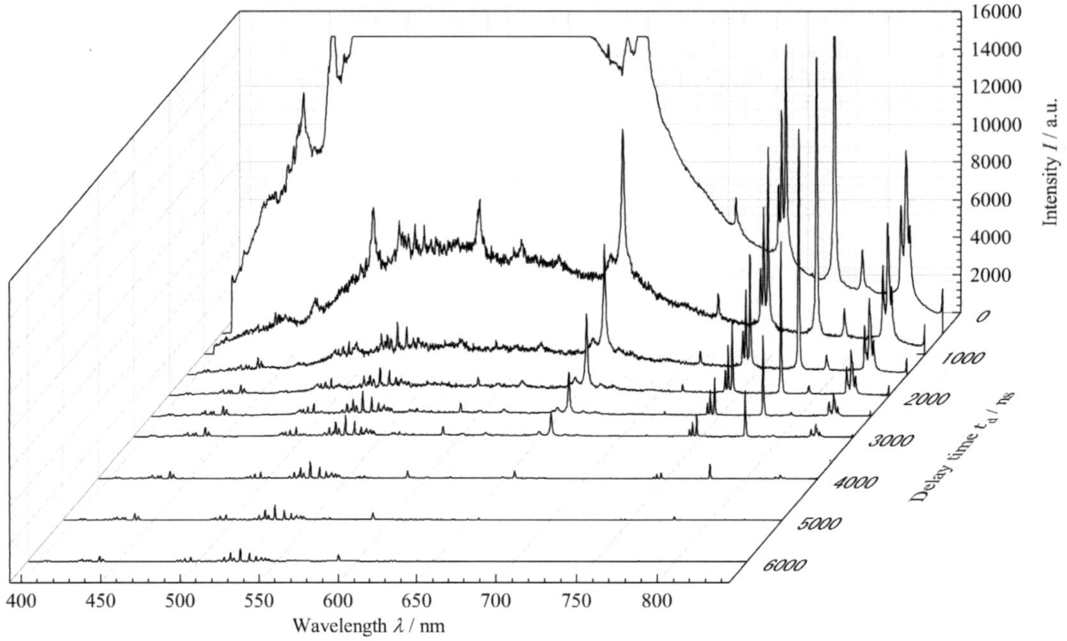


**Figure 6.** Trend of the spectrum as a function of the delay time for a power density of $\Psi = 2.83$ GW·cm$^{-2}$.

Figure 7 illustrates the emission intensity $I_{\lambda = 777.27}$ of the oxygen line at $\lambda = 777.27$ nm depending on the power density and the delay time. It is easily visible, that the emission intensity reaches the saturation of the spectrometer for the first 500 ns after the laser pulse is fired. For all power densities, the decrease of the emission intensity shows the same trend, but it is shifted to higher delay times for higher power densities. For the lowest power density, the decrease of the emission intensity starts immediately.

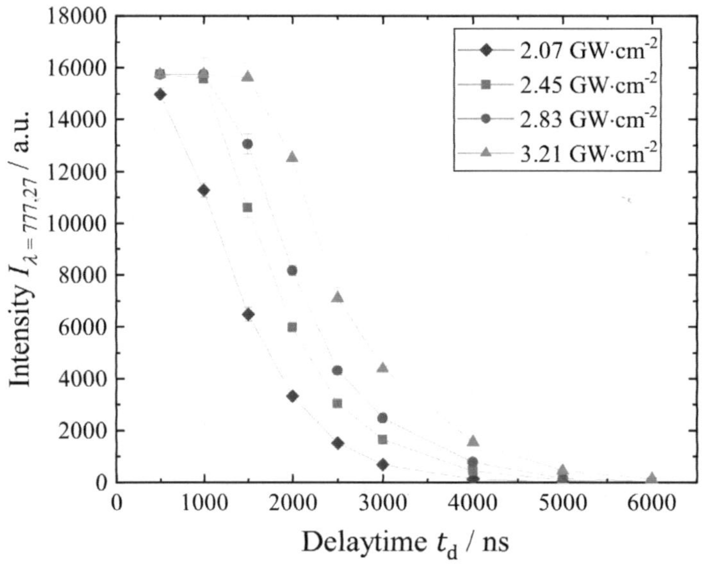


**Figure 7.** Trend of the emission intensity of the oxygen element line at $\lambda = 777.27$ nm as a function of the delay time $t_d$ and the power density $\Psi$.

Figure 8 shows the reference spectrum for $\Psi = 2.83$ GW·cm$^{-2}$. The reference spectrum was calculated from $n = 35$ spectra as described in equation (2). The spectrum exhibits a marginal ratio of continuum emission and the emission of some discrete element lines. It is noticeable that the explicit visible element lines are elements contained in ambient air like oxygen, nitrogen and hydrogen. Element lines from the metallic substrate aren't explicitly visible. There are two possible explanations for this behavior. First, the samples were positioned behind the focal point which means that the laser beam had to pass the focal point of the beam waist. This can favor fail ignitions in the focal point. Fail ignitions occurred rarely and were not reproducible in the performed investigations.

Additional investigations with diffuse surfaces point at a second and more conclusive explanation. The ability for laser radiation to couple into material is dependent on the diffusivity of the surface [31]. There are bad conditions for the laser radiation to be absorbed by the material because of the high reflectivity of the sample surfaces. Thus, only a small amount of the substrate material can be ablated and the share of stimulated elements from ambient air is bigger. It results in a higher emission intensity of these element lines in comparison to the emission intensity of element lines from the substrate material.

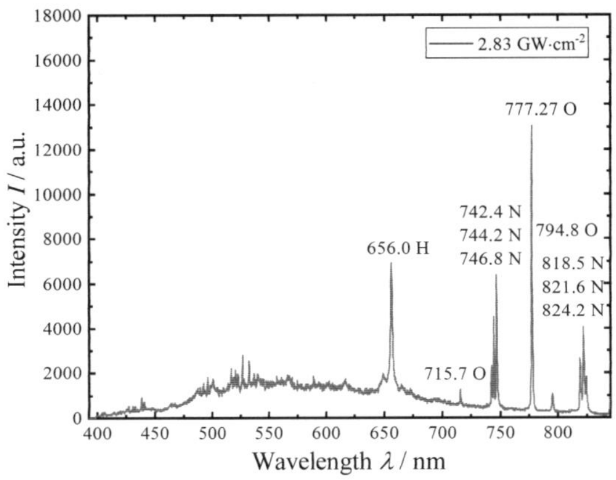


**Figure 8.** Reference spectrum for $\Psi = 2.83$ GW·cm$^{-2}$ with the identification of the element lines [32].

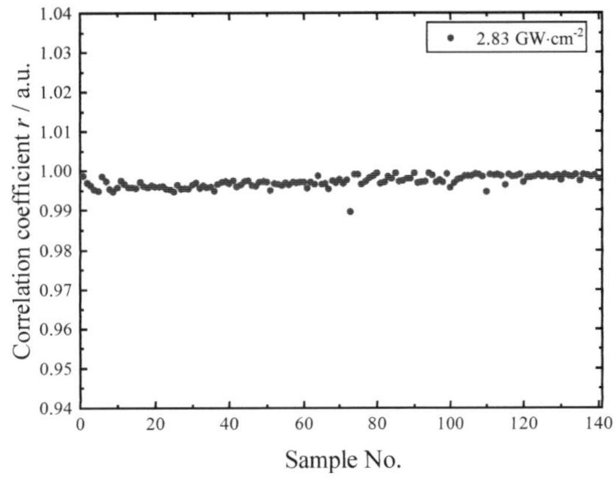


**Figure 9.** Correlation coefficient $r$ to reference spectrum for 140 single spectra, calculated for $\Psi = 2.83$ GW·cm$^{-2}$.

Further, reproducibility of the recorded spectra from cleaned sample surfaces was investigated as shown in figure 9. 140 spectra from cleaned sample surfaces were recorded and the correlation coefficients for the linear correlation with the reference spectrum were calculated. Average correlation coefficient amount $\bar{r} = 0.9972 \pm 0.0014$, which relates to a relative standard deviation (RSD) of RSD = 0.144 %. The lower 95 % confidence interval amounts to 0.997, the upper 95 % confidence interval amounts to 0.9975. It is clearly demonstrated for 140 single spectra, that the recorded spectra are reproducible. The results from LIBS measurement of cleaned samples surfaces are conform to the results from the surface energy and fluorescence measurements.

### *3.3. LIBS measurements of coated sample surfaces*

For the investigation of contaminated surfaces, ground and cleaned samples were coated with the forming lubricant Wisura Akalit CF10. The oil film was performed by means of a dip-coating process. Based on the variation of the coating velocity $v$ and the mixture of solvent and lubricant, 19 different film thicknesses $d_S$ were generated. The film thickness $d_S$ was measured using reflectometry at 35 sites on the sample for an average value of the film thickness. LIBS measurements of the coated samples were performed with the laboratory setup as shown in figure 3. On every sample, 35 LIBS

IOP Conf. Series: Materials Science and Engineering **480** (2019) 012006 doi:10.1088/1757-899X/480/1/012006

measurements were executed and the average spectra were calculated, analogous to equation 2. All spectra were recorded with power densities in the range of 2.45 GW·cm$^{-2}$ to 3.21 GW·cm$^{-2}$.

Figure 10 shows five average spectra for $\Psi = 3.21$ GW·cm$^{-2}$ and an increasing film thickness in a range between $d_S \approx 100$ nm and $d_S \approx 3100$ nm. The average spectrum for a film thickness of $d_S \approx 690$ nm is depicted in figure 11. It shows the identified element as well as molecule emission lines for an easier assignability. Taking a look at figure 10, it is obvious that the emission intensity of the visible element lines and molecule emissions change with the film thickness. In the wavelength range between $\lambda = 453.8$ nm and $\lambda = 474.4$ nm $C_2$ molecule emission [33] increases first and decreases for a film thickness higher than $d_S \approx 750$ nm. The other $C_2$ molecule emissions between $\lambda = 494.6$ nm and $\lambda = 517.5$ nm, $\lambda = 525.9$ nm and $\lambda = 563.8$ nm as well as between $\lambda = 593.2$ nm and $\lambda = 620.7$ nm show a similar behavior [33]. The hydrogen element line at $\lambda = 656.021$ nm, which was visible in the reference spectrum too, shows an equivalent behavior.

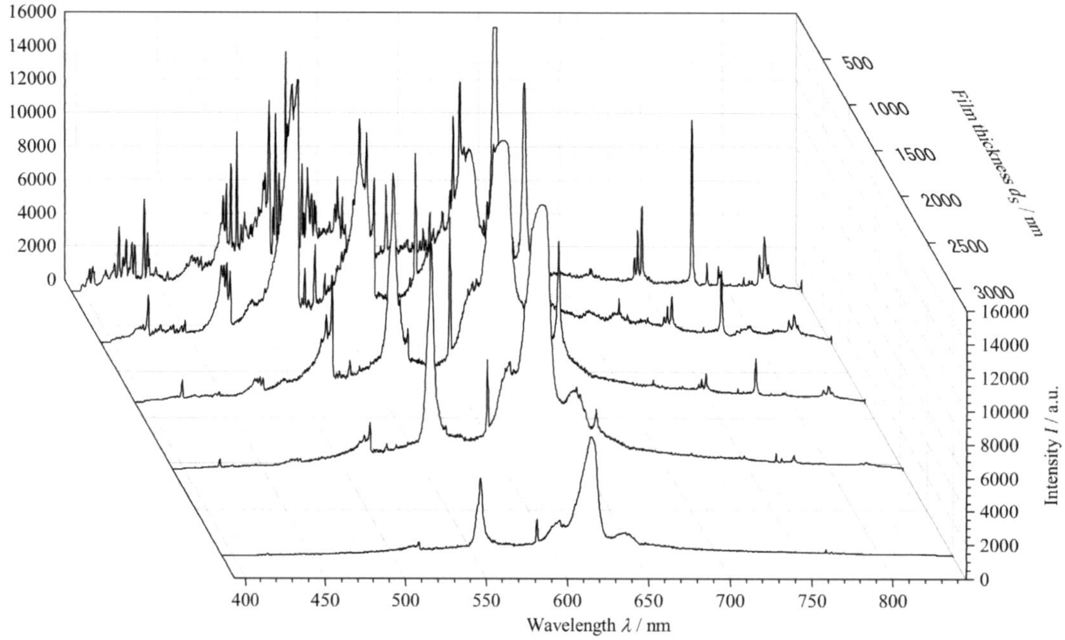


**Figure 10.** Appearance of the spectrum dependent on film thickness of Wisura Akalit CF 10 for power density $\Psi = 3.21$ GW·cm$^{-2}$.

The emission intensity of the oxygen element line at $\lambda = 777.27$ nm decreases with increasing film thickness, as shown in figure 12. The decrease in emission intensity is similar for all power densities but it is shifted to higher film thicknesses for higher power densities. Figure 13 shows the evolution of the $C_2$ molecule emission line at $\lambda = 516.073$ nm (see figure 11). The intensity reaches the saturation of the spectrometer for a film thickness between $d_S \approx 450$ nm and $d_S \approx 700$ nm. For a film thickness $d_S > 700$ nm, the emission intensity decreases. The begin of the decrease in emission intensity is shifted to higher film thicknesses with increasing power density $\Psi$.

This overall behavior can be explained in the following way: For a marginal film thickness, the power density is high enough to ionize the ablated material. The emission intensity of the $C_2$ molecule emission increases and the emission intensity of the oxygen element line decreases in this range of film thickness. With increasing film thickness, the power density is no longer high enough to ionize the ablated material completely. The overall intensity of the recorded spectra decreases as shown. The explanation is supported by the shift of the decrease of emission intensity to higher film thicknesses for higher power densities.

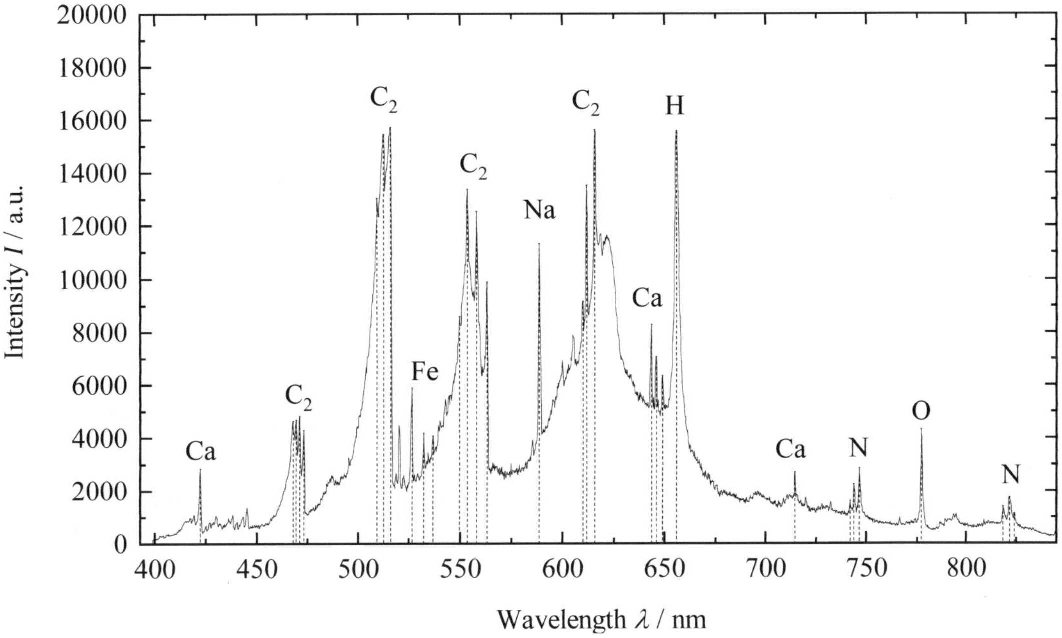


**Figure 11.** Average spectrum with marked element and molecule emission lines. The film thickness of Wisura Akalit CF 10 amounted $d_s \approx 690$ nm and the power density was $\Psi = 3.21$ GW·cm$^{-2}$.

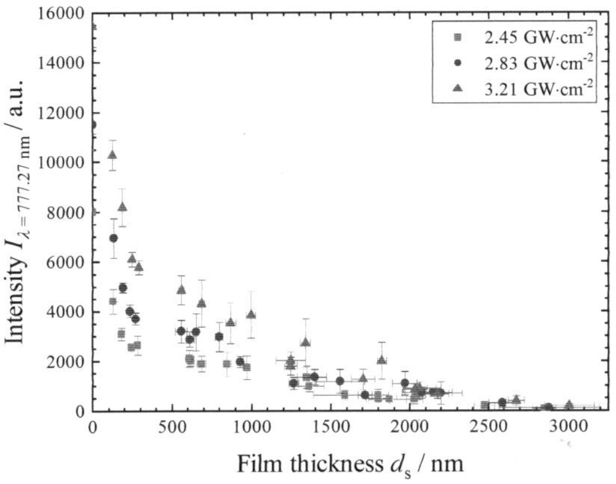


**Figure 12.** Trend of the emission intensity of the oxygen element line at $\lambda = 777.27$ nm dependent on film thickness $d_S$ of Wisura Akalit CF 10 and power density $\Psi$.

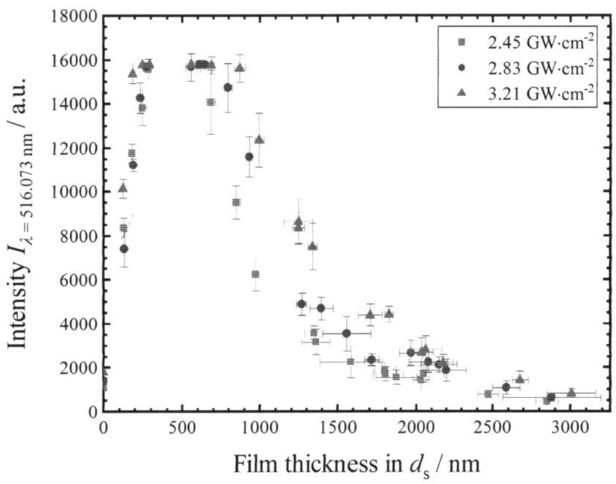


**Figure 13.** Trend of the emission intensity of the $C_2$ molecule emission line at $\lambda = 516.073$ nm dependent on film thickness $d_S$ of Wisura Akalit CF 10 and power density $\Psi$.

The linear correlation coefficient and the peak intensity ratio were used to evaluate the recorded spectra of the coated samples. The principle objective of the investigation is to identify how dependable spectra from coated samples can be distinguished from spectra of cleaned and uncoated samples. For the calculation of the correlation coefficient $r$, the reference spectra of the cleaned samples were used as a reference data set. The reference spectra of the different power densities were composed as shown in equation (2) from 35 single measurements. Equation (3) was used to calculate the peak intensity ratio. 35 LIBS measurements of every coated sample were recorded and the average correlation coefficient and peak intensity ratio were calculated.

Figure 14 shows the trend of the correlation coefficient $r$ for different power densities dependent on the film thickness. The highest correlation coefficient $r = 0.807 \pm 0.003$ was calculated for a film thickness of $d_S = 57.2 \pm 2.4$ nm at $\Psi = 2.45$ GW·cm$^{-2}$. In general, the correlation coefficient decreases with increasing film thickness of Wisura Akalit CF10. For thin film thicknesses up to $d_S \approx 450$ nm, the correlation coefficient $r$ decreases nearly linearly with a high gradient. In the range from $d_S \approx 500$ nm to $d_S \approx 1000$ nm, the correlation coefficient stagnates on one level. For a film thickness of $d_S > 1200$ nm, $r$ decreases slowly and approaches a value $r \approx 0.3$. Overall, a higher power density leads to a marginal higher value of the correlation coefficient.

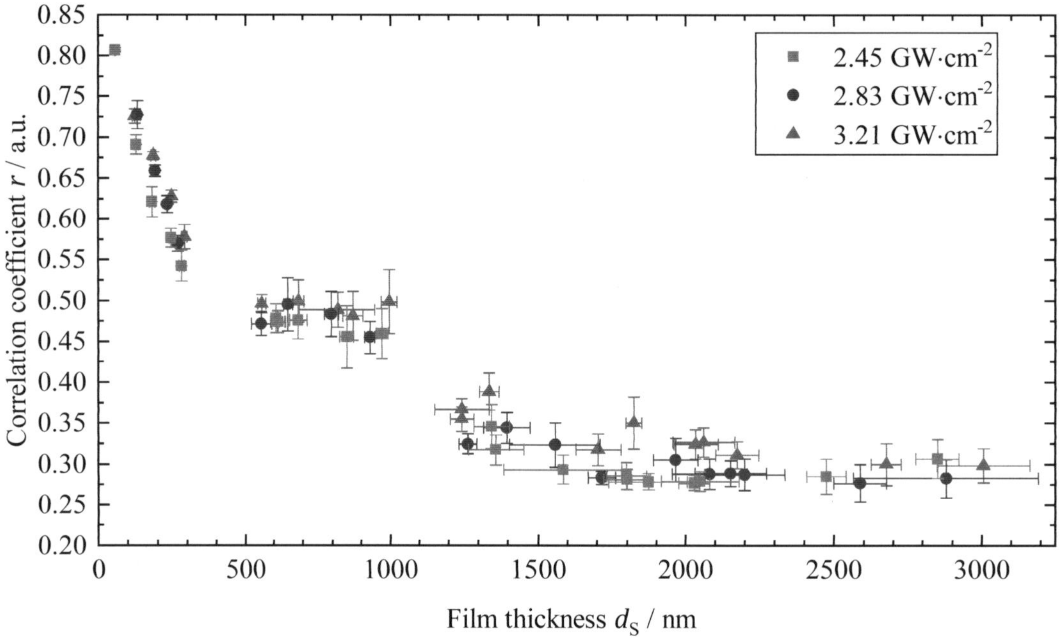


**Figure 14.** Correlation coefficient $r$ for increasing film thickness and three different power densities.

These three different stages can be explained with the trend of the $C_2$ molecule emission intensity, shown in figure 13. In the first film thickness region, the element line emission intensities rise rapidly and nearly linearly. Major changes occur in the element line emission intensities and therewith in the data set. In the second region, the element line emission intensity reaches the saturation of the spectrometer detector chip. Additionally, changes in the appearance of the spectra are very marginal. In the third film thickness region, the overall intensity decreases, but the appearance of the spectra changes in a marginal way, see figure 10.

Figure 15 depicts the trend of the peak intensity ratio of the hydrogen line at $\lambda = 656.021$ nm to the oxygen line at $\lambda = 777.27$ nm for the power density $\Psi = 3.21$ GW·cm$^{-2}$. The peak intensity ratio increases in the film thickness range up to $d_S \approx 600$ nm. Afterwards, the peak intensity ratio approaches to a value of $k \approx 4$ for a film thickness $d_S > 650$ nm. The trend of the peak intensity ratio can be described by a Boltzmann fit. The trend is explained with the course of the intensity of the hydrogen element line and the oxygen element line for increasing film thicknesses, which is depicted in figure 16. The emission intensity of the oxygen element line decreases rapidly in the film thickness region up to $d_S \approx 600$ nm. The emission intensity of the hydrogen element line reaches the saturation of the spectrometer for the same film thickness region. For $d_S > 600$ nm, the oxygen emission intensity and the hydrogen emission intensity decreases steadily with increasing film thickness and a constant peak intensity ratio is reached, see figure 15.

The investigation exhibit, that the linear correlation as well as peak intensity ratio are useable to distinguish spectra of coated samples from spectra of cleaned samples very easily. Additionally, it is shown that the values of the correlation coefficient $r$ and the peak intensity ratio $k$ can't be assigned to film thickness values directly. The correlation coefficient $r$ and the peak intensity ratio $k$ aren't explicit in the examined range of film thicknesses.

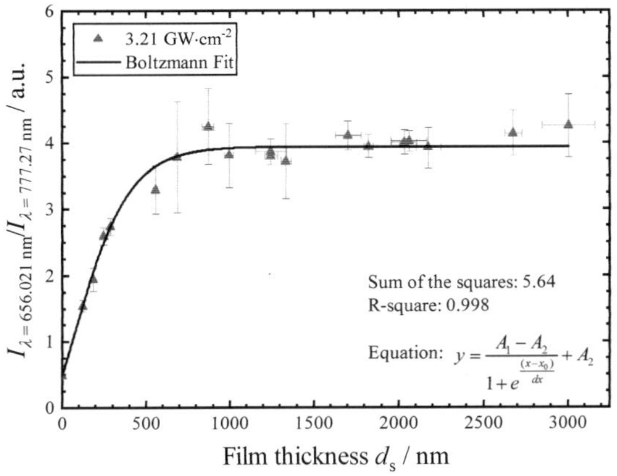


**Figure 15.** Trend of the peak intensity ratio $k$ of the hydrogen element line at $\lambda = 656.021$ nm to the oxygen element line at $\lambda = 777.27$ nm as a function of the film thickness $d_S$ for the power density $\Psi = 3.21$ GW·cm$^{-2}$.

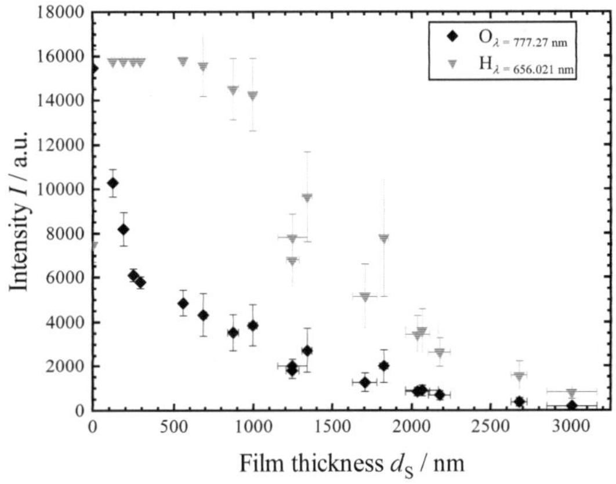


**Figure 16.** Trend of the emission intensity of the hydrogen element line at $\lambda = 656.021$ nm and the oxygen element line at $\lambda = 777.27$ nm as a function of the film thickness $d_S$ for the power density $\Psi = 3.21$ GW·cm$^{-2}$.

## 4. Conclusions

The investigation of the ground samples by means of surface roughness measurements using the profile method showed constant and low surface roughness values without any predominant direction. The examination of the surface energy and the fluorescence detector voltage showed a stable and reproducible cleanliness condition of the samples. The fluorescence detector voltage was measured on 115 samples with 35 measurement points on each sample. The surface energy was measured on three different sites on overall six samples.

The main objective of the investigation was to examine the possibility to locate contaminations on sample surfaces by laser-induced breakdown spectroscopy. In the first step, LIBS measurements were performed on cleaned samples and the impact of the delay time $t_d$ and the power density $\Psi$ were studied. The spectra of the cleaned samples show a strong continuum emission in a short first time range after the laser was fired. The appearance of continuum emission and the time range is well described in [26, 27]. For the next few microseconds, emission of discrete element lines from ambient air occurs. This behavior was also found in another investigation [32]. As expected, the intensity of the element lines decreased with increasing delay time. For increasing power densities, the curve progression looks the same but it is shifted to higher time values. The reference spectra of the cleaned samples were calculated for the examined power densities and correlation coefficients were determined for 140 recorded spectra from cleaned samples. The constant high and reproducible correlation values support the results from the surface energy measurements and the fluorescence measurements.

The samples were coated with Wisura Akalit CF 10 as a model contamination by a dip-coating process. Altogether, 19 different film thickness values were produced and the film thickness was determined using reflectometry. LIBS measurements of the coated sample surfaces were performed with constant delay times and three different power density values. Correlation coefficients in relation to reference spectra of uncoated samples and peak intensity ratios were calculated and investigated.

Summarized, the correlation coefficient *r* shows a decreasing amount with increasing film thickness values. For high film thickness values, the correlation coefficient approximates to a low and nearly constant amount. The peak intensity ratio *k* shows a similar behavior. First, the peak intensity ratio increases with increasing film thickness. Afterwards, the peak intensity ratio approximate to a nearly constant amount for higher film thickness values. The study shows that the linear correlation to reference spectra of cleaned and uncoated samples as well as the peak intensity ratio are suitable to detect lubricant contaminations on sample surfaces. Additionally, both described methods aren't directly suited to determine film thickness values in the investigated film thickness range.

**Acknowledgments**
The authors acknowledge the Institute of Materials Science and Engineering, Chemnitz University of Technology and the Daimler AG, Stuttgart, for supporting the research. We would also like to thank all colleagues from process development at Daimler AG, above all Mr. Stefan Elischer, for supporting the research.

**References**
[1] Fahrenwaldt H J and Schuler V 2006 *Praxiswissen Schweißtechnik: Werkstoffe, Prozesse, Fertigung* vol 2 (Wiesbaden: Vieweg)
[2] DVS – Deutscher Verband für Schweißen und verwandte Verfahren e.V. 2015 *Merkblatt DVS 3203-4 Laserstrahlschweißen von metallischen Werkstoffen – Nahtvorbereitung und konstruktive Hinweise* (Düsseldorf: DVS Media)
[3] DVS – Deutscher Verband für Schweißen und verwandte Verfahren e.V. 2009 *Merkblatt DVS 3214 - Unregelmäßigkeiten an Laserstrahlschweißnähten – Ursachen und Abhilfemaßnahmen* (Düsseldorf: DVS Media)
[4] Geiger M, Kägeler C and Schmidt M 2008 High-power laser welding of contaminated steel sheets *Product Engineering - Research and Development* **2** 235–240
[5] Müller M G 2002 *Prozessüberwachung beim Laserstrahlschweißen durch Auswertung der reflektierten Leistung* (München: Herbert Utz)
[6] Wolf M 2006 *Zur Phänomenologie der Heißrissbildung beim Schweißen und Entwicklung aussagekräftiger Prüfverfahren* (Bremerhaven: Wirtschaftsverlag NW)
[7] DVS – Deutscher Verband für Schweißen und verwandte Verfahren e.V. 1993 *Merkblatt DVS 2809 - Laserstrahlschweißen in Elektronik und Feinwerktechnik* (Düsseldorf: DVS Media)
[8] Behler K, Sokolowski W and Beyer E 1987 *Laser Optoelektronik in der Technik: Vorträge des 8. Internationalen Kongresses Laser 87 Optoelektronik* ed Waidelich W (Berlin, Heidelberg: Springer) pp 529-532
[9] Oberhausen M 2007 *Der Einsatz laserinduzierter Fluoreszenzmessungen zur Detektion geringster organischer Belegungen auf Oberflächen* (Saarbrücken: Universität des Saarlandes Produktionstechnik, Dissertation)
[10] Fenske M, Markus S and Hildebrandt H 2012 *Qualitätssicheres Vorbehandeln und Kleben durch den Einsatz optischer Emissionsspektroskopie - Safebond* (Bremen: Forschungsvereinigung Schweißen und verwandte Verfahren)
[11] Markus S 2008 *Die Laser Induced Breakdown Spectroscopy (LIBS) als Inline-Verfahren zur Detektion von Oberflächenkontaminationen im Bereich der Klebtechnik* (Aachen: Shaker)
[12] Noll R, Bette H, Brysch A, Kraushaar M, Mönch I, Peter L and Sturm V 2001 Laser-induced breakdown spectrometry - applications for production control and quality assurance in the steel industry *Spectrochimica Acta Part B* **56** 637–649
[13] Feierabend A 2015 *Recycling und Rohstoffe* vol 8, ed Thomé-Kozmiensky K J and Goldmann D (Neuruppin: TK Verl. Thomé-Kozmiensky) pp 619–628
[14] Vadillo J M and Laserna J J 2004 Laser-induced plasma spectrometry: truly a surface analytical tool *Spectrochimica Acta Part B* **59** 147–161

[15] Vadillo J M, Palanco S, Romero M D and Laserna J J 1996 Applications of laser-induced breakdown spectrometry (LIBS) in surface analysis *Fresenius' Journal of Analytical Chemistry* **355** 909–912
[16] Kratochvíl T, Černohorský T, Knotek P, Kalina L, Návesník J, Pouzara M and Zvolskáa M 2014 Fast determination of the surface density of titanium in ultrathin layers using LIBS spectrometry *Journal of Analytical Atomic Spectrometry* **29** 1806–1812
[17] Lentjes M, Dickmann K and Meijer J 2005 *Aktuelle Methoden der Laser- und Medizinphysik Remagen* ed Hartmann U (Berlin: VDE-Verlag) pp 288–293
[18] Kaelble D H 1970 Dispersion-Polar Surface Tension Properties of Organic Solids *The Journal of Adhesion* **2** 66–81
[19] Owens D K and Wendt R C 1969 Estimation of the surface free energy of polymers *Journal of Applied Polymer Science* **13** 1741–1747
[20] Rabel W 1971 Einige Aspekte der Benetzungstheorie und ihre Anwendung auf die Untersuchung und Veränderung der Oberflächeneigenschaften von Polymeren *Farbe und Lacke* **77** 997–1005
[21] Holz P, Lutz C and Brandenburg A 2017 Optical scanner system for high resolution measurement of lubricant distributions on metal strips based on laser induced fluorescence *Proc. of SPIE* 103292A-1 - 103292A-11
[22] Fischer N 2005 *Messungen geringster organischer Belegungen auf diffus reflektierenden Oberflächen mit einem mIR-faseroptischen Prüfkopf* (Saarbrücken: Universität des Saarlandes Produktionstechnik, Dissertation)
[23] Clint J H 1973 Partial Deposition of Mixed Monolayers onto Solid Surfaces *Journal of Colloid and Interface Science* **43** 132–143
[24] Derjaguin B 1993 On the Thickness of the Liquid Film Adhering to the Walls of a Vessel after Emptying *Progress in Surface Science* **43** 134–137
[25] Noll R 2012 *Laser-Induced Breakdown Spectroscopy: Fundamentals and Applications* vol 1, (Berlin, Heidelberg: Springer)
[26] Miziolek A W Palleschi V and Schechter I 2006 *Laser Induced Breakdown Spectroscopy* vol 1, (Cambridge: Cambridge University Press)
[27] Singh J P and Thakur S N 2007 *Laser-induced breakdown spectroscopy* vol 1, (Amsterdam, Boston, London: Elsevier)
[28] Lentjes M, Dickmann K and Meijer J 2007 Calculation and optimization of sample identification by laser induced breakdown spectroscopy via correlation analysis *Spectrochimica Acta Part B* **62** 56–62
[29] Khater M A 2013 Laser-induced breakdown spectroscopy for light elements detection in steel: State of the art *Spectrochimica Acta Part B* **81** 1–10
[30] Cremers D A and Radziemski L J 2013 *Handbook of laser-induced breakdown spectroscopy* vol 2, (Chichester: John Wiley & Sons)
[31] Hügel H, Dausinger F and Graf T 2007 *Laser in der Fertigung: Strahlquellen, Systeme, Fertigungsverfahren* vol 2 (Wiesbaden: Teubner)
[32] Zhang J 2015 Untersuchungen von laserinduzierten Plasmen in Bohrungen (LIPS) (Mittweida: Hochschule Mittweida, Fakultät Mathematik/Naturwissenschaften/Informatik, Masterthesis)
[33] Lebedev V, Rabchinskii M, Kozlyakov M, Stepanov D, Shvidchenko A, Nikonorov N V and Vul A 2018 Laser-Induced Breakdown Spectroscopy: an advanced method for analysis of nanocarbon materials chemical composition *Journal of Analytical Atomic Spectrometry* **33** 240–250

# Emergence and impact of $Al_2TiO_5$ in $Al_2O_3$-$TiO_2$ APS coatings

**A Richter[1,*], L-M Berger[2], S Conze[2], Y J Sohn[1] and R Vaßen[1]**

[1] Forschungszentrum Jülich GmbH, Institute of Energy and Climate Research, Materials Synthesis and Processing (IEK-1), 52425 Jülich, Germany
[2] Fraunhofer Institute for Ceramic Technologies and Systems, Fraunhofer IKTS, Winterbergstr. 28, 01277 Dresden, Germany

[*] e-mail: ale.richter@fz-juelich.de

**Abstract.** Despite numerous studies and decades of industrial application, there is still a lack of understanding about the formation and the impact of aluminum titanate ($Al_2TiO_5$) in $Al_2O_3$-$TiO_2$ thermal spray coatings. Especially the influence of the feedstock powder characteristics on the phase composition has only crudely been investigated so far. Therefore, in this work we have characterized commercial fused and crushed $Al_2O_3$-$TiO_2$ feedstock powders: three of them containing 13 wt.% $TiO_2$ and three containing 40 wt.% $TiO_2$. The effect of the varying phase compositions of the powders and their relevance on the deposition efficiency, the phase compositions, the porosity, and the hardness of the respective APS coatings is described in detail. While detrimental to the mechanical properties of 40 wt.% $TiO_2$ coatings, we have found an enhancement of the hardness for 13 wt.% $TiO_2$ coatings with a high $Al_2TiO_5$/$Al_6Ti_2O_{13}$ content in the feedstock powder. Furthermore, it was found that $Al_2TiO_5$ may reform during APS when sprayed from an $Al_2TiO_5$-free powder.

## 1. Introduction

Thermal spraying is a group of technologies that is very versatile regarding the coating material (predominantly oxides, metals and hardmetals) to produce coatings with a typical thickness in the range of 100-500 µm with a high deposition rate. Atmospheric plasma spraying (APS) is the most common thermal spray process for ceramic materials using Ar, $H_2$, and/or $N_2$ as plasma forming gases. In the case of oxides, as the most important ceramic material group, melting of the feedstock material during the spray process is common [1–3]. The coatings are then built up by individual splats of rapidly solidified spray powder particles. Changes of the chemical and/or the phase composition of the coating as compared to the feedstock occur due to chemical reactions with plasma-forming gases, burning products (in the case of flame spraying) or with the environment. When spraying $Al_2O_3$ coatings, for example, the thermodynamically stable $\alpha$-$Al_2O_3$ commonly solidifies in its metastable $\gamma$-$Al_2O_3$ modification, which has a significant impact on the properties of the coating [4,5]. Furthermore, titanium oxide feedstock may be reduced during thermal spray [6,7] producing coatings with non-ordered and ordered (Magnéli-phases) suboxides, which would affect the melting point of the particles [6,8] and the electrical [6,7] as well as the thermal conductivity [9] of the sprayed coating.

Several commercially available feedstock powder compositions of the $Al_2O_3$-$TiO_2$ system ($Al_2O_3$ with 3, 13, and 40 wt.% $TiO_2$) are commonly processed to coatings. However, in the $Al_2O_3$-$TiO_2$ system a compound, aluminum titanate ($Al_2TiO_5$), exists at about 44 wt.% $TiO_2$. It is thermodynamically stable in a temperature range of about 1280 °C up to its melting point at 1860 °C [10], but may also exist in a metastable state as long as the rates of diffusion are sufficiently low (typically up to about 800 °C). This compound is proposed to have a significant impact on the properties of materials of the $Al_2O_3$-$TiO_2$ system as its properties strongly deviate from the properties of both single oxides. For example its hardness and elastic modulus are much lower than that of $\alpha$-$Al_2O_3$. It does show some promising thermal properties, however, due to a pronounced isotropy in its thermal expansion coefficient, leading to an average thermal expansion coefficient close to zero in polycrystalline $Al_2TiO_5$ ceramics [11].

Furthermore, it has to be taken into account that the process of $Al_2TiO_5$ decomposition is strongly affected by the oxygen partial pressure [12] and the grain size of $Al_2TiO_5$ [13]. For the decomposition process a contraction of $Al_2TiO_5$ crystals in the *a* direction of the lattice and a concurrent expansion in the other two directions, *b* and *c*, at different rates was proven. However, the complex mechanism has not been fully understood yet. Depending on the oxygen partial pressure $Al_2TiO_5$ forms an Al-rich solid solution $Al_{2-x}Ti_{1+x}O_5$ of anosovite ($Ti_3O_5$) and $Al_2TiO_5$ [12,14,15,17] which oxidizes and decomposes to $Al_2O_3$ and $TiO_2$ as final products in air. Reformation of decomposed $Al_2TiO_5$ above 1280 °C has been described in air and in vacuum [16,18,19]. The formation of a solid solution has also been observed in the case of thermal spray coatings [20]. Depending on the spraying conditions, changes in the composition of the solid solution could be observed between the feedstock material and an APS coating. It has been found that the homogeneity of distribution of Al and Ti in the powder is a key issue for the formation of $Al_2TiO_5$ in feedstock powders and the corresponding coatings [20].

More recently, several authors described $Al_6Ti_2O_{13}$ as an additional phase in the $Al_2O_3$-$TiO_2$ system [21–24], which was proposed to exist in thermal spray coatings as well [25]. It can be assumed that the properties of this compound are closely related to the properties of $Al_2TiO_5$. All of the described phases are expected to affect the properties of the coatings and, therefore, need to be controlled during the spraying process.

APS $Al_2O_3$-$TiO_2$ coatings have found widespread use in industry for the functionalization of metallic substrates with a focus on mechanical and electrical properties. Furthermore, the machinability, wear resistance or conductivity of the coatings can be adjusted by changing the $TiO_2$ content. Current applications of $Al_2O_3$-$TiO_2$ coatings include alternator parts, heaters, valves, textile manufacturing tools, and wear parts [26,27]. Several studies have addressed the influence of feedstock powders of different composition on the properties of APS coatings in the past [25,28–36]. Most of these studies focus on the comparison between nanostructured powders and the clad powder Metco 130, which consists of coarse alumina particles cladded with sub-µm titania [28–32,36,37]. However, the most commonly used type of commercial $Al_2O_3$-$TiO_2$ feedstock, namely fused and crushed powders, which are often preferred due to their low production costs, are rarely investigated.

Despite the multitude of academic studies and a long history of application of APS coatings there are still issues with the reliability of the coating properties, which stem from insufficient attention to the feedstock powder properties. In this study we address this issue by taking a deeper look at the influence of the compounds from the $Al_2O_3$-$TiO_2$ system on the properties of the coatings with a special focus on the different commercially available fused and crushed feedstock powders. Controlling the phase contents in fused and crushed powders is not straightforward and strongly depends on the heating, mixing and the cooling conditions during fabrication. Therefore, we have selected six fused and crushed powders of two common compositions, characterized them in detail, and investigated the impact of their phase composition on the properties of APS $Al_2O_3$-$TiO_2$ coatings. Furthermore, we investigated the behavior of an experimental powder, obtained by heat treatment of a commercial powder with very high $Al_2TiO_5$ content, during the spray process.

## 2. Experimental and materials

### *2.1. Feedstock powders*

A selection of six different commercially available $Al_2O_3$-$TiO_2$ feedstock powders (three with a nominal composition of 13 wt.% $TiO_2$ and three powders with 40 wt.% $TiO_2$) was used in this study. For both compositions one powder was produced by blending separately fused and crushed $Al_2O_3$ and

$TiO_x$ powders. These powders will be denoted as "blend13" and "blend40" in the following. The other two powders for each composition were produced by jointly fusing and crushing of the individual oxides by different manufacturers and under different processing conditions. They will be referred to as "jA13/jA40" and "jB13/jB40" in the following. The phase composition of powder jB40 was modified by a thermal treatment at 1150 °C with a heating rate of 5 K $min^{-1}$ and 4 h holding time in a Nabertherm LH30/14 chamber furnace under flowing air.

All feedstock powders were characterized in detail. This included a characterization of the cross-section of each powder by a FEI Phenom First Generation scanning electron microscope (SEM) in backscattered electron (BSE) mode. The particle size distribution was determined by laser diffraction (Horiba LA-950-V2) and the density of the powders characterized via a helium pycnometer (Quantachrome Instruments Ultrapycnometer 1000). A Bruker D4 Endeavor was employed to determine phase contents of each powder via X-ray diffraction (XRD). Each diffraction pattern was recorded in a range of 2Θ=10..130° with a step size of 0.02° and a measurement time of 2 s per step in a simultaneous measurement range of 4°. Phase quantification was then performed by Rietveld refinement using the TOPAS Software [38]. The mass changes of the feedstock powders were determined by gravimetric tests in alumina crucibles (triple determination of mass gain using 2 g of feedstock powder) at 1300 °C and for 6 h holding time under a continuous air flow.

### *2.2. Coatings*

Atmospheric plasma spraying (APS) was used to produce coatings with thicknesses of 500 to 800 µm. In particular, an Oerlikon Metco F4 plasma torch with a nozzle diameter of 6 mm was used in an Oerlikon Metco Multicoat System. Each coating was sprayed with an $Ar/H_2$ gas mixture with flow rates of 40 and 10 slpm, respectively. The stand-off distance was set to 150 mm and the torch was operated at a constant current of 550 A, resulting in a torch power of about 40 kW. The surface speed of the torch was set to 500 mm/s in relation to a stationary substrate. Prior to spraying, each grit-blasted mild steel substrate (50x50x2 $mm^3$) was preheated to about 120 °C. During spraying, the substrate was cooled by two air jets. The feed rate of each powder was determined at the beginning of the spraying experiments. This feed rate was then used to estimate the deposition efficiency (*DE*) after measuring the mass gain of the substrate. All powders were fed with a constant Ar flow rate of 3.5 slpm.

Following the coating deposition, each substrate was cut using an ATM Brillant 221 saw. XRD measurements were performed on as-sprayed coatings similarly to those for the powders, but with a reduced measurement range of 2Θ=10…80° and a measurement time of 0.75 s per step. Materialographically prepared cross-sections were firstly characterized by confocal laser microscopy (LaMi) to determine the porosity of each coating and the coating thickness. Ten LaMi images were recorded at a magnification of x500 and then analyzed via ImageJ [39]. Furthermore, BSE SEM images were recorded using a Hitachi TM3000. The Vickers hardness of the coatings was determined on the cross-sections by using a Leitz Miniload 2 with an applied force of 3 N (HV0.3) for 30 s. At least five indentations were made per coating.

## 3. Results and discussion

### *3.1. Feedstock powders and the respective coatings*

A similar powder morphology of irregularly shaped particles was found for all powders. In combination with a high similarity between the size distributions of all powders as exemplified by the $d_{50}$ in Table 1, this allows for a direct comparison of their behavior in the spray process. However, differences were detected for the powder densities (Table 1). In fact, the densities were in agreement with the phase analysis results obtained via Rietveld refinement (Figure 1). In the following "$TiO_2$" implies the rutile structure. The powder blends exhibited a pycnometric density in between the crystallographic densities of $\alpha$-$Al_2O_3$ (4.0 g $cm^{-3}$) and $TiO_2$ (4.3 g $cm^{-3}$). In contrast, powders with a significant content of $Al_2TiO_5$ (crystallographic density 3.7 g $cm^{-3}$) showed a lower density. Overall the blend13 and blend40 powders consisted of $\alpha$-$Al_2O_3$ and $TiO_2$ or several $TiO_x$ phases. Although, a mass gain of blend40 powder of 0.4 % indicated a low O-deficiency. Contrary, the jointly fused and crushed powders jA and jB differed significantly in their oxidation behavior as well as in their phase compositions. The theoretic maximum $Al_2TiO_5$ content is about 30 and 90 wt.% for compositions with 13 wt.% $TiO_2$ and 40 wt.% $TiO_2$, respectively. While the jA powders contained an $Al_{2-x}Ti_{1+x}O_5$ solid solution [20], the jB powders contained significant amounts of $Al_2TiO_5$ and $Al_6Ti_2O_{13}$ [21,23]. Due to the inhomogeneous distribution of the metallic constituents, the $\alpha$-$Al_2O_3$ content in the jA powders is higher than for the theoretic compositions due to an incomplete reaction with $TiO_2$. A significant mass increase for the jA powders by oxidation in air is connected to the reaction of $Al_{2-x}Ti_{1+x}O_5$ to $Al_2TiO_5$ or $Al_2O_3$ and $TiO_2$ [20]. The compositions of the jB powders corresponded well to the theoretic compositions according to the results of the phase analysis if $Al_6Ti_2O_{13}$ is included.

**Table 1.** Mean diameter $d_{50}$ and density of the investigated $Al_2O_3$-$TiO_2$ powders.

| powder | 13 wt.% $TiO_2$ | | | 40 wt.% $TiO_2$ | | |
|---|---|---|---|---|---|---|
| | blend13 | jA13 | jB13 | blend40 | jA40 | jB40 |
| $d_{50}$ (μm) | 35 | 37 | 32 | 34 | 34 | 30 |
| pycnometric density (g cm$^{-3}$) | 4.0 | 4.0 | 3.9 | 4.1 | 4.0 | 3.7 |
| average mass gain (%) | 0.16 | 0.73 | 0.03 | 0.40 | 1.91 | 0.42 |

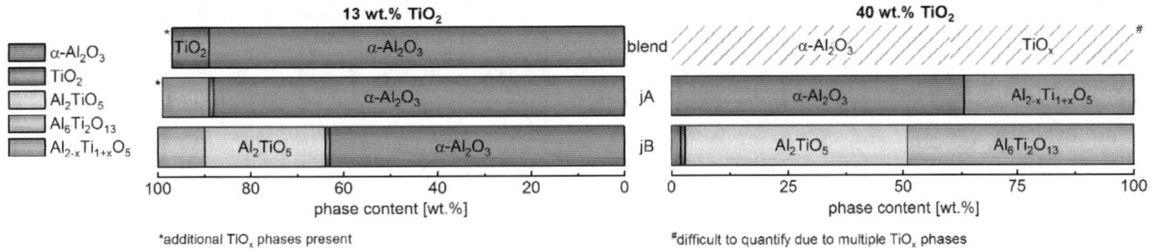


**Figure 1.** Phase compositions of the investigated feedstock powders according to Rietveld analysis.

The cross-sections of the powders are presented in Figure 2 a-c) and g-i) for the 13 and the 40 wt.% $TiO_2$ powders, respectively. A clear distinction can be made between the individual bright $TiO_2$/$TiO_x$ and the dark $Al_2O_3$ particles for the powder blends in Figure 2 a) and g). Surprisingly, a similar elemental contrast was observed for the jA powders in Figure 2 b) and h). Even though they were produced by joint fusion of $Al_2O_3$ and $TiO_2$, they are characterized by an inhomogeneous distribution of the metallic constituents. Both jA powders also showed the highest mass gains in the oxidation experiments in their corresponding groups. Only few powder particles show a homogeneous intermediate grayscale. The analysis results for the jA40 powder are consistent with the results of the earlier study [20]. In contrast, a homogeneous elemental distribution was found in both jB powders (Figure c) and i)). This corresponds well to the results of the phase analysis. Apparently, an insufficient intermixing of the metallic elements impeded the formation of $Al_2TiO_5$ for the jA powders, while for the jB powders a homogeneous intermixing of the elements facilitated its formation. Despite the decomposition of $Al_2TiO_5$ below 1280 °C [10], this phase can appear in the feedstock powder due to rapid cooling [40,41]. The formation conditions of $Al_6Ti_2O_{13}$ are still unclear. It is noteworthy that an inhomogeneous distribution of the metallic constituents in the jA powders correlates with a high deficiency of O relative to $Al_2TiO_5$, while for the jB powders the homogeneous distribution of Al and Ti and a low deficiency of O coincides with the co-existence of $Al_2TiO_5$ and $Al_6Ti_2O_{13}$. It appears that the presence of $Al_2TiO_5$ and $Al_6Ti_2O_{13}$ indicates a homogeneous distribution of the metallic elements and non-reducing conditions during the preparation of the powder.

IOP Conf. Series: Materials Science and Engineering **480** (2019) 012007 doi:10.1088/1757-899X/480/1/012007

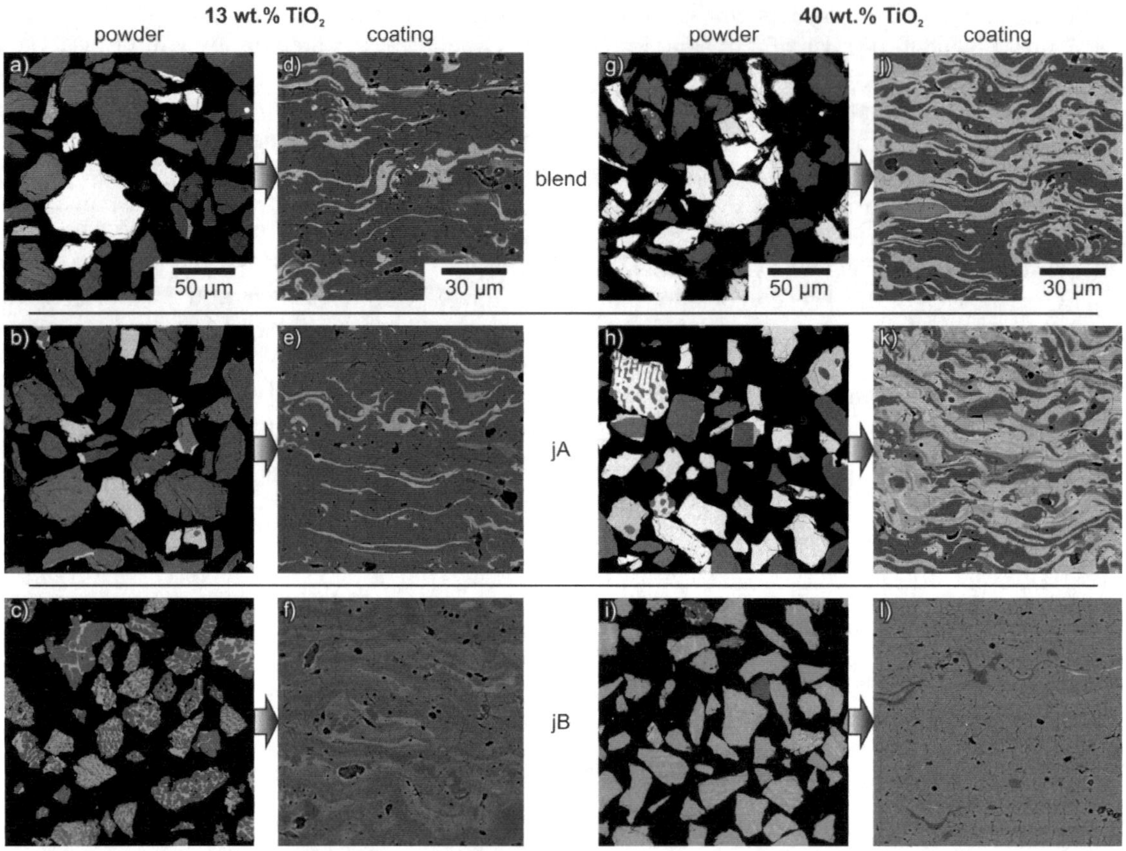


**Figure 2.** BSE SEM cross-sectional images of the feedstock powders with a)-c) 13 wt.% $TiO_2$ and g)-i) 40 wt.% $TiO_2$. d)-f) and j)-l): BSE SEM images of the respective APS coatings. The inserted scale bars are valid for all images in the respective column.

Cross-sections of the APS coatings produced from the various powders are presented in Figure 2 d)-f) and j)-l). Similarly to the feedstock powders, mostly dark or bright splats were observed in the microstructures of the coatings produced from the blended powders (Figure 2 d) and j)). Accordingly, the XRD patterns in Figure 3 indicate the presence of $\gamma$-$Al_2O_3$, $TiO_2$, and some $\alpha$-$Al_2O_3$. The formation of $\gamma$-$Al_2O_3$ in thermal spray coatings from $\alpha$-$Al_2O_3$ feedstock powders is well known [4,5]. However, no formation of $Al_2TiO_5$ or other compounds was detected in the coatings. In addition to $\gamma$-$Al_2O_3$, $TiO_2$, and $\alpha$-$Al_2O_3$, the coatings produced from powder jA exhibited some peaks that can be related to $Al_{2-x}Ti_{1+x}O_5$ (Figure 3), especially for jA40. With respect to the feedstock powders, the peaks related to this solid solution were slightly shifted to higher diffraction angles for the coatings, indicating an increased Ti content [20]. Similarly, the cross-sectional images in Figure 2 e) and k) show several splats with an intermediate gray scale between the $Al_2O_3$ and $TiO_2$ splats. So, in contrast to the blend powders, some changes in the phase composition of the jA powders occurred during APS, although these changes were probably restricted to powder particles containing both metallic elements. Lastly, the images of the coatings produced from the jB powders (Figure f) and l)) exhibited a low contrast between the individual splats, indicating a homogeneous distribution of the elements, similarly to the feedstock powders. Furthermore, only peaks related to $Al_2TiO_5$ were observed in the XRD pattern of the coatings produced from the jB40 powder (Figure 3). However, a clear distinction between $Al_2TiO_5$ and $Al_6Ti_2O_{13}$ is difficult from the XRD patterns of the coatings due to very broad peaks and a high similarity between the structures of these phases. Contrary to the jB40 coating, mostly $\gamma$-$Al_2O_3$ could be identified in the coatings sprayed from the jB13 powder with only minor peaks pointing at residual

$Al_2TiO_5$. At the same time, no peaks related to $TiO_2$ or $TiO_x$ could be identified. This may be due to Ti being incorporated in $\gamma$-$Al_2O_3$. This assumption is supported by a slight shift of the $\gamma$-$Al_2O_3$ peaks to lower diffraction angles (e.g. at 67°) which indicate an increase in its lattice constant. Transmission electron microscopy (TEM) investigations by Goberman et al. also indicated an incorporation of Ti in $\gamma$-$Al_2O_3$ for APS $Al_2O_3$-$TiO_2$ coatings [35,42]. Yang et al. have suggested the incorporation of Ti in an amorphous $Al_2O_3$ phase which is related to the amorphous "hump" at 2Θ=25..40° in the XRD diagrams of the coatings [29].

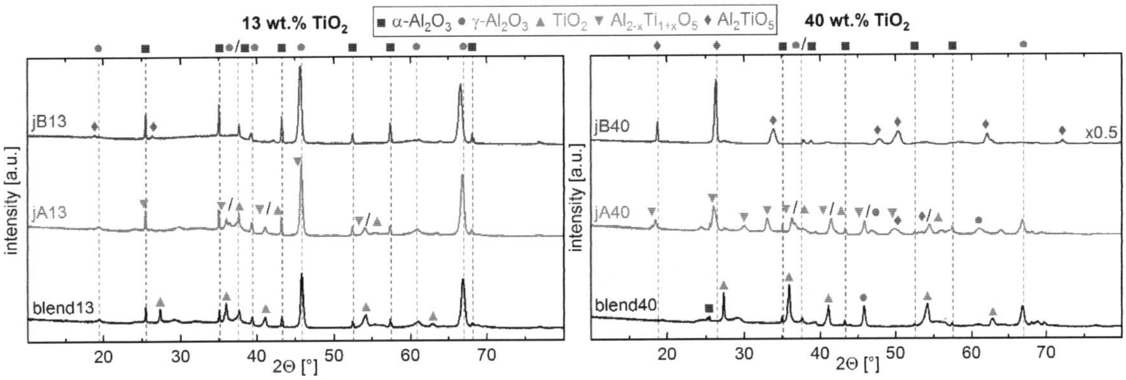


**Figure 3.** XRD diagrams of APS coatings sprayed by using the 13 wt.% $TiO_2$ feedstock powders (left) and 40 wt.% $TiO_2$ powders (right).

The *DE*, porosity and hardness of the APS coatings are presented in Figure 4. Low *DE*s were observed for the coatings sprayed from the blended powders. This can be related to the high melting point of $Al_2O_3$ at about 2050 °C which leads to insufficient melting of the pure $Al_2O_3$ particles in the plasma plume which will then bounce off the substrate/coating surface [3]. An improved intermixing of the metallic elements in each feedstock powder particle significantly decreases its melting temperature and thereby increases the *DE*. In the $Al_2O_3$-$Al_2TiO_5$ part of the phase diagram of the $Al_2O_3$-$TiO_2$ system this could lead to an up to 200 °C lower melting temperature. Accordingly, a higher *DE* was observed with the jB powders as compared to jA or the blend powders for both compositions. Surprisingly, the coatings sprayed with jB13 exhibited a slightly higher *DE* than the coatings sprayed with jB40. A possible explanation is the much lower thermal conductivity of $Al_2TiO_5$ (about 2 W (m K)$^{-1}$ [43,44]) phase in comparison to $\alpha$-$Al_2O_3$ (about 34 W (m K)$^{-1}$). Therefore, a lower melting degree is obtained due to impeded thermal transport through the particles despite a lower melting temperature.

The porosity of the coatings is also strongly affected by the degree of melting of the feedstock powder. This is especially evident for the 13 wt.% $TiO_2$ feedstock powders. Here, the coatings produced from blend13 and jA13 showed a significantly higher porosity than the coating from jB13 (Figure 4). With the 40 wt.% $TiO_2$ powders, however, a low and highly similar porosity was found for all coatings. Here, the phases with lower melting temperatures, namely $TiO_x$ and $Al_2TiO_5$, might have sufficiently promoted the densification of the coating.

The hardness of a coating is influenced by its porosity and phase composition. Therefore, the similar porosities of the coatings from the 40 wt.% $TiO_2$ powders allow for a comparison of the effect of the different phases. In particular, the occurrence of $Al_2TiO_5$ decreased the coating hardness distinctively (Figure 4). The highest hardness was measured for the coatings produced from blend40. Coatings prepared from feedstock powders with 13 wt.% $TiO_2$ showed hardness values higher than 1000 HV0.3 due to an increased amount of $\gamma$-$Al_2O_3$. Here, a slightly higher hardness was measured for the coating produced from the jB13 powder as compared to the other coatings. This might be related to the lower porosity of the coating, but could additionally stem from the distorted $\gamma$-$Al_2O_3$ phase as indicated by the XRD results in Figure 3. The latter assumption is also supported by spraying

IOP Conf. Series: Materials Science and Engineering **480** (2019) 012007 doi:10.1088/1757-899X/480/1/012007

experiments at an increased stand-off distance of 350 mm (not shown). Here, a similar hardness was reached for the jB13 coating despite an increase of the porosity to 9 %.

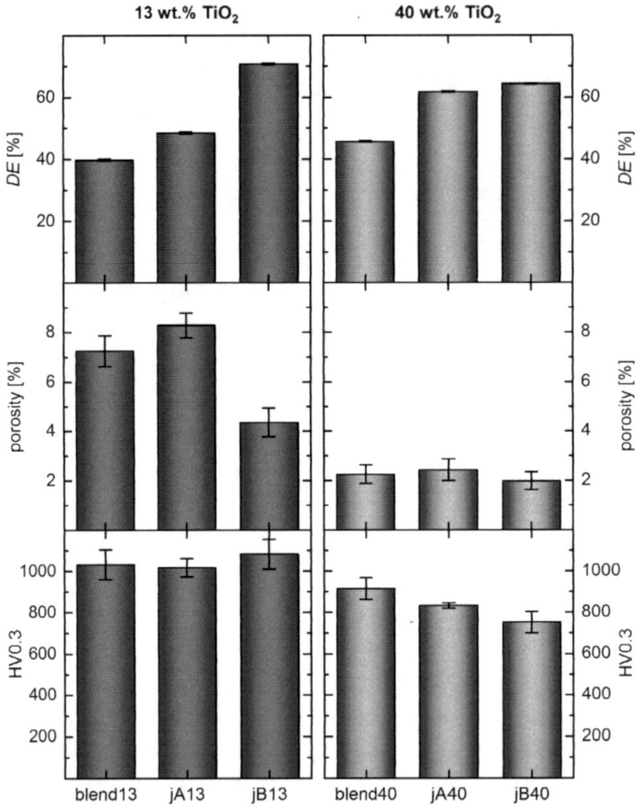


**Figure 4.** Deposition efficiency *DE*, porosity, and hardness HV0.3 of the APS coatings sprayed by using different feedstock powders for two $Al_2O_3$-$TiO_2$ compositions.

### 3.2. *(Re-)formation of $Al_2TiO_5$*

After annealing at 1150 °C for 4 h, the powder jB40, initially consisting of $Al_2TiO_5/Al_6Ti_2O_{13}$, solely contained $\alpha$-$Al_2O_3$ and $TiO_2$. The XRD patterns of the two powders are directly compared in the lower part of Figure 5 a) and show a full decomposition of $Al_2TiO_5/Al_6Ti_2O_{13}$. At the same time the morphology of the powder was preserved according to SEM images (not shown). Both powders were then used in APS and the XRD patterns of the corresponding coatings are presented in the upper part of Figure 5 a). Surprisingly, highly similar phase compositions were obtained for both coatings, indicating a complete reformation of the $Al_2TiO_5/Al_6Ti_2O_{13}$ during the spraying process. Also, the *DE*, the porosity, and the hardness were similar for both coatings. The small difference in the *DE* may be attributed to the lower thermal conductivity of $Al_2TiO_5$ in the jB40 powder. Here, the thermal properties of $Al_6Ti_2O_{13}$ are assumed to be similar to those of $Al_2TiO_5$. The higher porosity of the coating can be related to the development of a fine porosity within the feedstock powder due to the thermal decomposition. This porosity is then partially transferred to the coating. The reformation of $Al_{2-x}Ti_{1+x}O_5$ from an annealed powder (the original powder corresponded to jA40 in this work) has also recently been observed by Berger et al. for another APS process [20]. However, the $Al_{2-x}Ti_{1+x}O_5$ content of the feedstock powder and in the resulting coating were much lower in that case due to insufficient intermixing of the metallic constituents [20].

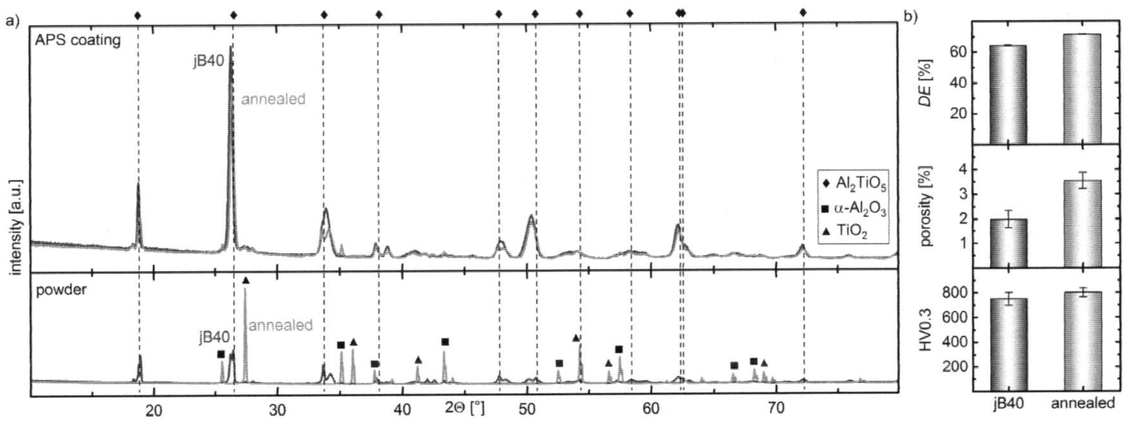


**Figure 5.** a) XRD diagrams of the powder jB (full), the same powder after thermal treatment (dotted), and the respective APS coatings. b) Deposition efficiency *DE*, porosity and Vickers hardness HV0.3 of the APS coatings produced from jB and the thermally treated powder.

## 4. Conclusions

In the current study we demonstrated the importance of controlling the phase composition in $Al_2O_3$-$TiO_2$ thermal spray coatings by a detailed characterization of the feedstock powder, especially with respect to the content of $Al_2TiO_5$ and $Al_6Ti_2O_{13}$. The presence of these phases in the jB powders indicates a homogeneous distribution of the metallic elements and non-reducing conditions during the preparation of the powder. Powders with a high content of these phases resulted in an improved sprayability, but severely decreased the hardness of $Al_2O_3$-40 wt.% $TiO_2$ APS coatings, while yielding a slight increase in hardness for $Al_2O_3$-13 wt.% $TiO_2$ coatings as compared to equivalent feedstock powders with inhomogeneous distributions of Al and Ti and low $Al_2TiO_5$/$Al_6Ti_2O_{13}$ contents.

Furthermore, we observed a full reformation of $Al_2TiO_5$/$Al_6Ti_2O_{13}$ during APS after a thermal treatment of the original $Al_2O_3$-40 wt.% $TiO_2$ feedstock powder, leading to a full decomposition of these compounds into $Al_2O_3$ and $TiO_2$. This illustrates that the formation of $Al_2TiO_5$ during spraying is possible if a sufficient intermixing of the single oxides is provided. Accordingly, it is paramount to be aware of the phase composition and phase distribution in any $Al_2O_3$-$TiO_2$ thermal spray feedstock powder and to control its manufacturing route to achieve full control of the phase composition and ultimately obtain reliable coating properties. Additional studies will focus on the localization of Ti in the 13 wt.% $TiO_2$ coatings and systematically determine the effect of varying $Al_2TiO_5$ contents on other coating properties, especially their thermal, electrical, tribological, and corrosion behavior. Furthermore, the deliberate stabilization of $Al_2TiO_5$ in the powders and coatings via additional oxides will be explored.

## Acknowledgments

This work was funded via the AiF (IGF: 19.471B / DVS: 02.109) by the German Federal Ministry of Economics and Technology within the framework of the program for promotion of "Industrial Joint Research" (IGF) of the Research Association on Welding and Allied Processes of the German Welding Society (DVS). The authors acknowledge the following members of the project advisory board for providing the feedstock powders: Ceram Ingenieurkeramik GmbH (Albbruck-Birndorf, Germany), H.C. Starck GmbH (Laufenburg, Germany), and Saint Gobain (Cologne, Germany). Furthermore, the authors would like to thank Frank Kurze and Sigrid Schwartz-Lückge for assistance with spraying the coatings and characterizing the feedstock powders.

**References**

[1] Fauchais P 2004 Understanding plasma spraying *J. Phys. D. Appl. Phys.* **37** R86–108
[2] Sampath S and Herman H 1996 Rapid solidification and microstructure development during plasma spray deposition *J. Therm. Spray Technol.* **5** 445–56
[3] McPherson R 1981 The relationship between the mechanism of formation, microstructure and properties of plasma-sprayed coatings *Thin Solid Films* **83** 297–310
[4] McPherson R 1980 On the formation of thermally sprayed alumina coatings *J. Mater. Sci.* **15** 3141–9
[5] Stahr C C, Saaro S, Berger L M, Dubský J, Neufuss K and Herrmann M 2007 Dependence of the stabilization of α-Alumina on the spray process *J. Therm. Spray Technol.* **16** 822–30
[6] Berger L-M 2004 Titanium oxide - new opportunities for an established coating material *Thermal Spray 2004: Advances in Technology and Application: Proceedings of the International Thermal Spray Conference, 10-12 May, 2004, Osaka, Japan* pp 934–45
[7] Sharma A, Gouldstone A, Sampath S and Gambino R J 2006 Anisotropic electrical conduction from heterogeneous oxidation states in plasma sprayed $TiO_2$ coatings *J. Appl. Phys.* **100** 114906
[8] Lewinsky Y 1997 *O-Ti p-T-x Phase Diagrams*
[9] Ok K M, Ohishi Y, Muta H, Kurosaki K and Yamanaka S 2018 Effect of point and planar defects on thermal conductivity of $TiO_{2-x}$ *J. Am. Ceram. Soc.* **101** 334–46
[10] Freudenberg B 1987 *Etude de la Reaction a l'etat Solide $Al_2O_3$ + $TiO_2$ -> $Al_2TiO_5$ Observation des Microstructures* (EPFL)
[11] Thomas H A J and Stevens R 1989 Aluminium Titanate - A Literature Review - Part 1: Microcracking Phenomena *Br. Ceram. Trans. J.* **88** 144–51
[12] Low I M, Oo Z and O'Connor B H 2006 Effect of atmospheres on the thermal stability of aluminium titanate *Phys. B Condens. Matter* **385–386** 502–4
[13] Low I M and Oo Z 2010 Effect of Grain Size and Controlled Atmosphere on the Thermal Stability of Aluminium Titanate *AIP Conference Proceedings* vol 1202 pp 27–31
[14] Sperisen T and Mocellin A 1991 On structures of mixed titanium aluminium oxides *J. Mater. Sci. Lett.* **10** 831–3
[15] Buscaglia V 1998 Decomposition of $Al_2TiO_5$ and $Al_{2(1-x)}Mg_xTi_{(1+x)}O_5$ Ceramics *J. Am. Ceram. Soc.,* **81** 2645–53
[16] Low I M and Oo Z 2008 Reformation of phase composition in decomposed aluminium titanate *Mater. Chem. Phys.* **111** 9–12
[17] Asbrink G and Magnéli A 1967 X-Ray Studies on Some Mixed Oxide Systems of Pseudobrookite Structure *Acta Chem. Scand.* **21** 1977
[18] Low I M and Pang W K 2013 In situ diffraction study of self-recovery in vacuum decomposed $Al_2TiO_5$ *J. Aust. Ceram. Soc.* **49** 48–52
[19] Low I M and Oo Z 2008 In situ diffraction study of self-recovery in aluminum titanate *J. Am. Ceram. Soc.* **91** 1027–9
[20] Berger L M, Sempf K, Sohn Y J and Vaßen R 2018 Influence of Feedstock Powder Modification by Heat Treatments on the Properties of APS-Sprayed $Al_2O_3$-40% $TiO_2$Coatings *J. Therm. Spray Technol.* **27** 654–66
[21] Norberg S T, Hoffmann S, Yoshimura M and Ishizawa N 2005 Al6Ti2O13, a new phase in the $Al_2O_3$-$TiO_2$ system *Acta Crystallogr. Sect. C Cryst. Struct. Commun.* **C61** i35–8
[22] Hoffmann S, Norberg S T and Yoshimura M 2005 Structural models for intergrowth structures in the phase system $Al_2O_3$-$TiO_2$ *J. Solid State Chem.* **178** 2897–906
[23] Berger M H and Sayir A 2008 Directional solidification of $Al_2O_3$-$Al_2TiO_5$ system *J. Eur. Ceram. Soc.* **28** 2411–9
[24] Göbbels M and Boström D 1997 Kristallzüchtung und Struktur von "$Al_6Ti_2O_{13}$" *Berichte der Dtsch. Mineral. Gesellschaft* **9** 128

[25] Vargas F, Ageorges H, Fournier P, Fauchais P and López M E 2010 Mechanical and tribological performance of $Al_2O_3$-$TiO_2$ coatings elaborated by flame and plasma spraying *Surf. Coat. Technol.* **205** 1132–6
[26] Vuoristo P 2014 Thermal Spray Coating Processes *Comprehensive Materials Processing* vol 4, ed D Cameron pp 229–76
[27] Di Girolamo G, Brentari A, Blasi C and Serra E 2014 Microstructure and mechanical properties of plasma sprayed alumina-based coatings *Ceram. Int.* **40** 12861–7
[28] Kear B H, Kalman Z, Sadangi R K, Skandan G, Colaizzi J and Mayo W E 2000 Plasma-sprayed nanostructured $Al_2O_3$/$TiO_2$ powders and coatings *J. Therm. Spray Technol.* **9** 483–7
[29] Yang Y, Wang Y, Tian W, Yan D, Zhang J and Wang L 2015 Influence of composite powders' microstructure on the microstructure and properties of $Al_2O_3$-$TiO_2$ coatings fabricated by plasma spraying *Mater. Des.* **65** 814–22
[30] Vicent M, Bannier E, Benavente R, Salvador M D, Molina T, Moreno R and Sánchez E 2013 Influence of the feedstock characteristics on the microstructure and properties of $Al_2O_3$- $TiO_2$ plasma-sprayed coatings *Surf. Coatings Technol.* **220** 74–9
[31] Wang Y, Tian W, Zhang T and Yang Y 2009 Microstructure, spallation and corrosion of plasma sprayed $Al_2O_3$-13%$TiO_2$ coatings *Corros. Sci.* **51** 2924–31
[32] Lima R S, Moreau C and Marple B R 2007 HVOF-Sprayed coatings engineered from mixtures of nanostructured and submicron $Al_2O_3$-$TiO_2$ powders: An enhanced wear performance *J. Therm. Spray Technol.* **16** 866–72
[33] Matikainen V, Niemi K, Koivuluoto H and Vuoristo P 2014 Abrasion, Erosion and Cavitation Erosion Wear Properties of Thermally Sprayed Alumina Based Coatings *Coatings* **4** 18–36
[34] Wang M and Shaw L L 2007 Effects of the powder manufacturing method on microstructure and wear performance of plasma sprayed alumina-titania coatings *Surf. Coatings Technol.* **202** 34–44
[35] Goberman D, Sohn Y H, Shaw L, Jordan E and Gell M 2002 Microstructure development of $Al_2O_3$-13wt.% $TiO_2$ plasma sprayed coatings derived from nanocrystalline powders *Acta Mater.* **50** 1141–52
[36] Zois D, Lekatou A, Vardavoulias M, Panagiotopoulos I and Vazdirvanidis A 2008 A comparative microstructural investigation of nanostructured and conventional $Al_2O_3$ coatings deposited by plasma spraying *J. Therm. Spray Technol.* **17** 887–94
[37] Jordan E H, Gell M, Sohn Y H, Goberman D, Shaw L and Jiang S 2002 Fabrication and evaluation of plasma sprayed nanostructured alumina–titania coatings with superior properties *Mater. Sci. Eng. A* **301** 80–9
[38] Bruker AXS 2008 TOPAS V4: General Profile and Structure Analysis Software for Powder Diffraction Data
[39] Schneider C A, Rasband W S and Eliceiri K W 2012 NIH Image to ImageJ: 25 years of image analysis *Nat. Methods* **9** 671
[40] Naghizadeh R, Rezaie H R and Golestani-fard F 2009 The influence of composition, cooling rate and atmosphere on the synthesis and thermal stability of aluminum titanate *Mater. Sci. Eng. B Solid-State Mater. Adv. Technol.* **157** 20–5
[41] Hoffmann S, Norberg S T and Yoshimura M 2006 Melt synthesis of $Al_2TiO_5$ containing composites and reinvestigation of the phase diagram $Al_2O_3$–$TiO_2$ by powder X-ray diffraction *J. Electroceramics* **16** 327–30
[42] Gell M, Jordan E H, Sohn Y H, Goberman D, Shaw L and Xiao T D 2001 Development and implementation of plasma sprayed nanostructured ceramic coatings *Surf. Coatings Technol.* **146–147** 48–54
[43] Papitha R, Suresh M, Das D and Johnson R 2013 Effect of micro-cracking on the thermal conductivity and thermal expansion of tialite ($Al_2TiO_5$) ceramics *Process. Appl. Ceram.* **7** 143–6
[44] Freudenberg B 1991 Aluminum Titanate *Concise Encyclopedia of Advanced Ceramic Materials* pp 20–2

IOP Conf. Series: Materials Science and Engineering **480** (2019) 012008 doi:10.1088/1757-899X/480/1/012008

# Investigation of HVOF-ID spraying with WC-CoCr -15+5 µm feedstock powder

**W Tillmann, C Schaak*, L Hagen and M Dildrop**

TU Dortmund University, Institute of Materials Engineering, Germany

* e-mail:christopher.schaak@udo.edu

**Abstract.** High velocity oxygen fuel (HVOF) spraying of WC-Co(Cr) with different chemical compositions, different powder size fractions, and different mean carbide sizes is a well-established research field for outer diameter (OD) applications. These coatings are typically applied as wear protective layers for different types of industries. Current demands for internal diameter (ID) coatings lead to a great interest in HVOF-ID spraying. This field of application necessitates a special spray gun equipment and spray powders with particle size fractions smaller than 20 µm. At the same time, the process control concerning both the spray gun configuration and the use of fine powders leads to new challenges which differ from those of OD HVOF spraying. In this study, HVOF-ID spraying using a WC-CoCr 86-10-4 (-15+5 µm) feedstock with a mean WC particle size of 400 nm is investigated with respect to the resulting coating properties. A statistical design of experiments (DoE) is utilized to enable a systematic analysis of various process parameter settings along with their interaction on the microstructural characteristics as well as the deposition efficiency (DE). Based on the results, a desirability-based multi-criteria optimization is carried out in order to produce adequate coating properties. The obtained knowledge about the spray system enables to realize dense WC-CoCr coatings with a porosity of approximately 1 %.

## 1. Introduction

Thermal spraying processes have significant disadvantages regarding the coating of inner surfaces such as cylindrical parts with small diameters or hard-to-reach undercuts when compared to other coating techniques, such as dip coating techniques, electroplating, or chemical vapour deposition. The reasons for this are mainly the process characteristics of thermal spraying processes, such as the perpendicularly directed particle spray jet, the long stand-off distance (SOD), and the large build-up of common spray guns. Current developments in research and industry aim to solve these problems with new spray gun concepts for ID applications. One research focus is prompted by the automotive industry with the cylinder liner surface coating and bore coating for combustion engines to lower the friction and wear [1,2]. Typical spray systems for this purpose are atmospheric plasma spraying (APS), twin wire arc spraying (TWAS), plasma transferred wire arc (PTWA) spraying, and rotating single wire (RSW) spraying [1,2,3]. In contrast, there is a lack of research and development in the field of high velocity oxygen/air fuel spraying with regard to ID applications. Nevertheless, high performance HVOF/HVAF-ID systems are already available for small IDs up to 43 mm [4]. To realize the coating of small IDs, the spray gun has to be smaller compared to conventional spray guns while, at the same time, the SOD has to be shorter. In addition, the compact design of the spray gun is associated with a lower combustion energy when compared to conventional spray guns. Consequently, the lower flame enthalpy leads to a reduced heat input into the spray particles while under many

IOP Conf. Series: Materials Science and Engineering **480** (2019) 012008 doi:10.1088/1757-899X/480/1/012008

circumstances (i.e. depending on the process parameters to be used, or the geometry of the parts to be coated) the spray particles are simultaneously subjected to a reduced dwell time. Hence, it is necessary to use spray powders with particle sizes smaller than 20 μm as feedstock [4,5]. Besides $Cr_3C_2$-NiCr coatings, WC-Co(Cr) coatings have already been established as appropriate candidates in the field of thermal spray technology to protect tribologically stressed surfaces against wear. In terms of HVOF spraying, typical agglomerated and sinterted WC-Co(Cr) feedstocks feature agglomerate sizes larger than 15 μm or 20 μm (lower limit value), respectively. Fine powder particles smaller than 20 μm with submicron carbides (WC < 1 μm) tend to overheat during spraying, which in turn favors the oxidation and decarburization of carbides. For WC-Co(Cr), decomposition phenomena can lead to the formation of $W_2C$ eta carbides such as $(W,Cr)_3Co_3C$, as well as Co(W,C) [6,7]. Moreover, the incorporation of Cr favors the formation of an intermetallic CoCr phase. Such phase transformation processes can provoke the formation of brittle phases, and thus reduce the metallic character of the binder [7,8]. As verified for WC-Co coatings, which were tribologically stressed during technological testing, the decomposition of WC results in a degradation of the mechanical properties [9]. With respect to the process control (i.e. injection of the feedstock), powders with small agglomerates exhibit a poor flowability. By applying conventional powder flowability characterization techniques (e.g. Hall Flowmeter), tests will fail with the result that the powder seems to have no flowability. Hence, customized powder feeder systems are necessary to provide a constant flow rate without demixing. During spraying, fine particles tend to follow the gas flow trajectories [10]. Due to the gas flow deviation close to the surface of the substrate, the so-called bow shock effect [11], which is known for cold gas spraying, could affect the particle impact velocity as well as the deposition efficiency.

In this study, a systematic analysis using a DoE is carried out to understand the influence of various spray parameters on the resulting microstructural characteristics as well as the DE. For this purpose, a fine-structured WC-CoCr 86-10-4 powder is sprayed utilizing a specialized HVOF-ID equipment.

## 2. Experimental details and methods

An agglomerated and sintered WC-CoCr 86 10 4 powder with an agglomerate size distribution of -15+5 μm and a WC Fisher sub-sieve size (FSSS) of 0.4 μm was used as feedstock (**Figure 1**). According to the manufacturer, the feedstock features a bulk density between 4.3 – 5.4 g/cm$^3$.

| **Durum Durmat 135.035** | | | |
|---|---|---|---|
| Powder type | | WC-CoCr 86 10 4 | |
| Particle size | | -15 + 5 μm | |
| WC particle size: | | 0.4 μm FSSS | |
| Bulk density (ISO 3923-2): | | 4.3 – 5.4 g/cm$^3$ | |
| Morphology: | | Mostly spherical | |
| **Chemical composition (in Wt.-%)** | | | |
| C | Co | Cr | W |
| 5.31 | 9.98 | 3.98 | balance |
| **Size fraction (CILAS)** | | | |
| | 15 μm | 4.51% | |
| | - 15 + 5 μm | Balance | |
| | - 5 μm | 0.66% | |

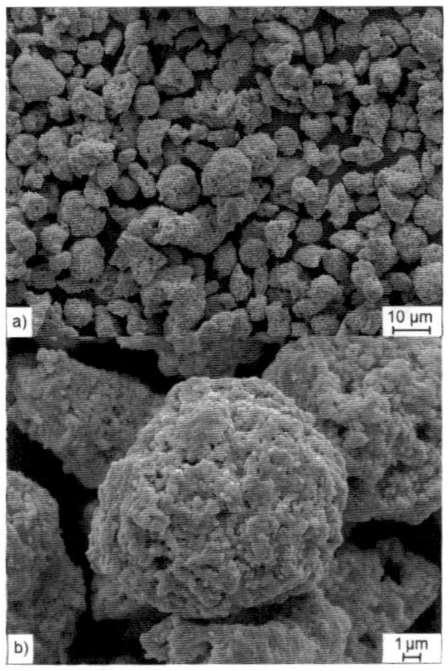


**Figure 1.** Powder specifications Durmat 135.035, chemical composition, size fraction, and SEM-images of the WC-CoCr particles in two different magnifications (a/b).

Within this study, the HVOF-ID spray gun IDCoolFlow Mono (Thermico, Germany) was utilized in order to conduct the spraying experiments. The HVOF-ID spray gun devices (e.g. IDCoolFlow) are described elsewhere [12]. In addition, the CPF2 twin-powder powder feeder system was used, enabling the feeding of fine-sized powder materials into the process due to the fluidization principle.

The spray gun was mounted on an industrial robot (type IRB4600/60, ABB, Switzerland) to ensure a robot-controlled movement within the experiments.

In this study, five independent variables were subjected to a central composite design (CCD) with 27 runs, including a single center run. In addition, two center runs were added to the design in order to reduce the residual variance of the model. This approach has already been used to develop thermally sprayed coatings [13]. The level settings considered in the CCD are based on the findings obtained from preliminary tests.

In terms of the CCD, the kerosene level (KL), oxygen level (OL), powder feeding rate (PFR), stand-off distance (SOD), and track pitch (TP) of the meander-shaped spray path were selected as independent variables. The level settings of the independent variables were converted to -1 (low), 0 (center), and +1 (high), whereas the so-called star points $-\alpha$ and $\alpha$ were converted to -1.664 and 1.664 to compute orthogonality within the CCD. The hydrogen level (HL), the backside cooling pressure, as well as handling parameters such as the spray angle, gun velocity, and amount of overruns were kept constant. The spray parameter settings are summarized in **Table 1**.

**Table 1.** Parameter settings for the CCD.

| Independent variables | Central Composite Design | | | | |
|---|---|---|---|---|---|
| | -1.664 | -1 | 0 | +1 | 1.664 |
| KL [l/h] | 2.87 | 3.4 | 4.2 | 5.0 | 5.53 |
| OL [l/min] | 220.07 | 240 | 270 | 300 | 319.93 |
| SOD [mm] | 58.36 | 65 | 75 | 85 | 91.64 |
| PFR [g/min] | 26.68 | 30 | 35 | 40 | 43.32 |
| Y-step [mm] | 2.33 | 2.50 | 2.75 | 3.00 | 3.17 |

| Constant parameters | | | |
|---|---|---|---|
| Spray angle [ °] | 90 | HL [l/min] | 70 |
| Gun velocity [mm/s] | 30.000 | Backside cooling pressure [psi] | 30 |
| Overruns | 25 passes | Powder carrier gas flow [l/min] | 10 |

In terms of the statistical analysis, the Vickers microhardness (MH), porosity (Po), mean roughness depth (Rz), and deposition efficiency (DE) were set as response variables. Statistically significant influencing factors and factor combinations were determined by means of significance tests of the null hypothesis $H_0$ as demonstrated in [13]. Hence, p-values were calculated based on a t-distribution with n - 1 degrees of freedom (where n is the number of experiments) according to t-statistic. Within this study, the significance level was set to 10% ($\alpha = 0.1$). The software Statistica (Statsoft, Oklahoma) was employed for regression and graphical analyses of the data obtained. Afterwards, optimized spray parameter settings were identified based on multi-criteria optimization, using the Derringer's desirability function [14].

In the spraying experiments, the coatings were deposited on rectangular (70 x 50 x 10 mm) C45 steel (1.0503; 258 ± 4 HV0.3) substrates. Prior to the coating deposition, the substrates were grit blasted with corundum using a grit size of F100 according to FEPA (blasting pressure: 3.5 bar; blasting angle: 45°; stand-off distance: 100 mm) and cleaned in an ultrasonic ethanol bath. Afterwards, the substrates were pre-heated to approximately 100 °C for 30 minutes. As verified by tactile measurements, the substrates exhibit an average roughness Ra of 1.42 ± 0.11 µm after grit blasting.

The response variables were obtained from metallographic investigations of the produced WC-CoCr coatings. Initially, the produced WC-CoCr coatings were cut and metallographically prepared (cross-section) using various grinding and polishing steps. Cross-section images were taken by means of light-microscopy (LM) and scanning electron microscopy (SEM). The MH was measured at the cross-section with a M400 hardness tester (Leco, Germany) applying a load of 2.942 N. Five indents per coating were used to calculate the average for MH. The Po of the coatings was optically measured at five different positions at the cross-section using a light-microscope with integrated software

AXIOPHOT (Carl Zeiss, Germany). The DE was calculated based on the measured weight of each sample before and after the coating deposition, considering the robot path over the sample as well as the powder feed rate for each individual experiment. Rz was measured with a tactile measuring device (Hommel Tester T-1000, Hommelwerke, Germany). For selected samples, the phase composition was analyzed by means of X-ray diffraction (XRD) using an Advanced D8 diffractometer (Bruker, Massachusetts) with Fe Kα radiation. For the experiments, the step width and exposure time were set to 0.05° and 3 s for each step. The measurements were conducted at an angle of incidence of 5 degrees. For selected spray parameter settings, the in-flight particle temperature and velocity were measured with an Accuraspray-g3 device (TECNAR Automation, Canada). The measuring position was varied to simulate different SODs.

**Table 2** summarizes the factor levels of the independent variables, with the corresponding results taken from metallographic examinations. The statistical design allows to estimate the linear (L) and quadratic (Q) effects and also the interactions between the variables.

**Table 2.** Level settings of the independent variables and results for MH, Po, Rz, and DE.

| | **Independent variables (factor levels)** | | | | | **Response variables (mean values)** | | | |
|---|---|---|---|---|---|---|---|---|---|
| **Run** | **KL [l/h]** | **OL [l/min]** | **SoD [mm]** | **PFR [g/min]** | **TP [mm]** | **MH [HV0.3]** | **Po [%]** | **Rz [µm]** | **DE [%]** |
| 1 | 1 | 1 | -1 | -1 | 1 | 976 | 2.1 | 21.0 | 38 |
| 2 | -1 | 1 | -1 | 1 | 1 | 1006 | 1.9 | 20.7 | 27 |
| 3 | 0 | 0 | 0 | 0 | 1.664 | 1027 | 2.9 | 20.0 | 32 |
| 4 | 1 | 1 | -1 | 1 | -1 | 997 | 3.1 | 23.2 | 44 |
| 5 | -1 | -1 | 1 | 1 | 1 | 942 | 5.1 | 20.0 | 36 |
| 6 | 1 | 1 | 1 | -1 | -1 | 934 | 3.3 | 19.5 | 43 |
| 7 | 0 | -1.664 | 0 | 0 | 0 | 1009 | 4.7 | 21.3 | 46 |
| 8 | -1 | -1 | 1 | -1 | -1 | 971 | 3.9 | 20.1 | 42 |
| 9 | -1 | 1 | 1 | 1 | -1 | 973 | 2.7 | 19.6 | 29 |
| 10 | -1 | 1 | 1 | -1 | 1 | 1032 | 2.5 | 19.2 | 28 |
| 11 | 1 | -1 | 1 | -1 | 1 | 971 | 3.9 | 19.6 | 52 |
| 12 | -1 | 1 | -1 | -1 | -1 | 1022 | 2.4 | 19.4 | 32 |
| 13 | -1.664 | 0 | 0 | 0 | 0 | 906 | 3.0 | 19.9 | 30 |
| 14 | 1 | -1 | 1 | 1 | -1 | 1074 | 3.6 | 21.6 | 50 |
| 15 | 0 | 0 | 0 | 0 | -1.664 | 999 | 2.9 | 21.3 | 38 |
| 16 | 1 | -1 | -1 | -1 | -1 | 969 | 1.3 | 21.6 | 43 |
| 17 | 0 | 0 | -1.664 | 0 | 0 | 1012 | 2.7 | 20.2 | 35 |
| 18 | 1 | 1 | 1 | 1 | 1 | 1078 | 2.7 | 19.5 | 38 |
| 19 | 1 | -1 | -1 | 1 | 1 | 993 | 4.2 | 20.8 | 50 |
| 20 | 0 | 0 | 1.664 | 0 | 0 | 946 | 3.5 | 18.6 | 38 |
| 21 | 1.664 | 0 | 0 | 0 | 0 | 954 | 3.5 | 19.7 | 52 |
| 22(C) | 0 | 0 | 0 | 0 | 0 | 985 | 3.2 | 21.1 | 39 |
| 23 | -1 | -1 | -1 | -1 | 1 | 919 | 3.4 | 21.6 | 41 |
| 24 | 0 | 0 | 0 | 1.664 | 0 | 948 | 3.3 | 21.9 | 37 |
| 25(C) | 0 | 0 | 0 | 0 | 0 | 953 | 2.9 | 19.5 | 37 |
| 26 | 0 | 0 | 0 | -1.664 | 0 | 976 | 2.7 | 19.2 | 31 |
| 27(C) | 0 | 0 | 0 | 0 | 0 | 974 | 2.7 | 19.8 | 36 |
| 28 | -1 | -1 | -1 | 1 | -1 | 903 | 5.6 | 21.6 | 39 |
| 29 | 0 | 1.664 | 0 | 0 | 0 | 971 | 2.9 | 19.1 | 28 |

## 3. Results and discussion

### *3.1. Main effects and interaction*

**Table 3** shows the effect estimates and the $p$-values for each response variable (MH, Po, Rz, DE). The independent variables affect the response variables positively or negatively, which does the sign of the estimate show. The significance level was set to 10 % (α = 0.1), and the significant effects are marked in bold. The line with (Q) and (L) refer to linear or quadratic effects.

**Table 3.** Effect estimates and *p*-values (α = 0.1) of the response variables.

| | MH [HV0.3] | | Po [%] | | Rz [µm] | | DE [%] | |
|---|---|---|---|---|---|---|---|---|
| | Est. | *p*-value [$10^{-4}$] | Est. | *p*-value [$10^{-4}$] | Est. | *p*-value [$10^{-4}$] | Est. | *p*-value [$10^{-4}$] |
| **Mean/Interc.** | 967.78 | 0.0 | 3.03 | 0.0 | 19.92 | 0.0 | 35.78 | 0.0 |
| **KL (L)** | 28.28 | **969.5** | -0.21 | 1703.3 | 0.37 | 2599.7 | 11.25 | **0.1** |
| **KL (Q)** | -25.72 | 1869.5 | 0.10 | 5501.7 | 0.00 | 9959.8 | 4.61 | **115.1** |
| **OL(L)** | 19.81 | 2244.5 | -1.22 | **0.2** | -0.80 | **303.7** | -9.57 | **0.4** |
| **OL (Q)** | 17.63 | 3514.5 | 0.50 | **163.2** | 0.33 | 3845.8 | 1.57 | 2993.7 |
| **SOD (L)** | 7.37 | 6373.3 | 0.48 | **92.6** | -1.26 | **31.8** | 0.94 | 4526.6 |
| **SOD(Q)** | 9.47 | 6094.2 | 0.02 | 8868.7 | -0.28 | 4526.3 | 1.11 | 4541.5 |
| **PFR (L)** | 11.42 | 4699.9 | 0.66 | **15.7** | 0.89 | **186.9** | 0.29 | 8114.6 |
| **PFR (Q)** | -2.55 | 8901.1 | -0.06 | 7331.3 | 0.56 | 1557.7 | -0.63 | 6681.9 |
| **TP (L)** | 11.51 | 4661.7 | 0.01 | 9574.4 | -0.59 | **889.9** | -1.72 | 1867.7 |
| **TP (Q)** | 34.09 | **921.0** | -0.12 | 4793.9 | 0.59 | 1367.8 | 0.10 | 9426.7 |
| **KL * OL (int.)** | -39.78 | **521.8** | 0.82 | **10.0** | 0.49 | 1979.0 | 1.13 | 4398.2 |
| **KL * SOD (int.)** | 6.57 | 7162.1 | 0.22 | 2096.7 | -0.25 | 5036.1 | 1.54 | 3000.8 |
| **KL * PFR(int.)** | 51.53 | **183.7** | -0.03 | 8470.4 | 0.23 | 5336.0 | 2.28 | 1390.6 |
| **KL * TP (int.)** | 1.50 | 9336.3 | 0.42 | **336.7** | -0.72 | **735.7** | 1.03 | 4788.7 |
| **OL * SOD (int.)** | -19.55 | 2952.1 | -0.03 | 8586.5 | -0.28 | 4430.5 | -1.42 | 3343.4 |
| **OL * PFR (int.)** | 1.05 | 9535.1 | -0.73 | **20.7** | 0.35 | 3536.4 | 0.11 | 9384.8 |
| **OL * TP (int.)** | 32.23 | 1020.8 | -0.58 | **77.9** | 0.19 | 5967.0 | -2.64 | **929.0** |
| **SOD * PFR (int)** | 18.30 | 3251.0 | -0.64 | **44.0** | -0.03 | 9259.0 | -2.27 | 1405.8 |
| **SOD* TP (int.)** | 8.57 | 6364.4 | 0.19 | 2895.5 | -0.13 | 7186.7 | -0.89 | 5402.7 |
| **PFR * TP (int.)** | 8.73 | 6306.4 | -0.25 | 1638.8 | -0.72 | **760.5** | -1.22 | 4031.4 |

Based on the effect estimates in **Table 3**, the MH is influenced by KL, TP, and the interactions KL*OL and KL*PFR. Po is mainly affected by OL, SOD, PFR, and the interactions KL*OL, KL*TP, OL*PFR, and SOD*PFR. The interaction KL*OL directly refers to the Lambda (λ) value (air–fuel equivalence ratio). The explanation for the observed relations is similar for all effects. Changes in the OL or KL result in differing flame temperature and gas velocities. Hence, these changes directly influence the powder heating and acceleration in the flame jet. Due to thermally induced microstructural changes, e.g. formation of brittle phases or the decomposition of WC to $W_2C$, the coating MH can be affected [15]. A high OL with a high KL combined with a high λ value result in a large total amount of combustion media (TAOCM), that leads to an increase in the particle velocity and reduction of the particle temperature. The impact velocity on the surface influences the compaction of the WC-CoCr coating and hence the porosity [16]. Further, the MH correlates with the Po. Based on a low heating, solid particles can re-bounce or are integrated into the coating and particles that are molten to a too high degree can burst during the impact. Both effects lead to a higher Po due to imperfections in the WC-CoCr coating. A similar effect occurs when PFR is increased or when SOD is varied.

The correlation between the PFR and Po as well as Rz depends on the higher mass flow of WC-CoCr particles [17]. To melt more particles, a higher combustion energy is necessary. If the flame energy is not increased parallel to the higher powder feed rate or limited due to the process control, the particles are not well heated and remain solid. Furthermore, Rz is affected by the named variables (OL, SOD, TP). A low particle velocity results in a rougher surface due to a lower compaction of a single sprayed layer.

With respect to the commercial use of the HVOF ID coating process, the DE is a very important value. As shown in **Table 2**, the DE fluctuates strongly within the range of 27 to 52 %. Based on **Table 3**, the DE is affected by KL, OL, and hence λ. **Figure 2** shows the main effect plots for the DE.

A higher λ means an oxygen excess during combustion. This excess is based on a reduced fuel level or an increased oxygen level. With higher λ, the combustion energy and the combustion chamber pressure are decreased [18]. The effect of λ is similar to the effect of the PFR: the particles are not sufficiently heated or melted to adhere on the substrate and to form a coating.

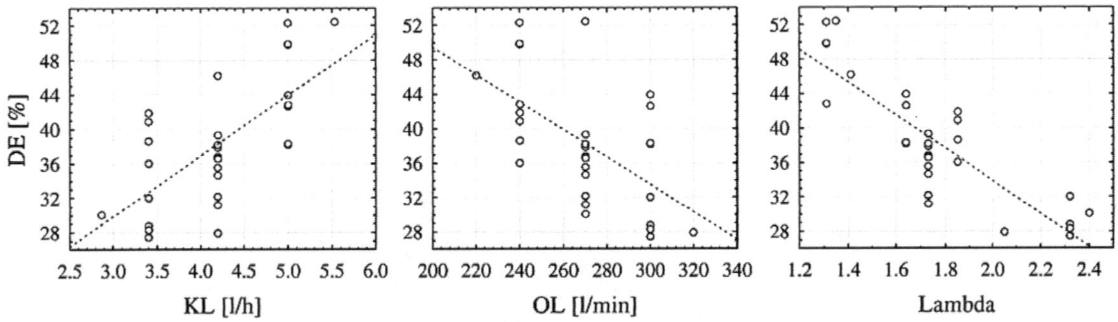


**Figure 2.** Effect plot for the deposition efficiency (DE).

Based on the results in **Table 3**, the response variables can be fitted by a second order model (1), enabling a correlation of the response variables to the independent variables [19].

$$Y_i = b_0 + \sum_{i=1}^{n} b_i x_i + \sum_{i=1}^{n} b_{ii} x_i^2 + \sum_{\substack{i,j=1 \\ i<j}}^{n} b_{ij} x_i x_j + \varepsilon \qquad (1)$$

where $Y_i$ is the predicted response, and $x_i$ and $x_j$ are independent variables that influence the response variable Y while $b_0$ is the intercept and $b_i$ is the linear coefficient. Furthermore, $b_{ii}$ is the quadratic coefficient and $b_{ij}$ is the linear-by-linear interaction between $x_i$ and $x_j$, where i ranges from 1 to 5. The residual standard error ε is the square root of the residual sum of squares divided by the residual degrees of freedom. Moreover, the data allows the generation of 3D-surface plots, showing the effect of each interaction of the independent variables on the response variables. To support the consistency and readability of the text, the 3D-surface plots are not given at this point. The regression models for the MH, Po, Rz, and DE are given within the SI data (**SI data, Figure S1**).

### *3.2. Optimization and verification*

Multi-criteria optimization, using the Derringer's desirability function [14], was employed to find an optimum combination of the independent variables. In this respect, the response variables were transformed into in a common scale. The desirability d of each response variable is assigned a value between 0 and 1, corresponding to the optimization goal (desirability at a maximum or minimum), as shown in **Table 4**. The desirability of a response variable takes the value 0 when the response drops to a lower specification limit (LSL). The value 1 is defined when the response variable exceeds the upper specification limit (USL). Values in between the limits are also assigned a desirability, whereby the transformation is linear (i.e. between 0 and 1).

Following the desirability transformation, the desirability index *D* is calculated as shown in [13]. Within this study, *D* is calculated with the same weight for all criteria (response variables). With regard to the multi-criteria optimization, the theoretical optimum with a desirability index of 0.8833 is reached. The resulting optimized spray parameters to deposit WC-CoCr in combination with the HVOF-ID gun are shown in **Table 5**. Furthermore, the predicted values for the MH, Po, Rz, and DE are given.

IOP Conf. Series: Materials Science and Engineering **480** (2019) 012008 doi:10.1088/1757-899X/480/1/012008

**Table 4.** Parameters for desirability transformation.

| | MH [HV 0.3] | | Po [%] | | Rz [µm] | | DE [%] | |
|---|---|---|---|---|---|---|---|---|
| | MH (max) | *d (MH)* | Po (min) | *d (Po)* | Rz (min) | *d (Rz)* | *DE (max)* | *d (DE)* |
| LSL | 900 | 0 | 2 | 1 | 19 | 1 | 25 | 0 |
| USL | 1080 | 1 | 4 | 0 | 23.5 | 0 | 45 | 1 |

**Table 5.** Optimized spray parameters and predicted response values.

| KL [l/h] | OL [l/min] | SOD [mm] | PFR [g/min] | TP [mm] | *D* |
|---|---|---|---|---|---|
| 5.0 | 293 | 91.6 | 43.3 | 3.17 | 0.8833 |

**Predicted values**

| MH [HV0.3] | Po [%] | Rz [µm] | DE [%] |
|---|---|---|---|
| 1162 | 2.4 | 19.9 | 35.8 |

WC-CoCr coatings were produced, based on the calculations using Derringer's desirability function in order to verify the predicted values of the response variables. All other spray parameter settings were kept constant according to **Table 1**. Three substrates were coated for the verification experiments. As shown in **Table 6**, the results reveal a significant deviation between the predicted and experimentally determined values (green: positive deviation, red: negative deviation).

**Table 6.** Predicted response values and experimentally determined values.

| MH [HV0.3] | | | Po [%] | | | Rz [µm] | | | DE [%] | | |
|---|---|---|---|---|---|---|---|---|---|---|---|
| pred. | exp. | DV [%] | pred. | exp. | DV [%] | pred. | exp. | DV [%] | pred. | exp. | DV [%] |
| 1162 | 1051 | - 10.57 | 2.4 | 1.1 | - 119.27 | 19.9 | 14.4 | - 38.24 | 35.8 | 25.3 | - 41.42 |
| 1162 | 1046 | - 11.06 | 2.4 | 1.1 | - 113.39 | 19.9 | 13.1 | - 51.95 | 35.8 | 24.2 | - 47.91 |
| 1162 | 994 | - 16.92 | 2.4 | 0.5 | - 378.00 | 19.9 | 13.9 | - 43.31 | 35.8 | 25.0 | - 43.18 |

However, the measured values for Rz and Po are lower than the predicted data. Thus, an average for Po of 0.9 % is measured, whereas an average for Rz of 13.8 µm is determined. The MH and DE tend to be lower than predicted as well. Especially, the DE determined by the experiments demonstrates a distinct reduction. Thus, the experiments reveal an average for DE of 24.83 %, which corresponds with a reduction of approximately 44 % in average. It can be concluded that the mathematic regression models seem to be inaccurate, leading to significant deviations. The reasons are not yet clear. Nevertheless, cross-section analyses of the produced WC-CoCr coating using optimized spray parameters confirm the low porosity, macroscopically dense microstructure, and smooth surface (**Figure 3**). However, the cross-section images (**Figure 3a, b**) showing the coating microstructure at higher magnification, reveal a heterogeneous microstructure and distinct micro-porosity. With the calculated SOD of 91.6 mm the minimal coatable internal diameter can be calculated as 171.6 mm, based on the HVOF-ID gun dimensions, i.e. housing of the spray gun with a diameter of approximately 80 mm.

IOP Conf. Series: Materials Science and Engineering **480** (2019) 012008 doi:10.1088/1757-899X/480/1/012008

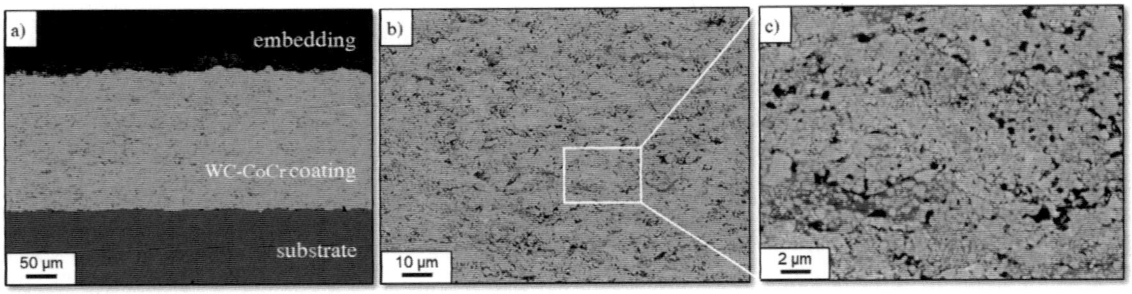


**Figure 3.** Cross-section of a WC-CoCr (-15+5 µm) coating deposited with an HVOF-ID spraying system using optimized spray parameters.

Based on the optimized spray parameter settings, further investigations on certain spray parameter variations were conducted in order to analyse the effect of changing the KL and TAOCM on the resulting MH. The cooling power and combustion chamber pressure (CCP) were measured as well. Furthermore, the in-flight particle velocity and temperature were determined. It was determined that a higher KL (which corresponds with a reduced λ, when OL and HL are kept constant) leads to a higher CCP and cooling power (**Figure 4**). As indicated by the increased cooling power during spraying, the spray particles are subjected to an increased heat input across the spray plume when they accelerate towards the substrate.

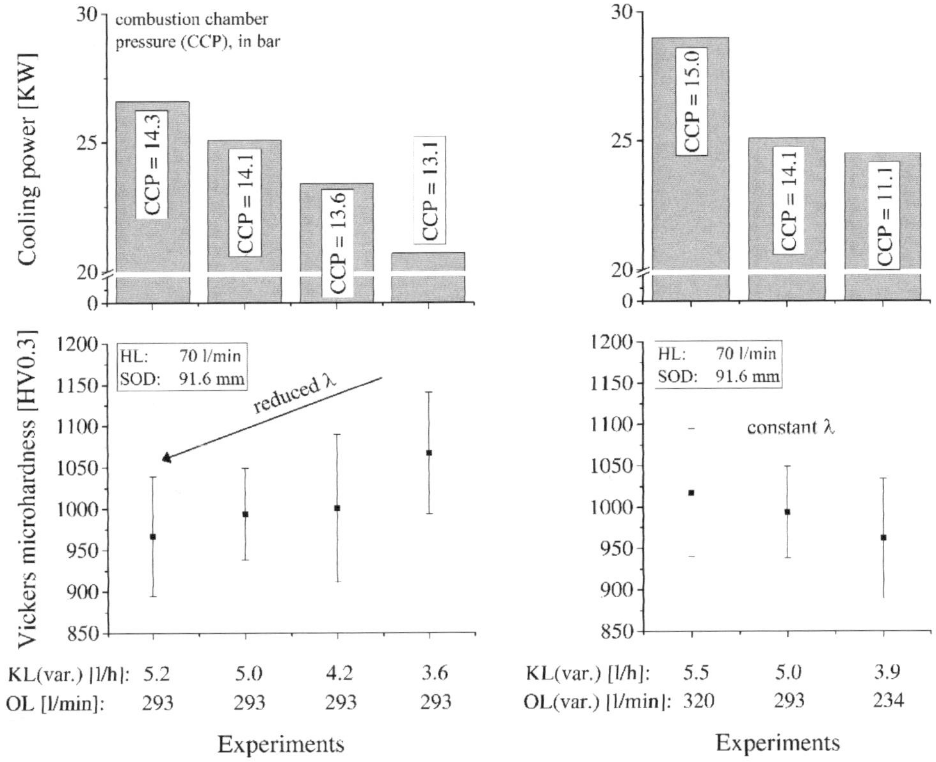


**Figure 4.** Results of selected spray parameter variations including the optimized spray parameters.

As demonstrated for higher KL (i.e. reduced λ), in-flight particle analyses (**Figure 5**) confirm the fact that the spray particles exhibit an increased temperature. At the same time, the velocity of the spray particles is slightly reduced. As a result, the spray particles have almost the same dwell time. It is assumed that the increased heat input could lead to more pronounced WC decomposition phenomena. XRD analyses (**Figure 6**) reveal that the WC-CoCr feedstock is mainly composed of WC, Co, as well as some traces of $Co_3W_3C$, Cr, and W. In contrast, the WC-CoCr coatings consist of WC, $W_2C$ (or $W_2C_{0.84}$), $Co_3W_3C$, and W. In addition, there is only a small amount of Co and Cr, which can be found within the experimental resolution. For a decreased λ, the XRD pattern indicates an increased amount of $Co_3W_3C$ and W. As discussed in [20,21], the hardness of the W-rich hard phases might decrease in the sequence of $W_2C > WC > Co_3W_3C$ (e.g. 29.4 GPa for $W_2C$, 16-22 GPa for h-WC, 23.5 GPa for WC, and less than 15.6 GPa for $Co_3W_3C$). As a result, the higher the content of $Co_3W_3C$ and W, while the amount of WC and $W_2C$ decreases simultaneously, the lower the microhardness. However, the data for MH deviates significantly, which might be related to the heterogeneous microstructure as shown previously (**Figure 3**).

Similar findings are observed for a change in the TAOCM. At a KL of 3.9 l/h and a OL of 234 l/min, the spray particles exhibit a temperature and velocity of 1931 ± 26 °C, and 536 ± 29 m/s, respectively. In contrast, a temperature of 1977 ± 6 °C and velocity of 620 ± 13 m/s is measured at a KL of 5.5 l/h and OL of 320 l/min. Hence, for a reduced TAOCM but at a same λ (KL: 3.9 l/h, OL: 234 l/min vs. KL: 5.5 l/h, OL: 320 l/min) the spray particles are subjected to an increased dwell time when compared to an increased TAOCM. Simultaneously, it is assumed that the spray particles sustain a high heat input, which is also indicated by the cooling power. This in turn, favors the phase transformation phenomena as previously described, leading to a possible degradation of the mechanical properties, indicated by a reduced microhardness (**Figure 4**).

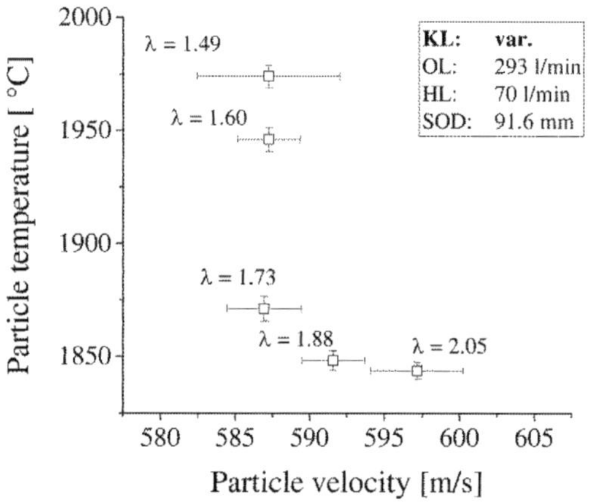


**Figure 5.** In-flight temperatures and velocities of WC-CoCr particles for different Lambda (λ) values.

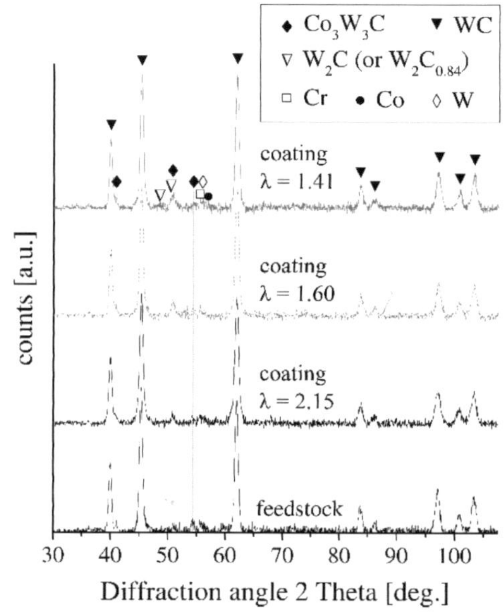


**Figure 6.** XRD analyses of the feedstock and different WC-CoCr coatings.

## 4. Summary and conclusion

In this study, a CCD experimental plan was used to describe the behavior of a HVOF-ID system in combination with a fine-structured, agglomerated and sintered WC-CoCr feedstock (-15+5 μm, WC particle size of 0.4 μm FSSS). A desirability-based optimization was conducted to determine optimal spray parameters (KL 5.0 l/h, OL 293 l/min, SOD 91.6 mm, PFR 43.3 g/min, TP 3.17 mm). The results obtained from the verification experiments differ from the predicted values based on a multi-criteria optimization using Derringer's desirability function. Hence, a low Po (0.9 %) and a reduced Rz

(13.8 µm) were achieved. Furthermore, the MH and DE were significantly lower than predicted. A reason for these results could be the inaccuracy in the mathematic regression model or in the desirability limit. Nevertheless, the systematic analysis by means of DoE is a promising approach to describe the system behavior of a HVOF-ID system, enabling to identify a usable SOD for ID applications. The SOD and dimensions of the spray gun determine the smallest coatable ID to achieve promising coating properties. For the present HVOF-ID system and WC-CoCr feedstock used within this study, the smallest coatable ID is 171.6 mm. Both the SOD and the combustion media, i.e. KL and OL, represent significant influencing factors as they affect the heat input into the spray particles, and thus, the phase transformation processes.

**Acknowledgments**
The results have emerged within the framework of the IGF Project No. 19.914 N with the subject of „Untersuchung der Einflussgrößen und prozess-technischen Randbedingungen auf die Schichtqualität beim Beschichten von rotationssymmetrischen Innenflächen mittels HVOF/HVAF". We thank the research association „Forschungsvereinigung Schweißen und verwandte Verfahren des DVS" and the Federation of Industrial Research Associations (AiF) for the promotion. The project was funded by the German Ministry of Economic Affairs and Energy via AiF within the framework of joint industrial research (IGF).

**References**
[1] Barbezat G and Herber R 2001 Durchbruch für Motorenbeschichtung *Sulzer Technical Review* **2** 8-11
[2] Bobzin K, Öte M, Königstein T, Dröder K, Hoffmeister H-W, Mahlfeld G and Schläfer T 2018 Development of novel fe-based coating systems for internal combustion engines *J. Therm. Spray Tech.* **27** 736–745
[3] Krauß B 2015 Trend: Beschichtete Zylinderlaufbahnen *VDI-Z Integrierte Produktion*
[4] Matthäus G 2018 Applications for HVOF and plasma coatings based on ultra-fine powder < 10 µm *Proceedings 11. HVOF Kolloquium Erding* 73-83
[5] Gutleber J, Molz R, He J, Weber C and Colmenares J 2017 New developments in HVOF spraying for internal diameter coatings *Proceedings International Thermal Spray Conference* 501-504
[6] Yuan J, Zhan Q, Huang J, Ding S and Li H 2013 Decarburization mechanisms of WC–Co during thermal spraying: Insights from controlled carbon loss and microstructure characterization *Mater. Chem. Phys.* **142** 165–71
[7] Berger L-M 2018 Hardmetal coatings – history and perspective *Proceedings 11. HVOF Kolloquium Erding* 93-100
[8] Gries B 2018 HVAF- Chance and challenge for users and for powder producers *Proceedings 11. HVOF Kolloquium Erding* 57-72
[9] Shipway P H, McCartney D G and Sudaprasert T 2005 Sliding wear behaviour of conventional and nanostructured HVOF sprayed WC–Co coatings *Wear* **259** 820–827.
[10] Marble F E 1963 Dynamics of a gas containing small solid particles Combustion and Propulsion (5th AGARDograph Colloquium) *Pergamon Press* 175-213
[11] Pattison J, Celotto S, Khan A and O'Neill W 2008 Standoff Distance and Bow Shock Phenomena in the Cold Spray Process *Surf. Coat. Technol.* **202** 1443-1454
[12] Bobzin K, Ernst F, Zwick J and Matthaeus G 2007 Analyse von Partikeleigenschaften beim Thermischen Spritzen von Mikropulvern. *Materialwissenschaft und Werkstofftechnik* **38** 149-154
[13] Tillmann W, Vogli E, Baumann I, Kopp G and Weihs C 2010 Desirability-Based Multi-Criteria Optimization of HVOF Spray Experiments to Manufacture Fine Structured Wear-Resistant 75Cr3C2-25(NiCr20) Coatings *J. Therm. Spray Technol.* **19** 392-408

[14] Derringer G and Suich R 1980 Simultaneous Optimization of Several Response Variables *J. Qual. Technol.* **12** 214-219
[15] Tucker R C 2013 *ASM handbook: Thermal Spray Technology* (ASM International)
[16] Oksa M, Turunen E, Suhonen T, Varis T and Hannula S-P 2011 Optimization and Characterization of High Velocity Oxy-fuel Sprayed Coatings: Techniques, Materials, and Applications *Coatings* **1** 17-52
[17] Xie M, Lin Y, Ke P, Wang S, Zhang S, Zhen Z and Ge L 2017 Influence of Process Parameters on High Velocity Oxy-Fuel Sprayed Cr3C2-25%NiCr *Coatings* **7** 1-10
[18] Baumann I T 2012 Hochverschleißfeste und konturnahe Werkzeugoberflächen durch Hochgeschwindigkeitsflammspritzverfahren Werkstofftechnologische Schriftenreihe **5**
[19] Weihs C and Jessenberger J 1999 *Statistische Methoden zur Qualitätssicherung und -optimierung in der Industrie* (Weinheim: Wiley VCH Verlag GmbH)
[20] Tillmann W, Hagen L and Schröder P 2017 Tribological Characteristics of Tungsten Carbide Reinforced Arc Sprayed Coatings using Different Carbide Grain Size Fractions *Tribology in Industry* **39** 168-182
[21] Li Y, Gao Y, Fan Z, Xiao B, Yue Q, Min T and Ma S 2010 First-principles study on the stability and mechanical property of eta M3W3C (M= Fe, Co, Ni) compounds *Physica B: Condensed Matter* **405** 1011-1017

IOP Conf. Series: Materials Science and Engineering **480** (2019) 012009 doi:10.1088/1757-899X/480/1/012009

# Investigation of low-pressure cold-gas dynamic spraying of polyamide-12 (PA12) on steel surfaces

**W Tillmann and J F Zajaczkowski***

TU Dortmund University / Institute of Materials Engineering

* e-mail: jonas.zajaczkowski@tu-dortmund.de

**Abstract.** Due to their good tribological properties and corrosive resistance, thermoplastic coatings are becoming increasingly important in industrial applications. Thermal spray technologies in general and flame spraying in particular belong to the most common application methods. However, the high thermal energy inevitably leads to the risk of thermal decomposition, especially in the case of materials such as polyamide-12, which have a small discrepancy between the melting and degradation temperatures. Furthermore, flame-sprayed coatings often are highly porous, reducing its effectiveness as a corrosive barrier. As an alternative approach, polyamide-12 coatings were applied to steel substrates using a conventional low-pressure cold-gas dynamic spray system. Various mixtures including ceramic powders ($Al_2O_3$) as well as different injection concepts were investigated and evaluated. The focus was put on the resulting coating morphology. Furthermore, the cold-gas spraying system was modified with a continuous temperature control for the gas temperature to enable a more precise adjustment based on the melting temperature of the material in order to reduce the risk of thermal degradation.

## 1. Introduction

The cold-gas dynamic spraying process, also simply known as cold spraying, is a relatively new spraying process developed in the 1980s by Alkhimov et al. [1] at the Institute of Theoretical and Applied mechanics of the Russian Academy of Science in Novosibirsk. Alkhimov et al. observed that under certain conditions in a supersonic wind tunnel experiment, tracer particles start depositing on surfaces rather than rebounding or eroding the surfaces. While not being the first to notice this effect, they used this observation to develop a new coating process with high deposition rates. With the cold spray process, it is possible to deposit various polymers, metal and metal alloy coatings, including metals such as copper or titanium, which at higher temperatures are, due to their high reactivity, very difficult to spray with conventional thermal spraying processes [2].

Although not entirely confirmed, it is generally assumed that the bonding of the particles in cold spraying is the result of the plastic deformation of the particles and the underlying layers/substrate, enabling a conformal contact, which combined with a high pressure, promotes a bonding of the particles [3]. The so-called critical velocity, which is necessary for this plastic deformation to occur, has been the topic of many studies in recent years [4–8].

Originally developed as a process that requires high pressures exceeding 20 bar, which are needed to obtain the high critical velocities of most metals, low-pressure cold-gas spraying (LPCGS) systems have been developed for the application of softer metals, which require lower critical velocities for a successful deposition [9]. In low-pressure cold-spraying, ceramic particles are often mixed with powders. These hard particles serve multiple purposes: They keep the nozzle clean by removing softer particles that might stick to the inner walls of the nozzle. Furthermore, they clean the substrate surface of impurities and oxides, activate the surface similar to grit blasting process, often rendering the conventional surface preparation unnecessary. They also have a densification effect by shot peening the already applied layers. Finally, hard particles often remain embedded in the coating, and hence reinforcing the coating [10, 11].

Mostly used for spraying metals, cold spray processes have been the subject of several studies for their ability to apply polymer coatings in the past few years because of their solid state deposition of particles, which reduces the risk of thermal degradation of the polymers. Alhulaifi et al. [12] simulated the cold spray process of polymers and concluded that unlike in usual cold spray applications, reaching a critical velocity is not always desirable, as the high sheer forces associated with higher impact velocities tend to remove previous layers. These simulations are backed with experiments from Xu et al. [13], who successfully deposited polyolefin particles with particle velocities as low as 100 m/s. Ravi et al. [14, 15] used a commercial cold spray system to successfully deposit ultra high molecular weight polyethylene (UHMW-PE) coatings on polymer as well as aluminum substrates. However, they noticed that a mixture of polymer and fine alumina particles was not only necessary for the aforementioned cleaning of the nozzle and the shot peening effect beneficial for the coating density and adhesion, but also played an important role for the cohesion of the coating because of intermolecular forces.

In this study, the viability of spraying polyamide-12 powder on steel substrates utilizing a commercial low-pressure cold-gas dynamic spraying system is investigated. Both pure powder and multiple mixtures with fine alumina particles were tested. Furthermore, the influence of the gas temperature and gas pressure was investigated. The results were analyzed in terms of both the coating topography and morphology.

## 2. Experimental and materials

The material used in this study was a pure polyamide-12 (PA-12) powder (Kern GmbH, Großmaischeid, Germany). The particle size distribution was measured via laser diffraction spectroscopy (S3500, Microtrac GmbH, Krefeld, Germany), resulting in the following values: D10 23 µm, D50 43.7 µm and D90 64.2 µm. Alumina particles were mixed with the thermoplastic powder to keep the nozzle clean during operation and to enable a shot peening effect (densification) of the polyamide particles. For this purpose, a commercial thermal spray powder Amperit 740 (H.C. Starck GmbH, Goslar, Germany) with a particle size distribution of -22+5 µm was used. The flowability of both powders and the respective mixtures was measured with a Revolution Powder Analyzer system (PS Prozesstechnik GmbH, Basel, Switzerland).

The coatings were applied utilizing a modified Dymet 413 spraying system (Dycomet Europe, Akkrum, Netherlands) operated with compressed air with a maximum pressure of 9 bar and a temperature of up to 650 °C. Usually, with this spraying system, the gas temperature can only be set using five separate settings. Because preliminary experiments showed no coating buildup without a preheating of the compressed air and severe thermal degradation of the thermoplastic powder occurred at the second temperature setting (gas temperatures around 350 °C), a more nuanced temperature control was needed. For this purpose, a potentiometer with a resistance range of 0 - 44 kΩ was installed into the circuit of the highest temperature setting. This allows stepless temperature control of the compressed air between 200 and 650 °C.

Because of instable feed rates and clogging problems when using the internal powder feeder unit of the Dymet 413 with light powders such as polymers, an external, conventional feed disk type powder feeder was utilized (Single-10C, Plasma-Technik AG, Wohlen, Switzerland). The carrier gas flow was minimized so that the powder injection was not forced, but still controlled by the Venturi effect-based suction of the nozzle.

The tests were carried out on samples with a 40 mm diameter and a thickness of 6 mm, made of carbon steel C45 (EN 1.0503). Prior to the spraying, the samples were grit blasted with corundum EKF-60 with a pressure of 4 bar at a distance of 100 mm and an with angle of 45°. Afterwards the samples were cleaned in an ultrasonic bath. The compressed air temperature, the process pressure and the volumetric mixing ratio of the powders were varied. These spray parameters are summarized in table 1. All coatings were applied with a gun movement velocity of 6000 mm/min and a y-step of 2 mm in a single pass with a spraying distance of 20 mm. The powder feeder was set to 20% feeder disk velocity. During all coating tests in this study, the samples were heated with a backside mounted heat gun GHG 600 CE (Bosch GmbH, Stuttgart, Germany) up to a front surface temperature of approximately 120 °C, which was measured with an optical pyrometer.

**Table 1.** Spraying parameters.

| Parameter | Unit | A | B | C | D | E | F[a] |
|---|---|---|---|---|---|---|---|
| Temperature | °C | 200 | 200 | 200 | 240 | 200 | 350 |
| Pressure | bar | 4 | 4 | 4 | 4 | 6 | 8 |
| Mixing ratio (PA-12 : $Al_2O_3$) | | 2:1 | 3:1 | 4:1 | 2:1 | 2:1 | 2:1 |

[a] modified powder injection

In addition to the typical radial injection of the powder before the nozzle, experiments were conducted with a radial powder injection behind the nozzle. The setup is shown in figure 1. This setup allowed the injection of pure polyamide-12 powder, which would otherwise quickly clog the nozzle in a conventional injection setup. In this setup, however, a higher temperature setting (setting 2, ca. 350 °C) and a high pressure (8 bar) were chosen to accommodate for the shorter time to heat up the particles and to transfer the kinetic energy.

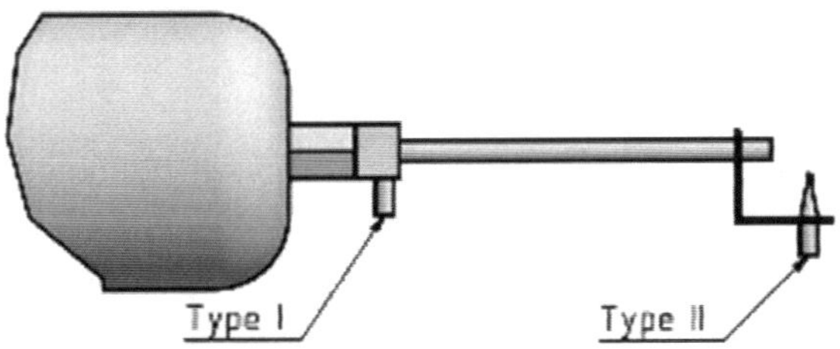


**Figure 1.** Radial injection of powder behind (type I) and in front of the nozzle (type II).

Some of the coatings were subsequently remolten at 180 °C for 2 hours to remove porosity. All coatings were metallographically prepared by creating cross-sections and examined by either light microscopy (BX51M, Olympus K.K., Shinjuku, Japan) or scanning electron microscopy (SEM) (JSM 7001F, Jeol Ltd., Akishima, Japan) with a BSD detector. The topology, including the roughness and waviness, of the coatings was measured with an optical profilometer InfiniteFocus (Alicona Imaging GmbH, Raaba, Austria).

## 3. Results and discussion

### *3.1. Effects of the powder mixture ratio*

The powder mixture ratio has a significant effect on the flowability of the mixtures, which is shown in figure 4 represented by the specific break energy, which is the energy needed for a powder to break apart and start an avalanche in the Revolution Powder Analyzer test [16, 17]. While pure alumina, which was supplied as a fused and crushed powder (figure 3), has a relatively high break energy because of the tendency of fine powders to agglomerate, the 2:1 mixture has an improved flowability compared to both pure powders. This improvement decreases with higher mixture ratios, showing no recognizable difference to the pure polyamide powder (figure 2) with the 4:1 ratio.

**Figure 2.** SEM image (SE) the polyamide-12 powder.

**Figure 3.** SEM image (BSD) of the alumina powder.

These differences in the flowability are visible in the resulting coatings exemplary displayed in the top views and topographical reconstructions (figure 5) and cross-sections (figure 6). Sample A, which was applied with a 2:1 mixture ratio, shows a macroscopically uniform coating. Sample B shows about the same coating thickness, although the coating is less uniform, with several uncoated spots visible. Sample C, with a volumetric powder mixture ratio of 4:1, shows a much thinner and very uneven coating. The coating thickness, roughness and waviness of the three samples are summarized in table 2.

**Table 2.** Thickness and topographical properties of samples coated with parameters A, B and C.

| Sample | Thickness [µm] | Roughness Ra [µm] | Roughness Rz [µm] | Waviness Wa [µm] |
|---|---|---|---|---|
| A | 438 ± 41.7 | 21.2 ± 4.1 | 107.2 ± 18.8 | 42.2 ± 26 |
| B | 388 ± 112.6 | 21 ± 3.4 | 95.1 ± 12.7 | 30.2 ± 9.9 |
| C | 265 ± 78.4 | 28.7 ± 3.8 | 126.5 ± 11.3 | 28.5 ± 9.4 |

IOP Conf. Series: Materials Science and Engineering **480** (2019) 012009 doi:10.1088/1757-899X/480/1/012009

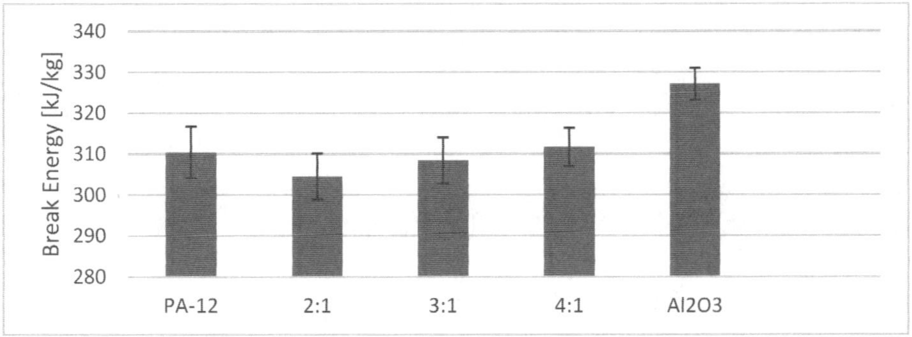


**Figure 4.** Specific break energy (in kJ/kg) for both pure powders and the respective mixtures.

**Figure 5.** Top view and topographical reconstructions (5x5 mm) of coatings produced with different powder mixtures; A: 2:1 ratio PA-12:$Al_2O_3$, B: 3:1, C: 4:1.

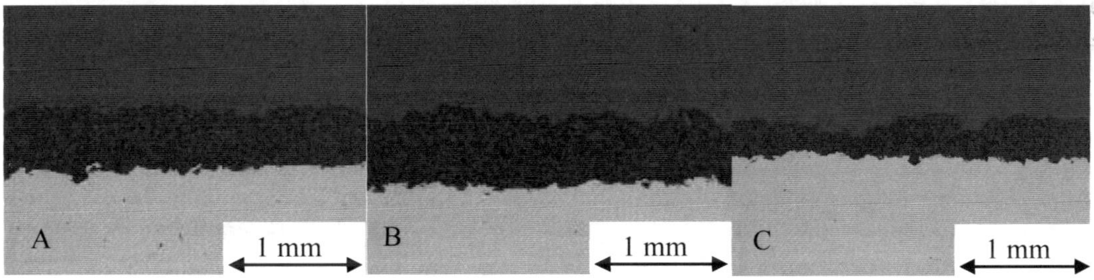


**Figure 6.** Cross-sections of coatings produced with different powder mixtures; A: 2:1 ratio PA-12:$Al_2O_3$, B: 3:1, C: 4:1.

A close up of the cross-sections of samples A and C reveal that the alumina particles remain embedded in the coating (Figure 7). Generally, the coatings are still highly porous, indicating that the fraction of alumina has no effect on the porosity of the resulting coatings and that the shot peening effect (hammering) known from other low-pressure cold-gas coatings is not as prominent in this coating.

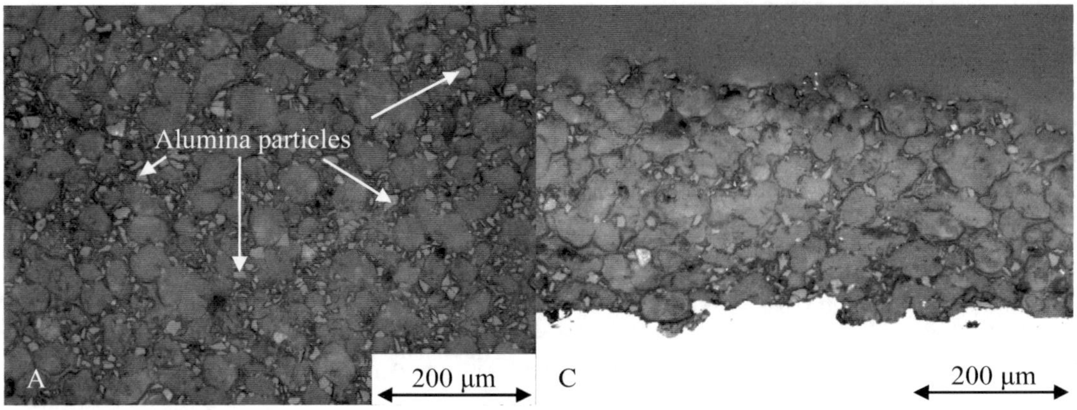


**Figure 7.** Cross sections of coatings produced with different powder mixtures; A: 2:1 ratio PA-12:$Al_2O_3$, C: 4:1.

### *3.2. Effects of the gas temperature*

Although a larger degree of partial melting of the thermoplastic powder and thus a better fusion between individual particles was assumed, the microstructures of the resulting coatings show almost no discernable differences with clear boundaries between the particles still being visible in both cross-sections. Figure 8 shows the cross-sections of the two samples A and D. They were both sprayed with almost identical spray parameters, the only exception being the process temperature at 200 °C for sample A and 240 °C for sample D respectively. At high powder mixture ratios, high temperatures lead to a clogging of the nozzle, as the polyamide particles already partially melt in the powder injector and start sticking to the wall of the nozzle.

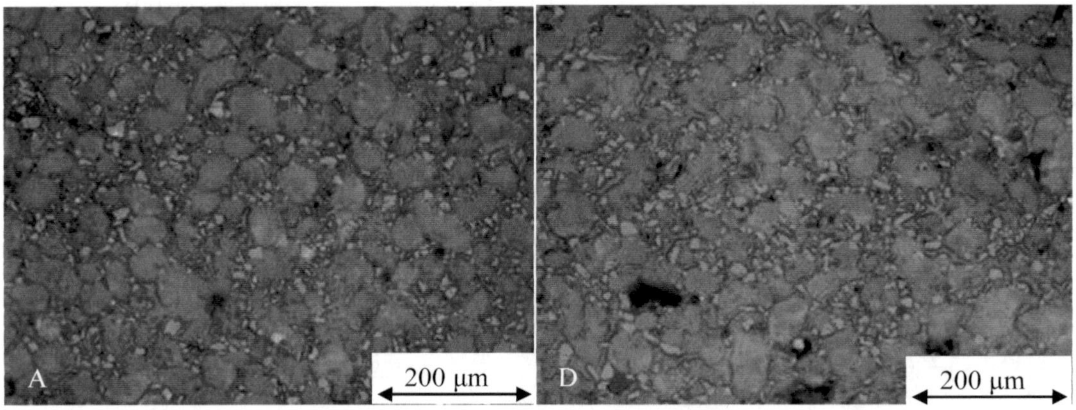


**Figure 8.** Cross-sections of coatings produced with different gas temperatures; A: 200 °C and D: 240 °C.

### *3.3. Effects of the gas pressure*

Figure 9 shows the cross section of a coating (sample E) sprayed with a low temperature setting with high pressure (6 bar). Compared to sample A, which was sprayed with the same temperature setting but employing a lower pressure (4 bar), several bonded particles are visible, indicating that because of the higher kinetic energy of the particles, the pressure has a significant influence on the cohesion and porosity of the coating, thus leading to an increased plastic deformation and compression heating. The downside of a higher pressure can be seen in the top view of the coating: due to the highly abrasive properties of the alumina particles, coupled with the gas temperature, which is still above the melting point of the thermoplastic, lower layers of the coating easily are blasted off the substrate. This effect can be reduced with lower alumina content.

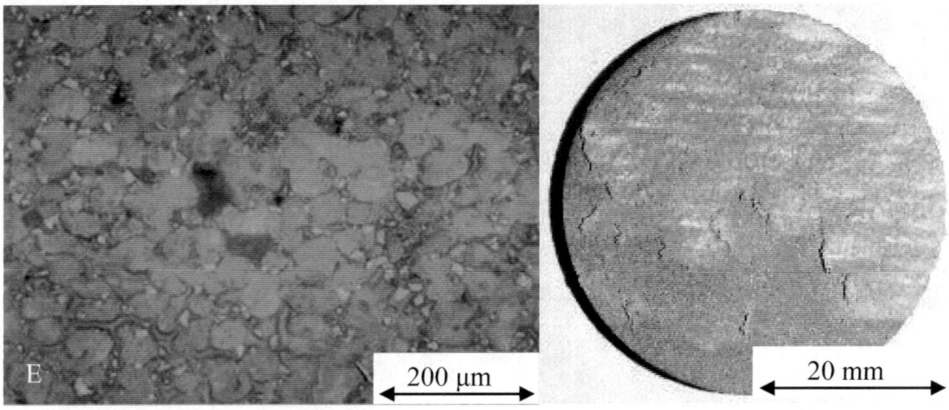


**Figure 9.** Cross-section (left) and top view (right) of a coating produced pressure of 6 bar.

### *3.4. Remelting of coatings*

To create pore free coatings, existing coatings sprayed with parameter A (mixture 2:1, low pressure, low temperature) as well as a rectangular sample coated with pure polyamide-12 (Sample F) by means of upstream radial injection (type II in figure 1, coatings shown in figure 10 before and after remelting) were heat treated at 180 °C for 2 h. Prior to remelting, the coating is very rough and highly porous. After remelting, the coating is smooth (roughness Ra 0.4 ± 0.1 µm, Rz 1.4 ± 0.7 µm) and transparent, indicating a very dense structure. Cross-section images (figures 11 and 12) of both remolten coatings show microstructures with no visible pores and with a good adhesion to the substrate. Since the defects in the microstructure of the coating generated with parameter A correspond in shape and size to the aluminum oxide particles, it can be assumed that these defects are not pores, but holes created in the cross sectioning process.

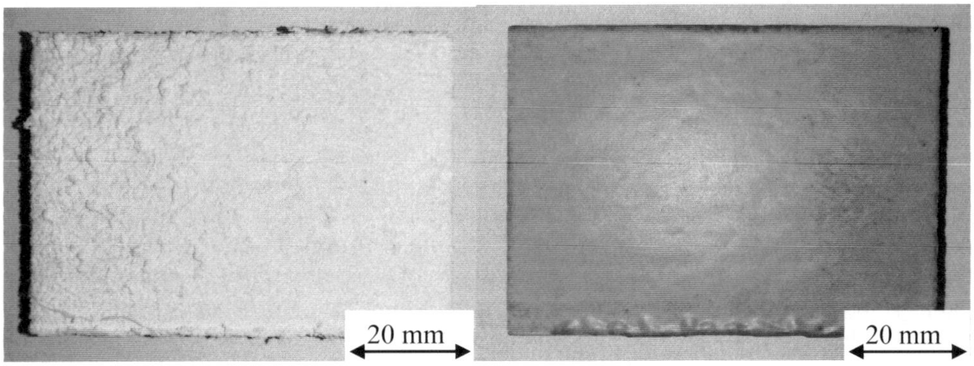


**Figure 10.** Sample F: Pure polyamide-12 coating before (left) and after (right) remelting.

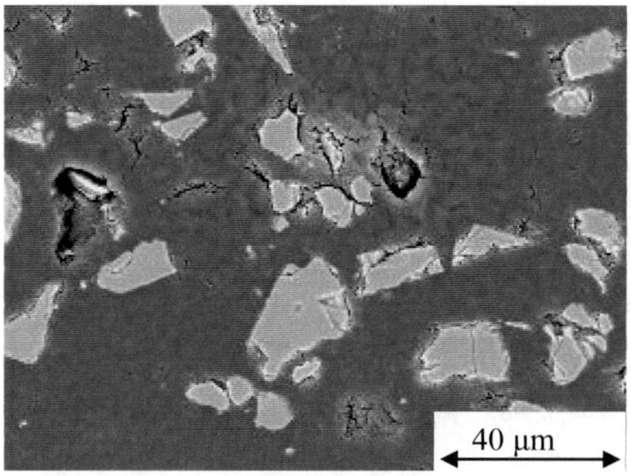


**Figure 11.** SEM image (BSD) of a cross-section of a remolten coating sprayed with parameter A.

**Figure 12.** Cross-section of a remolten pure polyamide-12 coating.

## 4. Conclusions

Polyamide-12 coatings were successfully applied by means of low-pressure cold-gas dynamic spraying. A variety of factors have been investigated and evaluated with regard to their effects on the microstructure of the coating. The investigation suggested that higher process pressures might help in reducing the coating porosity. However, because of the still comparatively high gas temperatures (above the melting point of the material), higher pressures also lead to a great erosion of the applied coating, hence necessitating further temperature control modifications in future studies.

By means of an upstream radial (type II) injection, pure polyamide-12 powder was applied to steel substrates without risking a clogging of the nozzle as it is the case with a conventional downstream (type I) injection. Subsequently, heat treating the coating at 180 °C lead to the production of a pore free polyamide coating that could be used both as a good corrosion barrier and a tribological coating.

Although currently being a two-step process with a need for more spraying parameter optimizations, the low-pressure cold-gas dynamic spraying process shows great potential to apply both dense reinforced and pure polyamide-12 coatings. Since the process only relies on pressurized air and electricity, which both can be generated carbon neutral, it is also preferable to conventional thermal spray processes with respect to the ever-increasing environmental requirements.

## References

[1] Alkhimov A P, Papyrin A N, Kosarev V F, Nesterovich N I and Shushpanov M M 1994 Gas dynamic spraying method for applying a coating *US5302414*

[2] Grujicic M, Zhao CL, Tong C, DeRosset WS and Helfritch D 2004 Analysis of the impact velocity of powder particles in the cold-gas dynamic-spray process *Materials Science and Engineering: A* **368** 222–30

[3] Wang Q, Spencer K, Birbilis N and Zhang M-X 2010 The influence of ceramic particles on bond strength of cold spray composite coatings on AZ91 alloy substrate *Surface and Coatings Technology* **205** 50–6

[4] Assadi H, Gärtner F, Stoltenhoff T and Kreye H 2003 Bonding mechanism in cold gas spraying *Acta Materialia* **51** 4379–94

[5] Dykhuizen R C and Smith M F 1998 Gas Dynamic Principles of Cold Spray *J Therm Spray Tech* **7** 205–12

[6] Irissou E, Legoux J-G, Arsenault B and Moreau C 2007 Investigation of Al-$Al_2O_3$ Cold Spray Coating Formation and Properties *J Therm Spray Tech* **16** 661–8

[7] Raletz F, Vardelle M and Ezo'o G 2006 Critical particle velocity under cold spray conditions *Surface and Coatings Technology* **201** 1942–7

[8] Schmidt T, Gärtner F, Assadi H and Kreye H 2006 Development of a generalized parameter window for cold spray deposition *Acta Materialia* **54** 729–42

[9] Maev R G and Leshchynsky V 2008 *Introduction to low pressure gas dynamic spray: Physics & technology* (Weinheim: Wiley-VCH)

[10] Koivuluoto H, Lagerbom J, Kylmälahti M and Vuoristo P 2008 Microstructure and Mechanical Properties of Low-Pressure Cold-Sprayed (LPCS) Coatings *J Therm Spray Tech* **17** 721–7

[11] Koivuluoto H and Vuoristo P 2010 Effect of Powder Type and Composition on Structure and Mechanical Properties of Cu + Al2O3 Coatings Prepared by using Low-Pressure Cold Spray Process *J Therm Spray Tech* **19** 1081–92

[12] Alhulaifi A S, Buck G A and Arbegast W J 2012 Numerical and Experimental Investigation of Cold Spray Gas Dynamic Effects for Polymer Coating *J Therm Spray Tech* **21** 852–62

[13] Xu Y and Hutchings I M 2006 Cold spray deposition of thermoplastic powder *Surface and Coatings Technology* **201** 3044–50

[14] Ravi K, Ichikawa Y, Deplancke T, Ogawa K, Lame O and Cavaille J-Y 2015 Development of Ultra-High Molecular Weight Polyethylene (UHMWPE) Coating by Cold Spray Technique *J Therm Spray Tech* **24** 1015–25

[15] Ravi K, Ichikawa Y, Ogawa K, Deplancke T, Lame O and Cavaille J-Y 2016 Mechanistic Study and Characterization of Cold-Sprayed Ultra-High Molecular Weight Polyethylene-Nano-ceramic Composite Coating *J Therm Spray Tech* **25** 160–9

[16] Krantz M, Zhang H and Zhu J 2009 Characterization of powder flow: Static and dynamic testing *Powder Technology* **194** 239–45

[17] Spierings A B, Voegtlin M, Bauer T and Wegener K 2016 Powder flowability characterisation methodology for powder-bed-based metal additive manufacturing *Prog Addit Manuf* **1** 9–20

IOP Conf. Series: Materials Science and Engineering **480** (2019) 012010 doi:10.1088/1757-899X/480/1/012010

# Thermally induced morphology changes of wire arc sprayed copper and corrosion-resistant steel (316L)

**R Winkler, E Saborowski, G Paczkowski and T Lampke**

Materials and Surface Engineering Group, Institute of Materials Science and Engineering, Chemnitz University of Technology, D-09125 Chemnitz, Germany

**Abstract.** This contribution deals with the determination of the mechanical and thermal properties of thermally sprayed materials. To this end, a comprehensive characterization of copper (Cu 98.7) and corrosion-resistant steel (316L) is carried out by temperature-dependent tensile, three-point bending and caloric tests within a temperature range from 293 K up to 1173 K. For this purpose, thick coatings were manufactured by wire arc spraying. The resulting volume was further processed by electrical discharge machining into suitable test samples. Reference specimens from the corresponding solid materials were also tested for evaluating the morphological features of the coatings. During the calorimetric examinations, the thermally applied materials were exposed to temperatures of up to 1173 K. As a result, permanent property changes could be observed. To investigate these heat treatment effects, metallographic preparations and XRD analyses were performed. Based on the results, conclusions on the morphology of the thermally applied materials could be drawn.

## 1. Introduction

Modern material hybrids are used to reduce moved masses with the aim of increasing energy and resource efficiency. The thermal spraying process offers the possibility of producing metallic layer systems on thin metal, polymer and ceramic substrates. For example, thermoplastics can be coated with metals such as titanium (cold gas spraying [1]), copper and aluminum (wire arc spraying [2, 3], cold gas spraying [4]) as well as hard material layers, for example WC-Co (high velocity oxygen fuel spraying [5]). The resulting coatings are characterized by their, in comparison to the corresponding solid material, differing mechanical and thermal performance due to their porous and oxidized morphology as well as former spray particle surfaces that form inter-splat boundaries within the coating's microstructure. For this, the determination of the characteristics of thermally processed materials is often carried out by using miniaturized samples and specially adapted test equipment [6-9].

In this contribution, two different metallic coating materials, Cu 98.7 (05T Praxair Tafa) and 316L (85T Praxair Tafa), are characterized by means of their mechanical and thermal material parameters. To achieve this, a thick coating of each material is applied to a thin steel substrate. The resulting material volume is further processed by wire electrical discharge machining (EDM) to the final specimen geometries. The specimens are characterized by tensile tests, three-point bending and laser flash analysis (LFA).

## 2. Materials and Methods

*2.1. Wire Arc thermal spraying process*

For the present study, Cu 98.7 (solid wire: Cu 98.7 (wt.%); Sn Trace; Si Trace; Mn Trace) was selected as typical intermediate layer material, suitable for buffering residual stresses as well as thermal protection of the substrate. The second selected material 316L (solid wire: Fe Balance, Cr 17, Ni 12, Mo 2.5, Mn 2, Si 1, C 0.08, P 0.04, S 0.03 (wt.%)) serves as an example for wear protection applications. The result is a single-phase, homogeneous and economically producible coating. Table 1 shows the optimized spraying parameters for both materials.

**Table 1.** Parameters for wire arc spraying (Visu-Arc 350).

| | current (A) | voltage (V) | pressure (MPa) | traverse speed (m/s) | track spacing (mm) |
|---|---|---|---|---|---|
| Cu 98.7 | 100 | 28 | 0.33 | 0.5 | 5 |
| 316L | 100 | 46 | 0.33 | 3 | 8 |

Standardized test conditions are required to determine correct characteristic values. The miniaturization of the specimens is limited due to the specific features of the processed material. Hence, a correspondingly large volume of the coating must be produced for manufacturing of the specimens. The wire arc spray system Visu-Arc 350 (Oerlikon Metco) is used for the thermal spraying process. A 2 mm thick sheet of carbon steel (1.0038) is used as substrate. The automated coating application (robot KUKA KR 30) is carried out using a meandering track program on a surface of 200 mm x 100 mm. Layer thicknesses of approx. 50 µm to 100 µm are sprayed per transition. After each 5 layers, a 5 minute cooling phase is performed in air. The result is a 14 mm thick semi-finished product, consisting of 170 individual layers. The material reservoir for specimen preparation is taken from the middle part of the coating. The further machining of the specimens to their final dimensions is done by EDM.

*2.2. Mechanical and thermal properties*

The mechanical properties of the thermally processed materials are determined using tensile and three-point bending tests. The specimens have a cross-section of 12 mm$^2$ for the tensile tests (bone sample, measuring length 15 mm, width 4 mm, thickness 3 mm) and 20 mm$^2$ for the three-point bending tests (sample length 45 mm, width 10 mm, thickness 2 mm). The geometry allows the usage of standardized testing devices. The experiments are carried out in a climate chamber with a temperature range from 293 K to 518 K. Thus, a broad database is generated. The universal testing machine Zwick/Roell Allround-Line 20 kN is used for applying the necessary loads. The tensile tests are carried out with a crosshead speed of 1 mm/min and a pre-load of 10 N. The measurement of the elongation is achieved by using tactile sensors. Tensile strength, yield strength and Young's modulus (tensile modulus) are determined from the resulting stress-strain curve. The three-point bending tests are used for verification of the data obtained from the tensile tests. A crosshead speed of 1 mm/min, an abutment distance of 40 mm and a pre-load of 100 N are used. As a result, the Young's modulus (flexural modulus) is determined.

In addition to the mechanical tests, caloric material parameters are determined for characterizing the specific thermal properties of the sprayed materials. The required specimens are also produced by EDM from the semi-finished products. The thermal diffusivity is determined by laser flash analysis (LFA 427, Netzsch Gerätebau, measurement in argon atmosphere, cylindrical specimen ø 12.3 mm x 3.5 mm). Since the thermal diffusivity is temperature-dependent, this experiment is carried out in a defined temperature range from 293 K to 1173 K.

To describe the morphology of thermally sprayed materials, XRD analyses were performed as well as metallographic preparation. An X-ray diffractometer "D8 Discover" from Bruker with Co-Kα radiation was used. The samples were measured with spot focus, a 1 mm collimator and a LynxEye-XE detector. The measuring time per sample was 8.1 h.

## 3. Results and discussion

The tensile and bending test results of the Cu 98.7 and 316L specimens reveal fundamentally different mechanical properties of the thermally sprayed material in contrast to its solid equivalent. The solid copper (rolled, annealed) has the expected ductility with a distinctive elastic and plastic deformation. At 293 K, a tensile strength of 247 MPa (flexural strength 411 MPa) and a maximum elongation of 40% are achieved. In contrast, the thermally sprayed material shows brittle failure with a significantly reduced maximum elongation of 0.3% and a tensile strength of 118 MPa (flexural strength 202 MPa). At elevated temperatures, the maximum elongation increases drastically up to 0.9%. The solid material shows a significant plastic deformation under the development of a constriction, while the thermally processed material fails in brittle mode after low elastic elongation. The interfacial adhesion between the individual spraying particles is weakened by intervening oxides and pores. Hence, the particles slide along their adjacent interfaces when load is applied. Consequently, a clearly increased number of pores with enlarged volume occur in the fracture area.

Similar observations have been made for 316L. The values of flexural strength (410 MPa, brittle failure) are significantly reduced compared to the solid material (flexural strength 702 MPa, no fracture).

Figure 1 illustrates the Young's moduli for Cu 98.7 and 316L determined by the three-point bending tests for a temperature range of 293 K to 518 K. The value decreases significantly with increasing temperature and differs widely between the tested solid and thermally processed material states. The Young's modulus is reduced to about 50 % (Cu 98.7 thermally sprayed: 48 GPa, solid: 103 GPa, 316L thermally sprayed: 95 GPa, solid: 196 GPa at 293 K).

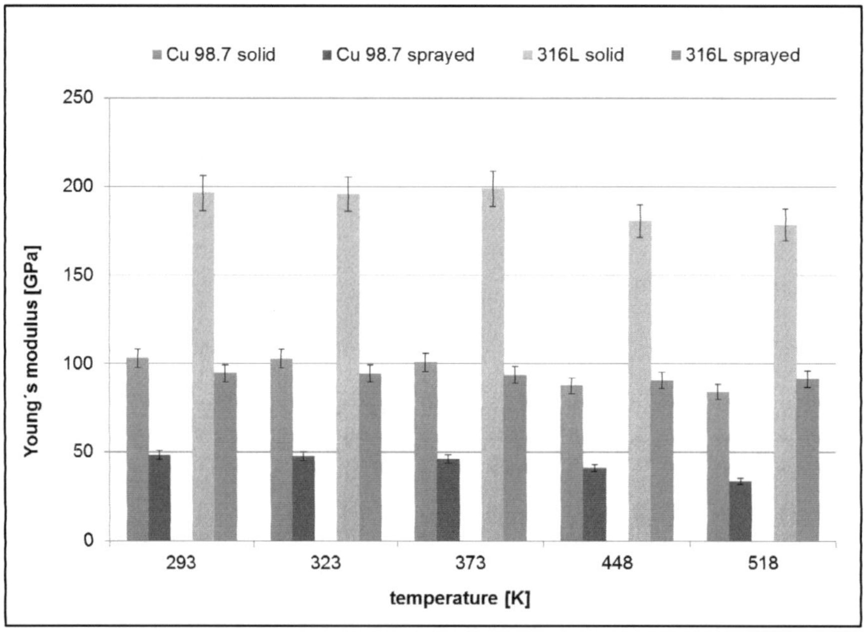


**Figure 1.** Young's modulus determined by bending tests, comparison between Cu 98.7 and 316L in solid as well as in thermally sprayed state.

Figure 2 shows the results of the LFA obtained for thermally sprayed copper. Specimen 1 was first heated at intervals of 100 K from 373 K to 1173 K. The graph of the thermal diffusivity in the heating process is not linear and differs clearly from the reference material. Based on the curve illustrating the cooling process, a significant change in the morphology of the thermally sprayed copper is indicated.

Consequently, it can be assumed that sintering of the copper particles took place at elevated temperatures over 973 K. This effect is known for thermally sprayed coatings, and was used and especially evaluated in cold sprayed copper layers [10-12]. The lasting effect of the heat treatment is confirmed by the repetition of the test (Figure 2: specimen 1 rep.). To prevent the particles from sintering, the experiment was repeated up to a maximum temperature of 573 K for specimen 3. At lower temperature levels, the effect is much less pronounced, since heating and cooling curve are almost congruent. Thus, the determined thermal diffusivity represents the characteristic behavior of the thermally sprayed material (Cu 98.7 thermally sprayed: 40 to 50 $mm^2/s$, solid: 117 $mm^2/s$ at 293 K).

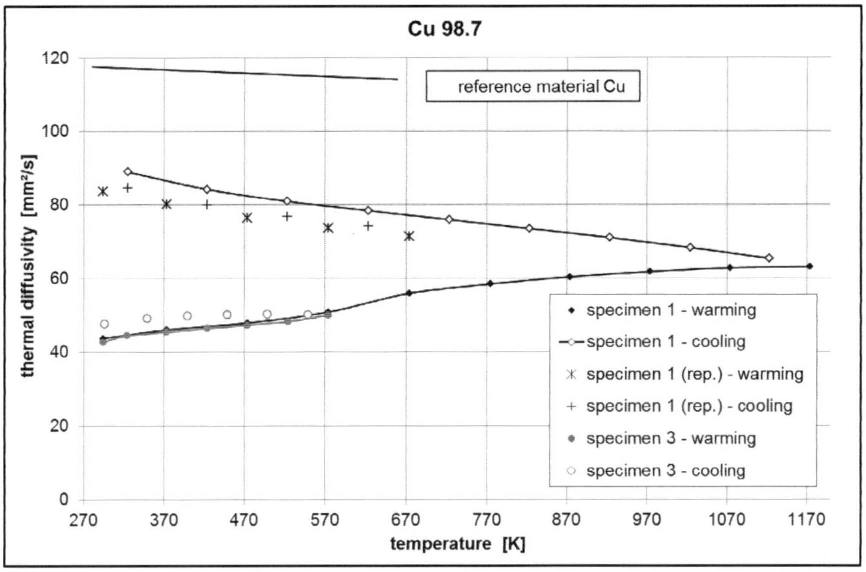


**Figure 2.** Thermal diffusivity of thermally sprayed Cu 98.7 at different heating conditions.

For the thermally sprayed 316L, a similar behavior can be observed in the test. Figure 3 shows the determined curves (test conditions according to Cu 98.7). Again, after the cooling process, a significant change in the thermal diffusivity is detectable. The graph is not linear in the heating process and also changes slightly during the cooling phase. By repeating the heating process of the specimen, the permanence of the change is confirmed (Figure 3: specimen 1 rep.). Another specimen (Figure 3: specimen 3) was heated up to 573 K. At this temperature, no significant change in thermal diffusivity could be observed (316L thermally sprayed: 1 $mm^2/s$, solid: X12CrNi18-8, 3.8 $mm^2/s$ at 293 K).

With the temperatures used in the test, no sintering can be assumed for steel-based materials. The effect may possibly indicate phase transitions. To clarify the effects of the heat treatment, a supplementary metallographic preparation and an XRD investigation are performed.

Figure 4 shows the micrographs of thermally sprayed copper for different heating conditions. In the initial state (as-sprayed, Figure 4a), single particles and lamellar oxidized intermediate phases can be observed. After heat treatment at 573 K (Figure 4b), there is no change in morphology optically detectable. However, as a result of the treatment with 1173 K (Figure 4c and d), significant changes can be identified. The number of visible particle boundaries is significantly reduced and the shape of the oxidized intermediate phases has changed to globular. Thus it can be assumed that sintering processes and phase transformations have taken place. On this basis, the strongly deviating caloric property profile of heat-treated Cu 98.7 can be explained in comparison to the as-sprayed state.

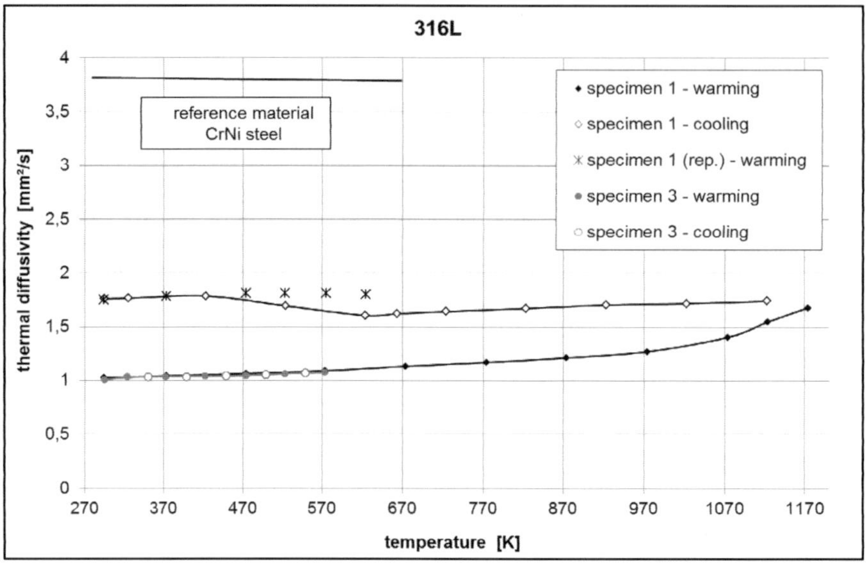


**Figure 3.** Thermal diffusivity of thermally sprayed 316L at different heating conditions.

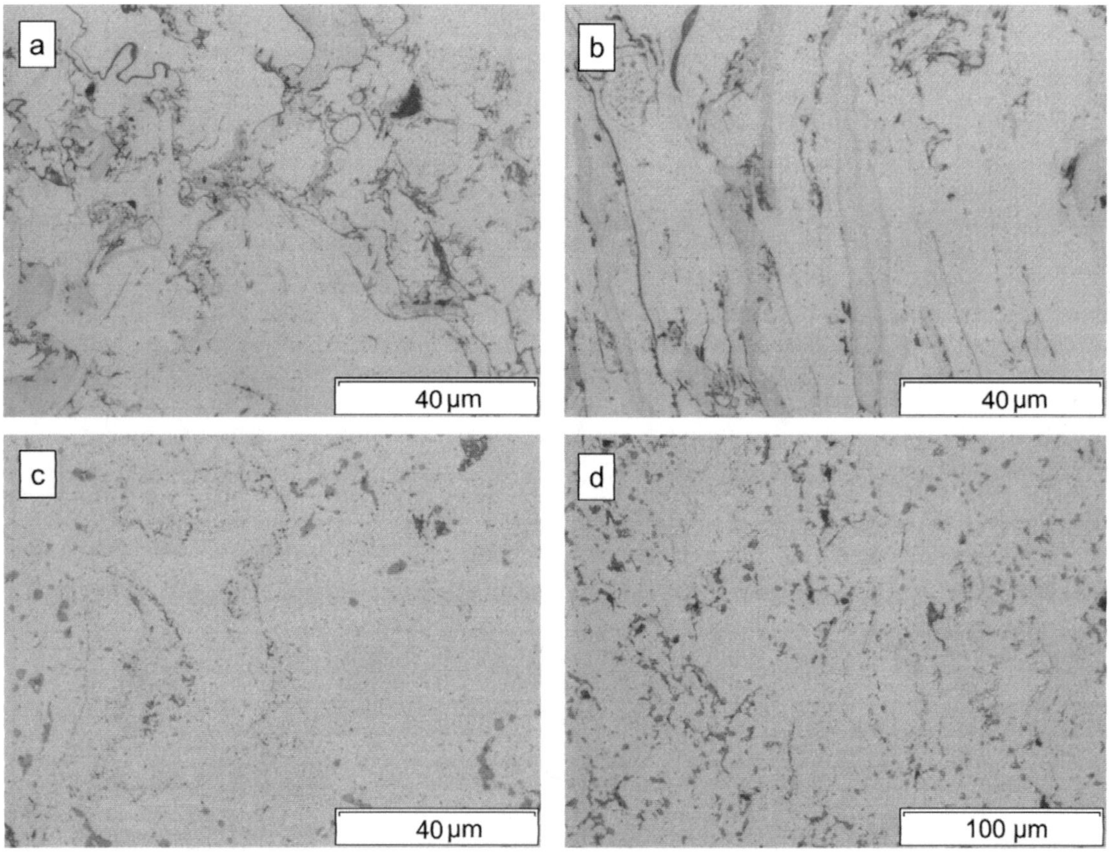


**Figure 4.** Thermally sprayed copper, as-sprayed (a), heat treatment 573 K (b), heat treatment 1173 K (c, d).

Figure 5 shows the results of the additionally performed XRD analysis for all heat treatment conditions (as-sprayed, 573 K, 1173 K) of the thermally sprayed copper. According to the diffractograms (Figure 5), copper as well as copper oxide (copper (I) oxide) can be detected. The diffractograms of the three different states indicate the same phase composition. The change of thermal diffusivity (Cu 98.7, specimen 1, Figure 2) is therefore due to the observed sintering effect.

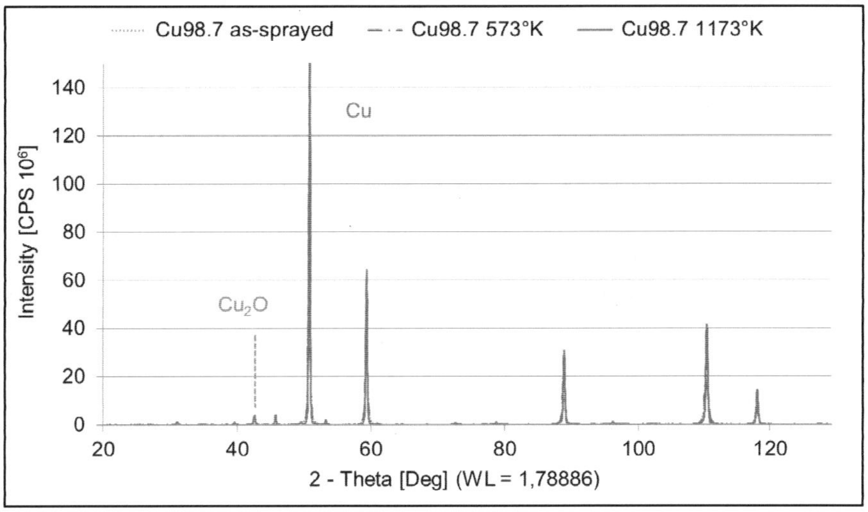


**Figure 5.** XRD analysis, diffractograms of thermally sprayed copper (Cu98.7) for different heat treatment conditions (as-sprayed, 573 K, 1173 K).

In Figure 6, the micrographs of thermally sprayed 316L for the three different heating conditions are displayed. The initial state (as-sprayed, Figure 6a) is characterized by a typical lamellar structure of deformed particles and oxidized intermediate phases. After heat treatment at 573 K (Figure 6b), the morphology has not changed similar to the copper sample. As a result of the treatment at 1173 K (Figure 6c and d), again optical changes occur. Especially in the intermediate phases, phase changes become optically visible. However, the well-defined particle boundaries remain. In accordance with the assumption for steel-based materials, sintering of the particles can thus not be concluded. A possible change in the phase composition is checked by the final XRD analysis.

Figure 7 shows the results of the performed XRD analysis for all heat treatment conditions (as-sprayed, 573 K, 1173 K) of the thermally sprayed 316L. The material consists of three main phases in the as-sprayed state, the austenitic stainless steel, ferrite and magnetite as well as additional oxidized phases. After heat treatment at 573 K, this composition remains almost unchanged. However, after heat treatment at 1173 K, most of the austenitic phase has been converted to ferrite and martensite. The amount of magnetite is also slightly reduced. The heat treatment was carried out under argon atmosphere, so no additional oxygen could be incorporated into the material. The phase transformation takes place as a consequence of diffusion processes and oxidation of the alloying elements, especially of the austenite-stabilizing nickel and manganese. The thermal diffusivity of iron (22.8 $mm^2/s$) differs significantly from unalloyed steel (<0.4% C, 12 $mm^2/s$ to 15 $mm^2/s$) or high-alloyed steel (X12CrNi18-8, 3.8 $mm^2/s$). It can therefore be assumed that the increase in thermal diffusivity is due to the effects of phase transformations.

The investigations show two typical mechanisms for property changes of thermally sprayed materials. The material-specific operating temperatures are of crucial importance and must be taken into account in order to avoid unwanted effects.

IOP Conf. Series: Materials Science and Engineering **480** (2019) 012010 doi:10.1088/1757-899X/480/1/012010

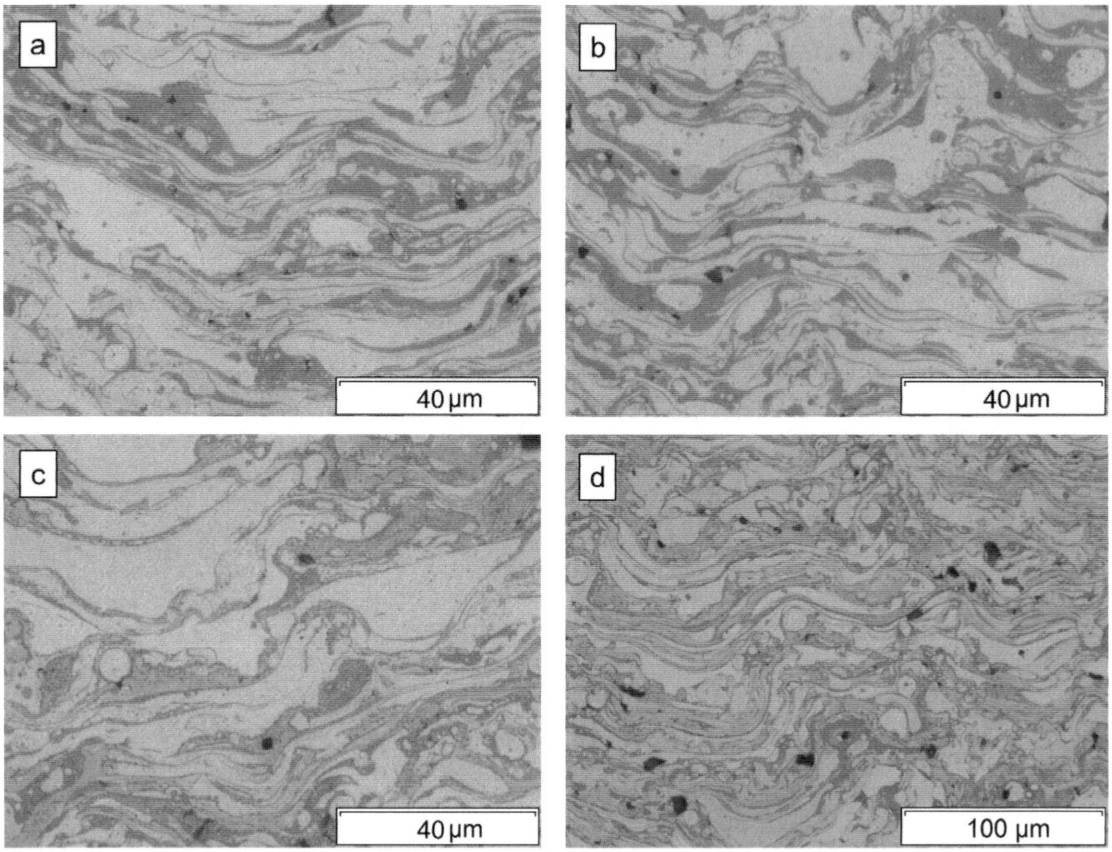


**Figure 6.** Thermally sprayed 316L, as-sprayed (a), heat treatment 573 K (b), heat treatment 1173 K (c, d).

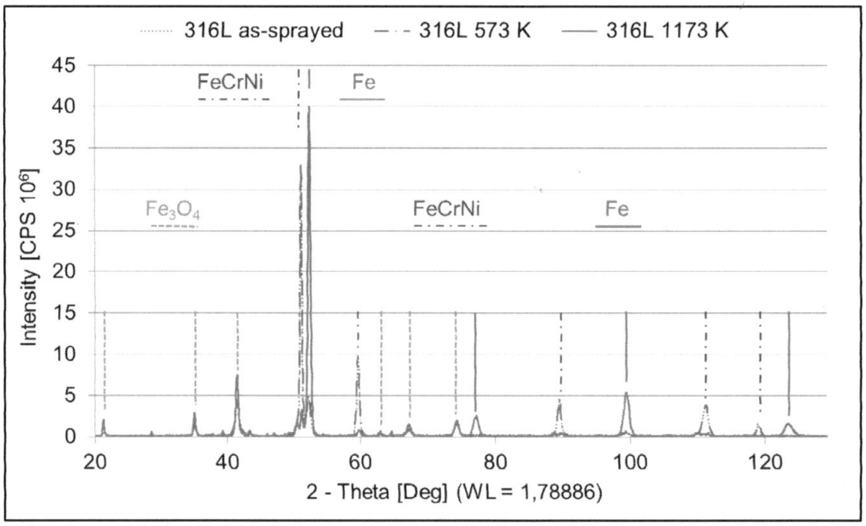


**Figure 7.** XRD analysis, diffractograms of thermally sprayed 316L for different heat treatment conditions (as-sprayed, 573 K, 1173 K).

## 4. Summary

Mechanical and thermal properties of wire arc sprayed Cu 98.7 and 316L have been studied. A novel approach for producing non-miniaturized specimens by EDM from thick coatings has been presented. The use of larger specimens reduces the influence of pores and oxides, which are considerably present in thermally sprayed material. Thus, error influences are reduced and the quality of the obtained data is increased. The mechanical test results revealed drastically reduced Young's moduli (Cu 98.7: 49%, 316L: 48%, measured at 293 K) of the thermally sprayed materials in comparison to their solid equivalent.

The laser flash analysis (LFA) was performed in a temperature range of 373 K up to 1173 K for both thermal sprayed materials. The graphs of the thermal diffusivity in the heating process showed atypical behavior especially at higher temperatures above 573 K, which indicated significant and permanent changes in the morphology of the thermally sprayed materials. At lower temperatures, the behavior also deviates from the solid materials but does not show any permanent changes.

To clarify the effects of the heat treatment, a supplementary metallographic preparation and an XRD investigation were performed.

For the tested copper (Cu 98.7), the micrograph after heat treatment with 1173 K revealed a significantly reduced number of visible particle boundaries. The shape of the oxidized intermediate phases changed to globular. Thus it can be assumed that sintering processes have taken place. The XRD diffractograms of the three different states showed no change in phase composition.

For the thermally sprayed 316L, a major phase change after heat treatment with 1173 K could be shown through metallographic preparation and XRD analysis.

The investigation shows fundamental property changes of thermally sprayed materials at elevated temperatures. The material-specific operating conditions are of crucial importance to avoid unwanted effects.

**Acknowledgements**

The IGF project No. 19.102 B / DVS No. 09.071 of the "Research Association on Welding and Allied Processes of the DVS" was funded by the "Federal Ministry for Economic Affairs and Energy" as part of the "Program for the Promotion of Industrial Community Research and Development (IGF/AiF)" on the basis of a resolution of the German Bundestag.

The authors would like to thank the AiF for the financial support of the project as well as the companies "Iwis Antriebstechnik GmbH & Co. KG" and "GfE Fremat GmbH & Co. KG" for the provision of demonstrators. We would also like to thank the members of the project support committee and the DVS who supported the project, as well as Dr. Ing. Rhena Wulf (TU Bergakademie Freiberg) for the performance of the caloric investigations.

## References

[1] Gardon M, Latorre A, Torrell M, Dosta S, Fernández J and Guilemany J M 2013 *Materials Letters* pp 97–99
[2] Liu A, Guo M, Gao J and Zhao M 2006 *Surface and Coatings Technology* pp 2696–2700
[3] Wielage B, Paczkowski G, Rupprecht C and Winkler R 2010 *Thermal Spray Bulletin* pp 148–152
[4] Zhou X L, Chen A F, Liu J C, Wu X K and Zhang J S 2011 *Surface and Coatings Technology* pp 132–136
[5] Ivosevic M, Knight R, Kalidindi S R, Palmese G R and Sutter J K 2006 *Surface and Coatings Technology* pp 5145–5151
[6] Gärtner F, Stoltenhoff T, Voyer J, Kreye H, Riekehr S and Koçak M 2006 *Surface and Coatings Technology* pp 6770–6782
[7] Bolelli G, Candeli A, Koivuluoto H, Lusvarghi L, Manfredini T and Vuoristo P 2015 *Materials & Design* pp 20–34

IOP Conf. Series: Materials Science and Engineering **480** (2019) 012010 doi:10.1088/1757-899X/480/1/012010

[8] Suffner J, Sieger H, Hahn H, Dosta S, Cano I G, Guilemany J M, Klimczyk P and Jaworskad L 2009 *Materials Science and Engineering* pp 180–186.
[9] Winkler R, Saborowski E, Paczkowski G, Grund T and Lampke T 2018 *Surface and Coatings Technology*
[10] Dietrich D, Wielage B, Lampke T, Grund T and Kümmel S 2012 *Advanced Engineering Materials* pp 275–278
[11] Hall A C, Cook D J, Neiser R A, Roemer T J and Hirschfeld D A 2006 *Thermal Spray Technology* pp 233–238
[12] Sudharshan Phani P, Srinivasa Rao D, Joshi S V and Sundararajan G 2007 *Thermal Spray Technology* pp 425–434

IOP Conf. Series: Materials Science and Engineering **480** (2019) 012011 doi:10.1088/1757-899X/480/1/012011

# Joining composite materials with reactive nickel-aluminium particles as an innovative additive in epoxy-based adhesives

**S Grohmann**[*], **M Friedrich and M F Zaeh**

Institute for Machine Tools and Industrial Management (*iwb*), Technical University of Munich, Boltzmannstrasse 15, 85748 Garching, Germany

[*] e-mail: sandra.grohmann@iwb.mw.tum.de

**Abstract.** Joints comprised of materials with different thermo-physical properties are often advantageous for lightweight components. However, the fabrication of such joints has become a key issue for joining technologies. An innovative approach to overcome the associated challenges is the use of reactive particles. Each reactive particle contains at least two reactants, which are able to undergo an exothermic, self-sustaining reaction after the ignition by an external energy input. Depending on the stoichiometric ratio of the reactants and the heating rate, temperatures up to 1500 K are reached within milliseconds for the nickel and aluminium system. In the automotive industry, bonding is state of the art for applications in electromobility. Yet, slow cross-linking of the adhesive increases cycle times and results in an initial low handling strength, which requires additional mechanical fixings. The benefits of using reactive particles in epoxy-based adhesives are shown in this paper. Varying quantities of reactive particles were integrated into two part epoxy adhesives and activated with microwave energy. Due to the promotion of the cross-linking of the epoxy resin and the reaction of the particles by the microwaves, curing times were significantly shortened. Joining aluminium and polypropylene specimens demonstrated the potential of the presented approach.

## 1. Introduction

Joining technologies play a key role in production engineering as they enable the use of composite materials for a resource-efficient lightweight design. A versatile and industrially established technology for joining components with different thermo-physical properties is adhesive bonding. However, processing reactive adhesives like epoxy resins, which are widely used due to their excellent temperature and environmental resistance, requires a certain period of time because of the curing reaction. In order to guarantee reasonable cycle times, additional mechanical fixings are often applied to overcome the initial low handling strength of the joint. Reactive particles, which were originally used in the field of combustion synthesis for fabricating high performance materials, represent a promising approach to reduce curing times. The exothermic reaction of each particle can be utilised as a tailored heat source to promote the cross-linking within the epoxy resin's network. By integrating reactive particles into adhesives, the advantageous characteristics of both constituents are combined in an innovative way.

### *1.1. Reactive particles in production engineering*

Combustion synthesis, also denoted as self-propagating high temperature synthesis, has become an important field of research in academia and industry since the discovery of the solid flame phenomenon by A. G. Merzhanov [1]. The fundamental reason for this development is the unique reaction characteristic of reactive systems. Two or more homogeneously mixed reactants undergo a self-sustaining, highly exothermic reaction after a defined energy input. As the term combustion synthesis implies, the main research focus is on the production of high performance materials, such as carbides, borides, nitrides, intermetallics, and composite powders [2]. The released energy allows the sublimation of impurities during the reaction and results in high-purity alloys. Another valuable approach is not to consider the reaction product, but the targeted use of the reaction heat. In propellants, the exothermic reaction of nickel and aluminium can be used to influence and stabilise the combustion reaction [3]. Furthermore, joining technologies, like welding, soldering, and bonding, require a controllable heat source, which can be provided by reactive systems.

In order to release the reaction heat in a defined way, the type of energy input to initiate the chemical reaction plays an essential role. Basically, two theoretical reaction modes need to be distinguished when studying the activation and reaction behaviour of reactive systems: the propagating and the simultaneous combustion mode [4]. A locally confined energy input into the reactive system results in the propagating mode and in the formation of a reaction front. Typical activation methods include for example a mechanical impact [5], a heated wire [6], or a laser beam [7]. The self-sustaining reaction front propagates through the reactive system until the reactants are fully converted into products. The simultaneous combustion mode occurs if the reactive sample is homogeneously heated to the ignition temperature, e. g. by a furnace [8] or electromagnetic radiation [9], and the combustion reaction is concurrently activated at each position within the reactive system.

Due to the high reaction rates and theoretical maximum adiabatic temperatures of up to 1910 K [10] for the nickel and aluminium system, an efficient energy transfer into the reactive system before and after the initiation is possible with electromagnetic radiation. Microwaves at a frequency of 2.45 GHz interact in an advantageous way with reactive systems and allow rapid and homogeneous heating [11].

Reactive powder materials, such as metallic particles, are commonly used in the field of combustion synthesis. For joining applications, a specific type of reactive systems has been developed. The so-called reactive multilayer systems consist of hundreds of alternating layers of the reactants [12]. However, due to their brittle nature as well as stability and handling issues [13] they are primarily suitable for joining planar or slightly curved surfaces. Reactive powder materials represent a more flexible alternative. As each particle contains at least two reactants, reactive particles can be used as an adaptable heat source. Figure 1 shows different intrinsic structures of reactive particles.

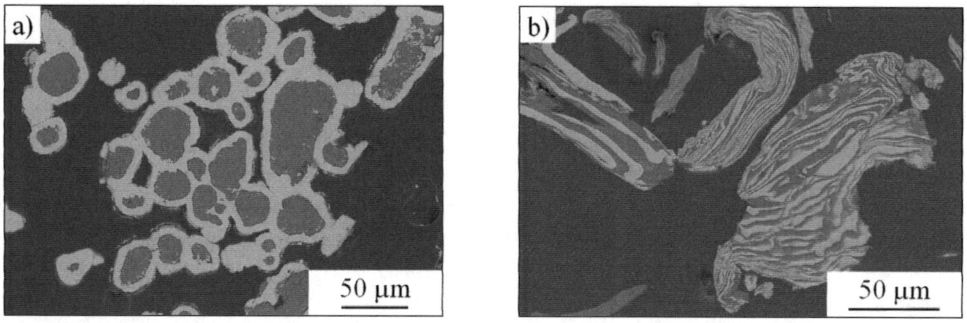


**Figure 1.** Scanning electron micrograph images of reactive particles with an intrinsic a) core-shell and b) lamellar structure.

Reactive particles with a lamellar structure are usually synthesised via ball milling [14]. A less common method is to shred reactive multilayer systems. Electroless [15] and electrochemical [16] plating processes of metal particles are efficient routes to obtain core-shell structures. As the size, shape, and stoichiometric ratio of the reactants are adjustable through the synthesis and significantly influence the activation and reaction behaviour and thereby the released energy, reactive particles provide a customisable heat source for a wide range of joining applications.

*1.2. Interactions between adhesives, additives, and microwaves*

Epoxy resins are used in various ways in production engineering, e. g. as surface coating, fibre-reinforced composite, and as highly relevant adhesive [17]. In this paper, the term epoxy resin refers to both the prepolymer with the reactive epoxy groups as well as the cured resin. Reactive epoxy groups, also denoted as oxirane, are able to form three-dimensional molecular networks with good mechanical strength as well as high temperature and excellent environmental resistance [18]. The curing agent plays a decisive role as it does not only initiate the reaction, but also determines the reaction mechanism and the resulting characteristics of the cured resin. Depending on the formulation of the single or two part epoxy resin, heat curing is used to initiate or additionally promote the chemical reaction. The major challenges of processing epoxy resins are long curing times, non-uniform polymerisation, and supplementary post-curing in order to complete the cross-linking of uncured areas.

An alternative and industrially established approach is using microwaves for curing polymers. Reduced thermal gradients and a homogeneous fast energy transfer directly into the matter are major advantages of microwave processing compared to the conventional heat transfer modes via conduction, convection, and radiation in an oven [19]. However, effective microwave curing is a demanding challenge as it depends on the formulation of the resin, the uniformity of the electromagnetic field, the operation and dimensions of the microwave oven, and the arrangement of the sample [17]. Nevertheless, comparative studies on thermal and microwave curing with different curing agents demonstrated the enhancement of reaction rates and a higher degree of curing when using microwaves [20]. Moreover, reaction mechanisms strongly depend on the operation mode of the microwave (pulsed or continuous [21]) and deviate from conventional heating [22].

The formulation of the epoxy resin and the resulting dielectric properties, particularly their change during the polymerisation [19], are of great importance for microwave curing as they determine the interaction with the electromagnetic field. Furthermore, the integration of additives into the epoxy resin is a sophisticated possibility to further enhance the absorption of electromagnetic energy. Microwave curing of epoxy resins filled with nanoscaled ferrites [23], silicon carbide nanoparticles [24], multi-wall carbon nanotubes [25], and fly ash [26] confirmed an efficient energy transfer into the composite and resulted in a considerable shortening of the curing time compared to conventional heating. In addition, energy can be directly transferred into the joint without damaging temperature-sensitive joining partners. Consequently, electromagnetic waves as well as additives have a positive effect on the curing process. As the use of reactive particles as an innovative additive in epoxy resins and the curing with microwaves have not yet been studied, this research aimed at investigating the resulting interactions.

## 2. Experimental setup and materials

In order to evaluate the cause-effect relationships between the reactive particles, the adhesive, and the microwave energy, various studies were conducted. First, curing times of different two part epoxy resins with varying quantities of reactive particles were investigated. Further knowledge about the influence of different quantities of other additives in two part epoxy resins and microwave power was gained by using differential scanning calorimetry (DSC) and thermogravimetric analysis (TGA). The findings obtained by that formed the basis for the preparation and the analysis of lap joints.

*2.1. Preparation and analysis of the epoxy resin*

The two part epoxy resin LOCTITE EA 3425 (Henkel AG & Co. KGaA) was applied with a manual dispensing gun and mixing tubes. Additives included reactive nickel-aluminium particles with a lamellar structure (Indium Corporation), manufactured by cold rolling and subsequent shredding of the multilayer system, aluminium powder (New Materials Development GmbH, 6.9 µm, 99.7 % purity), and carbon black powder (P-1500, PENTACARBON GmbH). All materials were used as received.

The thermal analyses were carried out with instruments and crucibles by TA Instruments. DSC measurements with the thermal analyser Q2000 required Tzero aluminium pans and Tzero hermetic aluminium lids as sample holders. The heating rate was adjusted to 10 K/min and the maximum temperature to 433.15 K. For TGA measurements, platinum sample pans (100 µl) were used as crucibles in the thermal analyser Q5000SA. The heating rate was again set to 10 K/min and the maximum temperature to 873.15 K. Argon was used for both analysis methods.

### *2.2. Experimental setup for microwave curing*

Initial curing experiments with reactive particles and two part epoxy resins were performed in a domestic microwave oven (HMT75M421, Robert Bosch Hausgeräte GmbH), which is widely used in scientific research [17] and is hereafter referred to as configuration I. For further studies, an experimental setup, which is denoted as configuration II, with a magnetron operating at a frequency of 2.45 GHz (MH2000S-215BB, Muegge GmbH) and a continuous output power was used. All samples were placed on the rectangular waveguide as illustrated in Figure 2.

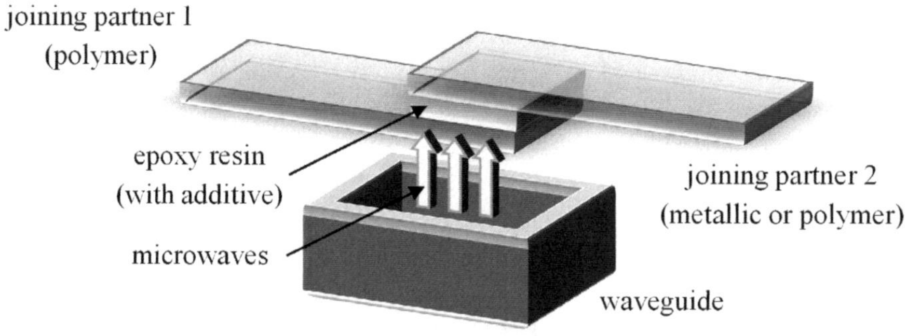


**Figure 2.** Scheme of the microwave curing process in configuration II.

In order to guarantee a homogenous microwave energy input into the adhesive and the additives, the joining partner facing the rectangular waveguide should not significantly interact with the electromagnetic field as this may result in unintended absorption or thermal damage. Therefore, lap joints with an overlap length of 12.5 mm consisting of a metallic and a polymer sheet as well as of two polymer sheets were studied. Dimensions of all test specimens for joining experiments were 100 x 25 x 2 mm according to DIN EN 1465.

### *2.3. Preparation and analysis of lap joints*

Polypropylene (PP-H, Technoplast v.Treskow GmbH) was chosen as the polymer joining partner due to its bondability with two part epoxy resins and a relatively low interaction with microwaves. Moreover, the surface treatment of polypropylene is relatively simple compared to chemical etching processes. After rinsing with isopropyl alcohol and ultrasonic cleaning, an atmospheric-pressure plasma, generated by the Openair FG5001 plasma generator (PlasmaTreat GmbH), was used for the surface treatment of the polypropylene samples. Suitable settings of the plasma generator were determined by contact angle measurements (Contact Angle System, DataPhysics Instruments GmbH) and are listed in Table 1.

Aluminium sheets (EN AW-2017-A, Bikar-Metalle GmbH) were prepared by rinsing with isopropyl alcohol and ultrasonic cleaning. The distance between the joining partners was ensured by a stainless steel wire (0.20 mm, Advent Research Materials Ltd.) and a bracket. This device was made out of Teflon (PTFE virginal unfilled, KTK Kunststofftechnik Vertriebs GmbH) to avoid unintentional interaction with the electromagnetic field and adhesion of the epoxy resin.

**Table 1.** Parameters for the plasma treatment of polypropylene sheets.

| parameter | value |
|---|---|
| travel speed of the nozzle | 15 m/min |
| distance | 10 mm |
| number of passes | 5 |

The analysis of the manufactured lap joints was carried out on the basis of tensile strength tests with the single-screw tensile testing machine Z100 (ZwickRoell GmbH & Co.KG) and a speed of the crosshead displacement of 60 MPa/s.

## 3. Results and discussion

In the following sections, three different studies are presented. In the first test series, the potential of reactive particles as an advantageous additive in two part epoxy resins was evaluated. Second, the influence of different additives, of varying quantities of the former, and of microwave power is discussed on the basis of the thermal analysis results. Finally, suitable parameter settings were derived from the previously generated knowledge and used for joining.

### *3.1. Influence of reactive particles on curing times*

The initial feasibility study was carried out in configuration I with varying quantities (0 wt.-%, 5 wt.-%, 10 wt.-%, 15 wt.-%) of reactive particles. The microwave power was set to 90 W, 180 W, and 360 W. Curing times were determined on the basis of five samples by indentation tests performed at regular intervals at the edges and in the centre of the sample. The results are shown in Figure 3.

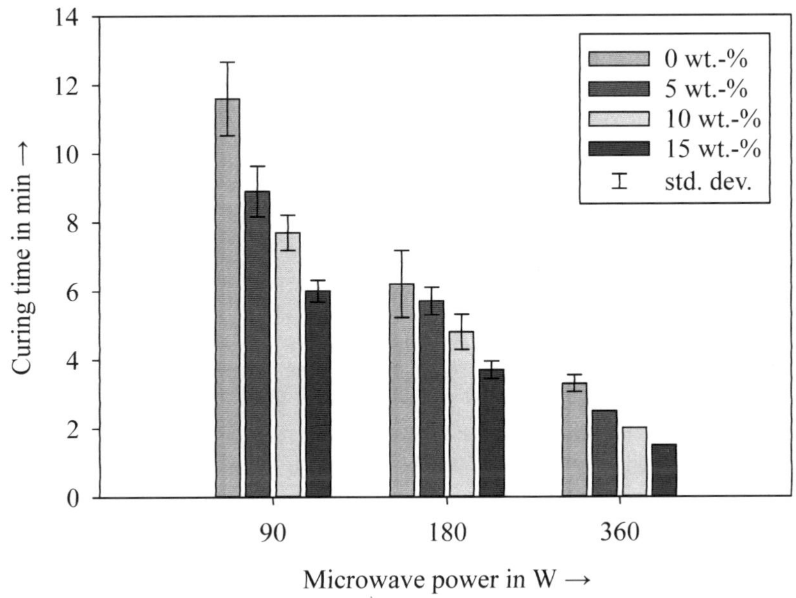


**Figure 3.** Curing times depending on the microwave power and the quantity of reactive particles.

It becomes obvious that both increasing microwave power and the presence of reactive particles significantly reduced curing times. As a reference, curing the two part epoxy resin LOCTITE EA 3425 at room temperature takes 24 hours according to the technical data sheet of the manufacturer. The values of the standard deviations demonstrate the good repeatability and range between 1.1 and 0. Due to the inverse temperature profile [27], curing always started in the centre of the sample. A higher microwave power than 360 W resulted in a change of colour of the adhesive and in a significant thermal damage. Furthermore, partial sedimentation of the reactive particles could be observed at longer curing times, which should be avoided for homogeneous cross-linking. This might have also been the reason for reacted and unreacted particles within the epoxy resin, which were detected with scanning electron microscopy as illustrated in Figure 4. However, it is important to note that it is not intended to release the maximum possible energy of the reactive particles since this would severely damage the epoxy resin. Therefore, the energy input and the heating rate, which are determined by the microwave power, should enhance exothermic intermetallic reactions, such as solid state diffusion, and phase transformations only to a limited extent.

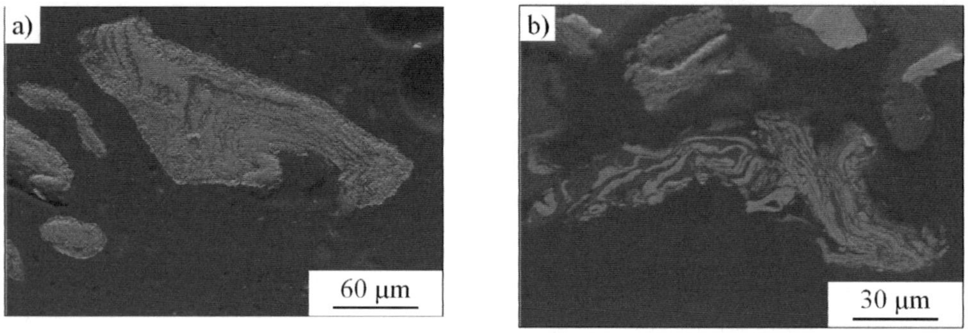


**Figure 4.** Scanning electron micrograph images of a) reacted particles with a relatively homogeneous nickel aluminide structure and b) unreacted particles with lamellas consisting of nickel (light grey) and aluminium (dark grey).

*3.2. Results of thermal analyses*

The second series of experiments included DSC and TGA measurements in order to investigate the polymerisation of the epoxy groups as well as the degree of cross-linking within the polymer network. Figure 5 shows the DSC curves for three different additives (10 wt.-%) compared to an unfilled reference on the basis of two samples.

The detected heat flow clearly indicates the exothermic polymerisation of the two part epoxy resin in all tested samples. Moreover, it can be derived that the three additives influenced the released enthalpy of the reaction. A comparison between the DSC curves reveals that the highest enthalpy of reaction is obtained for the unfilled reference sample, which does not contain any additional additive. The second highest enthalpy of reaction is observed for reactive particles. Apparently, these absorb a certain amount of energy, provided by the DSC oven, but do not significantly contribute to the overall enthalpy of the cross-linking reaction. Due to the relatively low heating rate, it is assumed that primarily solid state diffusion processes took place.

The samples for the TGA measurements were all cured in configuration II. In order to define the minimum and maximum microwave power, when operating with a continuous mode, curing times of unfilled reference samples were compared at first. By this means, thermal damage of the epoxy resin should be prevented. The microwave power was varied between 100 W and 200 W with 25 W intervals and curing times were tested as described before.

As in configuration I, curing times could be significantly reduced with increased microwave power. Curing at 100 W took approximately 16 min, while a microwave power of 200 W required 3 min for hardening. When comparing the results of configuration I (12 min at 90 W, 6 min at 180 W) with configuration II (16 min at 100 W, 3 min at 175 W), it can be assumed on the basis of the state of the art (see [17]) that the mode of operation had an influence on the polymerisation. However, curing at 360 W was not possible as significant thermal damage appeared at microwave powers above 200 W. Therefore, 200 W was chosen as maximum curing power for the epoxy resin LOCTITE EA 3425.

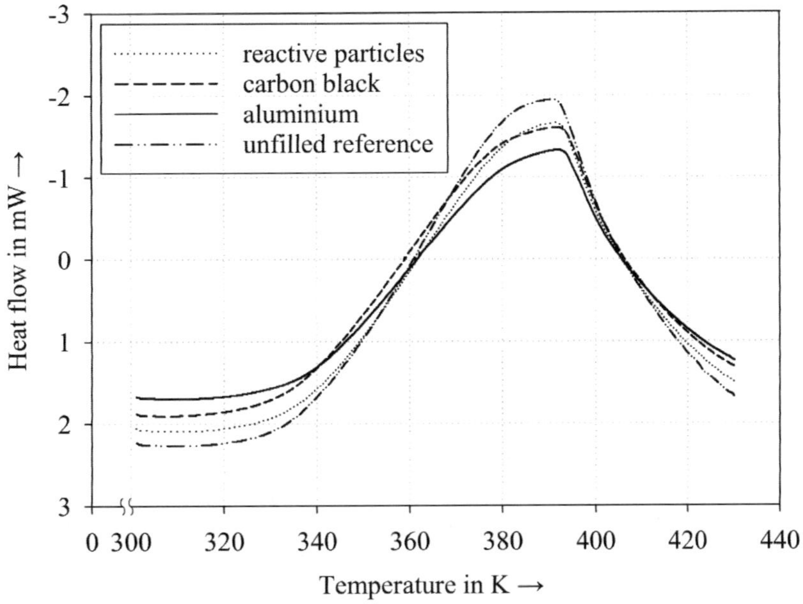


**Figure 5.** DSC curves of the two part epoxy resin with reactive particles, aluminium, and carbon black as additives and an unfilled reference.

TGA measurements of reference samples without additional additives were used to examine if post-curing at room temperature appeared a few minutes, 24 hours, and 48 hours after microwave processing in configuration II.

Almost identical TGA curves and degradation temperatures indicated that the samples were fully cured. The last set of experiments focussed on the interaction with the electromagnetic field and the quantity of the additive. Samples with 5 wt.-%, 10 wt.-%, and 15 wt.-% were cured at 200 W for 3 min. Figure 6 shows the corresponding TGA curves on the basis of two samples.

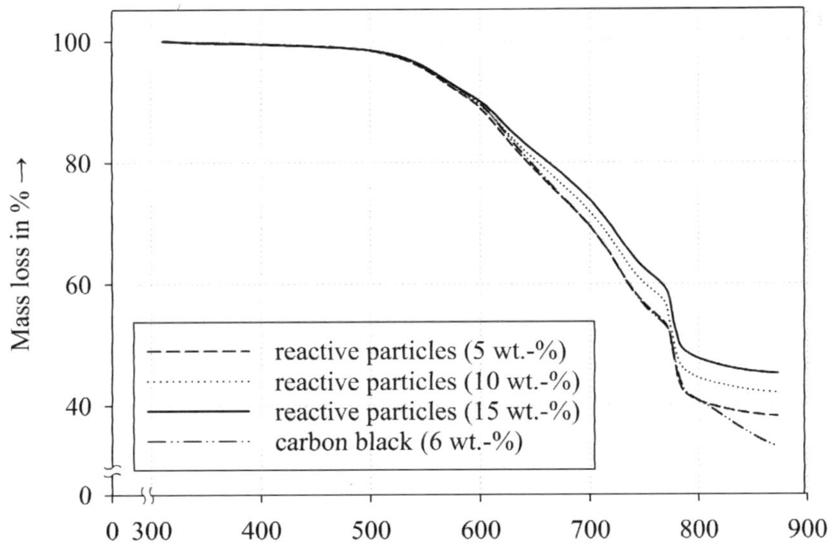


**Figure 6.** TGA curves of the two part epoxy resin with varying quantities of reactive particles and carbon black cured at 200 W for 3 min.

Due to the strong interaction of carbon black with the microwaves, samples with quantities above 10 wt.-% were significantly damaged. Therefore, the quantities for carbon black were reduced (3 wt.-%, 6 wt.-%, 10 wt.-%) in order to avoid evaporation of chemical components of the formulation. In configuration I, microwave processing with higher quantities of carbon black and microwave power was possible (see Figure 3). This result again highlights the significant influence of the microwave chamber and the operation mode. In contrast to reactive particles, TGA curves for carbon black were very similar to each other.

With aluminium or reactive particles as additive, no thermal damage was observed during the curing process, which could also be confirmed by the corresponding TGA curves. Identical quantities of the two metallic additives led to comparable TGA curves, which increasingly converged with reduced quantities within the epoxy resin. The comparison of the degree of cross-linking reveals that the samples with reactive particles exhibit lower degradation temperatures, which could be attributed to a too strong (pre-damage) or a too weak interaction with the electromagnetic field.

As a conclusion, it can be stated that the combination of microwave processing and reactive particles as additives enhances polymerisation and reduces curing times.

*3.3. Results of tensile strength tests*

As it was expected that the aluminium sheets would absorb and reflect a certain amount of the introduced microwave energy, first experiments aimed at scrutinising the transferability of the previously obtained outcomes to joining.

With the same settings as in the aforementioned studies (200 W for 3 min, configuration II) it was not possible to cure a reference sample consisting of an aluminium sheet, a polypropylene sheet, and the epoxy resin. Increasing the curing time and the microwave power led to the curing of the adhesive. Though the heating of the aluminium specimen took place to such an extent that a clear distinction between the interaction of the epoxy resin with the electromagnetic field and curing through thermal conduction was not possible anymore. In order to unambiguously evaluate the influence of the microwaves on the epoxy resin, lap joints of two polypropylene sheets were prepared and analysed. Nevertheless, the microwave power had to be adjusted to 300 W to obtain a curing time of 20 min. Figure 7 displays the obtained results of the strength tests on the basis of three samples.

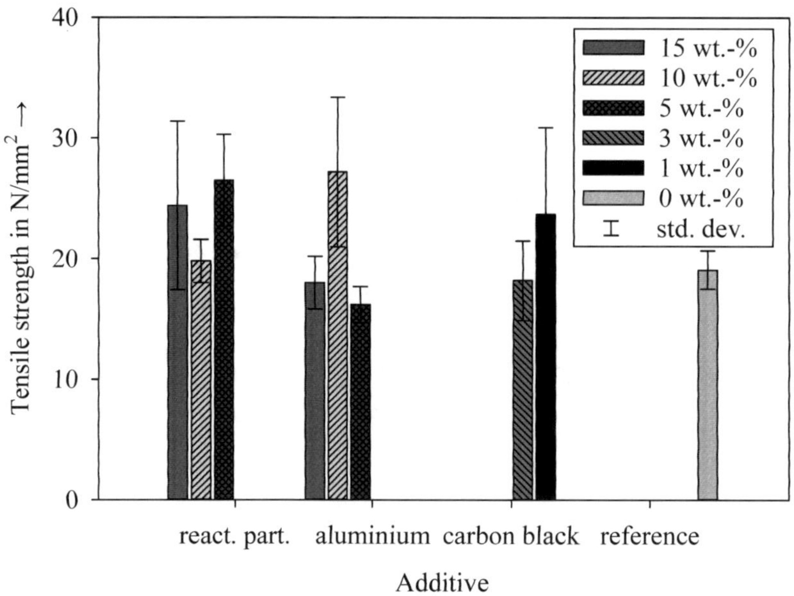


**Figure 7.** Tensile strength of polypropylene lap joints with varying quantities of reactive particles, aluminium powder, and carbon black as additives.

In contrary to the metallic particles, carbon black strongly interacted with the electromagnetic field above a quantity of 5 wt.-% and led to a relevant damage after a few minutes. For comparable results, the amount was reduced to 1 wt.-% and 3 wt.-%. The samples obtained by that were curable without defects.

Adhesive and cohesive fractures, hybrid forms, and failure of the base material were observed after the strength analyses. The high standard deviations impede a clear conclusion, which additive and which quantity is advantageous for joining. Additional studies should be carried out to investigate the cause-effect relationships and to further confirm the potential of reactive particles as additive in epoxy resins.

## 4. Conclusions

Epoxy resins are widely used in academia and industry due to their excellent thermo-mechanical characteristics. However, their long curing times represent a major disadvantage. Elaborated studies have shown that microwave processing and the use of additives are promising approaches to significantly reduce curing times and to influence the polymerisation in an advantageous way. Reactive particles, which undergo a self-sustaining exothermic reaction after an external energy input, represent an adaptable heat source for thermal joining applications. As they are also able to interact with electromagnetic fields at a frequency of 2.45 GHz, the use of reactive particles with a lamellar structure as a new, innovative additive in a two part epoxy resin was studied. The integration of reactive particles led to a considerable reduction of curing times. DSC and TGA measurements with no or further additives (aluminium and carbon black powder) revealed that the operation mode of the microwave oven strongly influenced the interaction between the epoxy resin, the additive, and the electromagnetic field. On the basis of these findings it was possible to fabricate composite joints. As the joining partners also interacted with the electromagnetic field to an essential extent, further studies on microwave processing of epoxy resins and reactive particles are required to enable an industrial usage.

## Acknowledgements

This research is kindly funded by the Deutsche Forschungsgemeinschaft (DFG, German Research Foundation) – 3376550608. The authors would also like to thank DELO Industrie Klebstoffe GmbH & Co. KGaA and Henkel AG & Co. KGaA for the sampling and the friendly support.

## References

[1] Merzhanov A G 1994 Solid Flames: Discoveries, Concepts, and Horizons of Cognition *Comb. Sci. Technol.* **98** 307–336

[2] Rogachev A S and Mukasyan A S 2014 *Combustion for Material Synthesis* (Hoboken: Taylor and Francis)

[3] Breiter A L, Mal'tsev V M and Popov E I 1990 Means of modifying metallic fuel in condensed systems *Combustion, Explosion, and Shock Waves* **26** 86–92

[4] Moore J J and Feng H J 1995 Combustion synthesis of advanced materials: Part I. Reaction parameters *Prog. Mater Sci.* **39** 243–273

[5] Reeves R V, Mukasyan A S and Son S F 2010 Thermal and Impact Reaction Initiation in Ni/Al Heterogeneous Reactive Systems *J. Phys. Chem.* C **114** 14772–14780

[6] Hardt A P and Holsinger R W 1973 Propagation of gasless reactions in solids-II. Experimental study of exothermic intermetallic reaction rates *Combust Flame* **21** 91–97

[7] Andrzejak T, Shafirovich E and Varma A 2007 Ignition mechanism of nickel-coated aluminum particles *Combust Flame* **150** 60–70

[8] Mukasyan A S, White J D E, Kovalev D Y, Kochetov N A, Ponomarev V I and Son S F 2010 Dynamics of phase transformation during thermal explosion in the Al–Ni system *Physica B: Condensed Matter* **405** 778–784

[9] Karimbeigi A, Zakeri A, Aboutalebi M R and Sharifi P 2013 Study on microwave synthesis and characterisation of bulk nickel aluminides *J. Mater. Sci. Technol.* **28** 86–91

[10] Varma A and Lebrat J-P 1992 Combustion synthesis of advanced materials *Chem. Eng. Sci.* **47** 2179–2194

[11] Rosa R, Veronesi P, Leonelli C, Bonamartini Corradi A, Ferraris M, Casalegno V, Salvo M and Han S H 2010 Microwave Activated Combustion Synthesis and Compaction in Separate E and H Fields *Advances in Science and Technology* **63** 197–202

[12] Weihs T P, Knio O, Reiss M and Blobaum K J 2001 Reactive multilayer structures for ease of processing and enhanced ductility *US Patent* US20010046597A1

[13] Boettge B, Braeuer J, Wiemer M, Petzold M, Bagdahn J and Gessner T 2010 Fabrication and characterization of reactive nanoscale multilayer systems for low-temperature bonding in microsystem technology *Journal of Micromechanics and Microengineering* **20** 64018

[14] Hadjiafxenti A, Gunduz I E, Tsotsos C, Kyratsi T, Doumanidis C C and Rebholz C 2010 Synthesis of reactive Al/Ni structures by ball milling *Intermetallics* **18** 2219–2223

[15] Schreiber S, Theodossiadis G D and Zaeh M F 2017 Combustion synthesis of reactive nickel-aluminum particles as an innovative approach for thermal joining applications *IOP Conference Series: Materials Science and Engineering* **181** 12008

[16] Schreiber S and Zaeh M F 2018 Electroplating of aluminium microparticles with nickel to synthesise reactive core-shell structures for thermal joining applications *IOP Conference Series: Materials Science and Engineering* **373** 12008

[17] Tanrattanakul V and SaeTiaw K 2005 Comparison of microwave and thermal cure of epoxy-anhydride resins *J. Appl. Polym. Sci.* **97** 1442–1461

[18] Ellis B1993 *Chemistry and technology of epoxy resins* (London, New York: Springer Science+Business)

[19] Thostenson E T and Chou T-W 1999 Microwave processing *Composites Part A* **30** 1055–1071

[20] Wei J, Hawley M C, Delong J D and Demeuse M 1993 Comparison of microwave and thermal cure of epoxy resins *Polym. Eng. Sci.* **33** 1132–1140

[21] Fu B and Hawley M C 2000 Comparative study of continuous-power and pulsed-power microwave curing of epoxy resins *Polym. Eng. Sci.* **40** 2133–2143

[22] Navabpour P, Nesbitt A, Degamber B, Fernando G, Mann T and Day R 2006 Comparison of the curing kinetics of the RTM6 epoxy resin system using differential scanning calorimetry and a microwave-heated calorimeter *J. Appl. Polym. Sci.* **99** 3658–3668

[23] Sauer H M, Spiekermann S, Cura E and Lie L H 2004 Beschleunigte Klebstoffhärtung mit Mikrowellen und Nanoferriten *adhäsion KLEBEN & DICHTEN* **48** 12–14

[24] Rabby M R-E, Jeelani S and Rangari V K 2014 Microwave processing of SiC nanoparticles infused polymer composites *J. Appl. Polym. Sci.* **271285** 1–11

[25] Sung P-C, Chiu T-H and Chang S-C 2014 Microwave curing of carbon nanotube/epoxy adhesives *Compos. Sci. Technol.* **104** 97–103

[26] Chaowasakoo T and Sombatsompop N 2007 Mechanical and morphological properties of fly ash/epoxy composites using conventional thermal and microwave curing methods *Compos. Sci. Technol.* **67** 2282–2291

[27] Bykov Y V, Rybakov K I and Semenov V E 2001 High-temperature microwave processing of materials *J. Appl. Phys. D* **34** R55-R75

IOP Conf. Series: Materials Science and Engineering **480** (2019) 012012 doi:10.1088/1757-899X/480/1/012012

# Influence of talc in polypropylene foam cores of sandwich structures with skins made of thermoplastic prepregs

**N Loypetch[1,*], J Tröltzsch[2], D Nestler[1] and L Kroll[1]**

[1] Institute of Lightweight Structures, Technische Universität Chemnitz, Chemnitz, Germany
[2] Karl Mayer Technische Textilien GmbH, Chemnitz, Germany

[*] e-mail: nalin.loypetch@mb.tu-chemnitz.de

**Abstract.** Properties of sandwich structures with a core of thermoplastic foam depends on cell size and cell density (cell number). Nucleating agents have a benefit to generate cell nucleation, which affects an increase in the cell density. In this study, sandwich structures composed of a polypropylene (PP) foam core and organic sheet skins in-situ produced by injection moulding process. A chemical blowing agent was employed to generate cellular structure in the core. The influence of the varying nucleating agent content with the same chemical blowing agent content on density, cellular structure and mechanical properties of the sandwich structures were investigated. The talc contents had a slight influence on the density of the sandwich structures. Moreover, flexural modulus and flexural strength of the sandwich structures consisting of the foam core with 4 wt% of talc were higher than the foam core without talc. However, the utilization of these talc contents with PP and the chemical blowing agent had no significant effect on the reduction of the cell size and the increase of the cell density.

## 1. Introduction and motivation

Sandwich structures consist of a light core material locating between two highly-stiff skin layers [1]. It is a composite material which has been commonly used in many applications such as automobile, aerospace and building [2]. One of the advantages of the sandwich structures is the combination of different materials of which dominant properties are different. Consequently, this combination can lead to desire properties. Sandwich structures components can be made of various kinds of material such as polymer and metal. In addition, they can be produced in form of a hybrid material which is composed of different kinds of material [1]. The structure of core materials can be found in many forms such as honeycombs and foam structures [2]. In case of the skin layers, they can be made of metal alloys or fibre composites [2].

Thermoplastic is one of the materials used for sandwich structure manufacturing due to high recyclability [4,5]. Moreover, the use of same kind of thermoplastic for the skin layers and the core material can achieve the skin and core adhesion without the use of adhesive layer between these two layers studied by Mechraoui et al. [6], Menrath et al. [7] and Tröltzsch et al. [8]. In these three studies, PP was used as the core material and the matrix of the composite skin layers.

Thermoplastic foams can be produced using a blowing agent to generate cell structures in injection mould parts. Due to the generation of the cell structures, the material content can be reduced, which leads to the reduction of material cost. The blowing agent can be divided into chemical blowing agent (CBA) and physical blowing agent (PBA) [9,10]. The advantage of CBA over PBA is that it can be employed directly with a conventional injection machine. Conversely, the PBA requires the additional equipment as can be found in MuCell® technology [11,12].

Due to the mechanism of CBA, cell nucleation, gas diffusion and cell growth occurs [13], which affects the cell size and cell density in thermoplastic foams. The cell size and cell density have an influence on the mechanical properties of foam structures [14]. The effective mechanical properties of the sandwich structures can be achieved if their core layer consists of the low cell size and the high cell density. Large cells can lead to a defect which can cause the failure of the sandwich structure's core.

The generating of cell nucleation controlling a quantity of cells depends not only on homogeneous nucleation (self-nucleation) but also heterogeneous nucleation which is generated by a nucleating agent [11]. Moreover, the gas diffusion from the polymer matrix into the nucleated cells causes the cell growth. This should be controlled - otherwise, it will affect large cell size or cell coalescence. Talc is one of the widely-used nucleating agents to increase the cell density in the thermoplastic foams either amorphous [15] or semi-crystalline [16].

The effect of talc on the microstructure has been investigated. Talc particles influence the prevention of the cell growth in PP and increase the nucleation site [17]. This cause the increase in cell number and cell density. In the study of Ruiz et al. [18], the microstructure of the foam depends not only on the kind of CBA but also fillers. In comparison between neat PP and PP with 7 wt% talc, the cell density in PP with 7 wt% talc was higher than neat PP, when the CBA was sodium bicarbonate. However, the cell density in both PPs was nearly the same when the CBA was the mixing between citric acid and sodium bicarbonate. The final density of all specimens was approximately 0.6 g/cm$^3$ in every case. The relation between amount of blowing agent and the effect of talc was also studied by Kim et al [19]. They found that talc had a significant effect on the increase of cell density in PP at the $N_2$ of 0.4 wt%. The influence of adding talc content did not affect the cell size and cell density at some amount of blowing agent. It is, for instance, 5 wt% of talc in the study of Kaewmesri et al. [20]. The addition of talc was effective when the $CO_2$ was low (about 1-3 wt%). This study also found that at 5 wt% of $CO_2$ and 0.8 wt% of talc, the desirable properties could be achieved. Wong and Park studied the cell nucleation in polystyrene (PS) by using talc as a nucleation agent [21]. The result suggests that the cell nucleation can be improved by using large talc particle. When the talc content was between 2 to 5 wt%, this effect was reduced. However, even small talc content (0.2–2.4 wt%) led to increase of cell content in PP foams produced by extrusion process [22].

Moreover, talc has also an effect on the mechanical properties of thermoplastic. The increase of talc content affected the improvement of tensile modulus [23–25]. However, it affects also the decrease of ductility, which can be found by the reduction of elongation at break. These have the same result as the talc micro-composite from the study of Wang et al. [26]. However, the results from the study of Palutkiewicz and Postawa were different [27]. PP with talc as well as PP with talc and CBA had a lower tensile strength than compact PP. Moreover, the weight of PP with talc and CBA was the lowest. Talc has also the influence on the increase of flexural strength and flexural modulus, especially when the talc content was 30-50 wt%, good stiffness in PP occurred [28]. The increase of tensile and bending properties of PP was found in the work of Lapcik et al. However, the ductility decreased which can be seen in the reduction of fracture toughness [29].

A pre-impregnated fibre layer in thermoplastic is called prepreg [30]. It is a composite material which consists of many stacking layers. The prepreg can be divided into unidirectional continuous fibre prepreg and bi-directional fabric prepreg. The use of continuous fibre lamination leads to the higher modulus and strength than discontinuous fibre [30]. An organic sheet in form of the thermoplastic prepreg composed of continuous glass fibres and PP matrix can improve bending performance of sandwich structure when it is used for the skin layer. However, the presence of the skin layers affects the limitation of heat transfer. This leads to more time for cell growth, which increases the possibility of cell coalescence.

The influence of talc has been mostly investigated in case of the nucleating agent in thermoplastic foams as well as the filler in compact thermoplastics produced by many processing techniques.

IOP Conf. Series: Materials Science and Engineering **480** (2019) 012012 doi:10.1088/1757-899X/480/1/012012

Nevertheless, a study of nucleating effect from talc on foam core morphology of the sandwich structures produced by injection moulding is not widely performed. Therefore, the aim of this study is to investigate the influence the talc content on the density, microstructure and bending properties of the sandwich structures.

## 2. Experiments and materials

### *2.1. Materials*

*2.1.1 Core material.* PP (Moplen 501H) used to produce the core material of the sandwich structures was supplied by LyondellBasell. The melt flow rate measured according to ISO 1133 was 2.1 g/10 min. The talc masterbatch was produced by a twin screw co-rotational extruder (Noris Plastic ZSC 25/40D). The masterbatch was composed of 80 wt% of Moplen 501H and 20 wt% of talc (UM-A1) supplied by Universal Mills GmbH. The median and top cut particle size were less than 1.2 and 5.0 µm respectively. Endothermic CBA (Cell.mix a072) supplied by KCD Kunststoffe, Additive und Beratung GmbH was used to generate cell structure in the core material of the sandwich structures. The CBA masterbatch consisted of 17 % of carbonate, which generate $CO_2$ at decomposition temperature approximately 175 °C.

*2.1.2. Skin material.* Glass fibre reinforced PP prepregs were used for the skin layers of the sandwich structures. The unidirectional glass fibre reinforced layer was applied in each layer. The components and the fibre orientation in each layer of the prepregs are shown in Figure 1. They were produced by applying OLU-Preg® technology. The density of the prepregs determined according to DIN EN ISO 1183 was approximately 1.53 g/cm$^3$. The fibre volume fraction was approximately 38 %. The thickness of the prepreg was approximately 0.8 mm. Tensile modulus and tensile strength of the prepregs according to DIN EN ISO 527-5 were15.1 GPa and 299 MPa respectively. The components of the sandwich structures are shown in Figure 2.

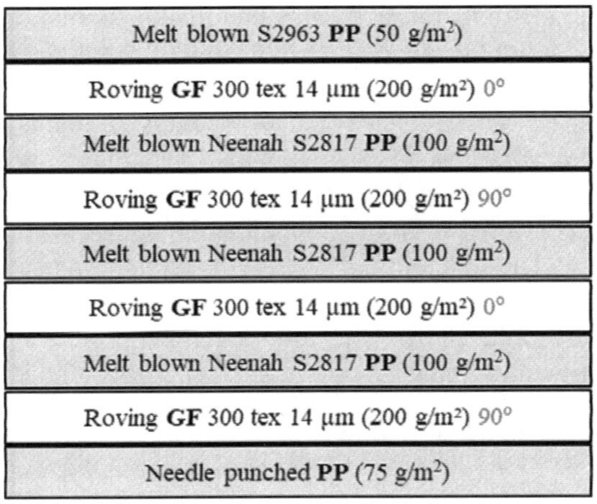


**Figure 1.** Components and fibre orientation in each layer of the unidirectional glass fibre reinforced PP Prepregs.

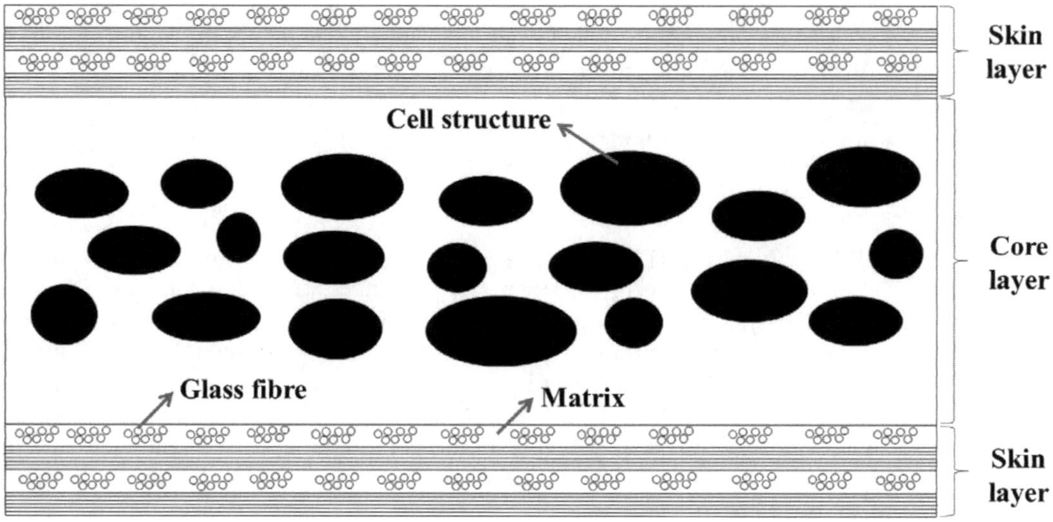


**Figure 2.** Components of sandwich structures.

*2.2. Experiments*

*2.2.1 Production of sandwich structures.* Injection foaming with low-pressure moulding technique was applied to produce the core material of the sandwich structures by using conventional injection moulding machine (KraussMaffei, KM-C80-380CX). A bolt-on plasticising unit of which the diameter of injection screw was 40 mm as shown in Figure 2. PP pellets were mixed with the CBA masterbatch of 4 wt% and the talc masterbatch at the content of 5, 10 and 20 wt% respectively so that the mixture with talc content of 1, 2 and 4 wt% can be achieved. The mixture was then introduced into a hopper. The height and width of the mould cavity were 100 and 250 mm respectively. The thickness of the injection-moulded specimens was adjusted at 4 mm. The continuous glass fibre reinforced PP prepregs were used as the skin layers of the sandwich structures. They were pre-heated in a vacuum oven (Heraeus Vacutherm) supplied by Thermo Scientific. The temperature was set at 140 °C for 2 minutes. The prepregs were applied on the mould surface with an adhesive tape as shown in Figure 3. The prepreg which was on the surface of moving-mould had a hole. The molten plastic can be injected pass through this hole into the mould cavity. The cylinder temperatures from the hopper to the nozzle were 50, 140, 160, 180, 195, 200, 205 and 210 °C respectively. The temperature of the hot runner was set between 200 and 210 °C. The injection speed was set at 110 $cm^3/s$. The mould temperature was set at 130 °C. After the molten plastic was injected into the mould cavity, the mould temperature was reduced to 70 °C. The list of the sandwich specimens is shown in Table 1.

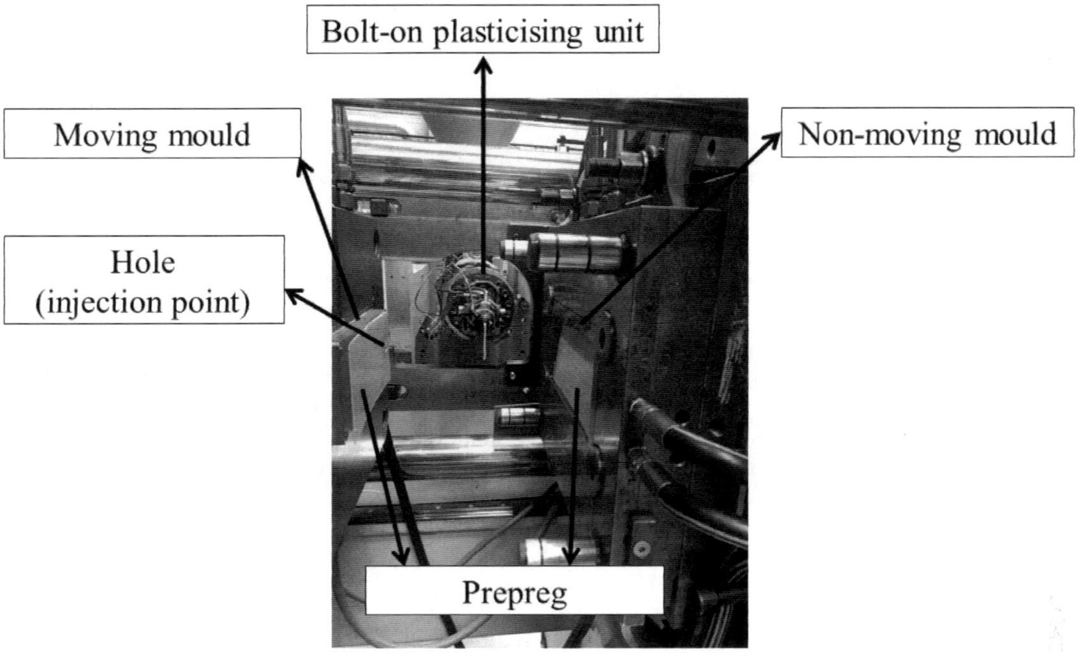


**Figure 3.** Injection mould machine component and position of the prepreg on the mould surface.

**Table 1.** List of the sandwich specimens.

| Specimen name | CBA (wt%) | Talc (wt%) | Skin layers |
|---|---|---|---|
| SW_F | 4 | 0 | Continuous glass fibre reinforced PP prepreg |
| SW_F_T-1 | 4 | 1 | Continuous glass fibre reinforced PP prepreg |
| SW_F_T-2 | 4 | 2 | Continuous glass fibre reinforced PP prepreg |
| SW_F_T-4 | 4 | 4 | Continuous glass fibre reinforced PP prepreg |

*2.2.2 Microscopy.* A specimen per kind of Sandwich structures were polished by a grinding-polishing machine (STRUERS TegraPol-15) and mounted by cold mounting technique. The height and width of the specimens were 10 mm. The microstructure was investigated by a light microscope (Axio Scope.A1) which supplied by Zeiss in panorama mode.

*2.2.3 Density measurement.* Five specimens of the sandwich structures were prepared for density measurement according to DIN EN ISO 1183. The specimens were weighed in air and in liquid which was alcohol with the density of 0.79 g/cm$^3$, respectively. The experiment was conducted by using a density test equipment and an analytical balance (KERN, ALS 220-4N).

*2.2.4 Bending test.* Three-point bending of ten sandwich structure specimens was performed according to DIN EN ISO 178. This test was performed by using a compression testing machine (Zwick/Roell Z5.0 TN). The applied preload was 2 N. The testing speed was 2 mm/min. The support span distance was 64 mm. The support span and the anvil had the radius of 5 mm.

## 3. Results and discussion

### *3.1. Microscopy*

The microstructures of the sandwich structures are shown in Figure 4 (a)-(d). These microstructures suggest that the continuous glass fibre reinforced PP prepreg (the skin layers) and the PP foam core (core material) can achieved good consolidation after the foam injection moulding, which also occurred in the work of Menrath et al. [7] and Tröltzsch et al. [8]. The SW_F had only the skin layers and the foam layers as shown in Figure 4 (a). In contrast, the sandwich structures with talc had the skin layers, compact layers and the foam layer as shown in Figure 4 (b)-(d). The connection of the cell structures in the middle of the sandwich structures occurred. It can be described that the lower temperature reduction rate at the middle area is due to distance to mould wall and the presence of the skin layers. Therefore, the cell structures in the middle area have more time to expand than the area near the mould wall.

Cell density in the foam core material of the sandwich structure with talc tended to increase with the increase of talc content from 1 wt% to 4 wt% as depicted in Figure 4 (b)-(d). However, the cell size in the PP foam core of the sandwich structures was observed without significant difference. The cell density of SW_F_T-1 and SW_F_T-2 was lower than SW_F. Nevertheless, SW_F_T-4 exhibited the cell density as nearly to same amount as SW_F. The reason of the variation in the cell size and the cell density may be described by the relationship between talc and blowing content [20].

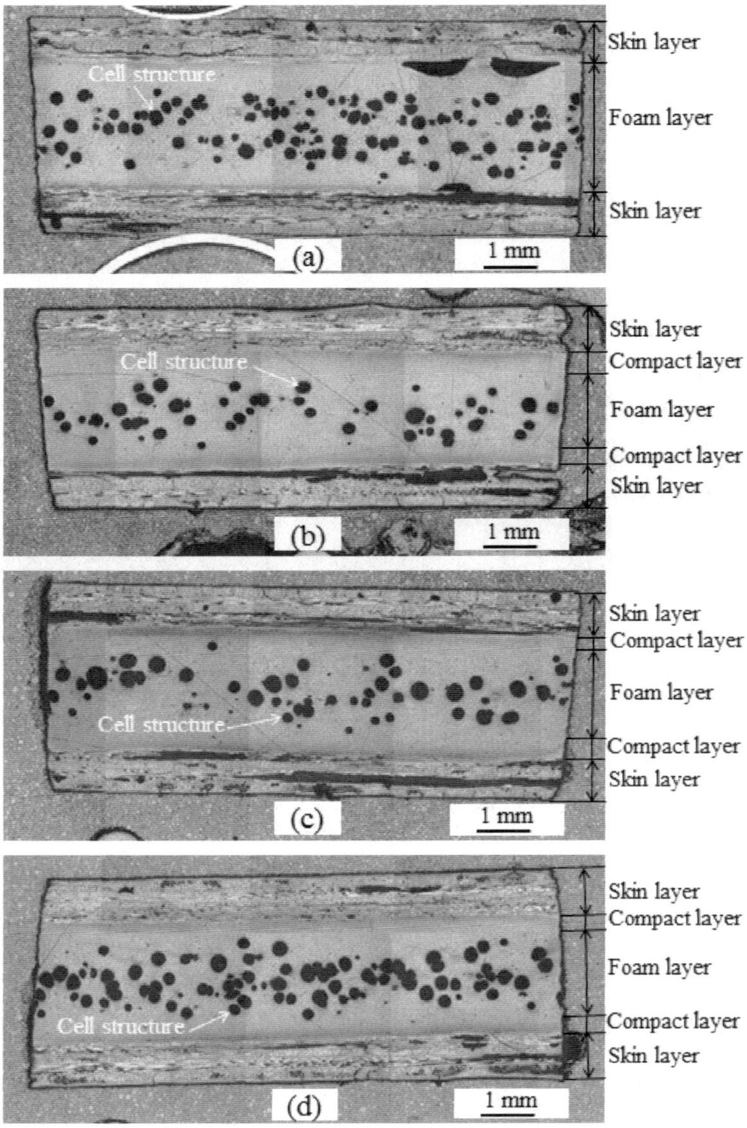


**Figure 4.** Microstructures of the sandwich structures
(a) foam core without talc (b) foam core with 1 wt% of talc
(c) foam core with 2 wt% of talc (d) foam core with 4 wt% of talc.

### *3.2. Density*

The result of the density measurement which is presented in Figure 5 reveals that the increase of talc content had slight effect on the reduction of the sandwich structure's density. SW_F_T-1 and SW_F_T-2 had a higher density than SW_F. In contrast, the density of SW_F_T-4 was lower than SW_F. The reduction of the density, when the core material of the sandwich structures was PP foam with talc, is consistent with the result of the part weight from the study of Palutkiewicz and Postawa [27].

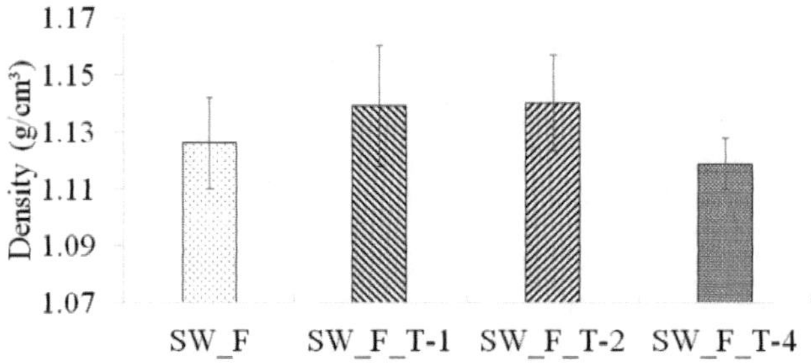


**Figure 5.** Density of the sandwich structures.

### *3.3. Bending properties*

The bending properties of the sandwich structures were demonstrated in Figure 6-8. Figure 6 indicates that the flexural modulus decreased when the sandwich structures consisted of 1 and 2 wt% of talc. However, the flexural modulus was slightly higher than SW_F when the content of talc was 4 wt%. Figure 7 demonstrates that the flexural strength significantly decreased when the content of talc in the sandwich structure was 1 wt%. Nevertheless, it increased when the talc content was 2 wt%. Moreover, the flexural strength led the higher value than SW_F if the talc content was 4 wt%. The increase of the flexural modulus and flexural strength of SW_F_T-4 was consistent with the study of Kant et al. [28] and Lapcik et al. [29]. In addition, when compared SW_F and SW_F_T-4, the compact layers which occurred in SW_F_T-4 had a positive improvement of the flexural modulus and the flexural strength. The flexural elongation at break presented in Figure 8 had the same tendency to the flexural strength. However, SW_F_T-4 could maintain the flexural elongation at break at nearly the same level as SW_F. This suggests that the presence of 4 wt% of talc in the foam core of the sandwich structures had no adverse effect from brittleness.

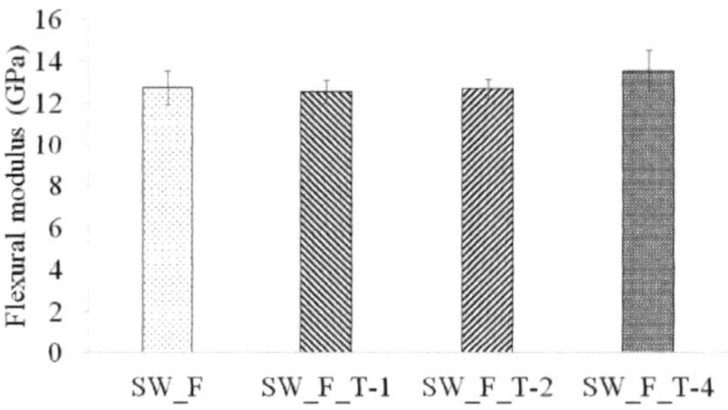


**Figure 6.** Flexural modulus of the sandwich structures.

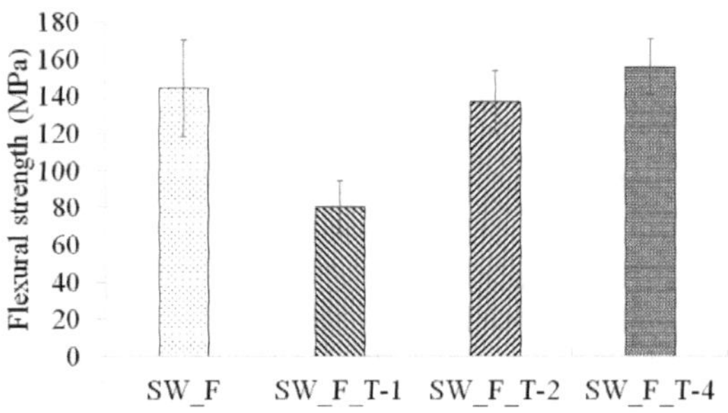


**Figure 7.** Flexural strength of the sandwich structures.

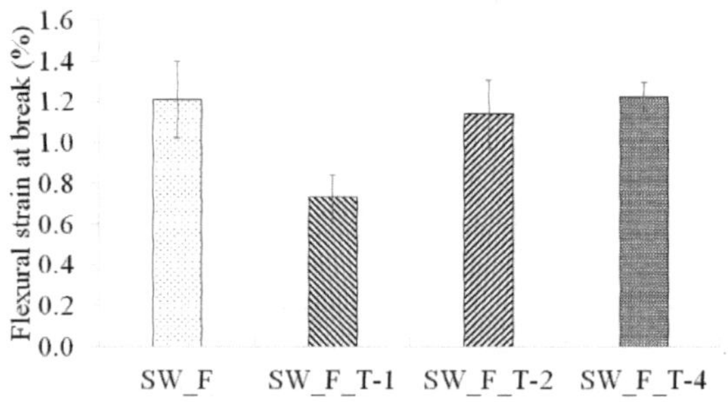


**Figure 8.** Flexural strain at break of the sandwich structures.

## 4. Conclusions

In this study, the sandwich structures composed of PP foam core, generated using CBA, and continuous glass fibre reinforced PP skin layers in the form of prepreg were investigated the influence of talc on their properties. The result of microscopy suggests that the presence of talc had no significant effect on the cell size and cell density of the cell structures in the core of the sandwich structures. However, talc affected the presence of compact layers in the sandwich structures. This may be caused by the cell growth prevention, which leads to the reduction of cell expansion under the skin layers. The slight reduction of the density of the sandwich structure occurred when the talc content was 4 wt%. In contrast, the density of SW_F_T-1 and SW_F_T-2 was higher than SW_F. In comparison between SW_F and SW_F_T-4, the slight increase of flexural modulus and flexural strength of the sandwich structures could be achieved by applying CBA and 4 wt% of talc in the core of the sandwich structure. Moreover, the flexural strain at break of SW_F_T-4 maintained nearly the same value as SW_F.

The use of the PBA to generate the cell structures is interesting to be conducted as a further study in order to compare the influence of PBA on the cell size and the cell density with that of CBA.

**Acknowledgments**

This work was supported by the German Research Foundation (DFG) within the Cluster of Excellence MERGE [Grant No. EXC 1075]. The financial support is gratefully acknowledged.

## References

[1] Nestler D, Jung H, Trautmann M, Wielage B, Wagner G, Drossel W-G, Drebenstedt C and Hipke T 2015 New sandwich structures consisting of aluminium foam and thermoplastic hybrid laminate top layers. *Mat. Sci. Forum* **825-826** 797-805

[2] Carlsson L A and Kardomateas G A 2011 *Structural and Failure Mechanics of Sandwich Composites* (Springer Science & Business Media) p 1

[3] Vinson J R 2005 Sandwich Structures: Past, Present, and Future In *Sandwich Structures 7: Advancing with Sandwich Structures and Materials*, ed O T Thomsen, E Bozhevolnaya and A Lyckegaard (Springer Netherlands) pp 3-12

[4] Corvaglia P, Passaro A, Manni O, Barone L and Maffezzoli A 2006 Recycling of PP-based Sandwich Panels with Continuous Fiber Composite Skins *J. Thermoplast. Compos. Mater.* **19** 731-745

[5] Grünewald J, Parlevliet P and Altstädt V 2017 Manufacturing of thermoplastic composite sandwich structures: A review of literature *J. Thermoplast. Compos. Mater.* **30** 437-464

[6] Mechraoui A, Riedl B and Rodrigue D 2011 Mechanical properties of polypropylene structural foams with fiber-reinforced skins *J. Cell. Plast.* **47** 115-132

[7] Menrath A, Henning F, Huber T, Roch A and Riess C 2014 Foam-injected sandwich panels with continuous-reinforced facings In Mechanical properties of polypropylene structural In *AIP Conference Proceedings* (AIP Publishing) pp 477–481

[8] Tröltzsch J, Helbig F and Kroll L 2016 Mechanical properties of polymer melt-impregnated fiber tape sandwiches using injection molding technology *J. Thermoplast. Compos. Mater.* **29** 1033-1046

[9] Maier C and Calafut T 1998 3 - Additives In *Polypropylene*, ed C Maier and T Calafut (William Andrew Publishing) pp 27-47

[10] Tolinski M 2009 Chapter 13 - Reducing Density: Polyolefin Foams In *Additives for Polyolefins*, ed M Tolinski (William Andrew Publishing) pp 195-204

[11] Kamal M R 2009 *Injection Molding: Technology and Fundamentals* (Hanser) p 282

[12] Osswald T A, Turng L S and Gramann P J 2008 *Injection Molding Handbook* (Carl Hanser Publishers)

[13] Mergenhagen T 2013 Chemical foaming agents in thermoplastics and thermosets In Blowing Agents and Foaming Processes (iSmithers Rapra Publishing)

[14] Altan M 2018 Thermoplastic Foams: Processing, Manufacturing, and Characterization In *Recent Research in Polymerization*, ed N Çankaya pp 117-137
[15] Leung S N, Wong A, Wang L C and Park C B 2012 Mechanism of extensional stress-induced cell formation in polymeric foaming processes with the presence of nucleating agents *J. Supercrit. Fluids* **63** 187-198
[16] Naguib H E, Park C B and Lee P C 2003 Effect of Talc Content on the Volume Expansion Ratio of Extruded PP Foams *J. Cell. Plast.* **39** 499–511
[17] Yetgin S, Unal H and Mimaroglu A 2014 Influence of foam agent content and talc filler on the microcellular and mechanical properties of injection molded polypropylene and talc filled *J. Cell. Plast.* **50** 563-576
[18] Reglero Ruiz J A, Vincent M, Agassant J F, Claverie A and Huck S 2015 Morphological analysis of microcellular PP produced in a core-back injection process using chemical blowing agents and gas counter pressure *Polym. Eng. Sci.* **55** 2465-2473
[19] Kim S G, Lee J W S, Park C B and Sain M 2010 Enhancing cell nucleation of thermoplastic polyolefin foam blown with nitrogen *J. Appl. Polym. Sci.* **118** 691–703
[20] Kaewmesri W, Lee P C, Park C B and Pumchusak J 2006 Effects of $CO_2$ and Talc Contents on Foaming Behavior of Recyclable High-melt-strength PP of microcellular PP *J. Cell. Plast.* **42** 405-428
[21] Wong A and Park C B 2012 The effects of extensional stresses on the foamability of polystyrene–talc composites blown with carbon dioxide *Chem. Eng. Sci.* **75** 49–62
[22] Naguib H E, Park C B, Lee P C and Xu D 2011 A Study on the Foaming Behaviors of PP Resins with Talc as Nucleating Agent *J. Polym. Eng.* **26** 565–588
[23] Guerrica-Echevarría G, Eguiazábal J I and Nazábal J 1998 Influence of molding conditions and talc content on the properties of polypropylene composites *Eur. Polym. J.* **34** 1213–1219
[24] Wang K, Bahlouli N, Addiego F, Ahzi S, Rémond Y, Ruch D and Muller R 2013 Effect of talc content on the degradation of re-extruded polypropylene/talc composites *Polym Degrad. Stab.* **98** 1275-1286
[25] Castillo L A, Barbosa S E and Capiati N J 2013 Influence of talc morphology on the mechanical properties of talc filled polypropylene *J. Polym. Res.* **20** 152
[26] Wang G, Zhao G, Dong G, Mu Y and Park C B 2018 Lightweight and strong microcellular injection molded PP/talc nanocomposite *Compos. Sci. Technol.* **168** 38-46
[27] Palutkiewicz P and Postawa P 2016 The investigation of selected properties of the porous moulded parts from talc-filled PP composites *J. Cell. Plast.* **52** 399-418
[28] Kant S, Urmila, J K and Pundir G 2013 Study of talc filled polypropylene-a concept for improving mechanical properties of polypropylene *Int. J. Res. Eng. Technol.* **2** 411-415
[29] Lapcik L, Jindrova P and Lapcikova B 2009 Effect of Talc Filler Content on Poly(Propylene) Composite Mechanical Properties In *Engineering Against Fracture* ed S Pantelakis and C Rodopoulos (Springer Netherlands) pp 73-80
[30] Mallick P K 2010 5 - Thermoplastics and thermoplastic–matrix composites for lightweight automotive structures In *Materials, Design and Manufacturing for Lightweight Vehicles* (Woodhead Publishing) pp 174-207

IOP Conf. Series: Materials Science and Engineering **480** (2019) 012013 doi:10.1088/1757-899X/480/1/012013

# Friction stir welding of glass fiber-reinforced polyamide 6: Analysis of the tensile strength and fiber length distribution of friction stir welded PA6-GF30

**S P Meyer*, B Jaeger, C Wunderling and M F Zaeh**

Institute for Machine Tools and Industrial Management (iwb), Technical University of Munich, Boltzmannstr. 15, 85748 Garching, Germany

* e-mail: stefan.meyer@iwb.mw.tum.de

**Abstract.** Friction stir welding (FSW) is a solid state welding process that is preferably used for aluminum alloys. In the study described here, FSW was applied to glass fiber-reinforced polyamide 6 (PA 6) and evaluated in terms of weld strength and fiber length distribution in the weld seam. For this purpose, the main effects on the tensile strength for friction stir welded specimens with a stationary shoulder were investigated. It could be shown that small tilt angles in combination with high contact pressures are advantageous. Using the optimum settings, a tensile strength of 50% of the base material strength could be achieved. Furthermore, an optical measuring method for large-volume fiber length measurement is introduced and evaluated. It has been demonstrated that the fibers shorten during the process. However, this has only a minor influence on the strength, while the influence of the tilt angle, the contact pressure, and the feed rate are significant.

## 1. Introduction

Friction stir welding (FSW) is a solid state joining technology and was patented in December 1991 by "The Welding Institute" (TWI) [1]. The main application is the joining of aluminum alloys, which are difficult to weld by using conventional fusion welding processes like laser beam welding [2]. Since FSW only softens and stirs the material, but does not melt it, high-strength compounds can be produced [3]. Due to this advantage, FSW is often used in sectors with a focus on lightweight constructions such as the aerospace or the automotive industry [3].

For lightweight construction plastics are also being used to an increasing extent. Fiber-reinforced plastics in particular offer the possibility to achieve high specific strengths with low densities. [4] However, joining fiber-reinforced plastics is still a challenge [5]. One problem is that during conventional welding processes (e.g. heating element welding) the fibers align parallel to the seam. This leads to reduced strength and introduces weak spot. [5–8]

Since the material flow in the process zone during FSW is different from conventional welding methods, it is assumed, that joining fiber-reinforced plastics using FSW increases the weld line strength due to an adapted reorientation of the fibers in the seam.

The aim of this work is to investigate the influence of different process and tool parameters on the tensile strength and fiber length distribution of glass fiber-reinforced polyamide 6.

## 2. State of the art

### *2.1. Conventional welding methods for plastics*

A variety of methods can be used for joining plastics. These include processes such as screwing, riveting, clipping, and adhesive bonding. However, these processes have considerable disadvantages with regard to the force propagation, which is often only possible at certain junctions. Also, additional mass is introduced through the mechanic joining elements (e.g. screws, rivets, or adhesives). These disadvantages can be avoided by welding [9]. In the following, various welding techniques for polymers will be discussed.

According to MENACHER [10], the welding processes can be classified corresponding to their heat input mechanisms. These include heat conduction (e.g. hot plate welding), convection (e.g. extrusion welding), radiation (e.g. infrared welding), friction via internal friction (e.g. ultrasonic welding), and friction by means of external friction (e.g. vibration welding and friction stir welding) [9;10] . Especially in industrial environments, hot plate welding and vibration welding are frequently employed. However, these methods have limited applicability for fiber-reinforced plastics due to a reduction of the weld strength compared to the strength of the base material [5; 7]. This relationship is defined by the welding factor *S*:

$$S = \frac{R_{m,weld}}{R_{m,base}}, \tag{1}$$

with $R_{m,weld}$ as maximum tensile strength of the welded sample and $R_{m,base}$ as maximum tensile strength of the base material.

In hot-plate-welding, the joining partners are heated by a contact surface until the joining partners plasticize in the contact area. The joining parts are then pressed together with the plasticized ends and thereby welded. The pressure applied to the joining zone causes a so-called squeeze flow, which presses the molten plastics out of the joining zone (see Figure 1). A more detailed description of the melt flow is given in section 2.2. In addition to the contact temperature *T*, the welding parameters include the contact pressure p and the dwell time *t*. [11]

With hot-plate-welding of various fiber-reinforced polymers welding factors between 0.42 and 0.83 can be achieved. For example, for a polypropylene (PP) with 25%[1] glass fiber-reinforcement, a weld factor of 0.42 was achieved [7]; with fiber reinforcements of 30%, strengths of 0.6 [12] and 0.61 [5] could be obtained. For polyamide (PA) with 30% glass fiber-reinforcement, a composite strength of 0.5 could be reached [12]. FIEBIG & SCHOEPPNER [5] achieved a weld factor of 0.83 for PA6-GF30. Here the fiber orientation within the weld seam was also examined and compared with the initial orientation of the fibers. It was found that a high amount of fibers were aligned parallel to the joining zone. FIEBIG & SCHOEPPNER [5] explained this based on the squeeze flow and the original fiber alignment.

In vibration welding, the components are moved relative to each other. In most cases one component is stationary, while the second component is set in motion with a frequency *f* and an amplitude *A* relative to it. Due to the high speed movements under a defined contact pressure, the components heat up and plasticize at the contact surface. A sudden deceleration and actuation of pressure causes the plastic material to solidify and the components to form a joint. Similar to hot-plate-welding, the squeeze flow causes material to be distributed parallel to the joining plane. [11]

For vibration welding of fiber-reinforced plastics, welding factors between 0.14 and 0.67 were achieved [5; 7]. GEHDE et al. [7] recorded a welding factor of 0.14 to 0.23 for polypropylene (PP) with 25% glass fiber reinforcement. FIEBIG & SCHOEPPNER [5] were able to obtain a welding factor of 0.56 for a PP with 30% glass fiber reinforcement and 0.67 for polyamide (PA) with 30% glass fiber-reinforcement. FIEBIG & SCHOEPPNER [5] also pointed out that the orientation of the fibers before the

[1] All weight data refer to weight percent.

IOP Conf. Series: Materials Science and Engineering **480** (2019) 012013 doi:10.1088/1757-899X/480/1/012013

welding process has a considerable influence on the strength of the joint. The paper showed, that the highest strength could be achieved for injection-molded test specimens [5]. KAGAN [13] investigated the influence of the fiber content on the relative weld strength. He showed that the welding factor for PA decreases with increasing fiber content. For unreinforced polyamide, a welding factor of almost 1 was achieved, while for PA with 45% glass fiber content the factor was only 0.39. [13]

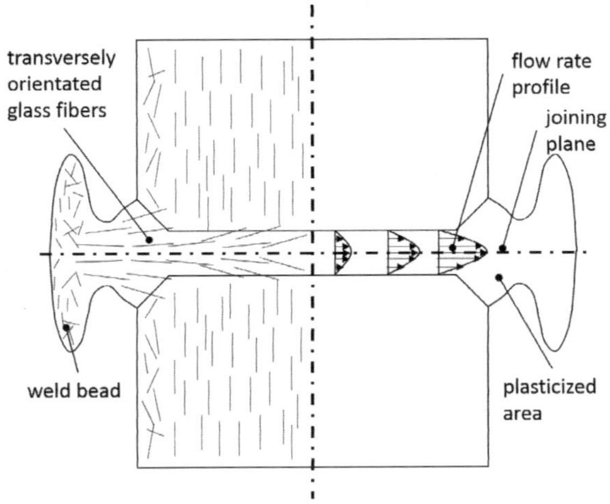


**Figure 1.** Schematic drawing of the reorientation of fibers in the weld zone due to the squeeze flow, based on [12].

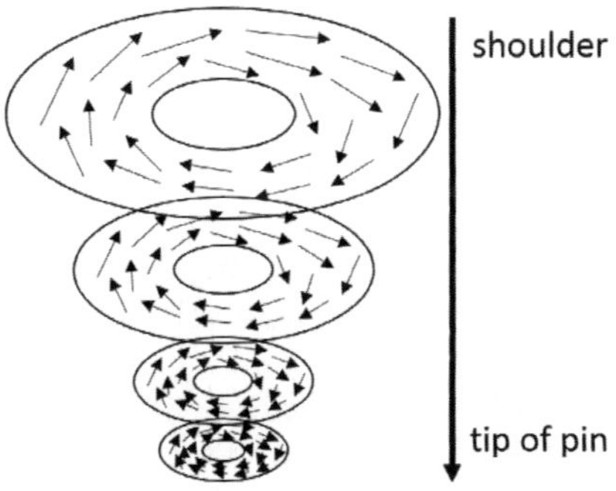


**Figure 2.** The vector field of the velocity around a probe is shown graphically for friction stir welding, based on [14].

### *2.2. Melt flow during welding*

One reason for the low welding factor of fiber-reinforced plastics compared to unreinforced plastics is the fiber orientation in the weld seam [5; 15]. As a result of the squeeze flow, the fibers are aligned perpendicular to the joining direction (see Figure 1). This results in a weld seam in which the reinforcing fibers are not aligned in the loading direction.

This means that the force is not transferred by the fibers in the welding zone. Rather, the fibers weaken the joint which leads to a reduced force transmission. [5–8; 15; 16]

Compared to the described conventional welding processes, the material flow during FSW is not perpendicular to the weld [17; 14; 18], but rather the material is transported around the pin [17].

ZHANG ET AL. [19] showed a numerical approach to describe the particle distribution in the FSW process on an Al 6061-T6 alloy plate. NANDAN ET AL. [14] depicts for friction stir spot welding (FSSW) that the material rotates around the pin, where the velocity on the relative position of the pin length (see Figure 2). SIMÕNES & RODRIGUES [20] presented the theory that the various thermo-mechanical properties of plastics lead to a differing material flow in the seam. One assumption is that the fibers in fiber-reinforced plastics are distributed differently during an FSW process compared to conventional welding techniques (e.g. hot plate welding or vibration welding). This distribution could lead to increased values of the welding factors when using FSW for joining fiber-reinforced plastics.

### *2.3. Influence of the fiber length on the tensile strength*

In addition to the fiber orientation, the fiber length has a major influence on the strength of the composite material [21; 22]. In order to take full advantage of the strength of the fibers in the reinforced polymer, the fiber length l has to be longer than the critical fiber length $l_c$:

$$l \geq l_c \tag{2}$$

The critical fiber length $l_c$ is defined by equation 3:

$$l_c = \frac{\sigma_{Fb} * d_f}{2 * \tau_b}, \tag{3}$$

with $\sigma_{Fb}$ as fiber breakage resistance, $d_f$ as fiber diameter and $\tau_b$ as shear strength of the matrix.

The critical fiber length describes the required length to transmit force through the interfaces until the complete fiber strength $\sigma_{Fb}$ is reached. Thus, fibers above the critical fiber length ensure that the complete fiber strength is exploited and in the event of an overload a fiber breakage or a combination of fiber pull-out and breakage occurs. Fibers shorter than $l_c$ cannot reach the full strength, meaning that the fiber-polymer composite does not reach the maximum possible strength. [22] During processing of discontinuous fiber-reinforced plastics, fibers are subject to thermal and mechanical stresses. These lead to fiber breakage and shortening during the process. A good overview of fiber breakage mechanisms, as well as a method for fiber length determination can be found in [21].

### *2.4. Friction stir welding for polymers*

Friction stir welding of thermoplastic materials is currently a popular topic of research and is not yet employed in industry. In the following, the most important findings regarding FSW of polymers are discussed.

STRAND [23] investigated the FSW process for unreinforced polypropylene (PP). For this purpose, a special tool with a "hot shoe" was developed, which warms the joining zone by using a heated shoulder and thus guarantees an improved heat input. In particular, the microstructure in the weld was examined and compared with that of the base material. The typical FSW onion-like rings could be found in the plastic component as well. Furthermore, it could be shown that a large pin diameter, a low feed rate, and a high heating capacity are advantageous. [23] The "hot shoe" concept introduced by STRAND [23] was also adopted and studied by other authors [24–27]. Using this method, MOSTAFAPOUR & TAFHIZAD [26] were able to join an unreinforced PA 6 with a welding factor of 0.9.

Based on these results, AYDIN [28] investigated the effect of preheated UHMWPE (ultra high molecular weight polyethylene) samples on the weld strength. It was found that the strength increases depending on the preheating temperature, with an optimum in the range of 70 – 90°C. [28]

Fundamental studies on friction stir welding of thermoplastic materials are primarily focused on polyolefins (PE and PP) [29–31]. It was shown that the process parameters rotational speed and feed rate have the greatest influence on the strength [30].

SADEGHIAN & BESHARATI [32] investigated the influence of the pin profile on the tensile strength of friction stir welded acrylonitrile butadiene styrene (ABS). However, both the conical and the cylindrical pin showed similar results.

The influence of the axial force on the weld strength of ABS was investigated on an FSW machine [33] and on a welding robot [34]. It was found that higher axial forces reduce welding defects. The feasibility of joining polymers by FSW with a robot was proven. [33, 34]

Besides polyolefins, polyamides are a frequently used group of polymers. Therefore, PANNEERSELVAM & LENIN [35] and HUSAIN ET AL. [36] investigated this group of materials. It could be demonstrated that there is an optimal rotational speed at which the specimens reach the maximum tensile strength [36]. Furthermore, [35] showed that a threaded pin reduces the weld defect susceptibility in polyamide specimens.

It has been demonstrated that a threaded conical pin also produces the best results for fiber-reinforced plastics (PP with 30% carbon fiber) [37; 38]. CZIGÁNY & KISS [39] demonstrated that the FSW process shortens the fibers in the plastic; nevertheless, fiber bridge bonds are formed. It is suspected that shorter fibers lead to the reduced strength. However, strength could increase due to the orientation of the fibers [39].

### *2.5. Fiber length measurement*

The fiber length measurement of short glass fibers is regulated in ISO 22314 [40]. The method is structured in the following steps:

1. Removal of the matrix via pyrolysis
2. Fiber distribution in a water bath by ultrasound
3. Manual measurement of (300±60) fibers with a microscope

However, this procedure is time-consuming. Therefore, GORIS ET AL. [21] developed an optical method based on a scanner solution and automatic image processing using a MATLAB algorithm. The first step, the removal of the matrix via pyrolysis, is executed according to the standard. Then, the fiber separation is performed by a fiber distribution system. For this purpose, the fibers are placed in a device (a type of funnel) and distributed over a glass plate using compressed air. The glass plate is then scanned and evaluated by a MATLAB algorithm. This method was shown to provide comparable results.

## 3. Objective and approach

The aim of the work described in this paper was to investigate the potential of FSW as a technology for joining fiber-reinforced plastics with a focus on the influence of the fiber length.

To achieve this, a process window for welding polyamide 6 with 30% glass fiber reinforcement was found. The produced specimens were tested for tensile strength and the fiber length distribution in the weld seam.

## 4. Experimental setup

### *4.1. Test rig and tools*

The experiments were conducted with an industrial robot (KUKA KR500-MT) using an FSW spindle. Furthermore, a patented DeltaN tool, i.e. a tool with a stationary shoulder, was used [41]. The FSW tool, which was used throughout the experiments, consisted of a flat shoulder with a diameter of $d_{S,1}$ = 20 mm and $d_{S,2}$ = 24 mm and a conical pin (cone angle $\delta_{Pin}$ = 10°, tip diameter $\delta_{Pin,tip}$= 4.7 mm, pin length $l_{Pin}$ = 5 mm, M8 thread).

### *4.2. Welding material and tensile test*

The tests were conducted on polyamide 6 plates with 30% glass fiber reinforcement (trade name: TECAMID 6 GF30 black) from Ensinger GmbH. The material had a thickness of 5.3 mm. Selected properties are given in Table 1.

**Table 1.** Selected mechanical and thermal properties of the examined polyamide 6 with 30% glass fibers [43].

| Mechanical properties | | | |
|---|---|---|---|
| Young´s modulus | $E$ | 5700 | MPa |
| Tensile strength | $R_m$ | 98 | MPa |
| Elongation | $\varepsilon$ | 4 | % |
| **Thermal properties** | | | |
| Glass transition temperature | $T_a$ | 49 | °C |
| Melting temperature | $T_m$ | 218 | °C |
| Specific thermal capacity | $c$ | 1.3 | J/(g*K) |
| Thermal conductivity | $\lambda$ | 0.41 | W/(K*m) |

The plastic plates were cut into 300 x 100 mm samples and welded together in butt joint configuration with a seam length of 250 mm. In accordance with DIN EN ISO 527-3 (specimen geometry 2), 20 mm wide sections were cut out and prepared for tensile testing (see Figure 3) [42]. The tensile strength $R_m$ of the produced specimens was measured with a tensile testing machine type 'Z020' from Zwick GmbH & Co. KG.

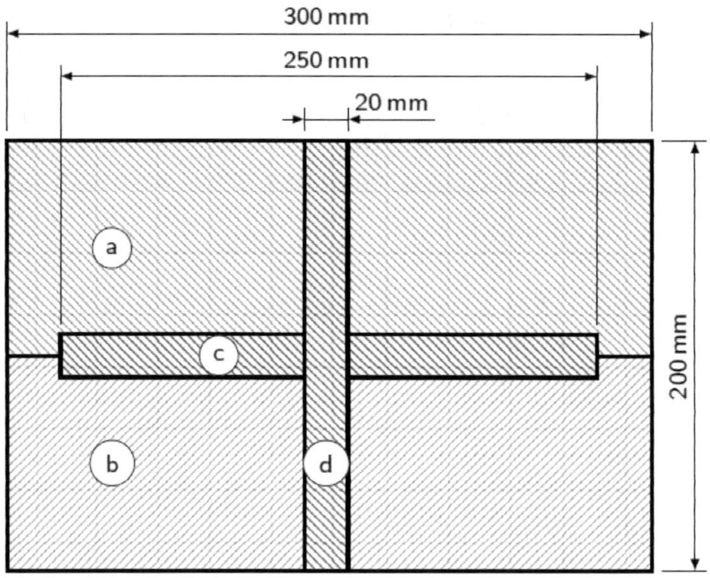


**Figure 3.** Schematic drawing of the joining partners (a and b) with indicated weld seam position (c) and tensile test specimen (d)

*4.3. Fiber length analysis*

In order to determine the fiber lengths in the weld seam, an optical measuring method was developed based on GORIS ET AL. [21] and SCHÄFER ET AL. [44] using a MATLAB algorithm. For this purpose, a 10 x 10 x 5.5 mm piece was cut out of the joining zone and ashed at (625±25) °C using a muffle furnace (model: WT-TM 18-S from WENDEL-TEC GmbH) according to DIN EN ISO 1172 [45]. The sample was held in a glow bowl and covered for the duration of the ashing (approx. 45 – 60 min) due to the risk of fire. During this process, the matrix material burned and exposed the fibers. Afterwards, the fibers were distributed on a glass plate via a fiber separation system similar to that of GORIS ET AL. [21]. For this purpose, the fibers were placed in the separation system and sputtered by compressed air. As a result, the fibers were separated and broadly distributed over the glass plate. The glass plate was then scanned in a lighting tent using a reflex camera (Canon EDS 60D) with a special close-up lens (Canon EF 50mm f/2.5 Compact Macro). To determine the magnification through the lens and to achieve the correction of the pixel length to the true length, a glass scale was placed on the glass plate during scanning. The dimensions of the high-precision glass scale were known and could therefore be used to convert the pixel length to the true length using the calibration factor $k$:

$$k = \frac{true\ length}{pixel\ length} \tag{4}$$

To evaluate the fiber length, the images taken were imported and evaluated using a MATLAB algorithm. First, the contrast of the image was increased using the *imadjust(x)* command (compare Figure 4a to Figure 4b). This command saturates the lower 1% and the upper 1% of all pixel values. This operation allows a better visibility of the fibers in the image. Next, the image was binarized using a global threshold method (ISO50% method) based on a gray value histogram (see Figure 4c). For this purpose, the gray value histogram of the image was determined and two peaks were detected due to the bimodal distribution of the values. The peak with the lower gray value was defined as background, while the peak with the higher gray value was identified as fiber. Then, the threshold value $TH_{global}$ was determined with equation 5:

$$TH_{global} = \frac{Gray\text{-}Value_{Material} + Gray\text{-}Value_{Background}}{2} \quad (5)$$

The gray scale image *g(x,y)* could then be converted with the threshold value $TH_{global}$ into a binary image *b(x,y)*:

$$b(x,y) = \begin{cases} 0 & for\ g(x,y) < TH_{global} \\ 255 & \geq TH_{global} \end{cases} \quad (6)$$

Thereafter, filter operations were applied to detect and eliminate errors such as pixel errors, agglomerations, and edge defects in the binary image. For this purpose, a filter was applied that separates and eliminates dust grains and other artifacts below a certain size. A second filter was used, which detects and deletes agglomerated objects. The third filter discerns objects that were located on the edge and thus were probably not completely in the field of view of the camera. These particles will be ignored. The last filter operation was used to detect overlapping objects. In this context, an ellipse was placed around each detected contiguous object and the aspect ratio was determined (see Figure 4d). Oblong fibers have a small aspect ratio, while crossing fibers have a relatively high aspect ratio and can therefore be detected and ignored. Through these filter operations, an image could be generated which can be analyzed based on the fiber length. In this image, contiguous objects were detected and measured using the Pythagorean approach (see Figure 4e). Then the length was determined by using the calibration factor from equation 4 and evaluated.

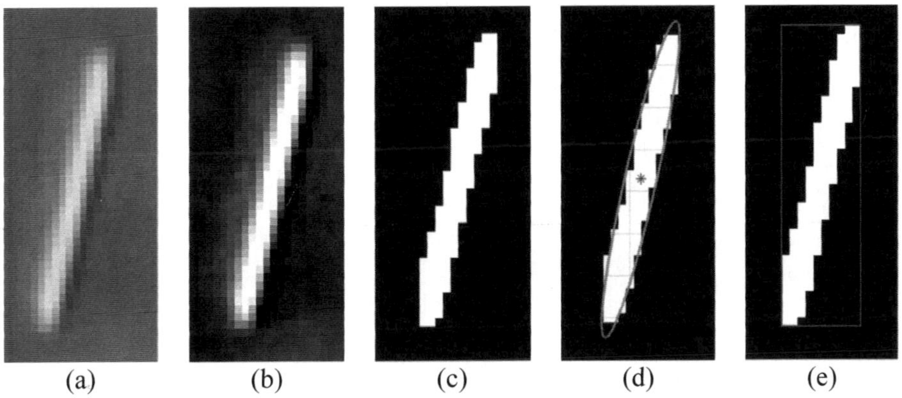


**Figure 4.** Exemplary detail display of a glass fiber during filter operations: (a) original gray scale image, (b) section after the contrast adjustment, (c) binary image, (d) ellipsis plot to determine the aspect ratio, and (e) length calculation by using the Pythagorean approach

## 5. Results and discussion

### *5.1. Process window*

As mentioned in the state of the art, the influence of the rotational speed has already been examined in detail. For this research, the rotational speed of 1000 1/min (first plasticization of the polymer) was gradually increased. At a rotational speed of 2000 1/min a uniform plasticization of the material could be observed. At the same time, the best seam surfaces appeared here. Subsequently, the rotational speed was set to 2000 1/min and only the parameters tilt angle $\alpha$, feed rate $v$, axial force $F_z$, and shoulder diameter $d_S$ were varied (see Table 2).

**Table 2.** Overview of the used parameters and results of the tensile tests including the base material (BM); the standard deviation was calculated from 5 samples.

| No. | Shoulder diameter $d_S$ in mm | Rotational speed $n$ in 1/min | Tilt angle α in ° | Downward force $F_Z$ in N | Feed rate $v$ in mm/min | Tensile strength $R_m$ in MPa | Standard deviation |
|---|---|---|---|---|---|---|---|
| 1 | 20 | 2000 | 2 | 2000 | 25 | 49.04 | 4.03 |
| 2 | 20 | 2000 | 1 | 1500 | 40 | 8.61 | 0.91 |
| 3 | 20 | 2000 | 1 | 1500 | 25 | 28.91 | 14.14 |
| 4 | 20 | 2000 | 2 | 1500 | 25 | 12.05 | 7.09 |
| 5 | 20 | 2000 | 2 | 2000 | 40 | 11.12 | 2.81 |
| 6 | 24 | 2000 | 1 | 2000 | 25 | 14.83 | 8.06 |
| 7 | 24 | 2000 | 2 | 1500 | 25 | 16.23 | 6.29 |
| 8 | 24 | 2000 | 1 | 1500 | 10 | 18.82 | 8.70 |
| 9 | 24 | 2000 | 2 | 2000 | 25 | 10.67 | 4.30 |
| 10 | 24 | 2000 | 2 | 1500 | 25 | 16.59 | 9.10 |
| 11 | 24 | 2000 | 2 | 2000 | 40 | 8.80 | 2.65 |
| 12 | 24 | 2000 | 1 | 1500 | 25 | 10.01 | 1.70 |
| BM | - | - | - | - | - | 98 | - |

In order to be able to evaluate the influence of the individual parameters, a statistical experimental design was developed. Two values were tested for each of the above mentioned parameters.

### *5.2. Analysis of the tensile strength*

Figure 5 shows the results of the tensile tests. For each value, 5 samples were evaluated. Except for samples 1 and 3, the strengths are low and the scatter of the values compared to the base material.

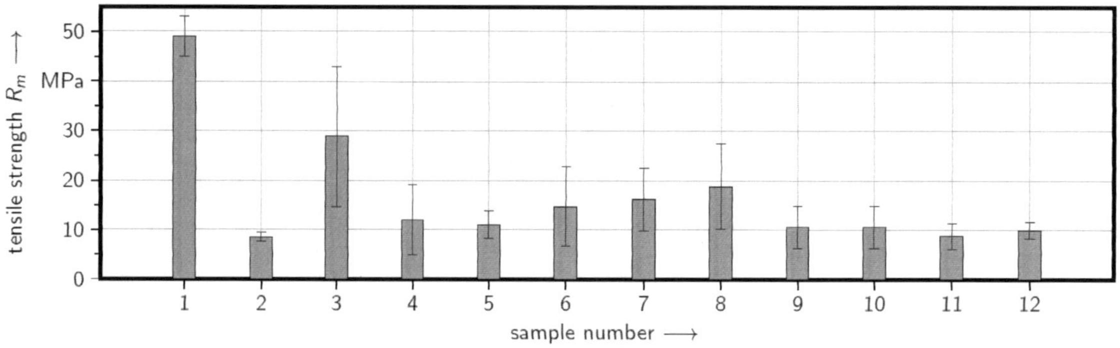


**Figure 5.** Bar chart of the results of the tensile tests and the corresponding standard deviation of 5 samples; the corresponding process parameters are listed in Table 2.

A statistical evaluation of the experiments shows that small shoulder diameters in combination with small tilt angles, low feed rates and high contact forces are advantageous (see Figure 6). Both smaller shoulder diameters and higher contact forces increase the contact pressure and thus the compression in the weld seam. Lower feed rates lead to an increased energy input and thus to a lower viscosity. This improves the dynamics of the material flow during the welding process. This effect can lead to fibers being carried along better and improve their orientation in the weld seam.

However, some effects seem to be contrary. A large shoulder diameter has advantages with a lower contact force and high feed rates.

The tilt angle is advantageous for low angles. This effect is independent of the contact force and the feed rate.

The influence of the shoulder diameter is significant. A comparison between sample 1 and sample 9 shows that only 1/5 of the strength is achieved, although only the shoulder diameter was varied.

The samples always broke on the retreating side.

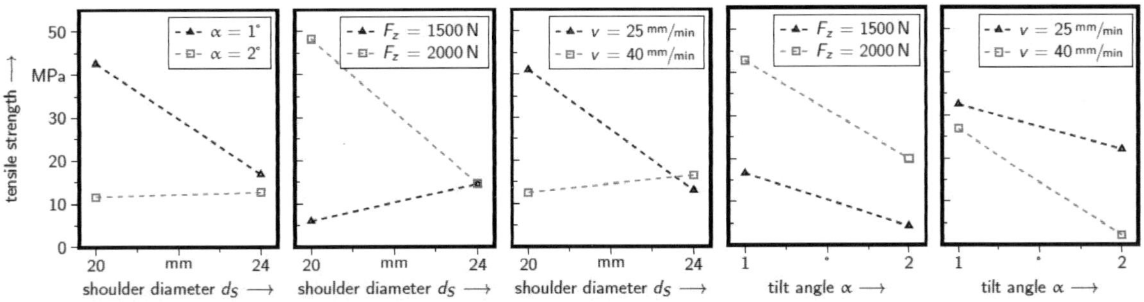


**Figure 6.** Statistical evaluation of the samples with regard to shoulder diameter and tilt angle.

### 5.3. *Analysis of the cross-sections*

The analysis of the cross-sections shows that the weld seam quality on the advancing side (AS) is significantly better (see Figure 7). On the retreating side (RS), shrinkage holes and pores were frequently formed. One possible reason for that could be the heat generation, which is lower on the RS. This means that the plasticized material cannot adhere to the base material at all or only with very limited adhesion. As a result, the joint is often discontinuous and the strength tends to be lower.

Comparing the micrographs of sample 1 (see Figure 7a) and sample 4 (see Figure 7b) shows that a reduced contact force $F_Z$ increases pore formation on the retreating side (RS). The effect of a higher feed rate is shown by a comparison from sample 1 to sample 5 (see Figure 7c). Besides a tunnel error in sample 5, the poor joint quality on the retreating side is visible. Additional pores can be seen on the advancing side, which further reduce the joint strength. By examining sample 1 and sample 9 (see Figure 7d), the influence of the larger shoulder diameter can be seen. A poor connection as well as a tunnel error is visible in sample 9.

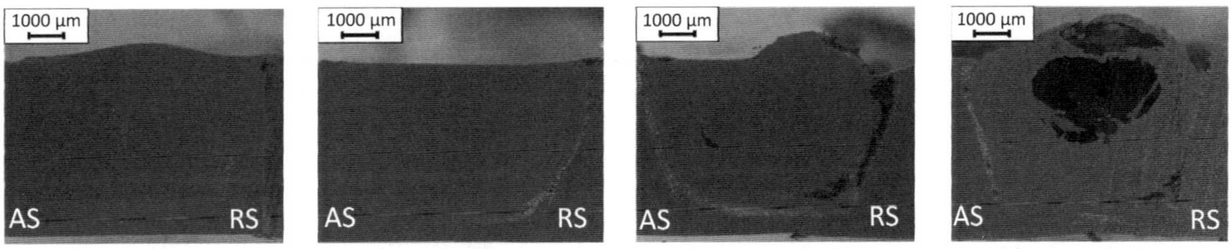


(a) sample 1 (b) sample 4 (c) sample 5 (d) sample 9

**Figure 7.** Micrographs of different samples with advancing side (AS) on the left side and retreating side (RS) on the right side of each cross section.

### 5.4. *Analysis of the fiber length*

The analysis of the fiber lengths of all samples shows that the fibers break during the process and shorten by 10 – 20% (see Figure 8). However, no correlation between the process parameters and the fiber shortening could be identified.

The distribution of the fiber lengths remains similar, compared to the base material. Both the base material and the welded samples show a normal distribution. The average fiber lengths are far below the critical fiber length. Fibers shorter than 0.3 mm are not detected by the measuring system due to the filter settings. Although this setting falsifies the measurement results, this adjustment is permissible as fibers smaller than 0.3 mm do not improve the strength.

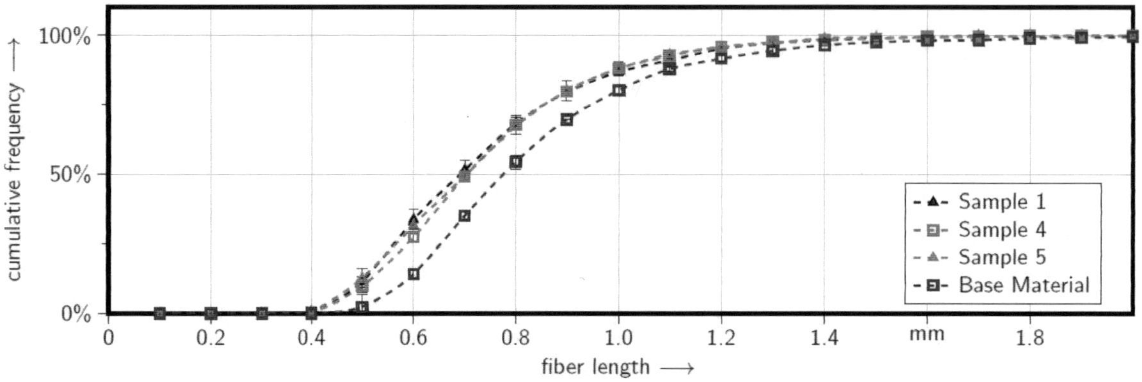


**Figure 8.** Comparison of the cumulated frequencies of the fiber lengths of different samples compared to the base material.

## 6. Conclusion and discussion

The influence of the process parameters on the joint strength of friction stir welded PA6-GF30 samples could be demonstrated in this study. It could be shown that a small tilt angle is to be preferred for FSW of plastics. Small feed rates, small shoulder diameters and high contact pressures are also advantageous. With the best parameter setting ($\alpha = 2°$, $F_z = 2000$ N, $v = 25$ mm/min and $d_S = 20$ mm) a maximum welding factor of 0.5 could be achieved. Furthermore, a method for large volume measurement of glass fibers was presented and evaluated. It could be shown that the FSW process breaks the fibers and shortens them. However, this shortening has no noticeable influence on the strength of the welded joint. This leads to the conclusion that the tensile strength is mainly influenced by the process parameters.

Thus it can be said that the process itself has to be controlled in a stable and reproducible way before the reinforcing fibres have a reinforcing effect. Furthermore, the fiber orientation in the joining zone should be investigated to check how the reinforcing fibers are deposited.

## Acknowledgments

This research and development project is funded by the German Federal Ministry of Education and Research (BMBF) within the Framework Concept "Research for Tomorrow's Production" (funding number 02P16Z010 – 02P16Z014) and managed by the Project Management Agency Karlsruhe (PTKA). The author is responsible for the contents of this publication.

## References

[1] Thomas W M, Nicholas E D, Needham J C, Murch M G, Temple-Smith P and Dawes C J Improvements relating to friction welding **WO 1993 01 09 35 A1**

[2] Mishra R S and Ma Z Y 2005 Friction stir welding and processing *Materials Science and Engineering: R: Reports* **1-2** 1–78

[3] Thomas W M and Nicholas E D 1997 Friction stir welding for the transportation industries *Materials & Design* **4-6** 269–273

[4] Lyu M-Y and Choi T G 2015 Research trends in polymer materials for use in lightweight vehicles *International Journal of Precision Engineering and Manufacturing* **1** 213–220

[5] Fiebig I and Schoeppner V 2016 Influence of the Initial Fiber Orientation on the Weld Strength in Welding of Glass Fiber Reinforced Thermoplastics *International Journal of Polymer Science* 1–16

[6] Bucknall C B, Drinkwater I C and Smith G R 1980 Hot plate welding of plastics: Factors affecting weld strength *Polymer Engineering & Science* **6** 432–440

[7] Gehde M, Giese M and Ehrenstein G W 1997 Welding of thermoplastics reinforced with random glass mat *Polymer Engineering & Science* **4** 702–714

[8] Potente H, Natrop J, Pedersen T K and Uebbing M 1993 Comparative Investigations into the Welding of Glass-Fiber-Reinforced PES *Journal of Thermoplastic Composite Materials* **2** 147–159
[9] Stokes V K 1989 Joining methods for plastics and plastic composites: An overview *Polymer Engineering & Science* **19** 1310–1324
[10] Menacher M Vibrationsschweißen von strahlenvernetztem Polyamid 66
[11] Baur E, Brinkmann S, Osswald T A, Rudolph N and Schmachtenberg E 2013 Saechtling Kunststoff Taschenbuch
[12] Bruessel A 1999 Fertigungstechnische und werkstoffspezifische Aspekte zum Fügen von Thermoplasten mittels Heizelement
[13] Kagan V 2003 Optimized Mechanical Performance of Welded and Molded Butt Joints: Part I — Similarities and Differences *Journal of Reinforced Plastics and Composites* **9** 773–784
[14] Nandan R, Roy G G and Debroy T 2006 Numerical simulation of three-dimensional heat transfer and plastic flow during friction stir welding *Metallurgical and Materials Transactions A* **4** 1247–1259
[15] Kamal M R, Chung Y-M and Gomez R 2008 Three-dimensional fiber orientation in vibration welded joints of glass fiber reinforced polyamide-6 *Polymer Composites* **9** 954–963
[16] Fiebig I and Schoeppner V 2018 Factors influencing the fiber orientation in welding of fiber-reinforced thermoplastics *Welding in the World* **8** 751
[17] Seidel T U and Reynolds A P 2001 Visualization of the material flow in AA2195 friction-stir welds using a marker insert technique *Metallurgical and Materials Transactions A* **11** 2879–2884
[18] Kumar K and Kailas S V 2008 The role of friction stir welding tool on material flow and weld formation *Materials Science and Engineering: A* **1-2** 367–374
[19] Zhang H W, Zhang Z and Chen J T 2005 The finite element simulation of the friction stir welding process *Materials Science and Engineering: A* **1-2** 340–348
[20] Simões F and Rodrigues D M 2014 Material flow and thermo-mechanical conditions during Friction Stir Welding of polymers: Literature review, experimental results and empirical analysis *Materials & Design* 344–351
[21] Goris S, Back T, Yanev A, Brands D, Drummer D and Osswald T A 2017 A novel fiber length measurement technique for discontinuous fiber-reinforced composites: A comparative study with existing methods *Polymer Composites* **6**
[22] Fu S 1996 Effects of fiber length and fiber orientation distributions on the tensile strength of short-fiber-reinforced polymers *Composites Science and Technology* **10** 1179–1190
[23] Strand S Jan. 2003 Joining plastics - can friction stir welding compete? *Electrical Insulation Conference and Electrical Manufacturing and Coil Winding Conference* 321–326
[24] Bagheri A, Azdast T and Doniavi A 2013 An experimental study on mechanical properties of friction stir welded ABS sheets *Materials & Design* 402–409
[25] Mostafapour A 2012 A study on the role of processing parameters in joining polyethylene sheets via heat assisted friction stir welding: Investigating microstructure, tensile and flexural properties *International Journal of the Physical Sciences* **4**
[26] Mostafapour A and Taghizad Asad F 2016 Investigations on joining of Nylon 6 plates via novel method of heat assisted friction stir welding to find the optimum process parameters *Science and Technology of Welding and Joining* **8** 660–669
[27] Azarsa E and Mostafapour A 2014 Experimental investigation on flexural behavior of friction stir welded high density polyethylene sheets *Journal of Manufacturing Processes* **1** 149–155
[28] Aydin M 2010 Effects of Welding Parameters and Pre-Heating on the Friction Stir Welding of UHMW-Polyethylene *Polymer-Plastics Technology and Engineering* **6** 595–601
[29] Saeedy S and Givi M K B 2011 Investigation of the effects of critical process parameters of friction stir welding of polyethylene *Proceedings of the Institution of Mechanical Engineers, Part B: Journal of Engineering Manufacture* **8** 1305–1310

[30] Bozkurt Y 2012 The optimization of friction stir welding process parameters to achieve maximum tensile strength in polyethylene sheets *Materials & Design* 440–445
[31] Gao J, Shen Y, Zhang J and Xu H 2014 Submerged friction stir weld of polyethylene sheets *Journal of Applied Polymer Science* **22** n/a-n/a
[32] Sadeghian N and Besharati Givi M K 2015 Experimental optimization of the mechanical properties of friction stir welded Acrylonitrile Butadiene Styrene sheets *Materials & Design* 145–153
[33] Mendes N, Loureiro A, Martins C, Neto P and Pires J N 2014 Effect of friction stir welding parameters on morphology and strength of acrylonitrile butadiene styrene plate welds *Materials & Design* 457–464
[34] Mendes N, Loureiro A, Martins C, Neto P and Pires J N 2014 Morphology and strength of acrylonitrile butadiene styrene welds performed by robotic friction stir welding *Materials & Design* 81–90
[35] Panneerselvam K and Lenin K 2014 Joining of Nylon 6 plate by friction stir welding process using threaded pin profile *Materials & Design* 302–307
[36] Husain I M, Salim R K, Azdast T, Hasanifard S, Shishavan S S and Eungkee Lee R 2015 Mechanical properties of friction-stir-welded polyamide sheets *International Journal of Mechanical and Materials Engineering* **1** 241
[37] Ahmadi H, Arab N B M, Ghasemi F A and Farsani R E 2012 Influence of Pin Profile on Quality of Friction Stir Lap Welds in Carbon Fiber Reinforced Polypropylene Composite *International Journal of Mechanics and Applications* **3** 24–28
[38] Ahmadi H, Mostafa Arab N B and Ghasemi F A 2014 Optimization of process parameters for friction stir lap welding of carbon fibre reinforced thermoplastic composites by Taguchi method *Journal of Mechanical Science and Technology* **1** 279–284
[39] Czigány T and Kiss Z Friction stir welding of fiber reinforced polymer composites *Proceedings of the 18th International Conference on Composite Materials*
[40] Deutsches Institut für Normung e. V. DIN EN ISO **DIN EN ISO 527-3** Plastics - Determination of tensile properties
[41] Silvanus J, Forster E and Tauscher E Schweißwerkzeug zum Verbinden von wenigstens zwei Werkstücken, Schweißverfahren und Werkstück **DE 10 2011 106 506 A1**
[42] Deutsches Institut für Normung e. V. DIN EN ISO **DIN EN ISO 1172** Prepregs, Formmassen und Laminate
[43] Ensinger Ltd 20.02.2018 TECAMID 6 GF30 black - Stock Shapes
[44] Schäfer C, Meyer S P and Osswald T A 2018 A novel extrusion process for the production of polymer micropellets *Polymer Engineering & Science* 1391
[45] ISO Internationale Organisation für Normung. ISO **22314** Plastics -- Glass-fibre-reinforced products -- Determination of fibre length

IOP Conf. Series: Materials Science and Engineering **480** (2019) 012014 doi:10.1088/1757-899X/480/1/012014

# Concepts for interface engineering and characterization in composite hybrid structures

**J Serna[1], C Zinn[2], I Scharf [3], A Dittes[3,*], S-D Schwoebel[3], C Schmidt[1], D Meiners[1], M Schaper[2] and T Lampke[3]**

[1] Department of Polymer Materials and Plastic Engineering (PuK), Clausthal University of Technology, Clausthal-Zellerfeld, Germany
[2] Chair of Materials Science (LWK), Paderborn University, Paderborn, Germany
[3] Chair of Materials and Surface Engineering, Chemnitz University of Technology, Chemnitz, Germany

[*] e-mail: axel.dittes@mb.tu-chemnitz.de

**Abstract.** Multi-material design is a key technology for lightweight design and calls for combining different materials such as steels, light metals and composites like carbon fiber reinforced plastics (CFRP) into load-bearing structures. Due to the minor chemical affinity of such materials and contact corrosion aspects, the design of an appropriate interface is a current challenge. Aiming on highly performant single-stage manufacturing processes, the paper presents investigations on several surface treatment techniques like a laser pretreatment, the use of primers and a sol-gel coating. Beside investigations on the achievable joint strengths, the corresponding joining concepts to facilitate the in-mould assembly of hybrid structures are validated. In this case the intrinsic manufacturing methods of the Resin Transfer Moulding (RTM) and the Automated Fiber Placement (AFP) technology are presented. Furthermore corrosion protection measures for CFRP and aluminum are analyzed via immersion and salt spray tests.

## 1. Introduction

The use of lightweight structures offers nowadays the possibility to achieve a significant reduction of weight. In this context, a complete substitution of a monolithic material by one another is in general not adequate for the consistent use of the potential of lightweight structures [1, 2]. In contrast, the multi material design approach considers the entire structure and is based on a hybrid material combination. The concept of the hybridization of individual components into an integral structure gains therefore more importance and can basically be realized through two different methods. The utilization of consecutive manufacturing processes (post-mould assembly) such as adhesive bonding or screwing and riveting does not exploit the maximum lightweight potential either [1]. As an alternative concept, the hybridization within a single-stage process is most promising.

However, the utilization of intrinsic hybrids needs to meet essential requirements such as resource-efficient production, characterization and design of load-optimized hybrid components. In this context, suitable methods for the design of the hybrid interfaces are of particular interest. The paper presents different approaches to design and characterize the interface of intrinsic hybrids made out of the following material combinations:

- CFRP with duroplastic matrix and aluminum

- CFRP with duroplastic matrix and steel
- CFRP with thermoplastic matrix and aluminum

The dissimilar material combinations show varying coefficients of thermal expansion and different electrochemical potentials as well as chemically and mechanically strongly differing properties. Thus, adhesion and corrosion aspects need to be taken into account. Apart from the adhesion qualities of the resin, the surface pre-treatment is one of the most important steps to achieve a satisfying joint strength [3][4]. In terms of improving the adhesion on steel, various surface pre-treatment techniques, *e. g.* etching, anodizing, plasma spraying or blasting, were investigated in the recent years (cf. [5][6]). Several investigations, especially on titanium alloys, revealed that a laser pre-treatment of the metal surface represents a fast, easy and cost-efficient way to enhance the adhesion [7][8][9]. The laser pre-treatment leads to a nano structured and open-porous surface as well as fine nano cavities, which enable a mechanical interlocking between the metal surface and the resin. This ensures a satisfying bond strength [3].

Sol-Gel coatings provide excellent corrosion protection of aluminum and steel substrates. However, the coatings can additionally be modified to achieve chemical compatibilization in metal-FRP composites [10]. In this way, an interface concept combining corrosion protection and high bond strengths can be utilized.

Further, surface treatments already existing on the market, like anodizing, as well as the utilization of priming processes will be investigated in this paper for their suitability for single stage manufacturing of intrinsic hybrids. The characterization of the interface is conducted based on shear, shear-lap and double cantilever beam (DCB) tests. In addition, different corrosion protection measures are applied accounting to the increased corrosion potential of CFRP and aluminum due to the high electrochemical potential difference between the two counterparts.

## 2. Surface treatments for hybrid structures

### *2.1. Sol-Gel process generating adhesive bonds*

The concept of the interfacial design through a sol-gel intermediate layer particularly addresses an improved adhesion between the CFRP and the metal. For this purpose, the chemical composition of the sol-gel is adapted towards the selected polymeric matrix (polyamide 6) and the sol is applied onto the metal substrate (aluminum) via dip coating [10]. The properties of the sol-gel are adaptable by using a hybrid sol [11], which is a mixture of different organically modified silicates (ORMOSIL). This allows the change of the wetting behavior of the interface due to the selected chemical composition. The effect of polymerizing the sol into gel is shown in Figure 1.

Furthermore, the silicate layer acts as a dense physical barrier and thereby decouples the corrosion potentials of the metal and the CFRP. Thereby, contact corrosion between the noble CFRP (redox potential of +0.74 V) and less noble aluminum (redox potential of -1.67 V) is avoided. Hence, the challenge is to develop a formulation for a sol-gel coating on aluminum, which resists the forming and joining process and therefore can be applied before the hybridization step.

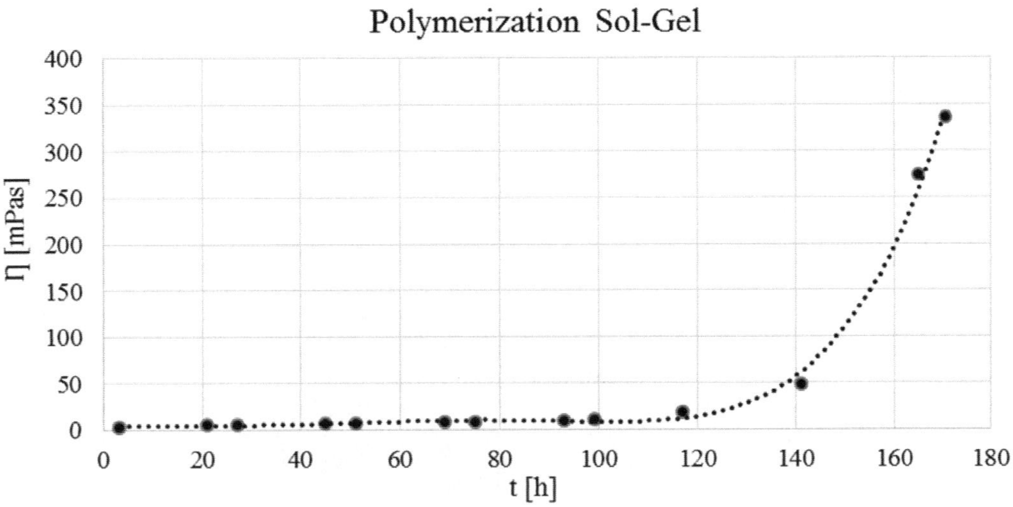


**Figure 1.** Polymerization of silicates – the mixture was applied after 3 days, the gel has a working life up to 5 days (20 °C, and SV-10).

The wettability of the fully crosslinked sol-gel coating can be further improved by an additional subsequent functionalization step in order to achieve an increased intermolecular interaction towards the polymer. For example, alkylated silicates are used to modify the surface in the desired manner [12]. In this way, the sol-gel layer gets a very oleophilic character promoting the connectivity towards multiple thermoplastic polymers. However, the challenge within the production of a suitable coating system is to achieve a maximum adhesion in combination of the coating and functionalization methods as well as the validation of the mechanical properties.

In order to characterize the mechanical adhesion properties of composites with different coating treatments, simple quasi-static shear tests were carried out. The composite sample geometry corresponds to the standard for the determination of the tensile lap-shear strength of bonded assemblies according to DIN EN 1465. The joining area was set to 25 x 12.5 mm using a force of 79 N during the melting time of 10 s. The shear tests were performed at a velocity of 1 mm/min on a Zwick universal tensile testing machine with a maximal nominal force of 20 kN.

Figure 2 summarizes the experimental results of the shear tests; the shear strength data are given for three different surface conditions. The untreated sol-gel coating can be taken as a reference. These results show that, on average, an increased shear strength occurs in the composite sample with an ORMOSIL coating treated with Trimethylchloro silane (Me3-Si-), whereas the shear strength of ORMOSIL treated with Octadecyltrimethoxy silane (n-C18-Si-) is hardly changed compared to the reference sample. Highest joint strengths are observed for ORMOSIL treated with Aminopropyltrimethoxy silane ($NH_2$-Si-). We note, that the shear strength data are considerably more scattered compared to the reference material. The interactions of the alkylated surfaces are based on van der Waals interactions. However, the surface of the octadecyl substituted sol-gel is less polar than the methyl substituted one. Thus, the interaction of the octadecyl towards the polar polyamide is less attractive and accordingly the joint strength is lower. The Amino functionalized surface interacts by hydrogen bonding with the polyamide. Thus stronger bonds are formed and a higher shear strength is achieved.

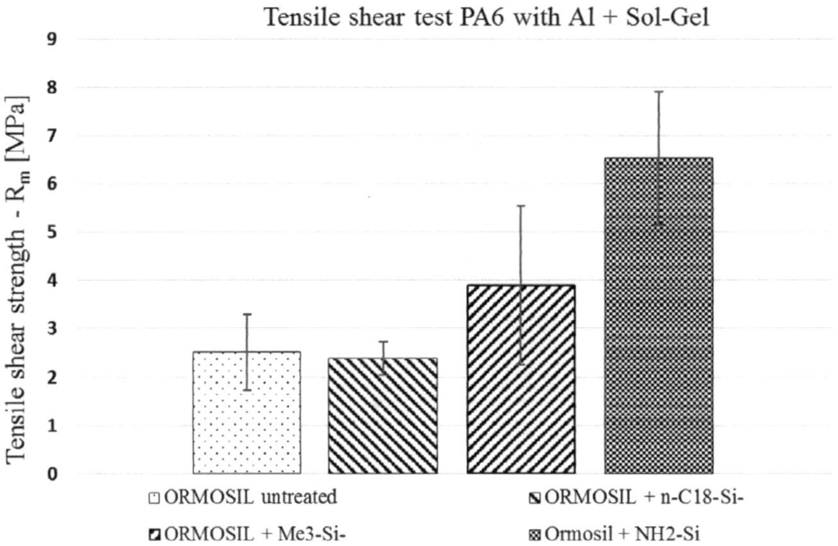


**Figure 2.** Overview of shear strengths of aluminum EN AW-6082 / PA6 composites in dependence of the coating condition. Sample size: Al 100x25x3 mm, PA6 70x25x4 mm, joining area 12.5 mm.

*2.2. Laser induced surface nano structuring for hybrid structures*

Beside the presented approach of improving adhesion through chemical bonding via a sol-gel coating process, a fast and efficient laser surface pre-treatment is used to improve the adhesion properties of hybrid structures based on a micro-alloyed steel (1.0548) as well as an aluminum alloy (EN AW-6082, 3.2315) and CFRP. As the laser-induced generated oxide layer cannot completely guarantee protection against (contact) corrosion, an additional glass fleece is utilized as an interfacial layer between the metal and the CFRP component. The intrinsic manufacturing of the hybrid structures is realized by using a modified Vacuum Assisted Resin Transfer Molding (VARTM) process. The matrix material used is a fast curing epoxy resin system (EPIKOTE 05475/ EPIKURE 5443, Momentive). Therefore, the manufacturing of a complex hybrid structural part is feasible in around 5 minutes.

**Surface pre-treatment**

Before starting the VARTM process, the steel and aluminum surfaces are completely pre-treated by laser irradiation. Therefore, a short pulse laser system (Powerline E Air 25, Rofin-Sinar, Nd:YVO4) with a wavelength of 1064 nm and a maximum power of 25 W is utilized. While structuring, the laser system operates in the pulsed mode and the laser beam is vertically positioned to and straightly focused onto the metal surface. Satisfying parameters, that means parameters which lead to a fine nano-structured and open-porous surface with various undercuts as well as with a good wettability and infiltration behavior, were determined by using scanning electron microscopy, atomic force microscopy, contact angle measurements and Focused Ion Beam investigations of fractured surfaces of shear tension specimens [13][14][15].

In this paper, the parameters leading to the best bond strength and corrosion behavior of steel-CFRP or, respectively, aluminum-CFRP specimens are shown. Constant parameters were the scan speed (800 mm/s), the hatch-distance (80 μm) and the repetition rate (1 time). The respective surface morphologies and the variable parameters are shown in Table 1.

**Table 1.** Parameter settings used for the laser pre-treatment of the steel and aluminum interfaces (left side) and resulting surface morphologies (right side).

| | Parameter settings | | Surface morphology |
| --- | --- | --- | --- |
| | Current in A | Pulse frequency in kHz | |
| Steel | 34 | 10 | 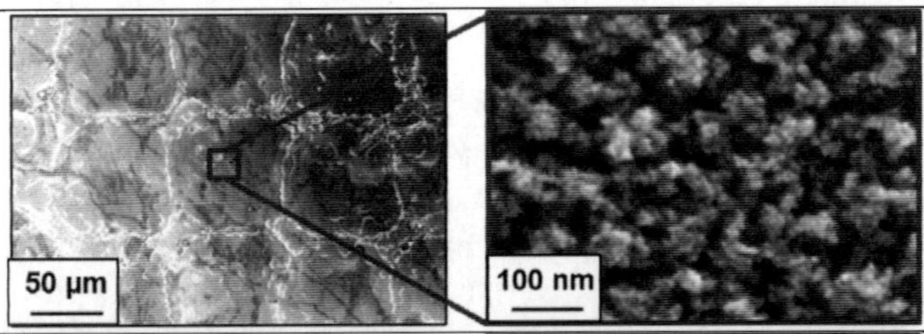  |
| Aluminum | 34 | 30 | 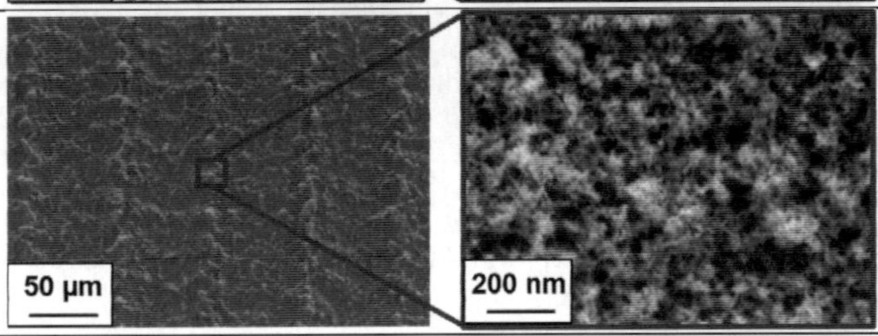  |

**VARTM manufacturing process**

The manufacturing process starts with the insertion of a metal sheet (steel or aluminum) into the tool cavity. Afterwards, six unidirectional carbon fiber textiles (HPT 320 C0, SGL) are placed, with respect to an alternating fiber orientation (0/90) °. In order to protect the metal component against (contact) corrosion, a 0.2 mm thin glass fiber fleece (JM SH35, Saertex) is inserted between the metal and the carbon fiber textiles. After fabricating a preform, closing and evacuating the tool cavity, which is pre-heated to 80 °C, the hybridization is started. The resin system is injected through four sprues in the upper tool cavity at a constant pressure of 0.5 MPa and for an injection time of 3 min by using a RTM unit (iJect touch 2.0, Wolfangel). The resin saturates the fiber preform completely and generates a firmly bonded connection between the CFRP component and the metal surface while the curing process. In case of the laser pre-treated metal surfaces, the fine nano cavities are filled with the low viscous resin so that an additional mechanical interlocking between the nano-structured metal surface and the resin occurs in the curing process. To ensure a non-porous hybrid component, the excessive resin, escaping from the riser, flows into a so named resin trap which is connected to a vacuum pump. The completely finished and cured hybrid structures is removed from the tool cavity after a curing time of 42 min. The entire manufacturing process of the hybrid structure is shown in Figure 3.

IOP Conf. Series: Materials Science and Engineering **480** (2019) 012014 doi:10.1088/1757-899X/480/1/012014

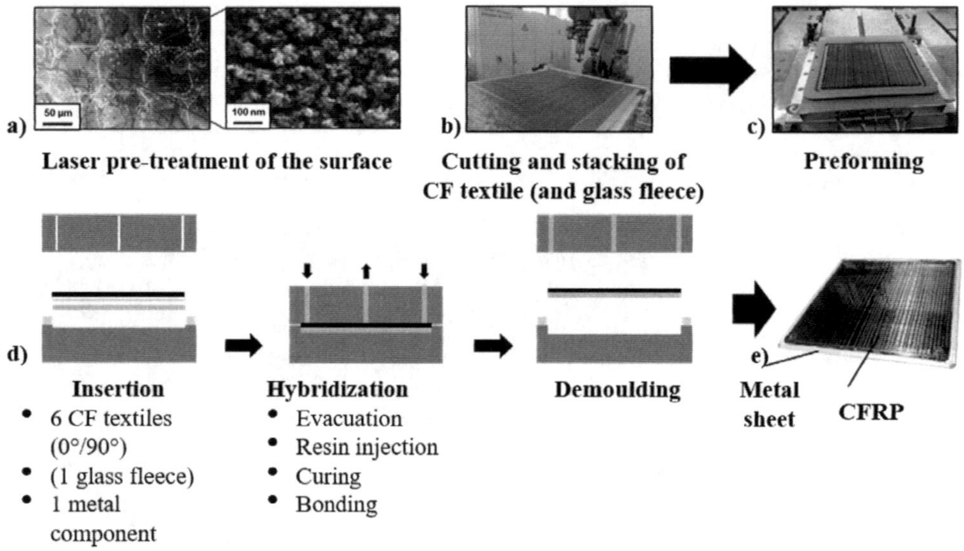


**Figure 3.** Intrinsic manufacturing process: a) laser pre-treatment, b) cutting and stacking of textiles, c) preforming, d) VARTM-process and e) finished hybrid structure.

**Mechanical and ageing properties**

After the intrinsic manufacturing process, shear tension specimens were cut out from the hybrid plates by waterjet cutting. Thereby, for each type (non-structured or laser pre-treated surface, with or without a glass fleece as an intermediate layer) and five specimens were cut out.

A part of the specimens was tested after a conditioning in normal climate (DIN EN ISO 291) and the other part was tested after an immersion test in a 5 % sodium chloride (NaCl) solution (DIN EN ISO 53287). The influence of the electrolyte on the specimens was tested weekly and, in total, for four weeks. All shear tests were carried out in accordance to the German standard DIN 65148 and by using a MTS 810 table top test system.

Figure 4 shows the results of the shear tests on the steel-CFRP specimens in dependency of the surface condition (non-structured reference (REF) or laser pre-treated), the insertion of a glass fleece (GF) and the condition (in the as received condition or after an ageing in an immersion test in a 5 % sodium chloride solution for 1, 2, 3 and 4 weeks). It can be seen that a laser pre-treatment of the steel surface has a beneficial influence on the shear strength.

In the as received condition, the shear stresses of the laser pre-treated specimens are at maximum four times higher than the shear stresses of the non-structured reference specimens. Furthermore, the insertion of a glass fleece has a positive effect on the shear stress. The non-structured reference specimens without a glass fleece reveal shear stresses around 8 MPa and the specimens with a glass fleece reach shear stresses around 14 MPa. It is assumed that the glass fleece lead to a homogenous interface thickness such that the applied stresses can be absorbed better by the interface. In addition, the glass fleece stabilizes, *i. e.* reinforces, the interface. The laser pre-treatment lead to shear stresses around 37 MPa (without a glass fleece) and 39 MPa (with a glass fleece). This improvement can be explained by the strong interlocking that occurs between the nano-structured steel surface and the resin during the infiltration and curing step of the VARTM procedure.

Regarding the development of the shear stresses while the four-week immersion test, it can be seen that the shear stresses of the non-structured reference specimens decrease with the duration of the test. In this case, the adhesion of the resin on the steel surface is as bad that the test solution can penetrate the interface and provoke crevice corrosion and, if the fibers of the CFRP component can get in contact with the steel, that means without a glass fleece in the interface, lead to contact corrosion. By using a glass fleece, the decrease of the shear stresses is extremely high, because the glass fleece absorbs the test solution penetrating the interface, whereby the shear strength is significantly reduced.

If a laser pre-treatment of the steel surface is performed, the shear stresses are constant throughout the entire test period of the immersion test. The firm interlocking between the laser-induced nano structure and the resin lead to a durably strong bonding strength. Higher shear stresses obtained after 2, 3 and 4 weeks of the immersion test can be explained by the specimens' position on the hybrid test plate, from which they were cut out. Close to the injection points, the laminate quality and therefore the interface quality is slightly better than in direct proximity to the riser.

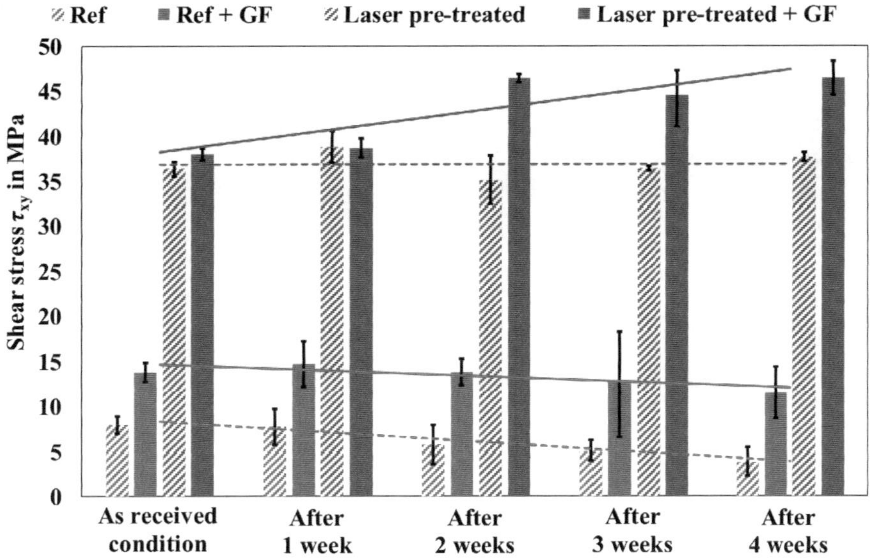


**Figure 4.** Shear strength of the steel-CFRP specimens in dependency of the surface condition, non-structured reference (REF) or laser pre-treated, the insertion of a glass fleece (GF) and the ageing condition, in the as received condition or after an immersion test in a 5 % sodium chloride solution for 1, 2, 3 and 4 weeks

Figure 5 presents the results for the aluminum-CFRP specimens. At first, it should be noted that no adhesion could be achieved for all non-structured aluminum-CFRP specimens. These specimens fail directly after the cutting process. By considering this information and the lower stiffness of the aluminum, which leads to a more flat force-path curve during the shear test, whereby the specimens fail at nearly the same path level than the steel-CFRP specimens, but naturally at lower maximum forces, the shear stresses revealed by a laser pre-treatment of the aluminum interface are very satisfying. In the as received condition, the specimens reach shear stresses around 30 MPa, without a glass fleece, and 26 MPa, with a glass fleece. It is assumed that the insertion of a glass fleece has no influence on the bonding strength and that the variation is caused by diversification resulting from the specimens' position on the hybrid test plate (see above).

As already seen in the results of the steel-CFRP specimens, the exposure of the aluminum-CFRP specimens to the 5 % Na Cl solution has no influence on the shear strength. The shear stresses are constant throughout the entire test period. The laser pre-treatment lead to a firm interlocking connection between the laser-induced nano structure and the resin. Furthermore, a defect-free interface area is generated.

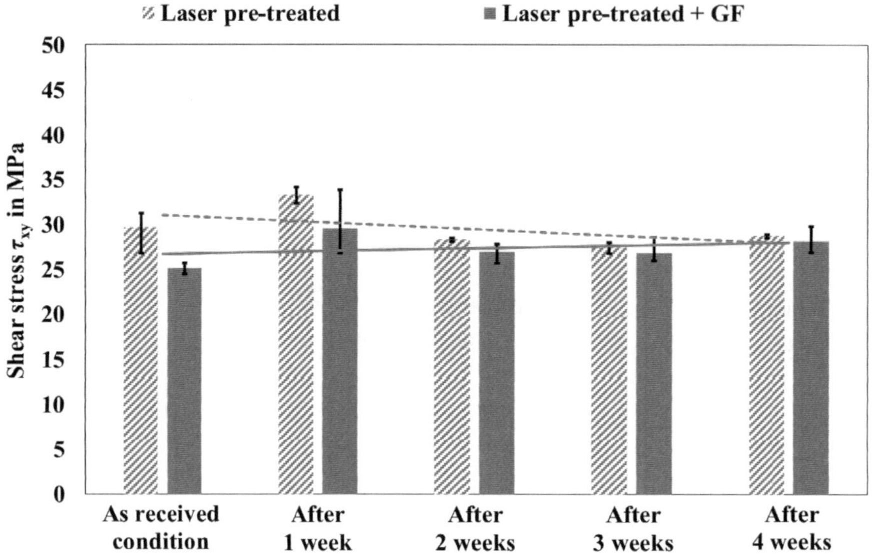


**Figure 5.** Shear strength of the aluminum-CFRP specimens in dependency of the surface condition, non-structured reference (REF) or laser pre-treated, the insertion of a glass fleece (GF) and the ageing condition, in the as received condition or after an immersion test in a 5 % sodium chloride solution for 1, 2, 3 and 4 weeks.

**Fracture and ageing behavior**

Regarding the fracture pattern of laser pre-treated steel- and aluminum-CFRP shear-tension specimens after the four-week immersion test, for both material combinations, without and with a glass fleece, no corrosion products can be detected. The failure behavior is cohesive/ interlaminar and the laser-induced nano structure, shining through the resin layer or the glass fleece, respectively, is intact. The golden-brown color of fractured surface of the laser pre-treated steel specimens is no indicator for rust, but the typical appearance of the laser-induced generated iron oxide layer. The typical color of an aluminum oxide layer is white. Therefore, it can be noted that the interface is not penetrated by the test solution. The interface area is completely intact and contact corrosion does not occur – neither for the specimens with an isolating glass fleece, nor for the specimens without a glass fleece, *i. e.* without an additional corrosion protection.

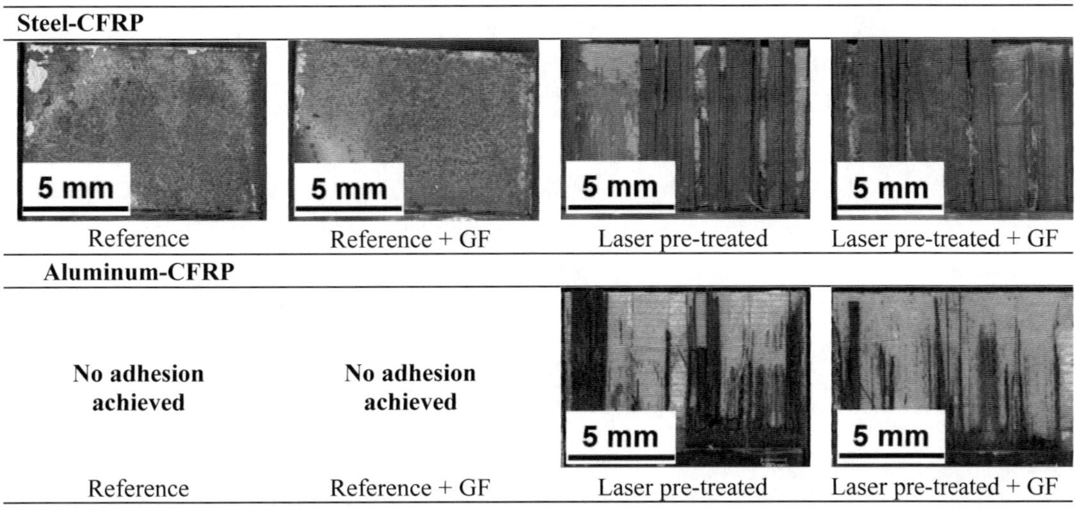


**Figure 6.** Fractured surfaces of the steel-CFRP and the aluminum-CFRP shear-tension specimens after a four-week immersion test.

A side-view of the interface area of small steel- and aluminum-CFRP plates after the four-week immersion test by using a digital microscope (Figure 7) shows that it is still recommended to utilize a corrosion protecting glass fleece as an intermediate layer. Especially the aluminum-CFRP specimens locally reveal more pitting corrosion, which is typical for aluminum due to its porous natural oxide layer [16], if no glass fleece is used.

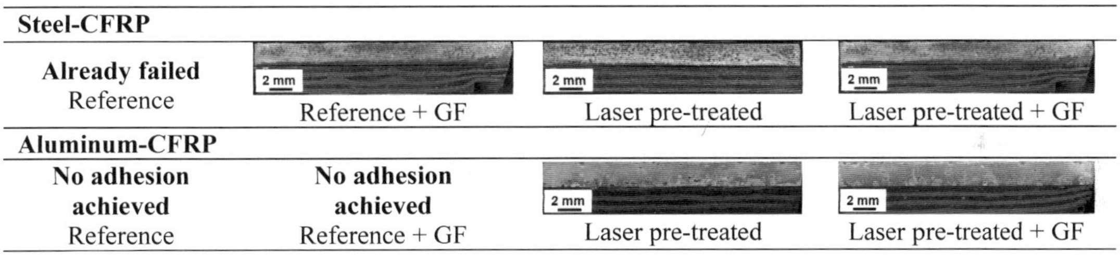


**Figure 7.** Side-view of the interface area of the steel- and aluminum-CFRP specimens after the four-week immersion test by using a digital microscope.

Confocal laser scanning microscopy of the above seen steel- and aluminum-CFRP plates illustrate that the use of a glass fleece as an intermediate layer is highly recommended (Figure 8). If no glass fleece is used, the aluminum component locally shows deep holes (blue colored areas) caused by pitting corrosion. These holes are deeper for the specimens, which have no glass fleece as an intermediate layer. For the steel plates, this effect is hardly visible, because steel prone to surface corrosion instead of pitting corrosion.

IOP Conf. Series: Materials Science and Engineering **480** (2019) 012014 doi:10.1088/1757-899X/480/1/012014

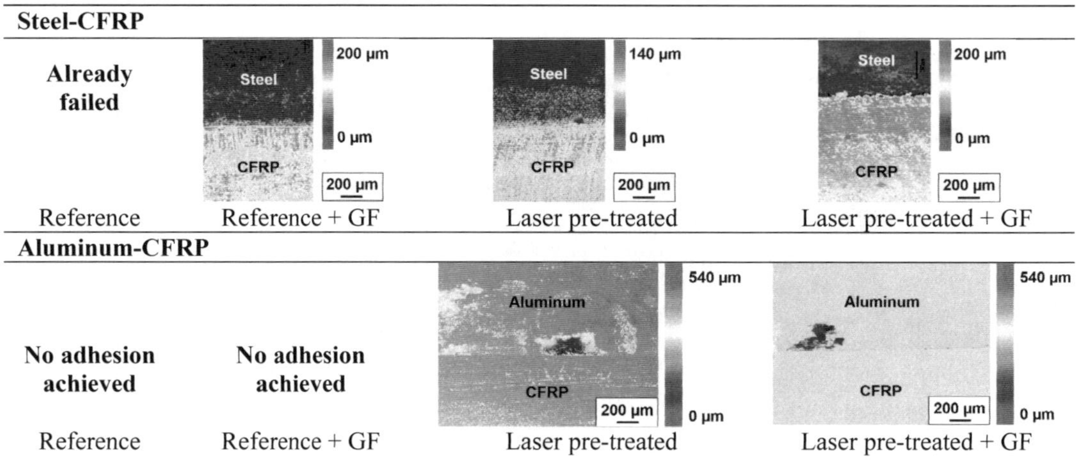


**Figure 8.** Side-view of the interface area of the steel- and aluminum-CFRP specimens after the four-week immersion test by using a Confocal Laser Scanning Microscope.

Summarizing, the laser pre-treatment of steel and aluminum surfaces offers a promising possibility to improve the bond strength and the ageing resistance of steel- and aluminum CFRP structures by creating a mechanical interlocked and durable connection. Despite, a glass fleece is recommended as an intermediate layer to completely exclude contact corrosion.

## 3. Surface treatment for an insert concept

The load transmission in thin-walled structures made of CFRP is often realized by joining methods from the metal construction such as bolt connections. However, such a distribution of loads does not exploit the potential of the carbon fiber design as the loads are only selectively introduced pointwise and the fibers are interrupted and damaged [1][17]. According to a fiber-fare design of load transmission, a novel insert which bases on several piled metal layers was developed, the so-called Multilayer-Insert (MLI). The multilayer design (Figure 9) allows a considerably increased the contact area between the metal and the CFRP for an optimal distribution of load to each CFRP-layer [18]. The insert is designed to be built up during the AFP (Automated Fiber Placement Process) and can be adapted to the fiber orientation and to the size and the geometry to meet the load requirements. Therefore, a separate module for automated insert placement (AIP) is developed and integrated in a fiber placement head [19].

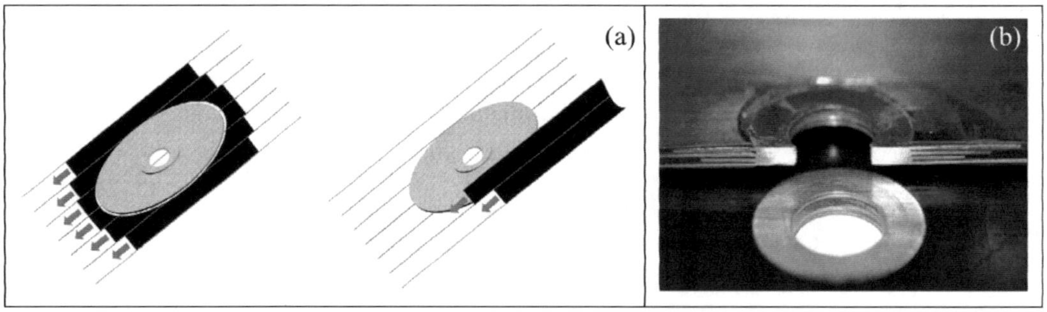


**Figure 9.** Design of a Multilayer-Insert: (a) conceptional setting and (b) exemplary specimen.

A central issue is in this case the surface pre-treatment of the aluminum layers. The pre-treatment has a strong influence on the process from both perspectives, the material science and production engineering, and must therefore be considered interdisciplinary. Here, the special requirements to avoid corrosion problems and guarantee a solid adhesion between aluminum

IOP Conf. Series: Materials Science and Engineering **480** (2019) 012014 doi:10.1088/1757-899X/480/1/012014

and CFRP are taken into consideration in [20]. Thereby, various coatings for aluminum that satisfy the previously mentioned requirements have been investigated. On the one hand, the coating should guarantee the adhesion between aluminum and CFRP and on the other hand, it should avoid the corrosion tendency between the two materials. In addition, the thickness of the coated aluminum must not exceed 125 μm to achieve the same thickness of a CFRP-Layer and avoid thickening in the laminate. Given that an aluminum layer has a thickness of 100 μm, the coating should not exceed 12.5 μm. Furthermore, the coating must overcome the AIP process without damage to guarantee the corrosion protection after the production. Table 2 shows the coatings and some of their properties performed that are investigated.

**Table 2.** Coatings for aluminum [21][22][23].

| **Coating** | **Used coating** | **Adhesion to aluminum** | **Corrosion protection** | **Thermal load capacity** | **Additional features** | **Coating thickness** |
|---|---|---|---|---|---|---|
| Anodized | Aluminum oxide $Al_2O_3$ | ++ | + | ++ | – High mechanical stress resistance (also for deformation)<br>– Coat properties flexible adaptable by process parameters (coating thickness, pore size, postcompaction, …)<br>– Coat electrically insulating | 10 μm |
| Chromium(III)-containing passivation | Surtec®650<br>trivalent Chromium oxide | + | o | ++ | – High suitability as adhesive surface for organical coats | ≈ 1 μm |
| Chromium-free-passivation | Nabutan®810 Titanium oxide | + | o | ++ | – High suitability as adhesive surface for organical coats | ≈ 1 μm |
| Epoxy primer | Cytec BR®127 epoxy basis | + | + | **O** | – Coat electrically insulating | 4 μm |
| Anodized + Epoxy primer | Aluminum oxide Cytec BR®127 | ++ | ++ | **O** | | 14 μm |
| Chromium(III)-containing passivation + Epoxy primer | Surtec®650<br>Cytec BR®127 | + | ++ | **O** | | 15 μm |
| Chromium-free-passivation + Epoxy primer | Nabutan®810<br>Cytec BR®127 | + | ++ | **O** | | 15 μm |

In the adhesion analysis, the energy release rates G1c are determined according to ASTM D 5528-01. In order to complete the adhesion analysis, tensile shear tests are carried out according to DIN 65148 where the interlaminar shear strength is determined. The corrosion tests are realized on the basis of salt spray tests according to DIN EN 60068-2-52 and the measurement of the open corrosion potential using a saturated calomel electrode.

IOP Conf. Series: Materials Science and Engineering **480** (2019) 012014 doi:10.1088/1757-899X/480/1/012014

**Adhesion Analysis**

The samples for the adhesion test are formed with a cross-ply laminate which consists of 0° and 90° layers. The laminate has a double-layer approach with four 0° layers in the center. One of the 0° layers is substituted by a coated aluminum layer, so the analyzed interface is exactly in the center of the laminate. The cross-ply laminate differs from the standard. This laminate approach is selected to reduce thermal stresses caused by the different thermal expansion coefficients of both materials.

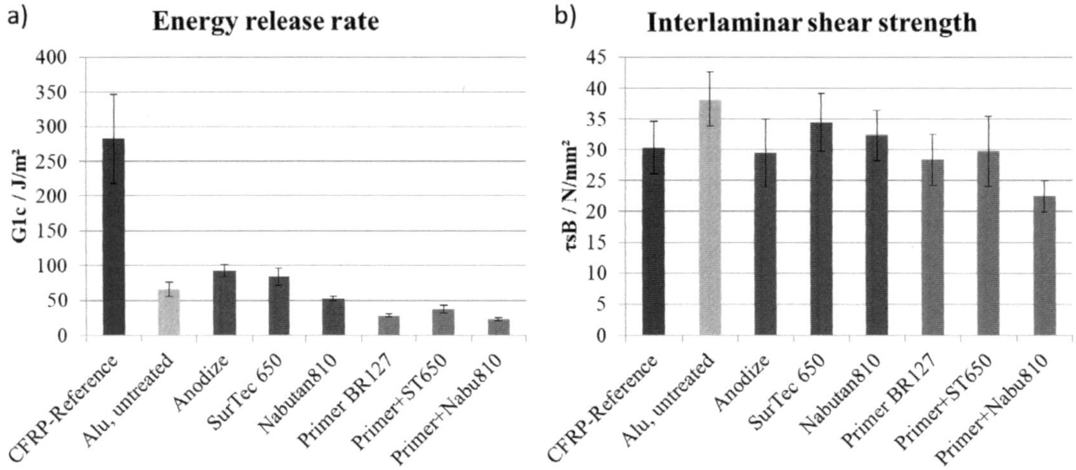


**Figure 10.** Results of the adhesion analysis.

It is shown in the adhesion analysis that the anodizing, the chromium (III)-containing and the chromium-free passivation provides the best adhesion properties (Figure 10). In the energy release rates, the values are significantly lower than the reference samples. Another illustration is shown in the interlaminar shear strength analysis. In this case, the chromium (III)-containing and the chromium-free passivation achieve higher values than the CFRP reference. This behavior is led back to the production of the samples. According to the norm DIN 65148, notches are cut in a way that they lead to stress concentration on the notch edges (Figure 11).

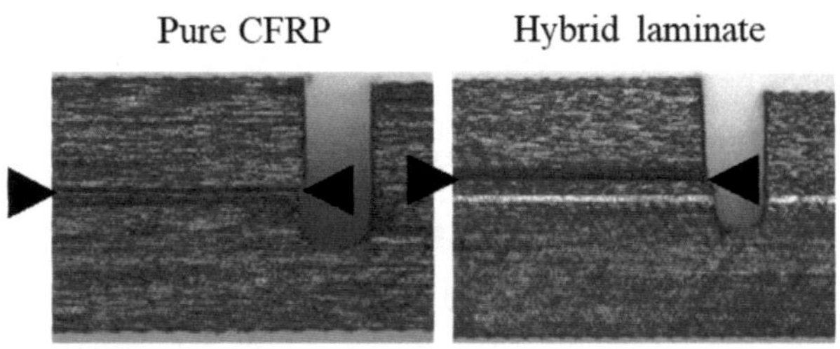


**Figure 11.** Failure behavior of shear-lap samples.

This is an already known issue in the shear-lap-tests. For this reason, the results can be accommodated only to a limited extent and permit only a qualitative statement. The values arising from the combination between anodizing and Primer BR®127 cannot be taken into consideration because of the long dwell time between both coating processes for logistic reasons. The adhesion properties of the anodizing process are reduced to a minimum after 48 hours. For completion within the framework of a further analysis, more samples have been coated. In this case, the dwell time was reduced to less than 48 hours. Additionally, the samples were differently analyzed in a shearing tool in which rectangular samples are sheared off from the midplane. Due to comparability the samples with untreated aluminum were analyzed with the same method. In this event, the values for the untreated aluminum are 33 N/mm². The combination of anodizing and Primer are 17 % lower than the reference with 28 N/mm² [14].

**Corrosion analysis**
The samples for the corrosion tests are manufactured according to the mentioned standard. In this case, aluminum plates with a thickness of 1 mm and a size of 110 x 100 mm are coated. A laminate in the size of 50 x 100 mm is built up on half of the area of the coated aluminum plate and cured in the autoclave. The other half of the coated aluminum can be used to observe and evaluate the corrosion effects. In Table 3 and Figure 12 the results of the corrosion tests are shown. The passivation coating showed signs of corrosion even though an adequate corrosion protection is generally expected. Only the anodizing has a sufficient corrosion protection.

**Table 3.** Observations after salt spray tests.

| Sample/Coating | | Observations after salt spray tests |
|---|---|---|
| Uncoated reference sample | - | – Very strong surface corrosion<br>– Lot of corrosion residue, especially in the Al-CFRP-contact area |
| Anodized | Aluminum oxide | – Sporadically pitting corrosion |
| Chromium(III)-containing passivation | Surtec®650 | – Surface corrosion<br>– Sporadically pitting corrosion |
| Chromium-free-passivation | Nabutan®810 | – Very strong surface corrosion<br>– Lot of corrosion residue, especially in the Al-CFRP-contact area |
| Epoxy primer | Cytec BR®127 | – Filiform corrosion on aluminum<br>– Filiform corrosion/infiltration of the protective layer in Al-CFRP-contact area |
| Anodize + Epoxy primer | Aluminum oxide + Cytec BR®127 | – No detectable corrosion |
| Chromium(III)-containing passivation + Epoxy primer | Surtec®650 + Cytec BR®127 | – Light filiform corrosion in Al-CFRP-contact area |
| Chromium-free-passivation + Epoxy primer | Nabutan®810 + Cytec BR®127 | – Filiform corrosion in Al-CFRP-contact area |

The results shown in Figure 10 indicate that in relation to improve the adhesion properties the passivation coatings and anodizing can come into consideration. In view of the corrosion analysis, (Table 3, Figure 11) only the anodizing and primer show adequate corrosion protection properties. For the use of aluminum inserts in CFRP, only anodizing as a coating can be considered because anodizing complies with both requirements, adhesion and corrosion protection.

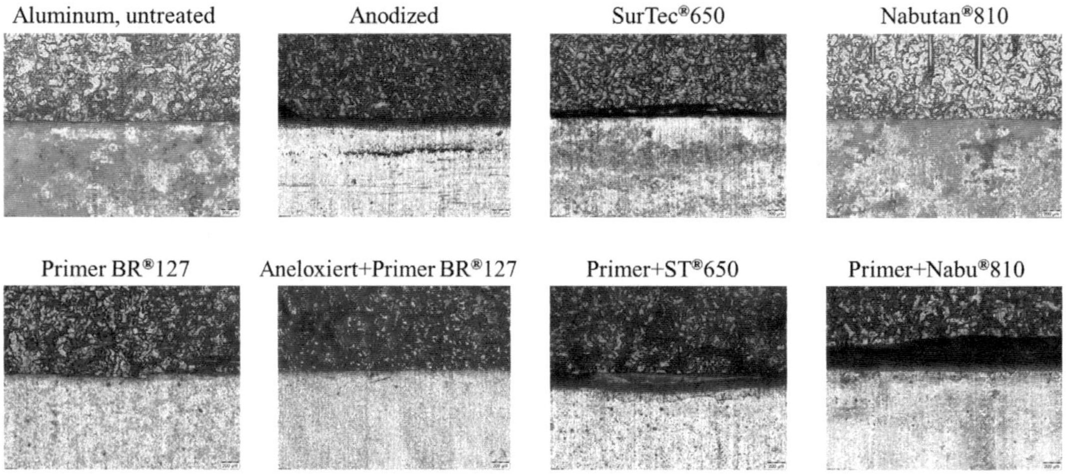


**Figure 12.** Results of the salt spray tests.

**Measurement of the open corrosion potential**

In order to measure the open corrosion potential, the aluminum plates with a thickness of 1 mm are coated in the same way like the samples for the salt spray tests. The difference is that the samples are not joined to CFRP, instead they will be used as an electrode in the galvanic cell. Finally, the results are contrasted and based on all the requirements, a recommendation for an appropriate coating is derived.

The measured electrochemical potential of the different coatings show the same tendency as already seen after the salt spray tests. It is unexpected that the electrochemical potential of SurTech650 is at the same level of the Primer BR®127, even though it showed weaker corrosion protection properties in the salt spray test (Figure 12).

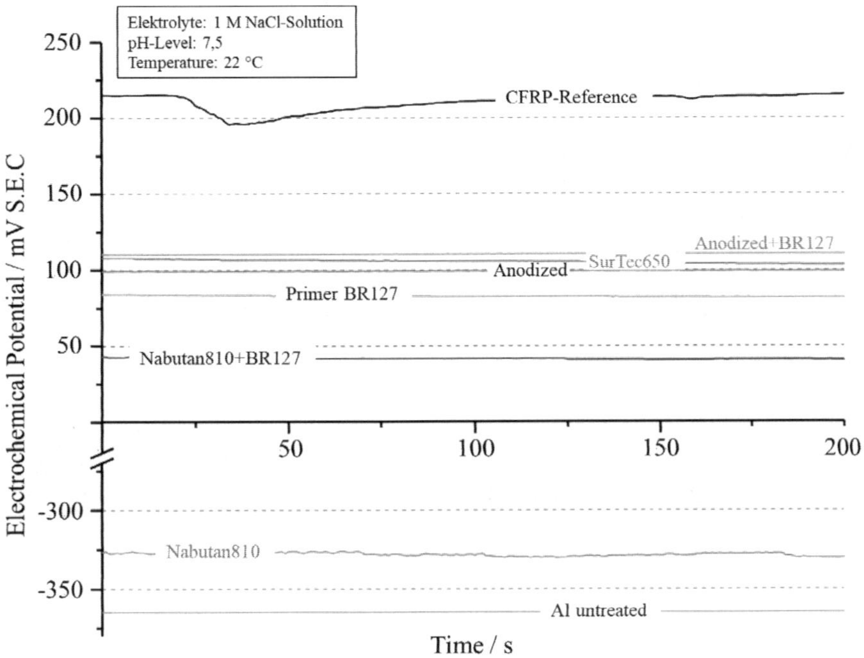


**Figure 13.** Open corrosion potential.

The measurement of the open corrosion potential shows that the anodizing causes a reduction of the electrochemical potential difference of 440 mV in comparison to the untreated aluminum. This result coincides with the results from the salt spray test in which no signs of corrosion were observed. The same behavior can be seen for the combination of anodizing and epoxy primer BR®127 and the epoxy primer BR®127. The electrochemical potential difference to CFRP is 110 mV and 120 mV respectively. A different behavior is noticed on SurTec®650, even though the electrochemical potential difference of 105 mV to CFRP is similar to the previously mentioned coatings, the corrosion protection in accordance with the salt spray test is not satisfactory (Table 3). The coating Nabutan®810 owns the highest electrochemical potential difference to CFRP and reflects a high tendency of corrosion in the results of the salt spray test. Through the additional application of the Primer BR®127, the electrochemical potential difference decreases but is still the highest. Thus, the corrosion protection remains insufficient as shown in the previous tests.

## 4. Conclusion

The paper investigates multiple interfacial design concepts for CFRP-aluminum and CFRP-steel hybrid intrinsic composites. In particular, the key aspects joint strength, corrosion protection as well as manufacturability by single-step joining processes were addressed. In this context, different manufacturing processes (RTM, AFP, hot pressing) were discussed. The sector of surface treatments offers a wide range of processes, whereby chemical (anodizing, sol-gel coating, chromium and titanium passivation), physical (laser pre-treatment) and functional (primers and overmolding by polyamide inter-layers) approaches for steel and aluminum were investigated. The results of this study can serve as a guideline for the production of hybrid structures. Depending on the production process and the material combination different methods to increase adhesion and avoid corrosion are shown and can be taken into account for the design of hybrid structures. An overview about the advantages, disadvantages and characteristics of the mentioned methods and processes is presented below.

By using a sol-gel based silicate coating, an increased adhesive strength of aluminum-polyamide 6 joints was achieved. The subsequent functionalization of the ORMOSIL coating by different silane coupling agents yield increased adhesion, while amino group bearing silane showed the highest joint strengths among the tested silane coupling agents. It counts as an advantage that the subsequent functionalization is easily adaptable towards other composites based on chemically different polymer matrixes. Beside improved adhesion, the ORMOSIL coating is promising to avoid corrosion due to its physical barrier properties. The applicable time span of the sol, determined by viscosity measurements accounts to 5 days and consequently suitability for production process chains is given.

A laser pre-treatment of steel and aluminum surfaces offers a promising possibility to improve, on the one hand, the bond strength and, on the other hand, the ageing resistance, *i. e.* the corrosion resistance, of steel- and aluminum CFRP structures. Due to the laser pre-treatment, a fine nano-structured and open-porous surface quality which is characterized by various undercuts is generated. During the infiltration process of the resin in the VARTM process, these small cavities are filled with the low viscous resin so that a firm mechanical interlocking between the resin and the metal surface, *i. e.* the laserinduced nano-structure, is generated. This strong connection lead to shear stresses which are up to 4 times higher than those of non-structured reference specimens. In addition, the connection is as firm and free of pores that the 5 % sodium chloride solution used in an immersion test is not able to penetrate the interface area so that the shear strength remains constant throughout a test period of four weeks. However, a glass fleece is recommended as an intermediate layer to completely exclude contact corrosion on cutting edges.

The process time of a surface pre-treatment by laser irradiation depends on the material, the laser system and the parameter settings, especially the scan speed, used. With the laser system utilized in this study, ablation rates of 0,6 $cm^2/s$ are possible. This short process time, the good automation capacity, practically no maintenance requirements and the excellent reproducibility qualify the laser pre-treatment for a large scale production.

The analysis of the surface pretreatments of the interface between aluminum and CFRP for AFP-processes shows that the combination of anodizing and the Primer BR®127 provides the best adhesion properties and corrosion protection. The electrochemical potential difference to CFRP is 105 mV and no signs of corrosion are observed in the salt spray tests. In further investigations, the transferability of the promising pre-treatment methods for different applications and complex structures should be investigated with reference to the adhesion and corrosion performance. Moreover, the pre-anodizing process can be optimized according to the properties needed. Through a parameter study, the adhesion properties can be improved. Furthermore, the shelf life of the combination between anodizing and primer should be analyzed. Finally, the chromium-free Primer BR®252 can be considered since it has been developed as an environment-friendly substitute of the Primer BR®127.

Summarizing, the selection of a surface pre-treatment method depends on the requirements of the respective application. A durably high joint strength and an effective corrosion protection are essential in the field of metal-CFRP hybrid structures. All presented methods lead to an increased joint strength and corrosion protection. By the laser pre-treatment and the anodizing or priming process higher shear strengths are achieved than by the sol-gel coating, but the last named procedure ensures superior corrosion protection. In many applications, the manageability of the pre-treatment procedure and the required processing time are key factors. In a large scale production, the pre-treatment process should be time efficient and practicable with a little manual effort. With process times of about 1 minute for coating itself and about 2.5 hours for subsequent heat treatment, the approach via a sol-gel coating shows an intermediate suitability, even though the process is highly automatable and convenient for large-scale production routes. The duration of the laser pre-treatment process strongly depends on the material, the surface area, the laser system and the parameters used. Nevertheless, the process time varies in a range of minutes and a complete corrosion protection can be easily realized by the insertion of an isolation layer in the interface area. The anodizing and priming processes constitute a fast as well as easy pre-treatment procedure and reveal high joint strengths and adequate corrosion protection, but require partly highly toxic and environmentally harmful process materials, which should be avoided nowadays concerning labor safety and environmental protection.

Finally, it can be stated that the selection of a pre-treatment method does not only depend on joint strengths, corrosion performance and efficiency, but also on ethic principles as well as national and international laws and directives.

**Acknowledgements**
The authors like to express their gratitude to the Deutsche Forschungsgemeinschaft (DFG) for the kind support of the interdisciplinary research cooperation within the DFG-Priority Program 1712 which enables basic investigations on "Intrinsic hybrid composites for lightweight structures".

**References**

[1] Schürrmann H 2007 *Konstruieren mit Faser-Kunststoff-Verbunden* (Berlin: Springer)
[2] Henning F, Bader B, Weidenmann K A 2011 Hybride Werkstoffverbunde *Handbuch Leichtbau* ed F Henning F and E Moeller (München: Hanser-Verlag)
[3] Kurtovic A, Brandl E, Mertens T and Maier H J 2013 Laser induced surface nano-structuring of Ti–6Al–4V for adhesive bonding *Int. J. Adhes. Adhes.* **45** 112–117
[4] Byskov-Nielsen J, Boll J V, Holm AH, Højsholt R and Balling P 2010 Ultra-high-strength micromechanical interlocking by injection molding into laser-structured surfaces *Int. J. Adhes. Adhes.* **30** 485–488
[5] Baburaj E G, Starikov D, Evans J, Shafeev G A and Bensaoula A 2007 Enhancement of adhesive joint strength by laser surface modification *Int. J. Adhes. Adhes.* **27** 268–276
[6] Mandolfino C, Lertora E, Genna S, Leone C and Gambaro C 2015 Effect of Laser and Plasma Surface Cleaning on Mechanical Properties of Adhesive Bonded Joints *Procedia CIRP* **33** 458–463
[7] Wang W X, Matsubara T, Takao Y, Yasuda K and Hayashi R 2015 Effects of curing conditions on the short time co-cure of CFRP–steel hybrid laminates *J. Reinf. Plast. Compos.* **10** 819–829
[8] Byskov-Nielsen J and Balling P 2009 Laser structuring of metal surfaces: Micro-mechanical interlocking *Appl. Surf. Sci.* **255** 5591–5594
[9] Montealegre MA, Castro G, Rey P, Arias JL, Vázquez P and González M 2010 surface treatments by laser technology *CM* **1** 19–30
[10] Kießling R, Ihlemann J, Riemer M, Drossel W-G, Dittes A, Scharf I, Lampke T, Sharafiev S, Pouya M and Wagner M. F-X 2018 A process and load adjusted coating system for metallic inserts in hybrid composites *Production Engineering* **12** 249-257
[11] Darwich S, Scharf I, Engelhardt G and Lampke T 2011 Sol-Gel Coating for aluminum Alloy: Self-Healing Characteristics *JEPT* **217** 3–13
[12] Mahadik S A, Parale V, Vhatkara S R, Mahadik D B, Kavale M S, Wagh P B, Gupta S and Gurav J 2013 Superhydrophobic silicia coating by dip coating method *Appl. Surf. Sci.* **277** 67–72
[13] Zinn C, Bobbert M, Dammann C, Wang Z, Tröster T, Mahnken R, Meschut G and Schaper M 2018 Shear strength and failure behaviour of laser nano-structured and conventionally pre-treated interfaces in intrinsically manufactured CFRP-steel hybrids *Composites Part B* **151** 173-185
[14] Zinn C, Schaper M, Serna J, Meiners D, Wang Z and Troester T 2016 Shear Edge Tests: a benchmark in investigating the influence of different surface pre-treatment methods on the shear stress of metal-CFRP hybrids manufactured by RTM process *Int. J. Automot. Compos.* **2** 244–71

[15] Wang Z, Bobbert M, Dammann C, Zinn C, Lauter C, Mahnken R, Meschut G, Schaper M and Tröster T 2016 Influences of interface and surface pretreatment on the mechanical properties of metal-CFRP hybrid structures manufactured by resin transfer moulding *Int. J. Automot. Compos.* **2** 272-298

[16] Zinn C, Wang Z, Tröster T and Schaper M 2018. Forming and corrosion stability of a laser pre-treated metal surface – Influence on the properties of metal-CFRP hybrid structure made by VARTM. Proceedings 3. Internationale Konferenz Hybrid Materials and Structures. 18.-19. April 2018. Bremen. 17-23

[17] Heißler H 1986 *Verstärkte Kunststoffe in der Luft- und Raumfahrttechnik* (Stuttgart: Kohlhammer Verlag)

[18] Hermann A, Kolesnikov B D, Pabsch A and Wilmes H 1999 Composite material with a reinforced connecting area PCT/DE1999/000790

[19] Schmidt C, Denkena B and Gross L 2016 Concept for automated production of CFRP-metal hybrid compounds integrated in an Automated Fiber Placement process *10th CIRP Conference on Intelligent Computation in Manufacturing Engineering - CIRP ICME '16* Ischia Italy

[20] Franzis R 2002 *Galvanic Corrosion. A practical guide for Engineers* (Weinheim: Wiley VCH)

[21] Bobzin K 2013 *Oberflächentechnik für den Maschinenbau* (Weinheim: Wiley VCH)

[22] Keller F, Hunter M S and Robinson D L 1953 Structural features of oxid coatings on aluminum *J. Electrochem. Soc.* **100** 411-419

[23] Ebnesajjad S, Ebnesajjad C 2014 Surface Treatment of Materials for Adhesive Bonding. Norwich

IOP Conf. Series: Materials Science and Engineering **480** (2019) 012015 doi:10.1088/1757-899X/480/1/012015

# Localized anodization of the aluminum alloy EN AW-7075 T6 by closed electrolytic free jet

**R Morgenstern[1,*], A Martin[2], N Lehnert[2], I Scharf[1], M Hackert-Oschätzchen[2], A Schubert[2] and T Lampke[1]**

[1] Institute of Materials Science and Engineering, Chemnitz University of Technology, 09125 Chemnitz, Germany
[2] Institute for Machine Tools and Production Processes, Chemnitz University of Technology, 09125 Chemnitz, Germany

[*] e-mail: roy.morgenstern@mb.tu-chemnitz.de

**Abstract.** Anodizing with closed electrolytic free jet enables the resource-efficient, local conversion and surface structuring of the high-strength alloy EN AW-7075 T6 aiming at the local improvement of the tribological behavior of functional surface areas. The objective of the present work was to confine the current distribution on the specimen's surface and to achieve a stronger localization of the coating oxide layer. For this purpose, the processing parameters working distance, voltage and process duration were varied systematically. Based on the refractive index of the porous oxide and laser scanning microscope imaging of the interference lines, the maximum thicknesses of the anodized spots were calculated. Moreover, the coating microstructure was investigated by scanning electron microscopy and the surface topography was measured with a stylus profiler. The results show that the reduction of the working distance to 100 μm enables a significantly improved localization of the coating growth at substantially lowered process voltages. Thus, the energy-efficient production of localized, wear-resistant oxide spots for tribological applications with local functionalization of surfaces through anodization by closed electrolytic free jet is proven.

## 1. Introduction

High-strength aluminum alloys are commonly used as lightweight materials for the construction of safety-related structures. However, due to their comparatively low hardness, they are susceptible to abrasive and adhesive wear. The anodic oxidation (anodization) is an established means to overcome this limitation and to open new fields of application [1]. Hard oxide ceramic conversion coatings can be generated in an electrochemical process using an oxalic acid bath at ambient temperature [2]. Conventionally, the whole workpiece is immersed into the electrolyte and connected as the anode. However, in many applications the tribological contact is limited to distinct surface areas, which solely need to be functionalized for wear protection. Hence, the limitation of the anodic oxidation process to highly stressed functional surfaces would allow for saving process energy and prolong the life cycle of the electrolyte reservoir. Furthermore, the dimensional tolerances of surfaces which are not anodized are not affected. Regarding modern light-weight construction, hybrid material concepts gain importance. Those include for example aluminum-polymer compounds or mixed constructions of aluminum and other metals. In this context, anodizing should be limited to the aluminum surface in order to avoid etching of other metals or damaging of polymer surfaces.

In conventional anodization the localization can be realized by masking of surfaces on which an oxidation is not desired with chemically resistant coatings that cannot be subverted by the electrolyte. These coatings have to be applied before and need to be removed after anodization in separate steps. As an alternative, local anodizing can be realized using sealed electrolytic cells. It was proved that micro capillary cells are suitable to generate anodized spots with defined diameters [2, 3]. According to an innovative approach of Kuhn et al., local anodizing can also be performed by using a closed electrolytic free jet [4]. The experimental setup of localized anodization using a continuous electrolytic free jet is extensively described in [4]. The process is closely related to electrochemical machining [5, 6], with the processing parameters such as electrolyte composition and voltage adapted to allow for surface passivation instead of material removal. At the beginning of this process, the surface passivation and coating growth is initiated at the impact area of the jet. With an increase in processing time the diameter of the oxide spot increases significantly since the surrounding surface is also wetted by the electrolyte and the current density distribution expands in lateral direction rather than passing the electrically insulating oxide layer at the center of the anodized spot. Consequently, the spots exhibit a rather low maximum oxide thickness of up to 500 nm in the center and a comparatively large diameter of approximately 2 mm [4].

Although it was proven that the electrolytic free jet technique is suitable to generate anodic oxide spots on high-strength aluminum alloys, localized surface structuring for tribological applications demands for a higher localization of the oxide growth. In order to meet this objective, the distance between the nozzle and the substrate surface was reduced from 2.5 mm to 0.1 mm. The resulting oxide thickness and the diameters of the oxide spots were measured in order to derive their influence on the localization of the coating growth during anodizing of the aluminum alloy EN AW-7075 T6 with closed electrolytic free jet.

## 2. Materials and methods

Sample sheets of the aluminum alloy AlZn5.5MgCu (EN AW-7075 according to EN 602) with the dimensions of 100 x 25 x 1.5 mm³ were used as substrate material. Grinding with SiC paper down to a grit size of 4000 and subsequent polishing with diamond suspension on polishing cloths were carried out in order to achieve smooth, shiny surfaces prior to anodizing. Due to the mechanical surface pretreatment, there was no necessity for etching prior to anodizing. Thus, the low surface roughness after polishing was not affected. The experimental setup used for anodizing with closed electrolytic free jet was already described in [4]. The used electrolyte was 6.1 g/l oxalic acid, which exhibits a conductivity of 63.1 mS/cm at 21.4 °C and a pH of 1.0. The experiments are conducted in potentiostatic mode at voltages of 20 V, 40 V and 60 V and at a working distance of 0.1 mm. Comparative experiments at voltages of 160 V, 230 V and 300 V and at a working distance of 2.5 mm were carried out to analyze the influence of the increased values. The current flow was recorded as a function of the process time.

In order to evaluate the surface topography, tactile measurements were conducted using a MarSurf LD 120 contour measuring unit (Mahr, Göttingen). An UV-confocal laser scanning microscope (LSM) VK-9700 (Keyence, Neu-Isenburg) was applied to measure the interference patterns at a constant wave length $\lambda$ of 408 nm. The maximum thickness of the anodized spots $d$ was calculated using equation (1) considering the number of the interference lines $z$ and the refractive index $n$ = 1,68 of the oxide layer, which was obtained from literature [7].

$$d = \frac{z*\lambda}{2*n} \qquad (1)$$

Prior to SEM investigations, the samples were subjected to 3-point-bending with a distance of 40 mm between the bearings up to a maximum deflection of 5 mm in order to generate cracks within the oxide layers. The microstructure was investigated by field emission scanning electron microscope NEON40EsB (FE-SEM, Zeiss,) at selected positions using a tilt angle of 55° to the electron beam.

## 3. Results and discussion

### *3.1. Electric current and electrical charge*

As already described by Sieber et al., potentiostatic anodizing in closed systems is usually characterized by a current decrease at the beginning due to the passivation of the surface [2]. In case of a compact barrier layer growth, the current can be expected to decrease to almost zero as time passes, because the oxide thickness increases on the entire surface. However, pore initiation and pore growth are typically observed in oxalic acid anodizing. Therefore, the current rises again in the following and eventually, an almost constant current level can be observed during the phase of steady pore growth [2]. As the entire coating thickness increases, a slight current decrease might occur due to the increasing length of the narrow pore channels and the associated increasing diffusion resistance. As can be seen in figure 1, there is only a slight current decrease during the first seconds of anodizing with closed electrolytic free jet, due to the passivation of the surface. No clear current minimum can be observed, because the current flow is not hindered to shift towards outer surface regions with a lower passive layer thickness and therefore a lower electrical resistance. After the initial decrease, the current remains at an almost constant level, which can also be explained by the initiation and growth of pore channels.

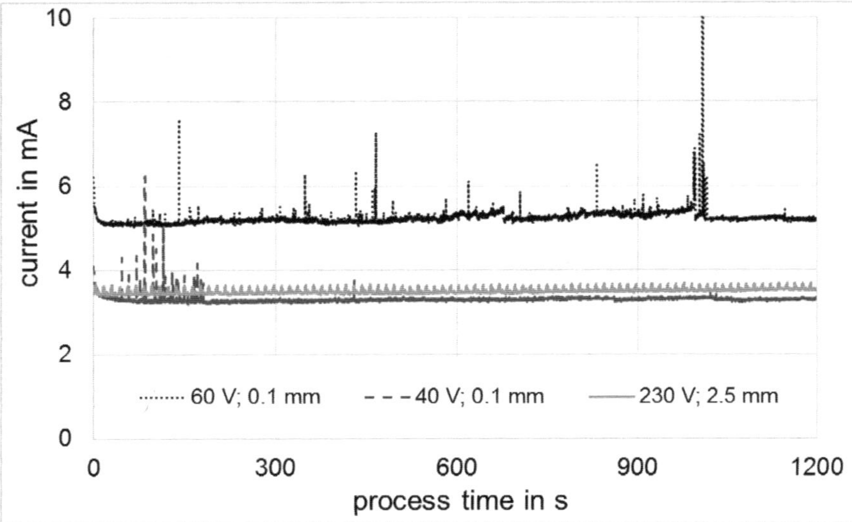


**Figure 1.** Current-time characteristics of anodizing with closed electrolytic free jet at 40 V and 60 V (work distance 0.1 mm) and 230 V (2.5 mm).

When comparing the current courses displayed in figure 1, it becomes obvious that the reduction of the working distance is a suitable means to achieve comparatively high anodizing currents at significantly reduced process voltages. In this particular case, approximately the same current values could be observed for the process voltages of 40 V and 230 V at working distances of 0.1 mm and 2.5 mm, respectively. Obviously, the electrical resistance of the electrolytic free jet is sharply decreased due to the reduced working distance. Furthermore, it is commonly observed in anodizing that the barrier layer thickness increases proportionally with the anodizing voltage. This effect could also contribute to an increased electrical resistance at high process voltages. However, when increasing the voltage from 40 V to 60 V at a working distance of 0.1 mm, the averaged current also increases by approximately 50 %. In this case, the electrical resistance seems to be identical.

IOP Conf. Series: Materials Science and Engineering **480** (2019) 012015 doi:10.1088/1757-899X/480/1/012015

### *3.2. Oxide thickness and microstructure*

Provided an almost constant process efficiency, the mass of the anodic oxide is directly proportional to the amount of charge transfer according to Faraday's law. Since the electric current differs only slightly over the process time, this proportionality also refers to the averaged current. Thus, it can be expected that the highest oxide mass is achieved using a working distance of 0.1 mm and a process voltage of 60 V. Furthermore, the oxide mass directly correlates to the average oxide thickness provided that the specific mass of the anodic oxides considering the density of alumina and the porosity is comparable. With these preconditions, it can be assumed that the average oxide thickness of the spots anodized with a process voltage of 40 V at 0.1 mm of working distance and 230 V at 2.5 mm of working distance are almost equal. Figure 2 represents chosen results of LSM imaging, where the detected interference lines are numbered. The maximum thickness of both anodized spots was calculated according to equation 1). Considering the total number of interference lines of $z = 16$ the maximum thickness amounts to approximately 2 µm in both cases.

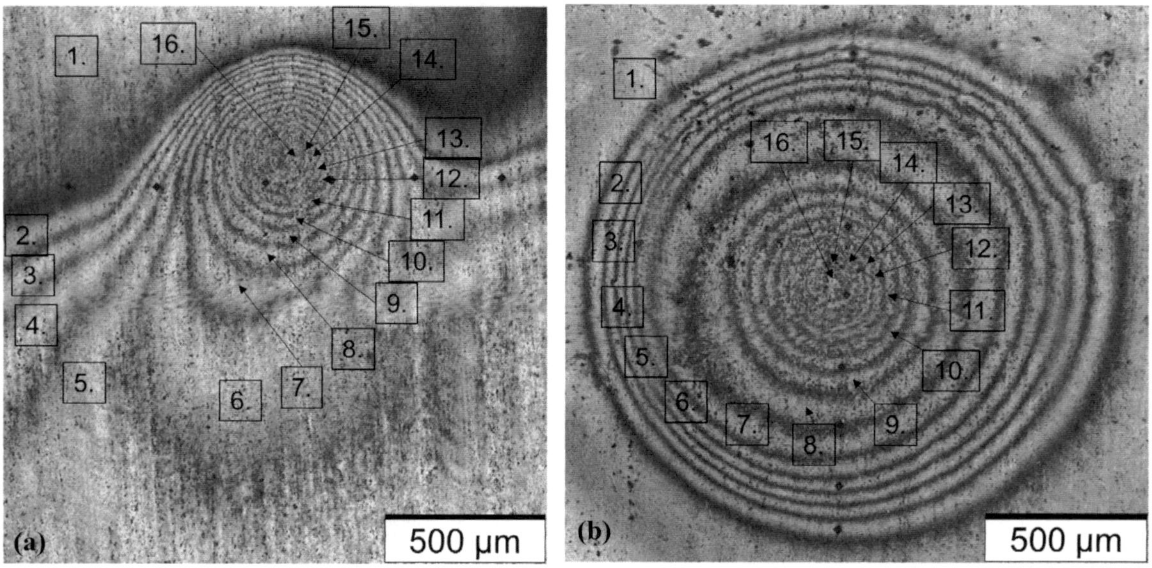


**Figure 2.** LSM images of spots with interference lines anodized with a processing time of 20 min at a) a voltage of 40 V and 0.1 mm of working distance and b) a voltage of 230 V and 2.5 mm of working distance

From the similar maximum thickness of both anodized spots a maximal oxide growth rate of approximately 100 nm/min was derived. However, the lateral distribution of the oxide spots varies significantly. While at 230 V the radius of the spot calculated from the second interference maximum to the center, exhibits approximately 750 µm, the radius of the spot anodized at 40 V amounts to only approximately 450 µm in horizontal direction (figure 2 a)). In addition, the circumferential shapes of the oxide spots differ significantly. While an almost rotation-symmetric spot can be recognized at 230 V, the results achieved at 40 V are characterized by an asymmetrical shape with a preferred extraction to the lower part of the image. The asymmetry can be explained by the preferential flow of the electrolyte towards this direction. The process is assisted by additional air flow to displace droplets resulting from the surface tension of the aqueous electrolyte in order to realize the free jet of electrolyte. Since the air flow is characterized by a preferential direction, the electrolyte is preferentially displaced in the same direction, which leads to the asymmetry especially at small working distance. At higher working distance the influence of the assisting air flow is obviously negligible.

A typical oxide microstructure of the anodized spot obtained at 230 V is shown in figure 3. As already derived from the time-dependent current characteristics of anodizing with closed electrolytic free jet in figure 1, the anodic oxide exhibits a porous microstructure. However, pore channels being regularly oriented perpendicularly to the substrate surface are hardly observable in figure 3. Similar highly non-conformal pore structures were already described by Morgenstern et al. regarding the immersion-based anodic oxidation of alloy EN AW-7075 in 0.8 mol/l oxalic acid at a process voltage of 60 V [8]. Moreover, Li et al. observed the growth of highly non-conformal pores during anodizing of high purity aluminum in 0.3 M oxalic acid at process voltages between 140 V and 400 V [9]. This type of competitive pore growth can be assigned to different dissolution rates at the ground of neighboring pore channels [10].

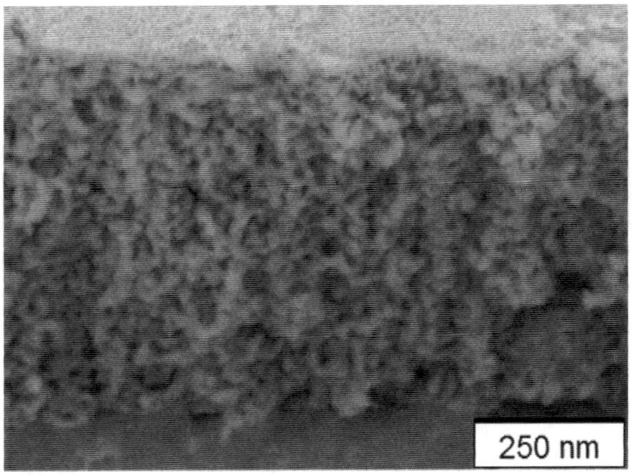


**Figure 3.** SEM image of the cracked oxide layer obtained at a working distance of 2.5 mm and a process voltage of 230 V after a processing time of 20 min.

It is known from anodizing in various electrolytes that the diameter of the pore-cells gradually increases with increasing anodizing voltage [11]. For deeper analyses into the applicability of the anodized spots for tribological applications, systematic investigations into influences of the applied voltages on the resulting pore sizes and the resulting hardness parameters are planned for future analyses.

### *3.3. Surface topography*

Apart from the hardness parameters the surface topography is also of great importance regarding both geometrical aspects, e.g. fitting tolerances, and the contact geometry in tribological systems. Typical cross-sectional profiles of anodized spots are presented in figure 4 a). It can be seen that flat elevations result from a large working distance of 2.5 mm. After anodizing with closed electrolytic free jet for 20 min at 230 V, the elevation height amounts to 0.4 µm at an elevation width of approximately 1.5 mm. The reduction of the working distance to 0.1 mm and process voltage to 60 V results in an almost similar elevation width, but at the same time to approximately twice the elevation height. A further reduction in processing voltage to 40 V keeping the working distance at 0.1 mm results in a significant decrease in diameter. Here, an elevation width of approximately 1.0 mm was detected, which represents a reduction to 67 % referring to the comparative results achieved with 60 V and 230 V. At the same time a comparatively large elevation height of 0.6 µm was realized, which is an increase to 150 % referring to the comparative result achieved with 230 V.

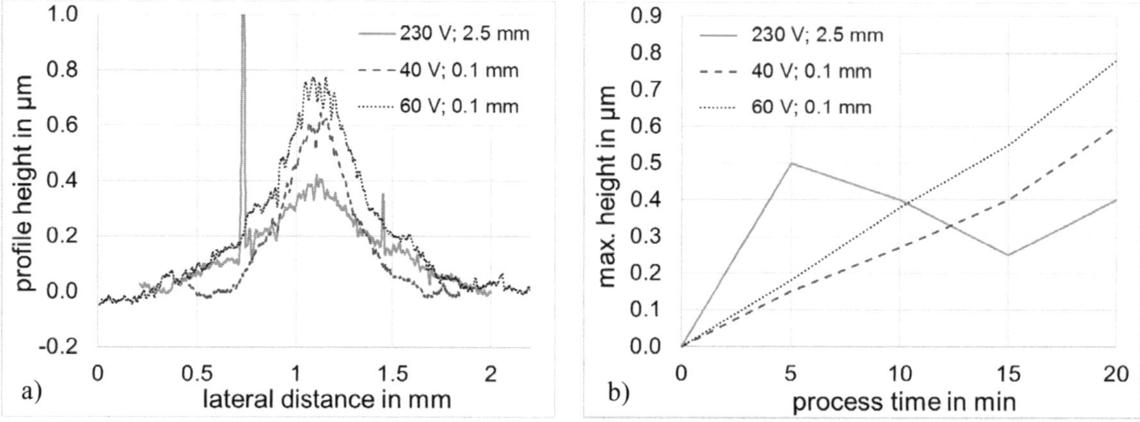


**Figure 4.** Results of tactile measurements: a) cross-sectional profiles of spots anodized with closed electrolytic free jet after an anodizing duration of 20 min and b) development of the maximum profile heights over the process time.

Apart from the topographical information, the time-dependent evolution of the maximum profile height enables us to draw conclusions about the mechanisms of oxide growth. As can be seen from figure 4 b), at the anodizing voltages of 40 V and 60 V, the maximum profile height increases almost proportionally with the anodizing time. Larger profile heights were achieved at a high process voltage of 230 V and a large working distance of 2.5 mm after a short anodizing time of 5 min. This implies a high oxide growth rate in the beginning of the anodizing process. However, the maximum of the analyzed profile height is already achieved after a process time of 5 min, afterwards a declining trend can be observed for increasing anodizing times. Considering the fact that the total oxide thickness increases further to approximately 2 µm after an anodizing time of 20 min as asserted in figure 2 b), this means that the anodic oxide growth proceeds towards the inside of the substrate material. The ratio between outwards and inwards growth rate depends on the ratio of oxide generation and oxide dissolution by the electrolyte.

As the diameter of the anodized spots increases with increasing working distance, the local current density in the center of the spot decreases. Hence, the local oxide growth rate decreases likewise. According to Nagayama and Tamura, the rate of field-assisted dissolution at the pore grounds depends on the local current density and might therefore not contribute to the intensified inwards growth [12]. The authors also state that the chemical dissolution increases with increasing porosity, due to the larger surface of the pore-walls [12]. However, it is not clear from the SEM investigations, whether the coating porosity is affected by the process voltage in anodizing with closed electrolytic free jet. For this reason, further research should be conducted in order to clarify the impact of the process conditions on the oxide porosity and the mechanical properties of the porous oxide layer.

## 4. Conclusion

The present work investigates the influence of the working distance and the process voltage on the localization of the anodic oxide growth in anodizing with closed electrolytic free jet. The results of optical interferometry show that comparable oxide thicknesses are achieved at considerably reduced process voltage from 230 V to 40 V, when applying a reduced working distance of 0.1 mm instead of 2.5 mm. The results of tactile probing indicate that for anodizing at reduced working distance and voltage an increased elevation height of the oxide spot along with a reduced elevation width is achieved at the same processing time. The reduction in width to 67 % and the increase in height to 150 % results in an increased aspect ratio of 225 %, which is a significant improvement regarding the aspired tribological application.

The results of the SEM investigations show that porous microstructures are generated. Further investigations should be performed in order to gain deeper information about the total porosity and the mechanical and tribological properties of the anodic oxide layers.

**Acknowledgments**
The authors gratefully acknowledge funding of this work by the German Research Foundation (Deutsche Forschungsgemeinschaft, DFG, projects LA1274/40-1 and SCHU 1484/20-1). The support by Dagobert Spieler, Elke Benedix, Steffen Clauß, Christel Gläser and Paul Clauß (all from the Institute of Materials Science and Engineering) is gratefully acknowledged.

**References**

[1] Wielage B, Alisch G, Lampke T and Nickel D 2008 *Key Eng. Mat.* **384** 263
[2] Sieber M, Morgenstern R, Kuhn D, Hackert-Oschätzchen M, Schubert A and Lampke T 2017 *IOP Conf. Ser.-Mat. Sci.* **181** 012044
[3] Nickel D, Dietrich D, Morgenstern R, Scharf I, Podlesak H and Lampke T 2016 *Adv. Mater. Sci. Eng.* 1374897
[4] Kuhn D, Martin A, Eckart C, Sieber, M, Morgenstern R, Hackert-Oschätzchen M, Lampke T and Schubert A 2017 *IOP Conf. Ser.-Mat. Sci.* **181** 012042
[5] Hackert-Oschätzchen M, Meichsner G, Zinecker M, Martin A and Schubert A 2012 *Precis. Eng.* **36** **4** 612
[6] Hackert-Oschätzchen M, Martin A, Meichsner G, Zinecker M and Schubert A 2013 *Precis. Eng.* **37** **3** 621
[7] Zekovic LD, Urosevic VV and Jovanovic B R 1986 *Thin Solid Films* **139** 109
[8] Morgenstern R, Scharf I and Lampke T 2018 *IOP Conf. Ser.-Mat. Sci.* **373** 012021
[9] Li Y, Ling Z, Hu X, Liu Y and Chang Y 2011 *J. Mater. Chem.* **21** 9661
[10] Chen S, Ling Z, Hu X, Yang H and Li Y 2010 *J. Mater. Chem.* **20** 1794
[11] Lee W, Ji R, Gösele U, Nielsch K 2006 *Nat. Mater.* **5** 741
[12] Nagayama M and Tamura K 1968 *Electrochim. Acta* **13** 1773

IOP Conf. Series: Materials Science and Engineering **480** (2019) 012016 doi:10.1088/1757-899X/480/1/012016

# Smart components by additive technologies

**I Dani***

Fraunhofer Institute for Machine Tools and Forming Technology, Germany

* e-mail: ines.dani@iwu.fraunhofer.de

**Abstract.** Smart components are key technologies for industry 4.0 as well as the Internet of Things. The application areas include e.g. systems for sustainable production, automotive, smart agriculture, and smart home. In order to integrate intelligence within a component, a sensor or actuator system typically needs to be embedded within the component itself during fabrication. Additive manufacturing has been extended to more and more industries and is now being used not only in toolmaking, but also in the series production of components. Geometric functionalization for conformal cooling is already state of the art in tool making. Despite the great potential of function-integrated parts, publications on sensor integration into parts fabricated by additive manufacturing are still rarely found. First approaches address multi-material printing by using insulating and conductive materials as well as the integration of smart elements during the printing process. Laser beam melting (LBM) is an additive manufacturing technology for geometrically complex parts based on the layer wise and selective melting of metal powders. As proof-of-concept of the developed integration concept, based on LBM, the functionalization of tools and medical implants will be presented.

## 1. Introduction

In the last years, a significant increase in applications of additive manufacturing technologies for production of direct metal parts occurred, especially in the medical and aerospace industry but also in engineering. Additive processes are characterized by outstanding geometrical flexibility and lack of tooling. Manufacturing of geometrically complex parts with additionally embedded sensors and actuators potentially enables the customization of products or real-time monitoring of engineering systems also in harsh environments. Additive manufacturing is especially suited for the production of smart structures with structurally integrated components due to its layer-by-layer fabrication process. Applications for sensor or actuator integrated smart parts include sophisticated biomechanical devices monitoring the interface between implant and bone and smart tools to monitor the process conditions or to predict maintenance. An extensive literature review on additive manufacturing of smart parts is contained in [1]. In particular, the powder bed based laser beam melting (LBM) allows a precise spatial energy input to fully melt metal powders to dense structures with mechanical properties comparable to those of conventionally manufactured parts. LBM enables a materially bonded embedding of functional components to obtain a maximum sensitivity of the adaptronic system. Further benefits include the complete encapsulation of the sensor within the metallic component and the utilization of far-field inductive energy transmission for wireless operation [2].

Tool and die making faces the challenge of producing molds for complex and high quality injection molded parts. The molds have to be equipped with a multitude of functions, such as locally defined heat conductivity, ventilation of the cavity, individual tempering options, toughness, strength, wear resistance, sliding, and demolding functions. By using additive manufacturing processes, some of these functions can be implemented in the smallest possible space. Approaches for linking fluid-carrying functions in mold inserts with the application of temperature and pressure sensors for process monitoring were developed and are described in this paper.

Potentially, tool manufacturing for different sheet metal forming technologies can also benefit from the advantages of LBM. The focus is on incorporating a conformal cooling system to reduce cycle time into a tool for press hardening. To compare the simulation results with the real forming process tool, integrated sensors are applied - the position and the type of integration in the tool are crucial aspects for reliable measurements. Design and manufacturing of the tools are explained in the paper and the results of first forming tests are presented.

The integration of sensors and actuators into artificial joints has been discussed for several years. Integrated sensors or actuators for monitoring of osseointegration, for detection of implant loosening or for in vivo implant load monitoring substantially improve the functionality of the implants. Especially for an acoustic stimulation, actuators placed on or near the surface are not able to create useable structure-borne acoustic waves. A structural embedding and materially bonded integration of the actuator is favored. A concept for integration of an encapsulated actuator system was developed and is presented.

## 2. Manufacturing concept and materials

### *2.1. Technology*

Laser beam melting is a very versatile additive manufacturing technology and in this paper is used as base process. The starting point is a 3D CAD model, which is virtually sliced into thin layers with a layer thickness of about 45 µm. Based on this data the physical part is built. A thin layer of metal powder, stored in a dose chamber, is uniformly distributed across the working area by a coater blade (Figure 1). The light emitted from an ytterbium fiber laser system at the wavelength of 1070 nm is directed by a galvanometer scanner across the powder layer according to the cross section of the part. The machine used, a M2 LaserCUSING® from Concept Laser, applies an island exposure strategy, i.e. the segments of each layer are irradiated in stochastical succession. Upon irradiation the powder melts to form a melt pool, covering not only the top powder layer, but also the already solidified layer below. After laser exposure of the first powder layer, the build platform is lowered, the next powder layer is applied and the exposure process starts again. These steps are iterated until the part is completed. Then, the unmelted powder is removed and the part is separated from the build platform and from the support structure.

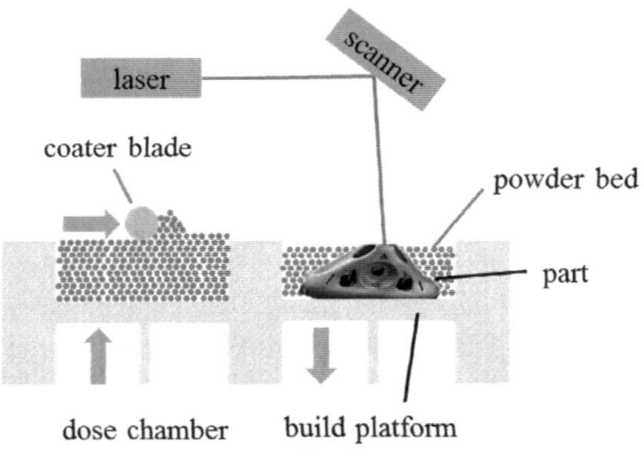


**Figure 1.** Scheme of the laser beam melting (LBM) process.

To integrate a sensor in a metal part, the outer part is manufactured by LBM (Figure 2). From the cavity, designed to house the sensor, the unmelted powder is removed and the sensor (incl. the wiring) transferred. Subsequently, the LBM process continues, i.e. the next powder layer is applied and selectively melted.

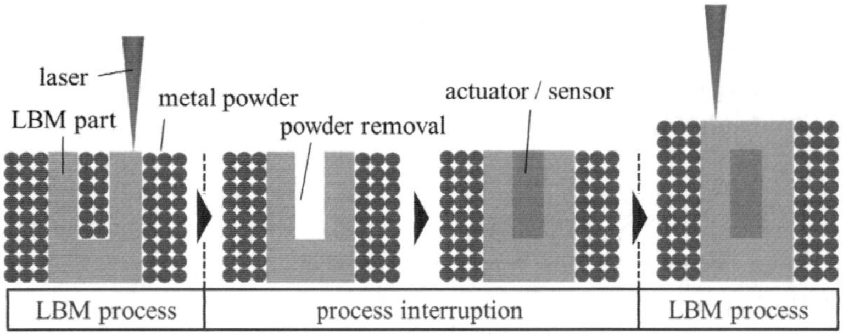


**Figure 2.** Principle of component integration into metal parts manufactured by LBM [3].

*2.2. Selection of Sensors and Actuators*

The sensors and actuators to be integrated in parts manufactured by LBM have to meet special requirements. The measuring element and the wiring must have a high temperature resistance in order to withstand the very high local thermal load during laser beam melting of the surrounding material and a subsequent heat treatment of the part. Further aspects such as low costs, small installation space, good availability and good accuracy are noteworthy. Sensors to be exploited include thermocouples, resistance strain gauges and capacitive differential pressure sensors. Industrial applications studied in this paper include tools for injection molding and for hot forming of metal sheets.

In the injection molding process the material is plasticized in an injection unit at temperatures of up to 300 °C; during injection into the mold the temperature reaches still about 250 °C. Thus, for the intended application of thermocouples in injection molding tools a measuring range from room temperature up to 250°C is required. Thermocouples were produced from a nickel-chromium wire connected to a nickel wire (type K) and insulated by a double layer of glass fiber. The measuring range is between −200 °C and +1250 °C.

To monitor the pressure in the mold cavity of the injection molding tool resistance strain gauges and capacitive differential pressure sensors can be applied. To avoid contact between the sensor surface and the mold, the pressure is measured indirectly via a membrane. The thickness of the metallic membrane is a crucial parameter and is determined by FE simulations. Main advantage of using the membrane and a fluid medium is that the pressure exerted on the surface of the pressure sensor is significantly reduced, allowing the application of a sensor with a smaller maximum operation pressure. Simulations show a pressure in the active fluid medium (hydraulic fluid) of max. 250 bar for a membrane thickness of min. 2.7 mm and an assumed pressure of 2000 bar at the injection nozzle. An electronic differential pressure transmitter (PU5401, ifm electronic GmbH) with a measuring range up to 250 bar was selected.

Also, the alternative measuring system, a strain gauge, is covered by a metal top layer, in this case with a thickness of 0.3 mm. If a force acts on the surface of the sample, the top layer and the strain gauge deform and the resulting voltage is amplified by a bridge circuit, at the same time compensating temperature deviations. For this integration method, the strain gauge has to withstand the temperature necessary to melt the surrounding steel, so only relatively expensive high temperature strain gauges are suitable. Inconel 600 encapsulated strain gauges (KHCD-5-200-G11 C2M, Kyowa) with a maximum operation temperature of 800 °C were selected.

For the medical implant application, a pre-fabricated actuator system described in [2] and [4] was used (Figure 3), ensuring a materially bonded connection between the actuator system and the surrounding implant, a complete encapsulation of the system, and a wireless energy and date transmission. The primary objective of the actuator system with a total weight of 2 g and size of 28 mm x 6 mm x 6 mm is the generation of oscillations between the two interfaces of the actuator with calculated forces in the millinewton range.

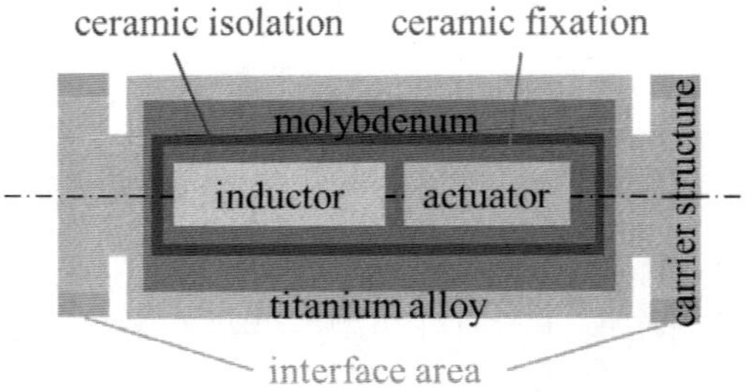


**Figure 3.** Pre-fabricated encapsulated actuator system.

### 2.3. *Materials*
The injection molding tool and the specimen for the pre-tests were additively manufactured by LBM from maraging steel powder (1.2709). The functional part of the hot forming tool was also manufactured by LBM from steel powder (1.2709) on a base part made from hot-working tool steel (1.2343). The medical implant was made from Ti6Al4V titanium alloy powder.

## 3. Results and discussion

### 3.1. *Smart Tools for application in injection molding*
In injection mold making, additive manufacturing technologies are increasingly applied for conformal cooling [5], avoiding inhomogeneous temperature distributions in the tool due to optimized cooling channel geometries. Application of conformal cooling channels is reported to reduce the cooling phase up to 50% [6]. Monitoring of the complex process of injection molding leads to increased controllability and effectivity of the process. Main parameters influencing the process include mold temperature and cavity pressure [7].

Application of an intelligent monitoring and control system by integration of thermocouples and pressure sensors into additively manufactured molds allows implementation of innovative and highly productive injection molding technologies. A mold concept adapted to additive manufacturing including conformal cooling channels, a near surface thermocouple, and a pressure sensor is shown in Figure 4. This paper describes a first approach for integration of the selected sensors into test specimen and their functional qualification.

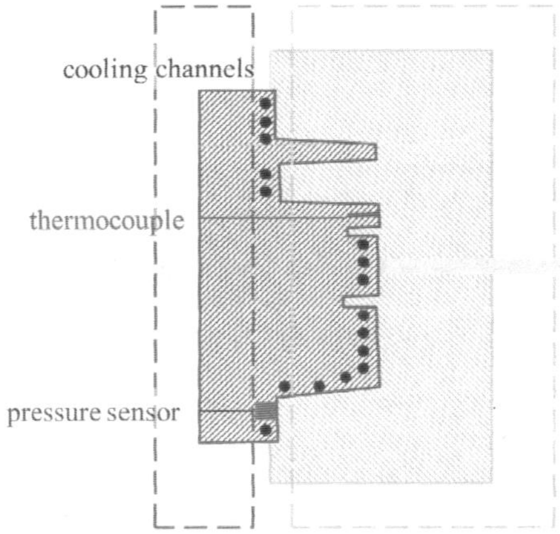


**Figure 4.** Mold concept adapted to additive manufacturing including conformal cooling channels, a near surface thermocouple, and a pressure sensor.

Thermocouples described above were threaded into additively manufactured steel specimens after loose powder material had been sucked off (Figure 5). Since the space conditions usually are very limited, different geometries of these specimens were tested (Figure 5a). At the lowest foot height (this sample is indicated with "*" in Figure 5), placement was possible only with the help of a threading aid. After threading, the specimens were finished with additively manufactured top layers with a thickness of 1 mm.

**Figure 5.** Integration concept for thermocouples: manufacturing of the base body, f.h. – foot height (a), insertion of the thermocouples (b), continuing of the build process (c), test specimen with integrated thermocouples (d).

The functionality of the sensors after integration was confirmed on a heating plate with a set temperature of 200 °C. After 10 minutes, the temperatures given by the thermocouples show values from 70 °C to 106 °C. Origin of this variation may be the inhomogeneous temperature distribution of the heating plate, deviations in the height of the top layer and in the bonding of the thermos wires to the surrounding material.

The integration concept for the strain gauge is similar to the concept used for thermocouples. The base part is prepared by LBM using steel powder. After finishing the cavity for the strain gauge, the manufacturing process is suspended, the powder sucked from the cavity and the strain gauge placed horizontally in the cavity, flush to the surface of the part. The cable is routed down the prepared channel. Finally, the next powder layer is applied and the LBM process continued until the top layer with a thickness of 0.3 mm is finished.

For the high temperature strain gauges only the integration concept was proved so far using dummies made from chromium-nickel steel. After additive manufacturing of the basic bodies, the strain gauges are welded in the LBM machine onto the basic body (Figure 6). Number and size of the welding spots were varied (Figure 6 c, Table 1). The distance between two adjacent welding spots is 1.5 mm. In the samples with smaller welding spot width (samples 1 and 3) the strain gauge is completely connected to the base body.

After welding, the additive manufacturing process is continued and the cover layer prepared. The cross-section of sample 3 demonstrates the successful integration of the strain gauge (Figure 6d).

**Figure 6.** Integration of strain gauges: manufacturing of the base body (a), insertion of the strain gauge dummies (b), variation of joining design (c), cross-section of an integrated strain gauge (d).

The integration of the capacitive differential pressure sensor using an active fluid medium is shown in Figure 7. To verify the integration concept housings were built by LBM. The membrane thickness of the as built housings varies from 3.7 mm to 4.1 mm. After separation from the build platform the defined membrane thickness of 2.7 to 3.2 mm is reached (Figure 7b). Each specimen is then threaded with a 6.35 mm (1/4 inch) bore diameter (Figure 7c). The flange surface is polished (Figure 7d). Then the active medium is filled into the cavity of the housing. Finally, the pressure sensor is screwed in. The pressure acting on the surface of the membrane of the housing is transformed by the medium into hydraulic pressure and measured by the pressure sensor.

**Table 1.** Overview on samples for embedding of strain gauges.

| | | Welding spot size | |
|---|---|---|---|
| Sample | Joining design according to Figure 4c | Length (μm) | Width (μm) |
| 1 | Type A | 500 | 250 |
| 2 | Type B | 500 | 500 |
| 3 | Type B | 1 000 | 250 |
| 4 | Type A | 1 000 | 500 |

For verification of the measuring principle the manufactured test specimen are characterized in a universal compounding machine (PROMESS) with force-distance monitoring. The force was varied from 10 kN to 40 kN for each sample and the voltage of the pressure sensor recorded. A nearly linear relationship between the applied force and the measured voltage was verified (Figure 7e).

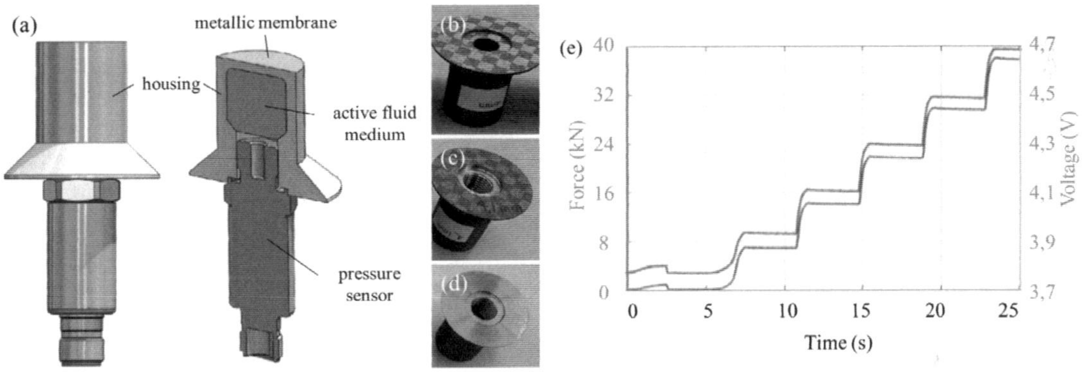


**Figure 7.** Design of the pressure sensor test specimen (a), housings: as built (b), threaded (c), polished (d), result of the force-distance monitoring: applied force and corresponding voltage in dependence of the time for a membrane thickness of 2.9 mm [8] (e).

### *3.2. Smart Tools – Embedding of thermocouples into hot sheet metal forming tools*

Automotive parts made from high-strength steels formed by press hardening offer enormous potential for lightweight design. By use of this technology, the sheet metal is heated above 900 °C, the austenitizing temperature, and subsequently quenched during the forming operation to below 200 °C, forming a martensitic microstructure.

A tool for forming of gear pans from coated heat-treatable boron steel (22MnB5) was designed and an innovative cooling system developed using thermal und forming simulations [3]. For comparison of the simulated results with the forming process the embedding of thermosensors was necessary to enable a close positioning near to the contour and therefore showing short reaction times.

To find the optimum position of the thermocouple in the tool preliminary experiments were conducted. The additively manufactured specimens contained two curved channels with large radii for feeding the wires into the tool. To encapsulate the thermocouple, a top layer is manufactured by LBM. Powder application immediately after the embedding is critical, because the measuring tips of the wires must not protrude the surface of the solidified part. Therefore, the channel end was recessed to integrate the thermocouples as flush as possible (Figure 8).

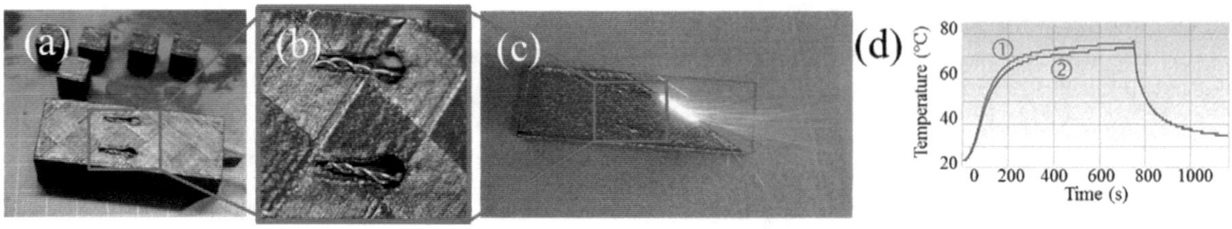


**Figure 8.** Integration of a thermocouple into a test specimen: base body with inserted thermocouples (a), detail of the thermocouples (b), additive manufacturing process of the top layer (c), temperature profiles of the two thermocouples inside the specimen (d) [3].

For proof of concept the temperature profiles of the two thermocouples inside the specimen were monitored on a heat plate. Small temperature differences between the sensors may be explained by the quality of the joint between the thermocouple and the additively manufactured part. Characterization of the joints shows a material bonding of the thermocouple to the surrounding material.

To produce the punch for forming tests, a thermocouple was integrated 3 mm below the contour of the punch of the tool (Figure 9). During additive manufacturing of the top layer, the temperature was recorded for functional qualification. The peaks in Figure 9d clearly show when the area above the thermocouple is exposed to the laser in each layer. The measured maximum temperature is 815 °C and well below the melting temperature of the steel (about 1400 °C), due to the short exposure time to the laser and the limited thermal conductivity of steel. During manufacturing, the part heats up to about 135 °C. To minimize the residual stress and to adjust the mechanical properties of the tool a heat treatment completed the manufacturing process.

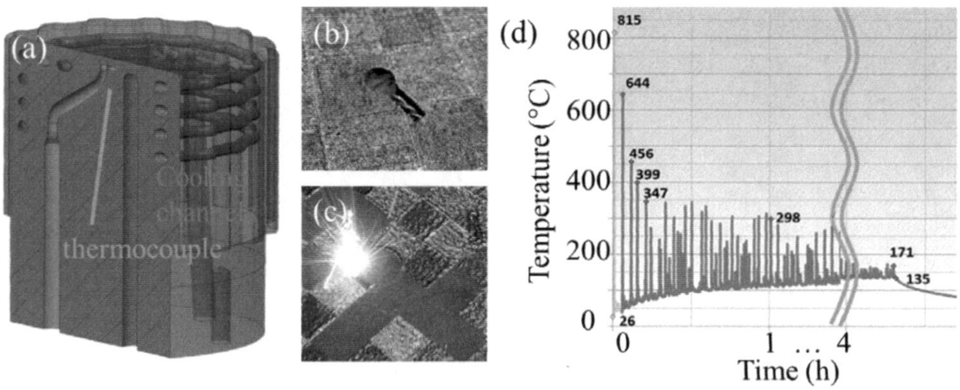


**Figure 9.** Design of the punch (a), thermocouple inserted into the punch (b), embedding of the thermocouple by LBM (c), measured temperature during additive manufacturing (d) [9].

To confirm the results of the simulation and of the pre-tests, extensive forming experiments were carried out under production-like conditions. Holding and cooling times were varied, starting at 10 seconds, a typical value for press hardening, and reduced to only 3 seconds. The thermocouple detected a tool temperature difference of just 10 K between the start of cooling at maximum tool temperature (closing of the tool) and minimum temperature at extraction of the part (opening of the tool). The entire process reached a steady state temperature after only 10 forming cycles (Figure 10). Using the optimized, additively manufactured tooling, the cooling time could be reduced by up to 70% from 10 s to 3 s, for this type of component.

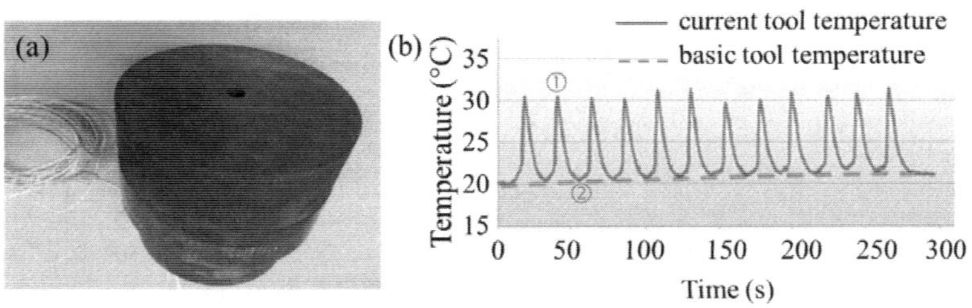


**Figure 10.** Manufactured punch with integrated thermocouple (a), temperature profile of the punch over 12 forming cycles with 3 s holding time, ① - closed forming tool, ② - opened forming tool (b) [3].

### *3.3. Smart implants*

Important attributes of endoprosthesis are biomechanical and material properties as well as the realization of a stable implant-bone-interface. The latter can be characterized by imaging technologies (e.g. X-ray) or functional diagnostics. However, these are not regular examinations and an implant loosening is often detected too late. A routine monitoring by means of an implant-embedded system would be useful. This paper describes the integration of a materially bonded, hermetically encapsulated actuator in a hip implant for a defined vibrational excitation within the implant and the surrounding bone for monitoring implant loosening and structural health. Variations in the implant-bone-interface lead to a shift in the natural frequencies, which can be detected by wireless data transmission.

The hip stem (Figure 11) was additively manufactured by LBM from Ti6Al4V titanium alloy powder. For this material, a heat treatment is necessary for adjusting micro-structural and mechanical properties and to relieve residual stresses. However, conventional post-process heat treatments cannot be applied due to the embedded temperature-sensitive actuator system. Therefore, in a first step only the base body of the hip stem including a cavity for incorporation of the actuator system is manufactured, cleaned and subsequently heat-treated. After the heat treatment, the alloy contains alpha and beta phases. The encapsulated actuator is then placed into the cavity flush to the top surface. Afterwards, the part is positioned in the LBM machine and the additive process is continued till the part is finished. After support removal, the cone for fitting the ball head and a mounting hole for the impactor was machined and the unmachined areas are shot-blasted. The finished part is shown in Figure 11b.

Computer tomography verifies the material bond between the actuator system and the hip stem (Figure 11c). Static properties of the completed hip stem correspond to the properties of the volume fractions of the heat-treated and the as-build parts.

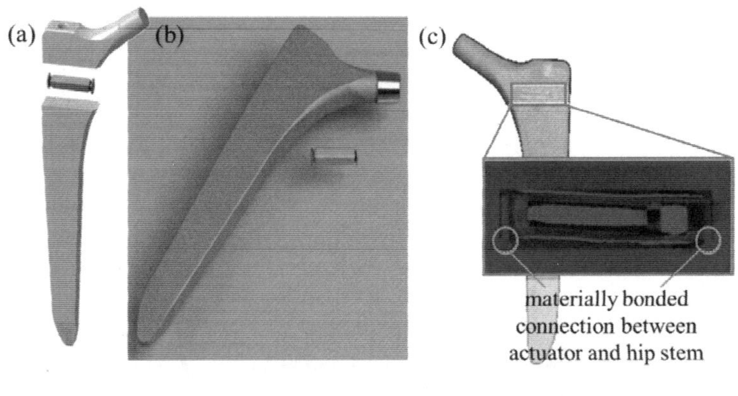


**Figure 11.** Principle of actuator integration (a), manufactured hip stem with embedded actuator system (b), computer tomography scan with detailed view of the actuator system (c) [2].

Dynamic measurements experimentally validated the functionality of the embedded actuator. A numerical modal analysis using a finite element model is utilized to identify the initially unknown natural frequencies. The first two modes for lateral and vertical bending of the hip stem are located at 2180 Hz and 3488 Hz, resp. (Figure 12). These frequency values are used for the excitation for the next investigations. The analysis of the dynamic response of the implant was carried out using a 3D laser scanning vibrometer to measure the surface velocities in three spatial directions. The experimentally determined frequencies inducing the largest displacement and velocity at 2135 Hz and 3416 Hz deviate slightly from the simulation values. The maximal surface velocity at the stem tip reaches about 15 µm/s and 7 µm/s, respectively [2].

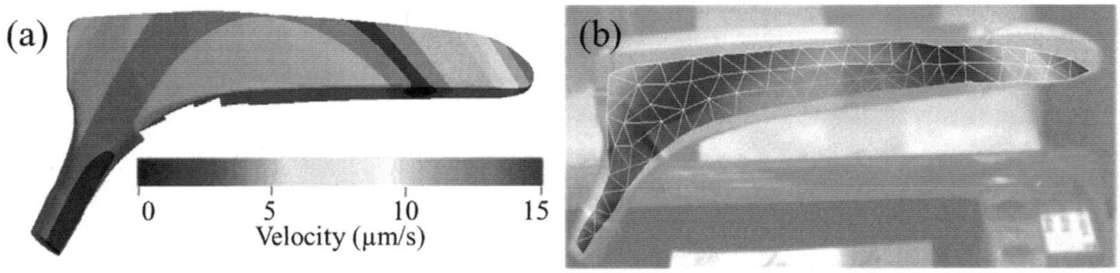


**Figure 12.** Result of the FE-model based modal analysis (a), functional qualification by laser scanning vibrometry (b).

## 4. Summary and conclusion

Additive manufacturing technologies constantly widen the scope of their economic viability and increasingly answer the requirements of smart devices. In this paper, a general methodology for the integration of sensors and actuators into metal parts was successfully applied. Due to the material bonding between the functional component and the surrounding material short reactions times and high sensitivity were ensured even under harsh process conditions. The direct embedding into the structural component enables monitoring of main process parameters near the surface of the tools without weakening the mechanical properties. Functional qualification of the embedded sensors and actuators verifies the usability of the integration concept.

For an industrial application of the integration principle, the reproducibility of the performance of the functional components and the automatization of the process chain have to be improved. Application-oriented development of smart products is intended together with machine manufacturers and end users.

## Acknowledgments

The results presented in this paper are based on several funded projects. The author sincerely acknowledges funding from the Manunet project HiperFormTool (02PN2000) by the German Federal Ministry of Education and Research (BMBF) within the Framework Concept "Research for Tomorrow's Production", from the project FunGeos by the BMBF under the grant mark 03ZZ0208A, and from the Fraunhofer Lighthouse Project "Theranostic Implants". The responsibility for the content of this publication lies with the author.

## References

[1] Lehmhus D, Aumund-Kopp C, Petzoldt F, Godlinski D, Haberkorn A, Zöllmer V and Busse M 2016 Customized Smartness: A Survey on links between Additive Manufacturing and Sensor Integration *Procedia Technology* **26** 284-301

[2] Töppel T, Lausch H, Brand M, Hensel E, Arnold M and Rotsch C 2018 Structural Integration of Sensors/Actuators by Laser Beam Melting for Tailored Smart Components *JOM* **70** 3 321-327

[3] Gebauer M, Stoll P, Müller B, Spierings A, Polster S, Feld T, Klinger M and Zurbrügg A 2016 High performance sheet metal forming tooling by additive manufacturing *Proc. of 6th International Conference on additive Technologies iCAT Nürnberg*, ed. I Drstvenšek and M Schmidt (Ljubljana: Interesansa-zavod) pp. 354-361

[4] Lausch H, Töppel T, Petters R, Gronde B, Herrmann M and Funke K 2016 Multi-material approach to integrate ceramic boxed temperature-sensitive components in laser beam melted structures for bio and other applications *Fraunhofer Direct Digital Manufacturing Conference DDMC* Berlin, ed. B. Müller (Stuttgart: Fraunhofer Verlag) pp. 261-268

[5] Spierings A B, Levy G, Labhart L and Wegener K 2011 Production of functional parts using SLM – Opportunities and limitations vol 5, *Innovative developments in virtual and physical prototyping : proceedings of the 5th International Conference on Advanced Research and Rapid Prototyping, Leiria, Portugal*, ed. P J Bártolo et al. (Boca Raton: CRC Press) pp. 785-790

[6] Hsu F, Wang K, Huang C and Chang R 2013 Investigation on conformal cooling system design in injection molding. *Advances in Production Engineering & Management* **8** 2 107-115

[7] Ogorodnyk O and Martinsen K 2018 Monitoring and control for thermoplastics injection molding. A review *Procedia CIRP* **67** 380-385

[8] Ye Y 2018 *Sensorintegration in generativ gefertigte Werkzeuge* Master Thesis (TU Braunschweig)

[9] Gebauer M 2017 Hochleistungswerkzeuge für die Blechumformung mittels Laserstrahlschmelzen *Proc. of the 14$^{th}$ Rapid.Tech Conference Erfurt*, ed. M Kynast, M Eichmann, G Witt (München: Carl Hanser Verlag) pp. 140-150

IOP Conf. Series: Materials Science and Engineering **480** (2019) 012017 doi:10.1088/1757-899X/480/1/012017

# WeldForming – a new inline process combination for the improvement of weld seam properties

**S Härtel[1,*] and T-E Adams[2]**

[1] Professorship Virtual Production Engineering, Institute for Machine Tools and Production Processes, Chemnitz University of Technology, 09107 Chemnitz, Germany
[2] Chair of Welding Engineering, Institute of Joining and Assembly, Chemnitz University of Technology, 09107 Chemnitz, Germany

[*] e-mail: sebastian.haertel@mb.tu-chemnitz.de

**Abstract.** The arc welding process reduces the components' prospective lifetime due to the high heat input and the resulting inhomogeneous crystalline structure. In order to compensate this disadvantage, subsequent treatment processes decoupled from the welding process are state of the art. This paper presents the novel process combination, WeldForming, which uses the heat generated during the welding process for a subsequent forming process in the seam area. The goal is to change the microstructure of the welded joint by using thermal forming processes right after the welding by inducing recrystallization processes in order to avoid the classic zone formation - coarse grain, fine grain and intercritical zones. The process development is based on numerical simulations. For the first time, a numerical model was developed in which a welding and a rolling process can be combined in a single simulation model, taking into account the microstructural changes. The necessary material model for the description of the microstructural changes (recrystallization behaviour) is based on thermophysical simulations. A novel approach was the development of a material model for the filler metal G4Si1 in its cast-like state, reflecting the microstructure developing during the solidification of the weld metal. By combining the knowledge gained from the thermophysical simulation and the numerical process simulation, a process window for the novel methodology of WeldForming could be identified. On the basis of this, a test setup was developed with which the functioning of WeldForming was proven.

## 1. Introduction

In industry, welding is one of the most frequently used methods to join metal sheets and other product forms. Welding is a process in which two or more parts are joined permanently at their contact surfaces by a suitable application of heat or pressure, or a combination of both. Often a filler material is added to facilitate metallic continuity [1]. Depending on the initial geometry and material or other problem specifications, different welding methods may be used. In most cases, the joining zone and an optional filler material are partially liquefied in order to achieve metallic continuity after solidification.

In fusion welding, energy input is integral to all welding processes, which can lead to extensive transformations and changes of the microstructure and properties of these materials. This applies in particular to steel in the area of the weld metal [2, 3]. The characteristics of heat input during the welding process will lead to temperature gradients and different cooling rates during the solidification of the weld pool. This can cause an irregular distribution of alloying elements (e.g., segregation) and various microstructural features after solidification may evolve. The resulting weld seam microstructure most closely resembles as a cast-like microstructure. The part of the base material which is altered by the weld thermal cycle is called heat-affected zone (HAZ). This zone is exposed to thermal cycles but is not melted. If high heat input welding, e.g. gas metal arc welding (GMAW) is applied to steels, the microstructure of the HAZ close to the fusion line becomes notably coarse and the toughness of this zone is strongly degraded [4]. Figure 1 shows the characteristic appearance of a weld seam in a cross-section.

The characteristic of the HAZ affects the mechanical properties of the weld seam. Depending on the welding process and the distance from the weld seam, microstructural imperfections may occur. This can in turn lead to undesirable properties of the whole assembly, e.g. the joint has inferior strength relative to the unaffected base material [5]. Experiments conducted on high-strength low-alloy steels by Mohandas [6] reveal the effect of the microstructural changes in the HAZ on the overall mechanical properties of the assembly.

When these steels undergo weld thermal cycles they exhibit softening in a certain region of the HAZ. The degree of softening in the HAZ is thereby a function of the weld thermal cycle, which is a characteristic of the welding process [7]. The softening characteristics also depend on the kinetics of the phase transformations of the steel and are a function of the chemical composition of the steel [8]. To reduce these negative effects of the welding procedure and restore properties, additional steps in the form of post-treatment processes (e.g., mechanical or thermal treatments) are typically necessary. The established processes are able to reduce residual stresses (stress relief heat treatment), notch sharpness, and near-surface weld seam defects (e.g., TIG melting) in the course of separate thermal post-processing steps. However, there is no targeted optimization of the weld seam microstructure available at present. All current post-processing processes are decoupled from the welding process and thus require subsequent processing steps. The introduction of the WeldForming process intends to make the additional decoupled post-treatment processes for welded components obsolete. The main advantage of the WeldForming process is the targeted adjustment of the microstructure and properties of a welded joint. The intention is to restore the mechanical properties of the joint by producing a base material-like microstructure all over the weld through recrystallization processes. This can be achieved through coupling a welding and a subsequent forming process, directly utilizing the heat from the welding process to initiate recrystallization. The thermal microstructure optimization by means of dynamic, metadynamic and static recrystallization will be stimulated by the forming process. In addition to the recrystallization processes, flattening of the weld seam will occur and increase the dynamic strength of the assembly by reducing critical notches. A schematic of the WeldForming process is shown in Figure 2.

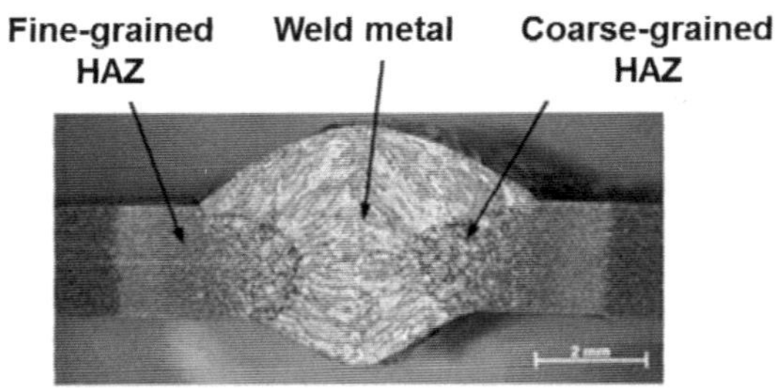


**Figure 1.** Macroscopic cross-section of a weld seam produced by gas metal arc welding.

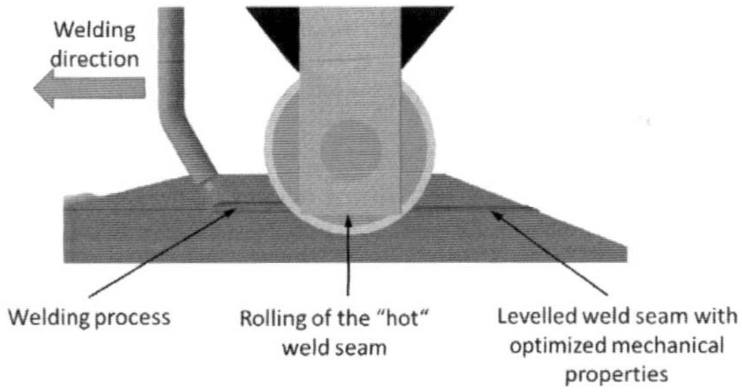


**Figure 2.** Schematic of the WeldForming process.

In the newly developed WeldForming process, the post-treatment of the weld seam is carried out inline - directly after the welding process in one process stage. This means that no additional process steps are necessary and an additional heating (as in a stress relief treatment) is obsolete. The heat of the welding process is directly utilized, thus resources in the form of energy and time are conserved.

IOP Conf. Series: Materials Science and Engineering **480** (2019) 012017 doi:10.1088/1757-899X/480/1/012017

## 2. Experimental and materials

### *2.1. Experimental setup*
In order to implement the WeldForming process experimentally, a water-cooled GMAW-welding torch was integrated into an existing Duo-Reversier cold rolling mill. The minimum possible distance, construction wise, between the welding torch and roll center is 130 mm, as shown in Figure 3. The welding speed in this setup is controlled, by adjusting the revolutions per minute of the rolling mill.

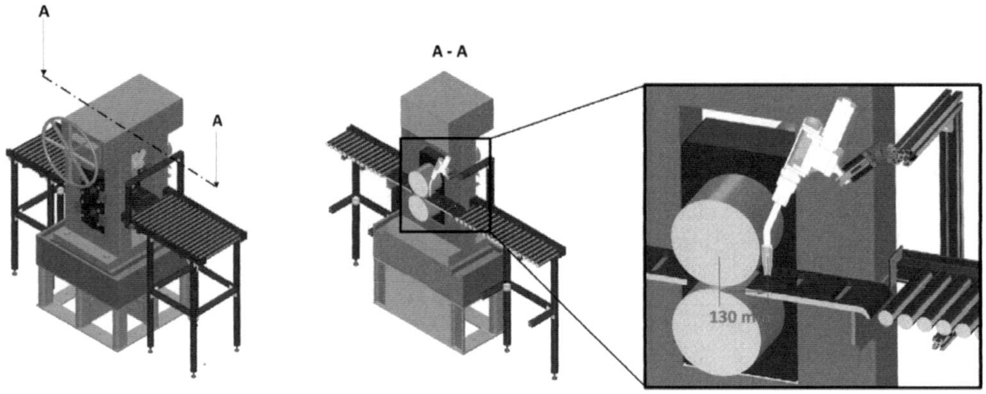


**Figure 3.** Experimental setup for the WeldForming process.

### *2.2. Welding tests*
The basis for the development of WeldForming is the welding process. Welding tests have been carried out in advance to determine a process window in which a stable welding process with a high weldreinforcement could be established.

The welding tests are carried out on the normalized steel S235 (1.0037) with an overmatching design of the welded connection by the additional filler material G4Si1 (1.5130). This means that the additional filler wire used for welding is higher alloyed than the base material. The two plates (thickness: 4 mm) are butt-welded together with a sheet-thickness-dependent seam geometry according to DIN EN 9692-1 and with a defined weld reinforcement and excess penetration. The reference welds are carried out on a FroniusTransPuls Synergic 5000 CMT R welding power source with a connected VR 700 CMT wire feeder. To ensure the required reproducibility of the results, the welding torch is guided by a six-axis articulated arm robot produced by Comau.

Figure 4 shows cross-sections of welding seams and the corresponding process parameters for different welding speeds of 0.7 m/min, 0.9 m/min and 1.4 m/min. It can be seen that at lower welding speeds of 0.7 m/min and 0.9 m/min, a higher weld reinforcement can be achieved. The welding arc burns longer at one point and more filler material can be melted per time unit. However, if the welding speed is too low, there is a risk of melt pool overheating, resulting in weld defects.

The cross-sections of the specimens with a welding speed of 0.7 m/min and 0.9 m/min show a comparable seam geometry regarding the weldreinforcement. However, the sample of the lower welding speed shows a lower penetration. This can be recognized by the flat seam flanks in the upper part of the sample. This means that less base material is melted and less thermal energy is introduced in the joint area.

The welding power is almost identical for the welding speeds of 0.7 m/min and 1.4 m/min with approx. 9.0 kW. A maximum welding power of 11.4 kW is achieved at a speed of 0.9 m/min.

To which extent the differences in the seam geometry, the welding power and the thermal in- and output influence the process will be numerically analysed in the presented paper.

| | |
|---|---|
| welding speed | 0.7 m/min |
| wire speed | 9.0 m/min |
| power | 8.6 kW |

| | |
|---|---|
| welding speed | 0.9 m/min |
| wire speed | 9.5 m/min |
| power | 11.4 kW |

| | |
|---|---|
| welding speed | 1.4 m/min |
| wire speed | 11.4 m/min |
| power | 9.0 kW |

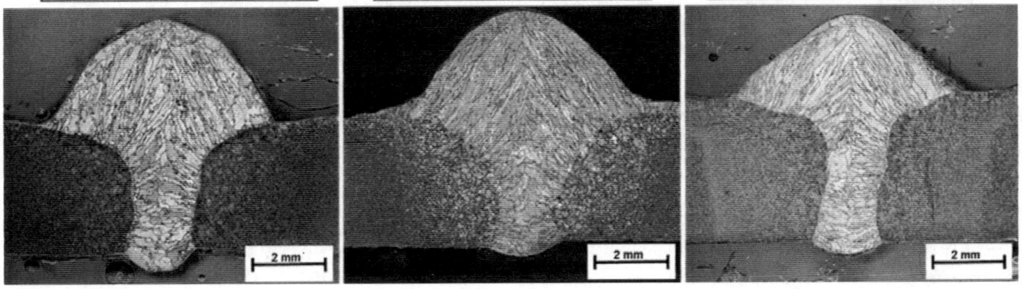


**Figure 4.** Geometry of the seam and the process parameters depending on the welding speed.

*2.3. Material characterization*

The aim of the WeldForming process is to transform the brittle, cast-like microstructure in the weld metal into a finer-grained and thus ductile structure. To predict this transformation numerically (see chapter 3.4), it is necessary to determine the flow curves and the recrystallization behaviour. For this, thermo-physical material tests were carried out by using the forming and quenching dilatometer DIL 805 A/D. The initial microstructure of the cylindrical samples has to be identical to the cast-like microstructure of the weld metal after the welding process. For this, an adjusted melting process of the filler material with subsequent cooling (cooling rate similar to the welding process) was performed (see [10]).

To determine the flow curves cylindrical upsetting tests were done. The dynamic recrystallization behaviour of the filler material in cast-like microstructure was also determined by upsetting tests. The static as well as the metadynamic behaviour were determined by relaxation tests using the forming and quenching dilatometer.

In the end, the entire recrystallization kinetics are modelled according to the JMAK theory (see [11]). Figure 5 shows an example of temperature depending flow curves and dynamic recrystallization kinetics for a strain rate of 1 $s^{-1}$.The hot flow curves were determined by compression tests up to a maximum plastic strain of 0.7. This is sufficient for the selected test conditions, since the steady-state range has already been reached and the yield stress remains almost constant for plastic strain higher than 0.6. The yield stress was assumed to be constant for plastic strains over the experimentally recorded range of 0.7. This is permissible according to [12], since the hardening and softening speeds in the steady-state range are almost identical.

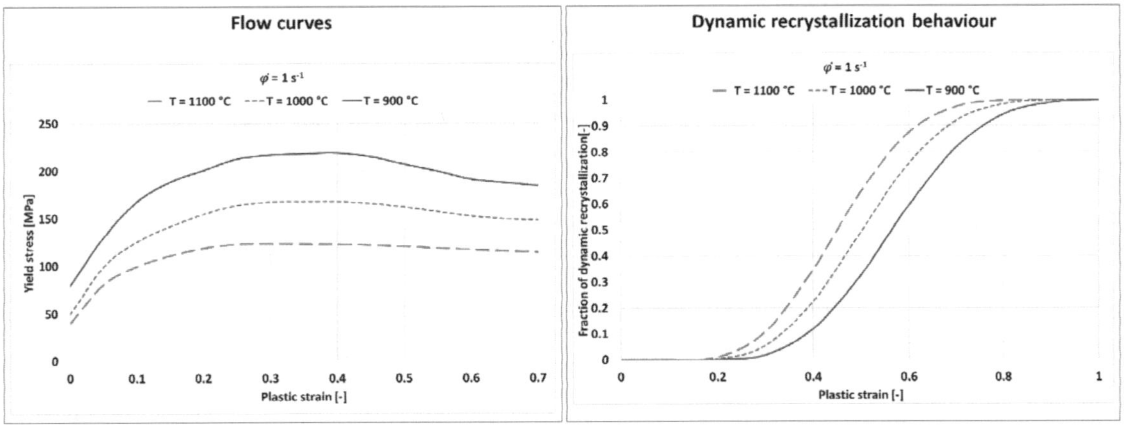


**Figure 5.** Flow curves and dynamic recrystallization for a strain rate of 1 $s^{-1}$ of weld metal.

IOP Conf. Series: Materials Science and Engineering **480** (2019) 012017 doi:10.1088/1757-899X/480/1/012017

## 3. Numerical process development

### *3.1. Principal approach*

In order to reproduce the novel methodology of WeldForming realistically, a numerical model was developed in which a welding and a forming process (hot rolling) can be combined in a single simulation model, taking into account the microstructural changes.For a realistic numerical simulation of the process in advance, the approach shown in Figure 6 is used. At the beginning, the welding process was numerically simulated and compared with the reference process (see chapter 3.2). In this way, a suitable heat source and the necessary parameters can be determined iteratively. Subsequently, a numerical simulation of a hot rolling process was performed, including a comparison with the reference test. This was done in order to determine suitable cross-linking parameters, friction models and coefficients, time increments and heat transfer coefficients (chapter see 3.3). On the basis of the validated individual processes, the rolling model was then extended by the heat source and the associated heat source parameters, so that the new WeldForming process can be reproduced realistically.

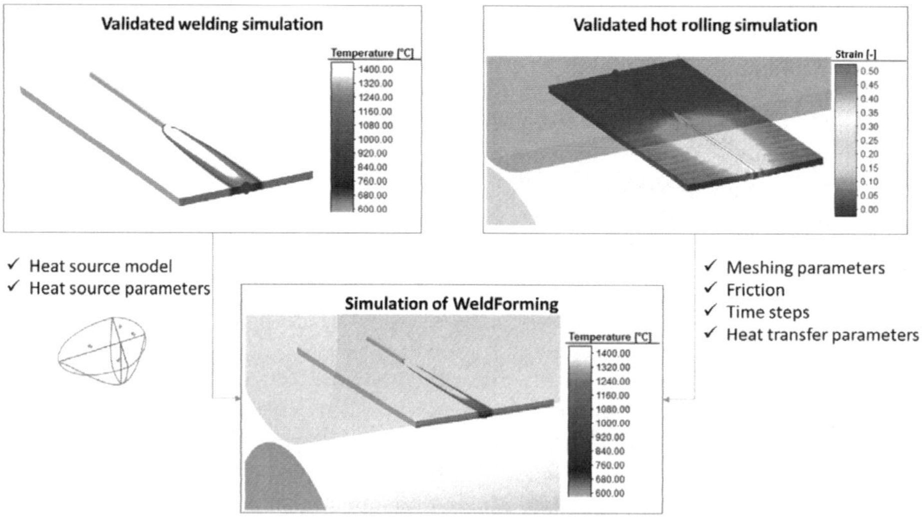


**Figure 6.** Procedure for the simulation of the WeldForming process based on validated single processes.

### *3.2. Simulation of welding process*

The simulations of the welding processes are carried out with the simufact.welding software tool. In order to ensure a realistic numerical representation of the welding process, reference tests were carried out (see chapter 2.2). On the basis of the tests, the weld seam geometry was analysed and transferred to the Finite-Element (FE) model. Furthermore, temperature measurements were carried out in order to obtain verification data for the simulation adjustment. These data are necessary to simulate the welding process using a simplified model for the heat source. For the arc-welding process used, a heat source model according to GOLDAK is recommended [9]. The welding process with a welding speed of 0.9 m/min is used as reference process. Figure 8 on the left shows the points where the temperature was measured by using thermocouples type K. By iteratively adjusting the geometric parameters for the heat source (shown in Figure 7), the temperature curve can be calculated realistically, as shown in Figure 8, right. Figure 7 shows the geometric parameters used for this purpose. The welding parameters of the reference process which are necessary for the simulation are shown in Figure 4.

IOP Conf. Series: Materials Science and Engineering **480** (2019) 012017 doi:10.1088/1757-899X/480/1/012017

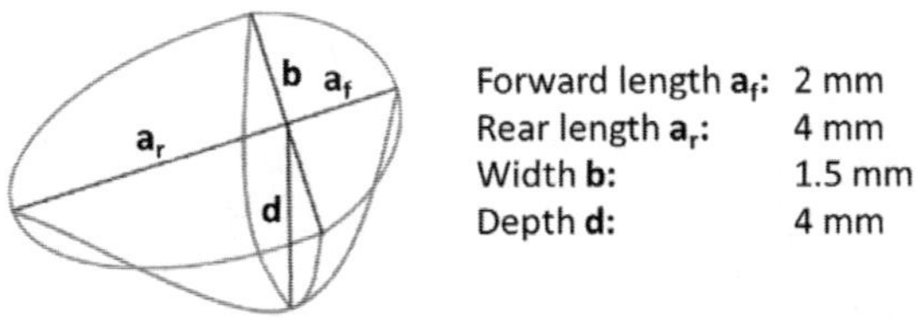


**Figure 7.** GOLDAKS double ellipsoid heat source model and used parameters.

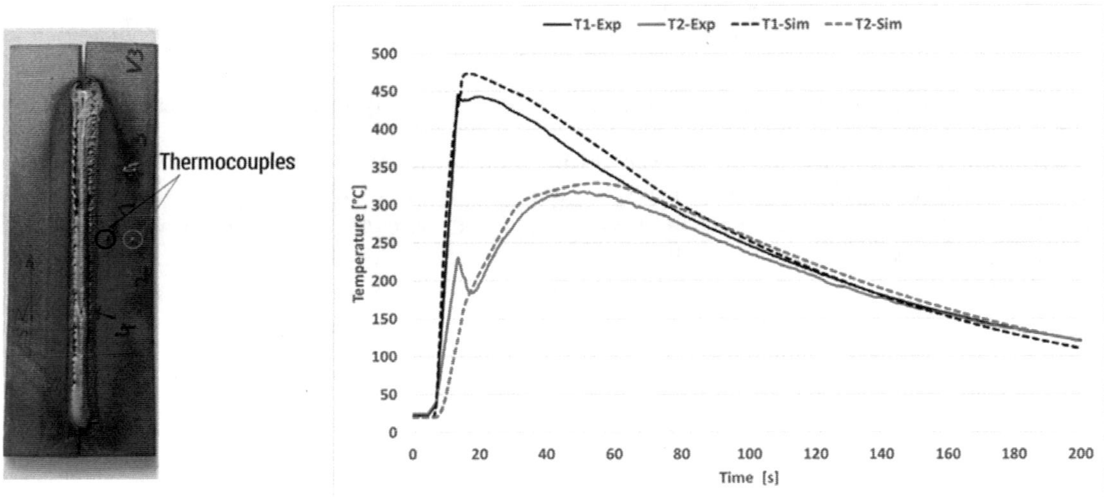


**Figure 8.** Measuring points for temperature measurement and comparison of measured and simulated temperature curves.

Advanced validation of the heat source model is necessary. With the validated model a realistic numerically prediction of the thermal heating during the simulation of WeldForming is possible.

### *3.3. Simulation of hot rolling*

The simulation of the hot rolling process was performed with the FE-tool simufact.forming. Figure 9 shows the structure of the simulation model for the hot rolling process. It can be seen that the roll intake and outtake areas are also modelled, since heat transfer from the hot sheet to the roll table takes place before, during and after the rolling process. The heat transfer coefficient between roll table and sheet is set to 2000 W/(m²·K) and between rollers and sheet 3000 W/(m²·K). The heat transfer to the environment was also considered with 50 W/(m²·K).

Figure 9 on the right also shows the modelled weld seam and the cross-linking of sheet and weld seam. The flow curves of the materials were implemented for the necessary forming temperatures (700 - 1200 °C in Δ 100 °C) and forming speeds of 1 $s^{-1}$ and 5 $s^{-1}$. The heating caused by plastic deformation was also considered. It was assumed that ca. 90 % of the plastic work are converted into heat. The validation of the hot rolling process was carried out by a comparison of the force-time curves calculated and measured in reference processes. Furthermore, the seam geometry after rolling was also compared to the simulation results. During the reference tests, the welded sheets (welding speed: 0.9 m/min; seam geometry see Figure 4, middle) were heated up to 1100 °C and1200 °C in the chamber furnace and then the entire weld sample was rolled to a sheet thickness of 3.9 mm.

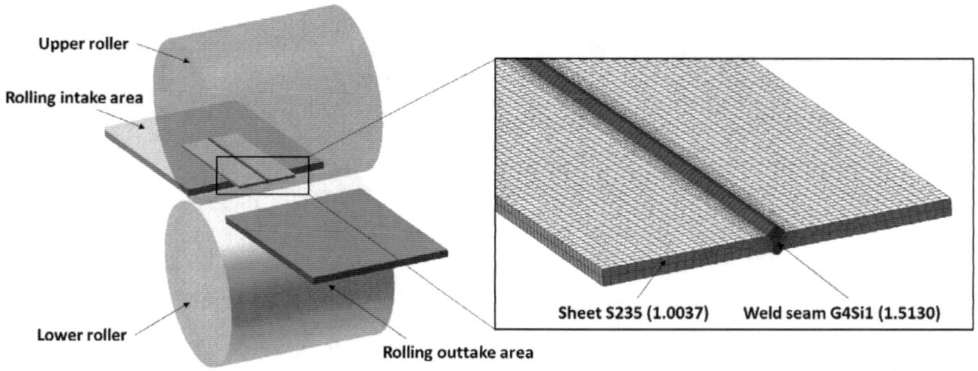


**Figure 9.** Simulation model for the hot rolling process (left) and modelled weld seam (right).

Figure 10 shows the comparison between the calculated and measured force-time curves at an initial sample temperature of 1200 °C and 1100 °C. It is apparent that both the qualitative and the quantitative force-time curves correspond very well and that the average deviation is less than 10 %. In addition to the comparison of the rolling force curves, the levelled seam was used as an optical comparison criterion. There is again a high qualitative agreement between simulation and experiment (see Figure 11). Finally, due to the constant comparison between simulation and experiments, a realistic simulation model for the hot rolling process could be developed.

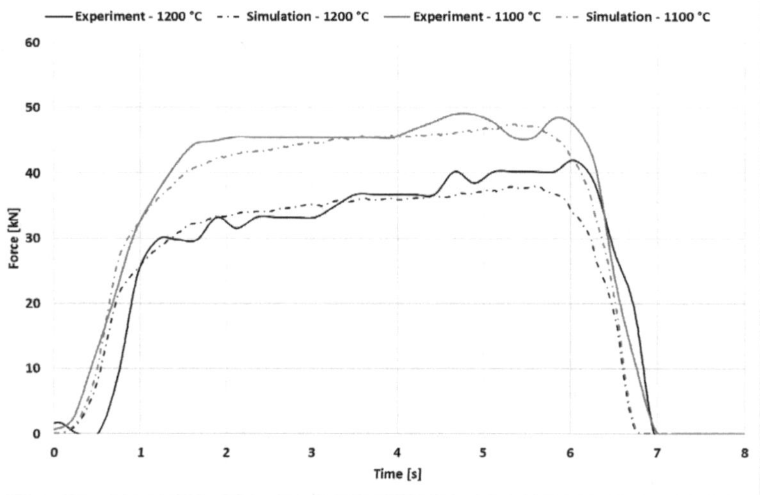


**Figure 10.** Comparison of simulated and measured force-timecurves for rolling temperatures of 1200 ° C and 1100 ° C.

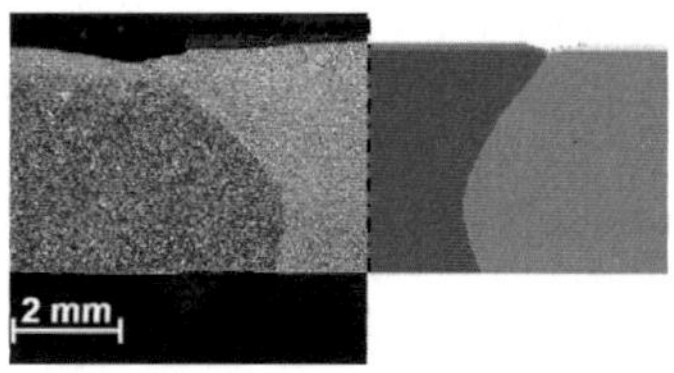


**Figure 11.** Comparison of the levelled seam after hot rolling at 1100 ° C.

### *3.4. Simulation of WeldForming*

The combined WeldForming simulation was done in the FE-tool simufact.forming. The basis is the FE model of the hot rolling process, which is extended by the validated GOLDAK heat source of the welding simulation. Since the FE tools simufact.forming and simufact.welding use the Marc-Solver, such an extension is possible.

The main goal of the WeldForming process simulation is the numerical determination of a process window, in which the WeldForming can be realized experimentally. The aim is to achieve a homogeneous microstructure by means of recrystallization in the welding zone. The recrystallization behaviour is essentially influenced by the plastic strain, the forming temperature and the strain rate.

Since the welding speed influences the seam geometry (and thus the plastic strain), the thermal in- and output (and thus the forming temperature) and the rolling speed (and thus the strain rate), the welding speed is an essential influencing factor in the microstructure homogenization by recrystallization. With the help of numerical simulations, the influence of the welding speed on the plastic strain, forming temperature and strain rate, as well as the resulting fraction of total recrystallization, will be analysed. The distance between the centre of the roll and the welding torch is set to the currently minimum possible distance of 130 mm.

Figure 12 shows the distribution of the plastic strain in the seam area. The evaluation area is located in the middle of the specimen, after a welding length of 100 mm. It can be seen that at a welding speed of 0.7 m/min and 0.9 m/min, a similar plastic strain is achieved by rolling of the weld seam, because the weld seam geometry before rolling/after welding is similar (see Figure 4). The rolling of the seam, which is welded with a speed of 1.4 m/min, leads to a lower plastic strain in the seam area, since the weld reinforcement is significantly smaller.

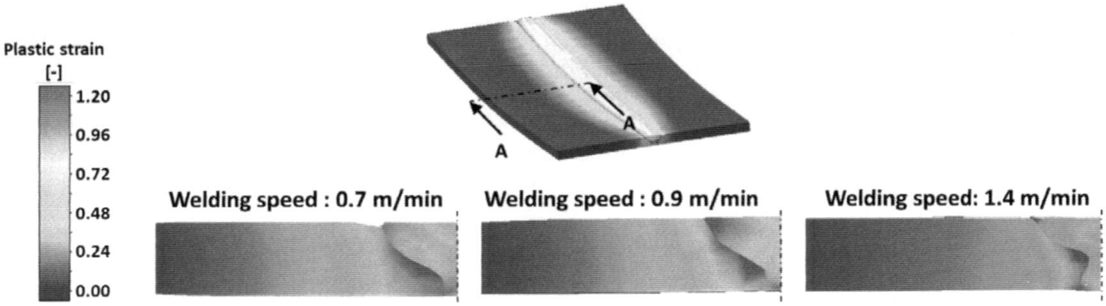


**Figure 12.** Comparison of plastic strain distribution after the WeldForming at different welding speeds.

The analysis of the occurring strain rates in the forming area shows that a higher welding speed leads to higher strain rates (see Figure 13). The strain rate has a significant influence on the dynamic recrystallization behavior. Because of that the recrystallization kinetics (see chapter 2.3) for a range of 0.5-5 $s^{-1}$was determined.

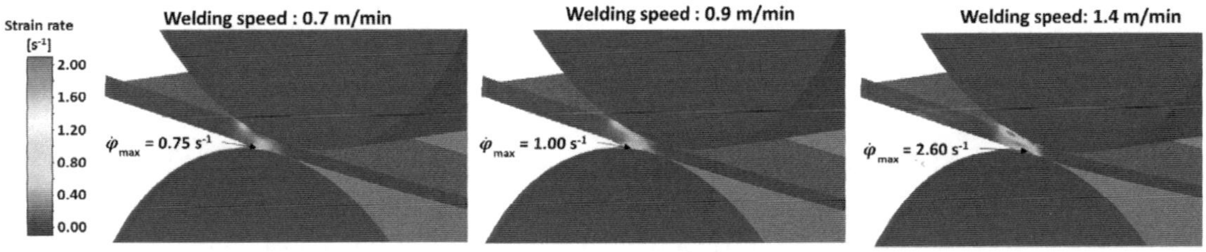


**Figure 13.** Comparison of strain rate distribution for WeldForming at different welding speeds.

The welding speed also influences the thermal in- and output in the seam area. On the one hand, a high welding speed means that due to a shorter process time less heat can be dissipated to the contacting tools and the environment. On the other hand, the higher welding speed results in less heat being introduced by the welding process, since the welding process time and power depend on the welding speed. Furthermore, the volume of liquefied material is smaller at higher welding speeds, so that less "hot" welding material and thus thermal energy is introduced. Figure 14 shows the temperature distribution during the process in a half-section. It can be seen that at a welding speed of 0.9 m/min a temperature of 890 °C is reached in the rolling gap, whereas for a welding speed of 0.7 m/min and 1.4 m/min, only 740 °C and 780 °C respectively are reached, which is significantly lower. The welding speed of 0.9 m/min thus represents the optimum parameters to achieve the highest possible temperature in the rolling gap.

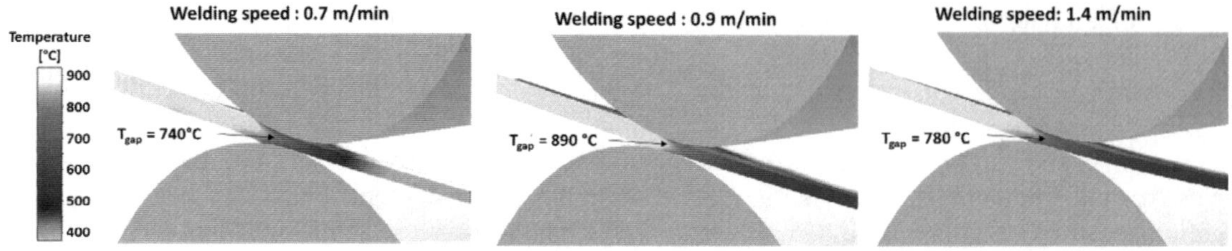


**Figure 14.** Comparison of temperature distribution for WeldForming at different welding speeds.

The parameters that were previously investigated separately have a significant influence on the recrystallization behaviour in the weld area. Using the material model developed on the basis of the JMAK theory, the total fraction of recrystallization can be calculated in the simulation. Figure 15 shows the total recrystallization behaviour at the investigated welding speeds. The highest fraction of total recrystallization is achieved at a welding speed of 0.9 m/min. According to this there is an optimum process window in which the combination of welding parameters and forming of the resulting weld seam lead to a maximum of recrystallization. A general statement that a high or low welding speed leads to the best possible result cannot be made.

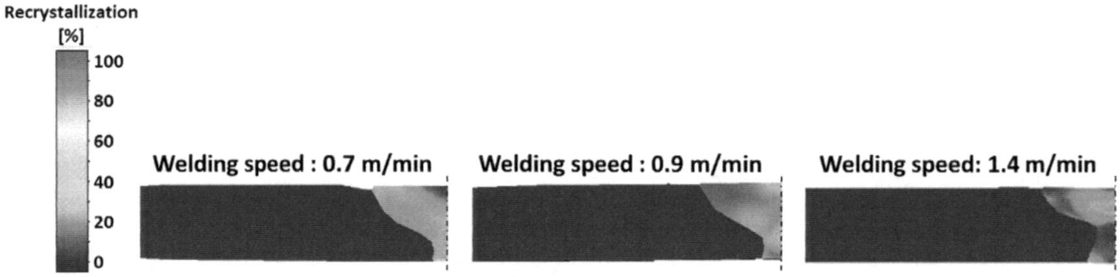


**Figure 15.** Comparison of total recrystallization after WeldForming with different welding speeds.

The simulation provides a parameter set with which the WeldForming process can now be implemented experimentally. The experimental results of the simulation are discussed in the following chapter.

## 4. Experimental results

On the basis of preliminary experimental and numerical investigations, a suitable welding speed could be determined at which a total recrystallization of approx. 75 % was numerically predicted. In order to validate this experimentally, the WeldForming process was performed with a welding speed of 0.9 m/min and the corresponding parameters (Figure 4, middle). The analysis of the microstructure in the seam area was done by light microscopy. The point of investigation is thereby identical to the simulation and located in the middle of the sample.

Figure 16 shows the real test setup as well as the microstructure as welded (Figure 16, left) and the microstructure after WeldForming. It can be seen that the typical cast-like structure of the weld metal can be transformed into a grain structure by means of recrystallization processes (Figure 16, right). Thus, the functional principle of the novel methodology of WeldForming is proven. However, it can also be seen that elongated grains caused by the rolling process are present in the microstructure. This suggests that the recrystallization process is not yet fully completed. This in turn correlates with the preliminary results of the numerical investigations, in which a total fraction of recrystallization of approx. 75 % was calculated.

**Figure 16.** Real experimental setup, microstructure as welded (left) and microstructure after WeldForming (right).

In addition, the intake and outtake temperatures (50 mm before and after the centre of the roll) were measured with a pyrometer. Figure 17 shows the measured inlet and outlet temperature as well as the numerically calculated inlet and outlet temperature. The difference between the measured and calculated temperatures is less than 15%, so it can be assumed that the temperature distribution in the component and its development during the process is reproduced realistically. Furthermore, the analysis of the inlet and outlet temperatures shows a rapid decline in temperature from 1130 °C to 440 °C as a result of the contact between the weld seam and the cold rolls of the rolling mill during the process. The high cooling rate prevents static recrystallization effects, which should follow the forming process, as the temperature falls below the required level needed for a static recrystallization. The required temperature to initiate the static recrystallization according to the modelling after the JMAK theory (see chapter 2.3) is approx. 650 °C, taking into account the previous forming conditions.

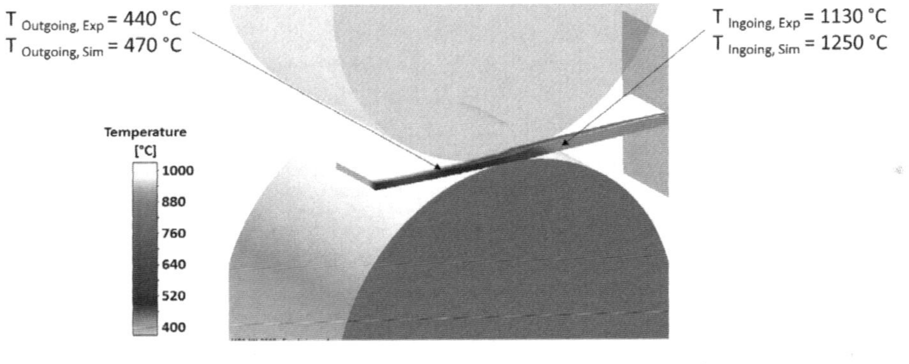


**Figure 17.** Measured and numerically determined in- and outgoing temperature.

## 5. Conclusions

For the first time a numerical process coupling of a welding and hot rolling process in one single simulation model was carried out, in the context of the present work. It is based on verified individual processes, taking into account the microstructural changes (recrystallization effects) in the welding area. On the basis of the simulation model, the influence of the welding speed on the degree of deformation, the forming temperature and the deformation rate was analysed to finally evaluate the recrystallization effects. The first assumption that a high welding speed improves the recrystallization behaviour (since less welding heat is dissipated due to the shorter process time) could be refuted. It was shown that at a welding speed of 0.9 m/min approx. 75% of the microstructure recrystallizes, whereas at a welding speed of 0.7 m/min approx. 50% recrystallizes, and at a welding speed of 1.4 m/min only approx. 30% recrystallizes. On the basis of the numerical findings, the welding process was implemented experimentally. It was proven that it is possible to transform the cast-like and brittle microstructure of the weld metal into a recrystallized grain structure as a result of hot forming. However, with the current machine concept it has not been possible to achieve a complete recrystallization. For this reason, the welding torch will be adjusted in future work to achieve a minimum distance between the centre of the roll and the welding torch of approx. 110 mm. According to numerical investigations, this distance is necessary in order to guarantee complete recrystallization in the seam area.

**Acknowledgements**
The authors gratefully acknowledge the funding by the German Research Foundation (Deutsche Forschungsgemeinschaft, DFG) within the project "Property improvement of weld seams due to an inline hot forming process", funding reference number: 326220175.

Acknowledgement also needs to go to the company simufact engineering GmbH for the support in the development of the FE model for the WeldForming process.

**References**
[1] DIN 8593-6 Manufacturing processes joining - Part 6: Joining by welding; Classification, subdivision, terms and definitions. September 2003
[2] Ashby M F and Easterling K E 1982 A first report on diagrams for grain growth in welds *Acta Metallurgica* **30** 1969–1978
[3] Ion JC, Easterling KE and Ashby MF 1984 A second report on diagrams of microstructure and hardness for heat-affected zones in welds *Acta Metallurgica* **32**1949 – 1962
[4] Kojima A, Yoshii K I, Hada T, Saeki O, Ichikawa K, Yoshida Y, Shimura Y and Azuma K 2004 Development of High HAZ Toughness Steel Plates for Box Columns with High Heat Input Welding *Nippon steel technical report* **90** 39-44
[5] Dilthey U 2005 Schweißtechnische Fertigungsverfahren 2 – Verhalten der Werkstoffe beim Schweißen (Berlin: Springer Verlag) p 124
[6] Mohandasa T, Madhusudhan Reddy G and Satish Kumar B 1999 Heat-affected zone softening in high-strength low-alloy steels *Journal of Materials Processing Technology* **88** 284-294
[7] Madhusudhan Reddy G and Mohandas T 1996 Ballistic performance of high strength low alloy steel weldments *Journal of Materials Processing Technology* **57** 23-38
[8] Lundin C D, Gill T P S and Qiao C Y 1989 Heat affected zones in low carbon micro alloyed steels *Proceeding on Recent Trends in Welding Science and Technology* p249.
[9] Hashemzadeh M, Chen B Q, Guedes and Soares C 2015 Numerical and experimental study on butt weld with dissimilar thickness of thin stainless steel plate. *The International Journal of Advanced Manufacturing Technology* **78** 319–330
[10] Adams T-E et al 2018 *DVS-Berichte DVS-Congress Friedrichshafen* vol 344 (Düsseldorf:DVS Media GmbH) pp 239 – 244
[11] Ullmann M 2014 *Rekristallisationsverhalten von geglühtem AZ31-Gießwalzband beim Warmwalzen* vol 358 (Freiberg: Freiberger Forschungshefte Reihe B Werkstoffwissenschaften)
[12] Hensel A and Spittel T 1978 *Kraft- und Arbeitsbedarf bildsamer Formgebungsverfahren* (Leipzig: Deutscher Verlag für Grundstoffindustrie)

IOP Conf. Series: Materials Science and Engineering **480** (2019) 012018 doi:10.1088/1757-899X/480/1/012018

# Pitting corrosion behavior of a laser hardened, high-alloyed steel

**T Mehner[1,*], P Landgraf[1], E Haack[2], I Scharf[1], T Grund[1] and T Lampke[1]**

[1] Chemnitz University of Technology, Institute of Materials Science and Engineering, Materials and Surface Engineering Group, D-09107 Chemnitz, Germany
[2] Inofex GmbH, Martha-Brautzsch-Str. 8, D-06108 Halle (Saale), Germany

[*] e-mail: thomas.mehner@mb.tu-chemnitz.de

**Abstract.** Laser hardening provides significant advantages in processing time, and energy consumption compared to conventional hardening processes, especially for small batches. In addition, the functional characteristics can be adjusted locally for different applications. However, the differences of the microstructural states in the laser-treated areas and the surrounding untreated base material may result in corrosion issues. In this paper, the corrosion behavior of the high-alloyed, laser-treated steel X153CrMoV12 (1.2379) used in the meat-processing industry is investigated in Hanks' balanced salt solution, a simulated body fluid, globally by immersion tests and locally by microelectrochemical investigations, depending on the measured temperature in the laser spot. Differences of the susceptibility to pitting, as well as correlation to thermodynamic calculations were found. Thus, the application-related corrosion behavior can be estimated numerically.

## 1. Introduction

Cutting tools in the meat-processing industry are often produced from high-alloyed steels. These materials provide several advantages, such as good food compatibility, non-toxicity, and hardenability. High hardness of knives is expected to be one of the most important properties and is usually correlated with long service life. However, during service, the knives are exposed to meat and the related corrosive media (e.g., blood), which can cause corrosion issues. Especially, pitting corrosion will occur in these media because the passive layer receives mechanical damage during cutting and is chemically attacked by $Cl^-$ ions.

For conventionally produced knives, annealed high-alloyed steel with low hardness is machined in order to obtain the basic shape of the knife with low tool wear. This is followed by hardening in a furnace with subsequent quenching and tempering. A high hardness of the cutting edge while maintaining good ductility of the knife's body can be achieved by local hardening with a laser. Especially for small batches, the laser hardening provides low machining time in combination with easy adaption to different shapes of the knives. Thus, a time- and energy-efficient machining is possible. However, chromium carbides are present (especially in the non-hardened base material) and the local chromium content is not sufficiently high for the formation of a stable passive layer. As a result, in the body and the transition zone to the laser-hardened areas, the knife is expected to be susceptible to corrosion. As laser hardening is a non-equilibrium process, the laser track's pitting resistance depends on the temperature in the laser spot ("laser temperature") in combination with the laser feed. In addition, the volume of the surrounding material impacts the heating and cooling rates, leading to a complex lateral and in-depth temperature distribution. This is of major importance as the thermodynamic behavior of some high-alloyed steels show specific characteristics that do not occur for low-alloyed materials. As an example, the solubilities of C, Mo, and Cr in the austenite during the austenitization of the material X153CrMoV12 (1.2379) are very sensitive to the temperature (see Figure 1 in section 3). Thus, when laser treated, it is expected that laser temperature and laser feed rate (laser travel speed) during the laser-hardening process will greatly impact the local chemical composition of the austenitic solid solution. In consequence, local differences of the corrosion resistance of the laser-treated areas are expected.

According to furnace hardening processes, the hardening temperature of X153CrMoV12 is set to 1050 °C. However, when applying laser hardening, the above-mentioned temperature distribution within the work piece results in hardness gradients and very low hardening depths when using a laser temperature of 1050 °C. In order to achieve sufficient hardness levels and hardening depths, it is usually recommend using a laser temperature that is 150 K above furnace hardening temperatures, i.e., 1200 °C for the selected base material [1]. However, this value as well as the heating and cooling rates strongly depend on the laser feed rate. Thus, even higher laser spot core temperatures might be necessary. The required adaptions of the laser-hardening process are barely investigated. Especially for high-alloyed steels, only insufficient data for laser hardening are available.

Besides hardness and hardening depth, further properties of X153CrMoV12 are scarcely considered in the literature [2]. Thus, in the present study, the corrosion behavior of this material is analyzed. By means of local current-density–polarization measurements with a microcapillary tube, the pitting potential is investigated for different laser temperatures and feed rates in order to determine the influence of the laser-hardening process on the pitting resistance. The experiments are supported by thermodynamic calculations.

## 2. Experimental and materials

The chemical composition of the material used for the production of the specimens is summarized in Table 1. Sheets with the size of 280 mm × 30 mm and a thickness of 8.5 mm were laser-hardened centered on the surface of the sheet along the long axis with a 3 kW diode laser. The laser hardening was performed with laser temperatures of 1050 °C, 1200 °C, 1280 °C, and 1350 °C each with a laser feed rate of 1 mm/s or 6 mm/s. Subsequently, the sheets were cut into specimens with the size of ≈10 mm × 30 mm × 8.5 mm and were grinded and polished according to standard procedures.

**Table 1.** Chemical composition of the used material X153CrMoV12 (main elements).

| Element | C | Cr | Mo | V | Mn | S | P | Fe | Other |
|---|---|---|---|---|---|---|---|---|---|
| Mass fraction (%) | 1.59 | 11.62 | 0.84 | 0.55 | 0.39 | 0.014 | 0.012 | 84.1 | 0.88 |

Two corrosion-testing methods were applied at room temperature utilizing the electrolyte Hanks' balanced salt solution (HBSS) [3], a simulated body fluid that can be used for emulating the corrosive media present during the cutting of meat. As first testing method, immersion tests were carried out for six days in 100 ml of the solution. The solution was replaced after 48 h and 96 h. Only the top surface was exposed to the HBSS. The bottom and the four sides were covert by lacquer. Three different samples were tested for each laser temperature. From this, the samples' general susceptibility to pitting was analyzed by optical microscopy with regards to the different areas: laser-hardened material (center line of the laser track), base material, and transition region. Potentiodynamic polarization tests were performed as the second testing method with the microcapillary technique using a capillary tube with a diameter of 800 µm, a potentiostat IMP83 PCT-BC (Jaissle Elektronik GmbH), and a potential scan rate of 1 mV/s. The stabilization of the open-circuit potential (OCP) took about 5–6 min before each measurement. At least three measurements were performed in the center of the laser track. Before the measurements were performed, the surfaces were metallographically prepared according to standard procedures and polished using an OP-S suspension (active oxide polishing suspension, Struers) directly prior to the measurements.

Thermodynamic calculations were done using the software JMatPro V7.0 (Sente Software Ltd.). Within this software, the CALPHAD method [4, 5] was used. Hereby, using a thermodynamic database, the phase equilibria are calculated for given chemical compositions and temperatures. As a result, the phase fractions and their individual chemical equilibrium compositions are determined. For the calculations in the present investigation, the chemical composition shown in Table 1 was used in the stainless steel mode. It is important to note that a thermodynamic equilibrium state is assumed in the CALPHAD method.

The pitting-resistance equivalent numbers (PRENs) were calculated based on the results of the thermodynamic calculations. For alloys without tungsten, DIN EN ISO 15156-3:2015 defines the PREN by

$$\mathrm{PREN} = 1 \cdot w_{\mathrm{Cr}} + 3.3 \cdot w_{\mathrm{Mo}} + 16 \cdot w_{\mathrm{N}} \tag{1}$$

with the mass fractions $w$ of Cr, Mo, and N, respectively. The nitrogen content of the alloy is not considered within the present investigation.

## 3. Results and discussion

Figure 1 shows the results of the thermodynamic calculations. In X153CrMoV12, austenite occurs in the temperature range of 793–1394 °C (Figure 1, left). Due to the ferrite-to-austenite phase transformation, there is a steep increase of the austenite mass fraction to about 82 % between 793 °C and 813 °C. Type $M_7C_3$ and $M_{23}C_6$ carbides constitute the remaining 18 %. These carbides decompose with increasing temperature and are fully dissolved above 1251 °C. At 1236 °C, the maximum of the austenite mass fraction (94 %) is reached. While the carbides are dissolved, the released C, Cr, and Mo atoms diffuse into the austenitic solid solution (Figure 1, right). The maximum equilibrium solubilities for these elements occur at 1233 °C (Mo), 1237 °C (C), and 1252 °C (Cr), respectively.

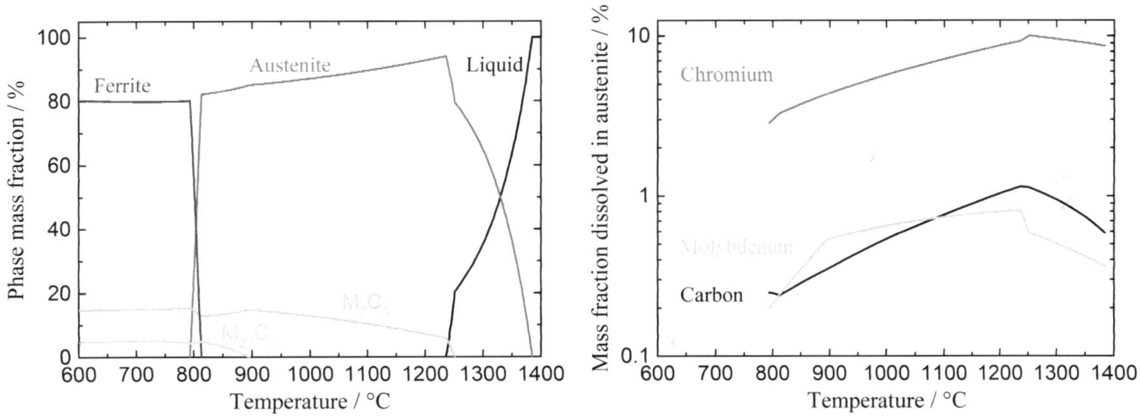


**Figure 1.** Equilibrium phase composition (left) and equilibrium mass fractions of Cr, Mo, and C that can be dissolved in the austenite (right) for the material X153CrMoV12 depending on the temperature (calculated using JMatPro).

Optical-microscopic images for laser-treated samples after immersion tests are shown in Figure 2. The zones with constant laser temperature are marked by red dashed lines. The width of transition zones and the remaining base material are marked in blue, and green, respectively. The laser track width is larger than the region with constant laser temperature due to quick self-quenching in the outward rim of the laser track. The transition region is characterized by a strong temperature gradient. The base material is not affected by the laser induced heat.

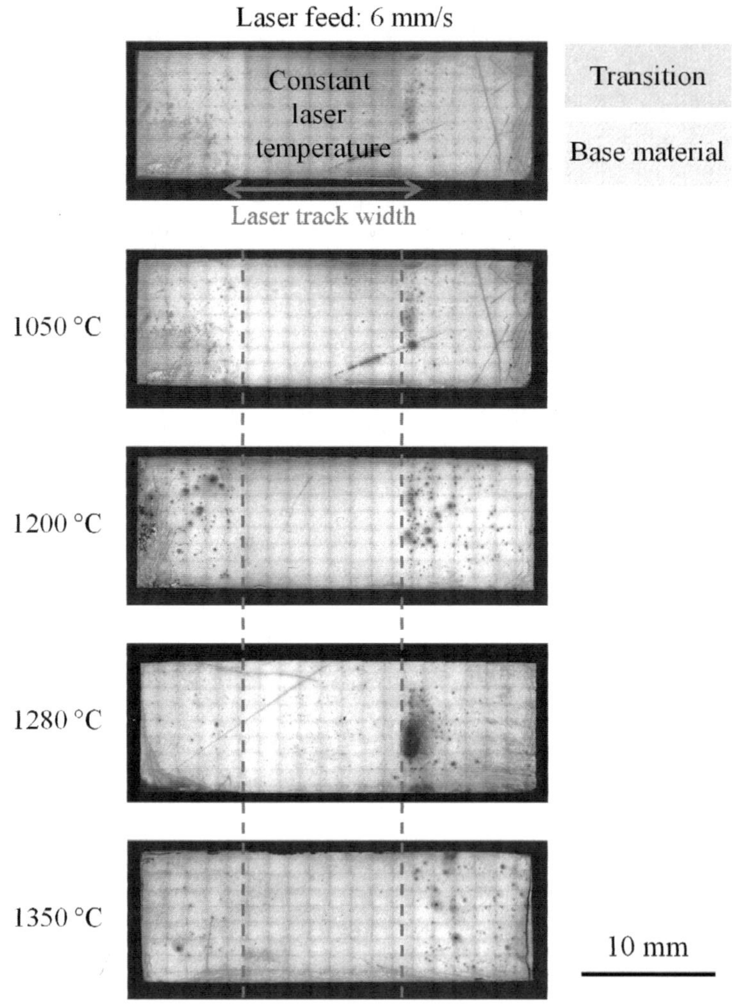


**Figure 2.** Optical-microscopic images of corroded samples (immersion tests in HBSS, laser feed 6 mm/s) after 6 d.

Within the laser track, corrosion pits are small, which indicates that pitting and repassivation occurs simultaneously within these areas. In contrast, in the transition zone, a large number of wide pits can be observed. Compared to that, the number of pits in the base-material region is lower, but pits are more developed than within the laser track. Within the temperature range that occurs in the transition zone, the precipitation of carbides or their incomplete dissolution (see Figure 1) is expected. Thus around the carbides, the mass fraction of Cr solved within the matrix is considerably below 12 %. As a result, the passive layer exhibits imperfections that promote the formation of large pits, while repassivation cannot occur. In the transition zone, this susceptibility to pitting is observed for all laser temperatures. However, it is shifted toward the laser track rim by about 1 mm for a laser temperature of 1350 °C. Similar results could be observed for the laser feed rate of 1 mm/s. As a result of the higher duration time of the material in the laser spot, visible oxidation products occurred in the center (2–3 mm) of the laser track for 1280 °C and 1350 °C.

To determine the optimum of the laser temperature with regards to pitting resistance, the pitting potential was determined in the center of the laser track by electrochemical measurements. Typical current-density–potential curves of the samples treated with different laser temperatures and feeds are shown in Figure 3.

IOP Conf. Series: Materials Science and Engineering **480** (2019) 012018 doi:10.1088/1757-899X/480/1/012018

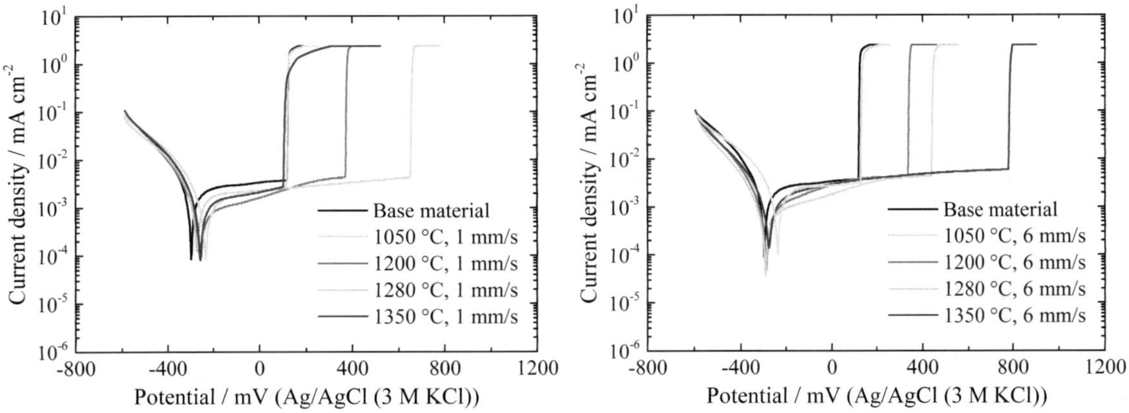


**Figure 3.** Current-density–potential curves of samples with different temperatures of the laser and laser feed rates of 1 mm/s (left) or 6 mm/s (right).

Due to the different solubility behavior of the alloying elements, the corrosion potential varies slightly for different laser temperatures. The pitting potential is clearly visible by the remarkably sharp increase of the current density. This is caused by the small area underneath the microcapillary tube (about 0.5 mm$^2$) that increases the method's sensitivity for the occurrence of pitting. The averaged pitting potentials are shown in Figure 4 along with the calculated PREN (from Equation (1) and Figure 1). As the PREN significantly changes in the temperature range of 1000–1400 °C, the pitting resistance is also strongly affected by the heat treatment temperature. The reason for this PREN change is the strong change of both the Cr and Mo solubilities in the austenitic matrix of the X153CrMoV12 material with increasing temperature.

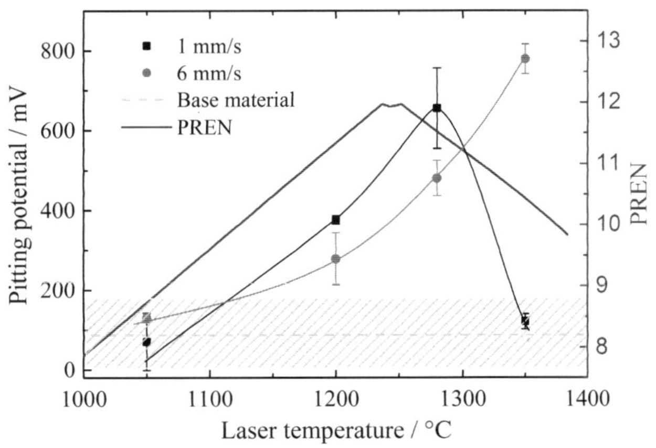


**Figure 4.** Pitting potentials for different laser temperatures and feed rates in comparison to the calculated PREN for equilibrium conditions. The black and red lines are shown to guide the eye and have no physical meaning. The error bars represent standard deviations.

For both laser feed rates, the temperature 1050 °C hardly affects the pitting potential, which is equal within the laser track and the base material. Only above about 1100 °C, the pitting potential exceeds the value of the base material. Using the laser feed rate of 1 mm/s, there is a maximum of the pitting potential (i.e., the most pitting-resistant state) at around 1280 °C. It slowly decreases towards lower temperatures and strongly towards higher temperatures. The pitting potential and the PREN exhibit similar curve shapes and their maxima occur at similar temperatures (around 1280 °C and 1240 °C, respectively). The minor difference of about 40 K is caused by the non-equilibrium conditions during the laser treatment. In the literature, shifts of Ac3 of up to 150 K compared to equilibrium conditions were reported [1, 6–8]. The resulting value depends on the heating rate and the time of the heat exposure. Due to the non-equilibrium conditions, the applied temperature has to be larger for the fast, local heat treatment by laser compared to a bulk heat treatment. This is related to the necessary diffusion of C, Cr, and Mo: The short-time heating process only allows for a short-time diffusion. However, when increasing the temperature, the diffusibility of the atoms increases [1]. As a result, the higher temperature compensates the shorter diffusion time and the resulting material state is similar to an equilibrium condition, but it occurs at a higher laser temperature. The steep decrease of the pitting potential for temperatures above 1280 °C is caused by the above-mentioned oxide formation on the surface in the center of the laser track. These oxidized areas are more susceptible to pitting due to related imperfections of the passive layer as well as differences of the electrochemical potential between small oxidic regions (hematite, magnetite) and the surrounding austenitic phases. For the low laser feed rate, the best pitting resistance of the laser track can be achieved in the temperature range of 1250–1300 °C. However, the oxidation in the center of the laser track limits the applicable laser temperature to about 1200 °C.

For a laser feed rate of 6 mm/s, the pitting potential steadily increases with increasing laser temperature within the considered temperature range. Due to the high feed rate, the treated samples do not reach the maximum pitting potential which leads to a steadily increased curve progression with increasing laser temperature. A shift of Ac3 as observed for a laser feed rate of 1 mm/s can be estimated to be at least 100 K. Hence using 6 mm/s, the heating and cooling rates apparently are in the same order of magnitude (> 1000 K/s) as described by Kostov [1]. The results prove that a high laser feed rate and a high laser temperature are advantageous for achieving good pitting resistance in the laser track for the material X153CrMoV12.

## 4. Conclusions

Pitting corrosion of the material X153CrMoV12 was studied in HBSS after laser hardening. The conventional hardening of this high-alloyed steel is usually conducted at 1050 °C. The thermodynamic calculations using the PREN indicate that the pitting resistance can be improved by increasing the hardening temperature to about 1250 °C. This is a consequence of significant changes of the behavior of the alloying elements regarding their solubility in the austenitic matrix in the temperature range of 800–1400 °C, which is much less pronounced in many other materials. Thus, the properties of X153CrMoV12 are very sensitive to the selected hardening temperature. Utilizing laser hardening with the related lateral and in-depth temperature gradients, the laser temperature for reaching the best pitting resistance strongly depends on the laser feed rate. A high feed rate results in large heating rates (non-equilibrium conditions) and to a shift of Ac3 of more than 100 K compared to conventional hardening. As a consequence, the usually assumed 150 K temperature increase required in laser-hardening processes does not result in the optimal pitting resistance, when using a high laser feed rate.

For a low laser feed rate of 1 mm/s, the pitting resistance in the center of the laser track is highest in the temperature range of 1250–1300 °C, but due to visible oxidized areas, a laser temperature of about 1200 °C should not be exceeded.

The most pitting-susceptible area is located in the transition region next to the laser track as the temperature is not sufficiently high for dissolving the initially present carbides. The related imperfections of the passive layer and differences in the electrochemical potential cause the observed strong pitting. As the temperature gradient is characteristic for the laser treatment, an appropriate heat treatment (dissolution of carbides) of the base material before laser hardening would result in an increase of the pitting resistance alongside the laser track.

Under consideration of the observed shift of about 40–70 K, the PREN can be used for predicting the laser temperature in order to gain good pitting resistance within the laser track when using a low laser feed rate. Thus, for minimized pitting, the determination of the PREN based on thermodynamic calculation is suitable for estimating required temperatures during laser treatments.

**Acknowledgments**
The authors gratefully acknowledge the financial support of the *Arbeitsgemeinschaft industrieller Forschungsvereinigungen* (AiF) in the framework of the program ZIM (project numbers ZF4116101DF5 and ZF4131901DF5).

**References**

[1] Kostov V 2014 *Untersuchungen zur zeitaufgelösten Spannungsentwicklung und Eigenspannungsentstehung beim Laserstrahlstandhärten am Beispiel des Stahls 42CrMo4* PhD thesis Karlsruher Institut für Technologie (in German)
[2] Oppenkowski A 2011 *Cryobehandlung von Werkzeugstahl* PhD thesis Ruhr-Universität Bochum (in German)
[3] Hanks J H and Wallace R E 1949 Relation of oxygen and temperature in the preservation of tissues by refrigeration *Proc. Soc. Exp. Biol. Med.* **71** 196–200
[4] Kaufmann L and Bernstein H 1970 *Computer Calculation of Phase Diagrams with Special Reference to Refractory Metals* vol 1 (New York: Academic Press)
[5] Bleck W and Moeller E 2018 *Handbuch Stahl: Auswahl, Verarbeitung, Anwendung* vol 1 (München, Carl Hanser Verlag) (in German)
[6] Schlicht H 1974 Beitrag zur Theorie des schnellen Erwärmens und schnellen Abkühlens von Stahl *Härterei-Techn. Mitt.* **29** 184–192 (in German)
[7] Miokovic T, Schulze V, Löhe and Vöhringer O 2003 Experimentelle Analyse und Modellierung des Kurzzeitumwandlungsverhaltens von Stählen am Beispiel von 42CrMo4 *Härterei-Techn. Mitt.* **58** 304–313 (in German)
[8] Pyzalla A, Bohne C Heiterkemper M and Fischer A 2001 Influence of laser rapid heat treatment on the corrosion resistance of the high nitrogen steel X30CrMoN15 1 + 0,3%N *Materials and Corrosion* **52** 99–105

IOP Conf. Series: Materials Science and Engineering **480** (2019) 012019 doi:10.1088/1757-899X/480/1/012019

# Neural network for prediction of hardness profiles for steel alloys after plasma nitriding

**J Pribbenow[1,*], M Mejauschek[2], P Landgraf[1], T Grund[1], G Bräuer[2] and T Lampke[1]**

[1] Chemnitz University of Technology, Materials and Surface Engineering Group
[2] Fraunhofer Institute for Surface Engineering and Thin Films

[*] e-mail: joerg.pribbenow@mb.tu-chemnitz

**Abstract.** Plasma nitriding is a state-of-the-art thermochemical treatment, which is used by the processing industry for hardening of the surface-near material volumes of tools and components. An artificial neural network, particularly backpropagation neural network, has been developed to investigate the hardness behavior of different steel types according to various process parameters. Therefor eleven steels with different chemical compositions were plasma nitrated with different concentrations of nitrogen at various temperatures. The hardness profiles were studied with indentation measurement. The experimental data were used to train the neural network with the Levenberg-Marquardt algorithm. For an optimal training, weight distribution, hidden neurons and gradients of neuron activation functions were varied, to allow simulations of hardness profiles for the different types of steel with deviant nitriding parameters.

## 1. Introduction

Plasma nitriding is a well-established process for the hardening of steel surfaces and the improvement of the chemical resistance. [1-3] Besides it is characterized by a very good reproducibility. Therefore, it is used for a wide range of products like tools, parts in mechanical engineering and in the food industry. [4-6] Depending on the application, the process parameters can be varied in a wide range to modulate the nitriding layer. Due to long process times, the determination of appropriate parameters to nitride new steels is a steady challenge for plasma nitriding companies. To reduce the experimental effort, an artificial neural network (ANN) can be used to investigate the hardness profiles and for the simulation of nitriding processes. [7-10] The investigation of the physical background of different hardening mechanisms, their effects on the microstructure and the resulting mechanical properties are not necessary, since their influences are implicated in the hardness profiles and hence in the mathematical analysis by the ANN.

## 2. Experimental and materials

The chemical composition of eleven selected steels was determined with X-ray fluorescence measurement (XRF) and is shown in Table 1. The alloys differ mainly in the concentration of the nitride forming elements like Cr, Mo, V, as well as the sum of alloying elements. The samples were cut to a height of 4 mm from a rod with a diameter of 35 mm. The grinded and polished samples were nitrated with various parameters which are presented in Table 2. The hardness profiles of the samples were measured three times with indentation measurement and a force of 250 mN to a depth of 300 µm.

**Table 1.** Chemical composition of investigated steels (XRF).

| Steel | Cr (%) | Mn (%) | Mo (%) | V (%) | Ni (%) | Co (%) | Ti (%) | W (%) |
|---|---|---|---|---|---|---|---|---|
| X3NiCoMaTi18-9-5 | 0 | 0 | 5.0 | 0 | 18.3 | 9.6 | 1.1 | 0 |
| X37CrMoV5-1 | 5.4 | 0 | 1.3 | 0.4 | 0 | 0 | 0 | 0 |
| X38CrMoV5-3 | 5.0 | 0 | 3.0 | 0.5 | 0 | 0 | 0 | 0 |
| X40Cr14 | 13.5 | 0 | 0 | 0 | 0 | 0 | 0 | 0 |
| X100CrMoV5 | 5.3 | 0 | 1.1 | 0.3 | 0 | 0 | 0 | 0 |
| X153CrMoV12 | 11.1 | 0 | 0.8 | 0.7 | 0 | 0 | 0 | 0 |
| 40CrMnMo7 | 1.9 | 1.5 | 0.2 | 0 | 0 | 0 | 0 | 0 |
| 45WCrV7 | 1.0 | 0.3 | 0.2 | 0 | 0 | 0 | 0 | 1.9 |
| 55NiCrMoV7 | 1.2 | 0 | 0.4 | 0.1 | 1.7 | 0 | 0 | 0 |
| HS6-5-2C | 4.7 | 0 | 4.8 | 1.9 | 0 | 0 | 0 | 6.2 |
| CPM REX M4 | 4.4 | 0.3 | 4.8 | 3.8 | 0 | 0 | 0 | 5.3 |

**Table 2.** Plasma nitriding process parameters.

| Nitrogen (vol.%) | Temperature (°C) | Pressure (pa) | Process time (h) | Pulse Duration (μs) | Pulse pause (μs) |
|---|---|---|---|---|---|
| 5, 10, 15, 20, 25, 80 | 480, 520 | 300 | 16 | 100 | 500 |

The ANN was designed with *Mathworks MATLAB* using the *Neural Network Toolbox*. It is a feed-forward neural network in which the Input Layer is connected by neurons via the Hidden Layer with the Output Layer as it is shown in Figure 1 and the data only move in this direction. To calculate the weights of the neurons backpropagation was used with tangent sigmoid transfer function inside of the hidden layers. For this so-called training of the ANN the Bayesian Regularization (BRA), Scaled Conjugate Gradient (SCGA) and Levenberg-Marquardt algorithm (LMA) were used to find the ideal optimization of the weights. The measured data has been divided into 70% training and 30% validation data. Only samples nitrated with 5, 15, 25 and 80 vol.% Nitrogen were used in combination with the parameters from the Input layer in Figure 1. Depending on each set of parameters three measured hardness profiles were used as target in the Output layer.

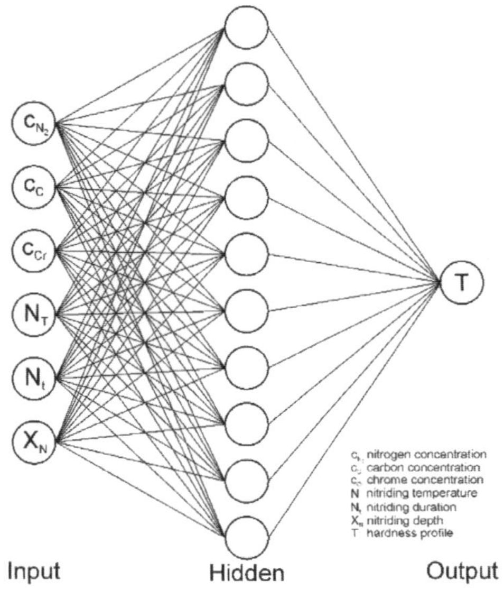


**Figure 1.** Schematic model of an ANN for prediction of the hardness profile of steels.

### 3. Results and discussion

For an optimal training of the ANN the number of Hidden Layers has been varied from 1 to 10 and the number of neurons therein from 1 to 10. The best approach was given by an ANN trained with three Hidden Layers, each with nine neurons. The LMA was very stable for this case what resulted in a high precision relating to the optimization of the weights. The result is a very good residuum R = 0.98 presented in Figure 2 as a combination of training and validation data. The correlation between simulated and measured data using the BRA and SCGA was insufficient. For that reason, both algorithms were excluded from the continuative investigations.

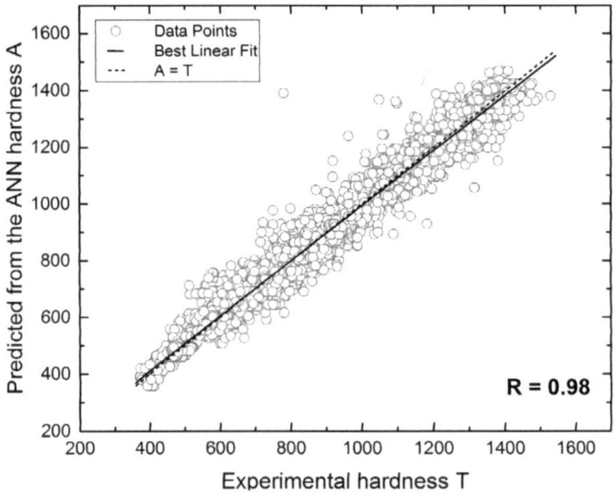


**Figure 2.** Correlation between the measured hardness profiles and analysis with the ANN. Training and validation data are summarized.

A graphical user interface was developed for an easy to operate application of the trained ANN and is presented in Figure 3 After choosing a given steel and different nitriding process parameters, which can differ from the experimental training data, the calculation of the hardness profile will be plotted.

IOP Conf. Series: Materials Science and Engineering **480** (2019) 012019 doi:10.1088/1757-899X/480/1/012019

The right hand side of Figure 3 shows averaged data points of three hardness profiles from 21 samples which were not part of the training compared to their simulated hardness data. The residuum R = 0.93 and can be considered as acceptable. The simulation and measurement is in a tolerable correlation except for some calculated hardness values in the range between 900 and 1200. In this context a large part of the measured hardness data points have values above their calculated ones. This is observable globally.

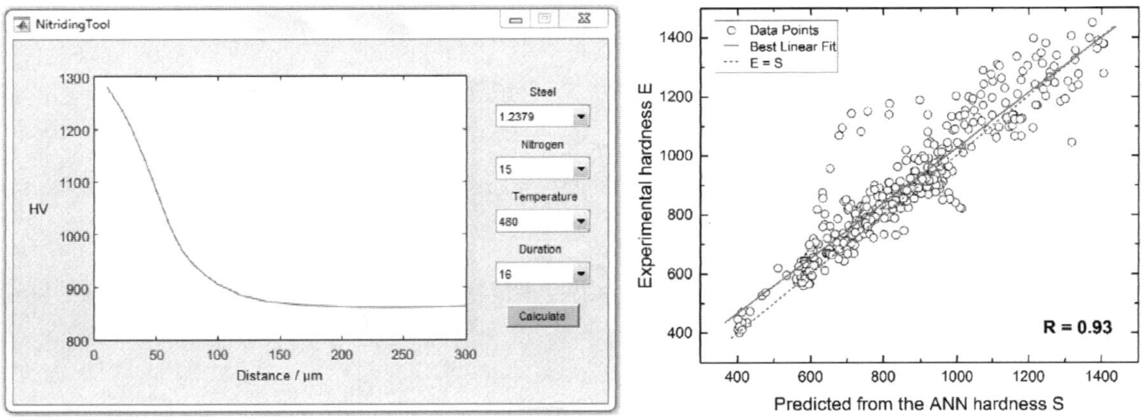


**Figure 3.** Graphical user interface for the simulation of hardness profiles and the sum of control measurements compared to the calculation of the ANN.

With a closer look on the hardness profiles it is possible to explain this deviation. On the left hand side of Figure 4 the simulated data coincide with the measured data. That case can be observed for the majority of the investigated samples. But there are some samples with a bigger divergence between simulation and experiment (Figure 4, 55NiCrMoV7 in the right hand side diagram). The reason can be found in the training data. Some of the nitrated steels show considerable fluctuations in their hardness profiles and are not suited for training without more measurements. The fluctuations of the hardness is not totally understood. The 55NiCrMoV7 steel has a low concentration of the nitride forming element Cr. In this case the hardness would be caused mainly through solid solution strengthening and not by precipitation hardening. During the fully automated measurement it is possible that the relative low content of nitrides will be measured with an above average frequency. This assumption will be supported by the fluctuations in the hardness profiles of the steel 40CrMnMo7 and 45WCrV7, which have a similar chemical composition like the 55NiCrMoV7 steel. To prove this assumption further investigations are necessary.

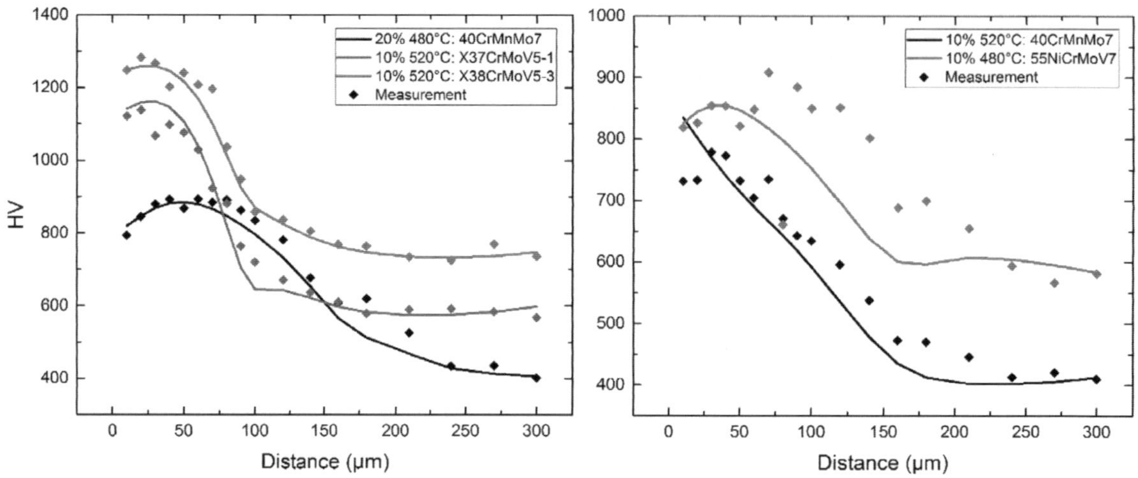


**Figure 4.** Comparison of the calculated and measured hardness profiles with a high correlation (right) and examples with a divergence (left) between calculated and measured data.

## 4. Conclusions

With the introduced ANN the possibility for a simulation of steel hardness profiles with various nitriding parameters is given with a good approach relating to the control experiments. The approach is hence considered as suitable to implement a bigger database, the work on it being already in progress. The future potential of the ANN-based prediction tool on nitriding process results for a variety of steel and nitriding parameters will hence grow with the growing range of parameters and materials. Using steels with other chemical compositions for the training of the ANN should likewise allow the simulation of deviant types.

**Acknowledgments**

The authors gratefully acknowledge the financial support of the Arbeitsgemeinschaft industrieller Forschungsvereinigungen (AiF), Germany (project numbers IGF 18741 BG /1 and 18741 BG /2).

## References

[1] Zhang Z L and Bell T 1985 Structure and Corrosion Resistance of Plasma Nitrided Stainless Steel *Surf. Eng.* **1** 131-136

[2] Larisch B, Brusky U and Spies H-J 1999 Plasma nitriding of stainless steels at low temperatures *Surf. Coat. Tech.* **116-119** 205-211

[3] Li C X and Bell T 2004 Corrosion properties of active screen plasma nitrided 316 austenitic stainless steel *Corros. Sci.* 1527-1547

[4] Menthe E and Rie K-T 1999 Further investigation of the structure and properties of austenitic stainless steel after plasma nitriding *Surf. Coat. Tech.* **116-119** 199-204

[5] Zeghni A E and Hashmi M S J 2004 The effect of coating and nitriding on the wear behaviour of tool steels *J. Mater. Process. Technol.* **155-156** 1918-1922

[6] Ochoa E A, Wisnivesky D, Minea T, Ganciu M, Tauziede C, Chapon P and Alvarez F 2009 Microstructure and properties of the compound layer obtained by pulsed plasma nitriding in steel gears *Surf. Coat. Tech.* **203** 1457-1461

[7] Malinovaa T, Malinovb S and Pantev N 2001 Simulation of microhardness profiles for nitrocarburized surface layers by artificial neural network *Surf. Coat. Tech.* **135** 258-267

[8] Filetin T, Zmak I and Novak D 2004 Nitriding parameters analized by neural network and genetic algorithm *J. Phys. IV France* **120** 355-362

[9] Zhechevaa A, Malinovb S and Sha W 2005 Simulation of microhardness profiles of titanium alloys after surface nitriding using artificial neural network *Surf. Coat. Tech.* **200** 2332-2342

[10] Paschke H, Nienhaus A, Brunotte K, Petersen T, Siegmund M, Lippold L, Weber M, Mejauschek M, Landgraf P, Braeuer G, Behrens B-A and Lampke T 2018 Adapted diffusion processes for effective forging dies *AIP Conference Proceedings* **1960** 040016

IOP Conf. Series: Materials Science and Engineering **480** (2019) 012020 doi:10.1088/1757-899X/480/1/012020

# Mechanical dissolution of copper additions in aluminium by Friction Stir Processing

**M Gnedel[1,*], A Zens[2], F Haider[1] and M F Zaeh[2]**

[1] Chair for Experimental Physics I, University of Augsburg, Universitätsstraße 2, 86159 Augsburg (Germany)
[2] Institute for Machine Tools and Industrial Management, Technical University of Munich, Boltzmannstraße 15, 85748 Garching (Germany)

[*] e-mail: maximilian.gnedel@physik.uni-augsburg.de

**Abstract.** Friction Stir Processing (FSP) can be used to locally modify properties in materials such as aluminium. Furthermore, the composition of the alloy can be changed by this technique. Intermixing specific micrometer-sized metal powders as well as metal foils helps to optimize both the microstructural stability during subsequent heat treatment, as well as the mechanical properties in general. Dispersing copper powder inside a matrix of AA1050 aluminium by FSP can produce homogenous solid solution of the two elements, if suitable processing parameters are used. A substantial hardening effect is shown for both effects, refinement of the grain structure and solid solution hardening. By calculating the diffusion length and using the results of investigations based on energy-dispersive X-ray spectroscopy (EDS), it can be shown that the dissolution is mainly mechanically driven and not due to diffusion. Furthermore, transmission electron microscopy (TEM) provides insights on the formation of intermetallic phases during FSP. The results may be used in future studies to evaluate the properties of such non-equilibrium alloys.

## 1. Introduction

Local enhancement of materials properties in light metals through a change in the microstructure and chemical composition is a topic with extensive research focus within physical metallurgy these days. In most cases, this is accomplished with the assistance of a high degree of plastic deformation, which is widely referred to as mechanical alloying. A variety of techniques can be applied for this purpose and many different metallic systems have already studied in order to evaluate if plastic deformation can help to create advantageous metallic composites or even highly supersaturated solid solutions.

### *1.1. Conventional techniques for mechanical alloying*

The first attempts at mechanical alloying were performed over 20 years ago by ball milling of metal powders with steel or ceramic spheres. At that time, the combination of copper and a variety of other elements was the primary focus [1–3]. However, this technique has continued to draw a large amount of attention and persists even today. A wide range of different material combinations have been tested and successfully brought to amorphisation, depending on the heat of mixing [4].

With high pressure torsion (HPT), extreme amounts of plastic deformation can be applied to metallic materials. As with ball milling, the focus of studies up to this point has been primarily copper-based systems, and depending on the composition of the base composition and on the relative strength of the components, composites similar to solid solutions have been obtained even for immiscible combinations [5,6] in equilibrium.

Accumulative roll bonding is also widely used to produce solid mixtures of metals under high plastic deformation. Often, aluminium is used as base material and, for example, mechanically alloyed with nickel as a harder alloying partner. Similar to the high pressure torsion experiments, the relative mechanisms of deformation to fracture are studied [7]. The Al-Cu system has gained much attention for processing with this technique; thus, the results of the conducted experiments can serve as reference values for the experiments done by FSP [8,9].

### *1.2. FSP for mechanical alloying*

It has been shown, that FSP is an effective technique for distributing additional alloying elements in various geometries like powder, wire or foil inside an aluminium matrix [10]. Especially for visualizing the material flow during the process, Cu foil pieces were used so far [11].

Mechanical alloying with powder is still the dominating method to integrate additional elements. Fe is for example a widely used alloying partner besides Cu [12][13].

Different goals are the driving forces for these studies, depending on the chosen element. Whereas in the Al-Fe system often the magnetic properties are studied, the addition of Zr can help to improve the microstructural stability [14]. Most elements are integrated to improve the mechanical properties like hardness or fracture toughness, which also holds for Cu [15–18]. Furthermore, the Al-Cu system, which is effectively hardenable by producing the Θ' phase ($Al_2Cu$) during annealing and together with the unique microstructure obtained by FSP is very promising for applications where high strength and light weight metal parts are needed. The local modification of structural parts by mechanical alloying with Cu is imaginable, but not yet studied.

Apart from the goal of producing an alloy with supreme mechanical properties, the mechanisms of intermixing during this process of partly diffusion, partly mechanical driven mixing have to be understood. Al-Cu as modelling system for soft/soft and partly miscible interaction is, besides the advantageous mechanical properties, an excellent system to examine this. It was already shown, that Cu powder can extensively be dissolved by FSP [19], but the crucial influencing variables are not yet identified.

## 2. Experimental and analytical set-up

### *2.1. Experimental set-up*

Figure 1 shows a schematic diagram of the experimental set-up, in which two EN AW-1050 sheets are aligned in butt-joint configuration. Copper powder, inserted in a groove in the middle of the sheets, is distributed in the weld nugget during processing. The thickness of the sheets was 8.0 *mm* and the dimensions of the groove where 1 *mm* x 0.5 *mm* filled with commercially pure copper CW008A powder with an average particle size of 3-4 *µm*. The probe performed either 2 or 4 passes through the material, for which the rotation direction was kept constant. Hereby, the advancing and retreating sides changed for each pass. The tool dimensions were 12 *mm* for the shoulder diameter, 5.0 *mm* for the base and 3.6 *mm* for the tip diameter of the conical probe, and 3.5 *mm* for the probe length. The shoulder plunge depth was 0.1 *mm* and a tool tilt angle of 2 ° was employed. The probe was threaded and furnished with three uniformly distributed flats. A CNC milling machine (Heller MCH250), which has been adapted to perform FSW processes, was used in position-controlled mode for all experiments.

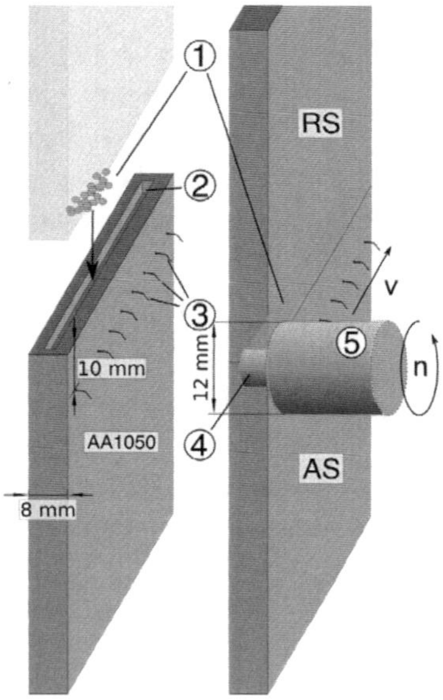


**Figure 1**. Schematic diagram of the sample production by a combination of Friction Stir Welding and Friction Stir Processing of the joint of two aluminium sheets with copper powder (1) in between. A groove (2) in the lower sheet is filled with commercially pure copper and the temperature during the process is measured by thermocouples (3). A rotating steel probe (4) with a shoulder (5) distributes the additional copper within the Aluminium matrix.

Nine Type K thermocouples with a size of 0.5 *mm* in diameter where placed inside holes 10 *mm* away from the joint. The holes were 0.6 *mm* in diameter and had a depth of 2 *mm*. The specimens for the hardness and compositional analysis where taken from the middle of the 286 *mm* long seam in order to ensure, that the analysis was preformed within an area in which the process was stable. The feed rate, rotation speed, and number of passes for each experiment can be found in Table 1.

**Table 1.** Process parameters for the production of the welding seams investigated in this work.

| Sample | Feed rate (*mm/min*) | Rotation speed (*RPM*) | Number of passes |
|---|---|---|---|
| 1 | 50 | 1500 | 2 |
| 2 | 50 | 2250 | 2 |
| 3 | 50 | 1500 | 4 |

### *2.2. Analytical set-up*
One specimen was extracted for each seam at approximately 143 *mm* from the starting point of the weld for hardness measurements and electron microscopy. The cross sections were prepared metallographically for scanning electron microscopy (SEM) with a final chemomechanical polishing using 40 *nm* silica particles. The specimens for transmission electron microscopy (TEM) were electrolytically thinned. High resolution SEM was executed using a ZEISS MERLIN microscope. A JEOL JEM-2100F allowed TEM with a resolution up to atomic scale.

## 3. Results and discussion

### *3.1. Structure*

The structural properties of the samples where investigated by light- and electron microscopy techniques, to study distribution, dissolved Cu fraction and possible precipitations formed during the process. An extensive study with the same experimental setup but coarser Cu powder and lower rotation rates preceded this study, but never led to particle dissolution.

Figure 2 shows a macrograph of a cross section of sample 2, which was polished, etched with Kroll's reagent and recorded with a light microscope. The cross section was also prepared at a second position approximately 2 *cm* further along the welding direction, which led to a nearly identical result. This shows, that the process is stable in the investigated region. Dissolved Cu is clearly visible as dark contrast, which is proven by energy dispersive X-ray (EDX) analysis (Figure 3-5). Few undissolved clusters can be identified as bright spots. The shape of the dissolved Cu resembles a marble cake, which already points to a complicated mechanical distribution during FSP. Sample 1 does not show any dissolution of Cu, whereas sample 2 and 3 both show rather homogenous distribution of dissolved Cu in the whole stir zone, which for sample 2 is remarkable, due to the execution of only two passes. A clearly defined border is visible at the bottom of the stir zone and the advancing side, which was investigated in detail, using electron microscopy.

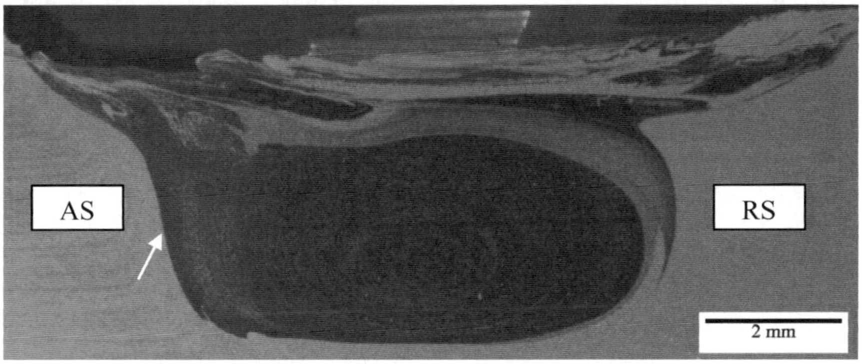


**Figure 2.** Macrograph of cross section of sample 2 taken with a light microscope after polishing and etching with Kroll's reagent. Advancing and retreating side of the second pass as well as the position of EDX analysis are marked.

In Figure 3 and 4 the edge region of sample 2 and 3 at the position marked in Figure 2 is shown, where Cu found by EDX analysis is displayed in red. Sample 2 shows a sharp edge, while in sample 3 an intermediate step with a width of approximately 200 $\mu m$ can be discovered. Undissolved Cu clusters with a size of up to 20 $\mu m$ are visible in both samples, which is also the maximum size of the base material Cu powder. The concentration of undissolved Cu particles in sample 2 is significantly lower, than in sample 3. The undissolved particles had a similar average diameter of 10.4 $\mu m$ for sample 2 and 11.3 $\mu m$ for sample 3, but the number of particles is up to twice as high for sample 3.

The sharp edge as well as the intermediate step in Cu concentration in the samples indicate complicated mechanical but laminar movement along the processing direction.

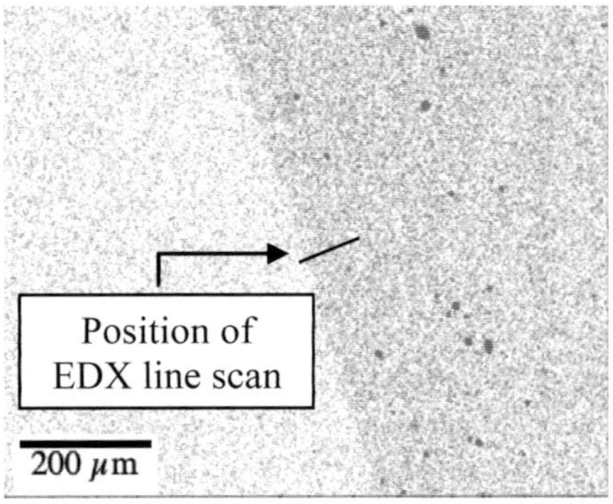


**Figure 3.** EDX mapping of the Cu concentration in sample 2 at the edge of the mechanically alloyed region. The position of the EDX line scan from Figure 5 is marked.

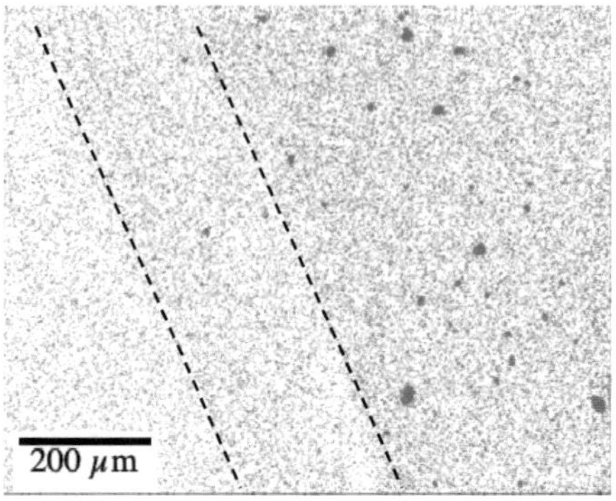


**Figure 4.** EDX mapping of the Cu concentration in sample 3 at the same position as in Figure 3.

To quantify the dissolved Cu content, an EDX line scan was carried out across the border region of sample 2. Its position is marked in Figure 3 and its result shown in Figure 5. The width of the region where the Cu concentration drops to 0 amounts to approximately 6 *μm*. Inside the mechanically alloyed region, the Cu concentration varies significantly between 0.2 and 0.34 *at%* along the measured line. More EDX spectra inside the stir zone were evaluated, which showed up to 0.73 *at%* for sample 2 and up to 0.56 *at%* for sample 3. This, together with the lower concentration of undissolved particles in sample 2 in comparison to sample 3 points out, that the higher rotation speed of 2250 *RPM* is more effective, then the 1500 *RPM* of sample 3, even though only two passes instead of four were executed. The strongly varying concentration and the sharp edge to the unalloyed region again suggest a complex laminar flow during dissolution. The strong concentration gradient accompanied by the changed material properties is furthermore interesting and could be exploited systematically.

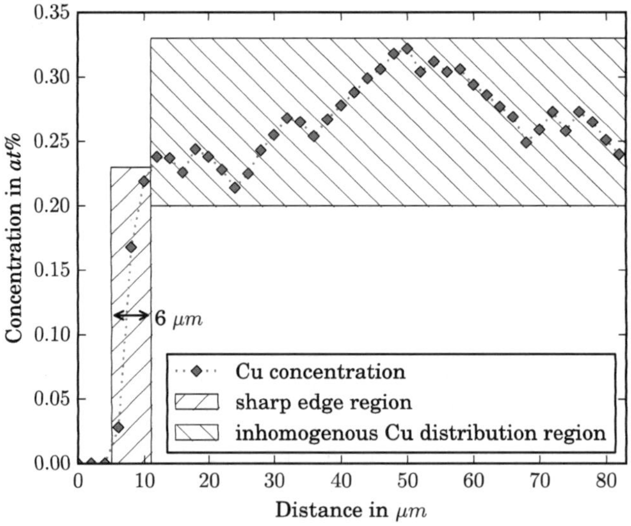


**Figure 5.** EDX line scan of Cu concentration of sample 2 at the position marked in Figure 2. The strongly fluctuation concentration profile inside the mechanically alloyed region as well as the clearly defined sharp edge is marked.

IOP Conf. Series: Materials Science and Engineering **480** (2019) 012020 doi:10.1088/1757-899X/480/1/012020

Sample 2 was subjected to high resolution transmission electron microscopy and showed intermetallic Θ'' precursor phases to the intentionally produced Θ' phase during age hardening of Al-Cu alloys. An image of one of these precipitations is shown in Figure 6. The formation of these phases at the elevated process temperature of approximately 400 *°C* inside the weld nugget (estimated from the temperature data measured by thermocouples) together with solid solution hardening are the reasons for the extreme hardness increase shown in Figure 7.

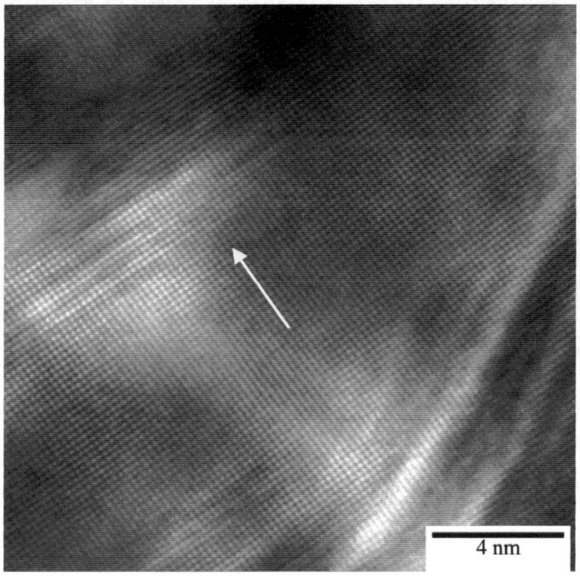


**Figure 6.** Θ'' phase formed during FSP visible via HRTEM and marked by arrow.

*3.2. Mechanical properties*

The findings of section 3.1 were believed to have strong impact on the mechanical properties of the mechanically alloyed and FSPed samples. Therefore, hardness mappings were recorded with 500 *mN* indentation force and 300 *µm* spacing between the indents and are shown in Figure 7. Sample 1 (Figure 7 (a)) shows no particle dissolution and a slight increase in hardness only around the particles, which did not get distributed effectively during FSP. Sample 2 (Figure 7 (b)) shows significant and reasonably homogenous hardening throughout the mechanically alloyed region up to values of 56 *HV*. Sample 3 (Figure 7 (c)) shows lower peak hardness values than sample 2 and the mechanically alloyed region has a more trapezoid shape.

The increased hardness values exactly match the contrast given by the Cu in the light microscopy images. Reasons for this enormous increase are solid solution hardening, formation of intermetallic precursor phases, as well as the unique microstructure with an average grain size of approximately 5 *µm* in the mechanically alloyed region (for all three samples).

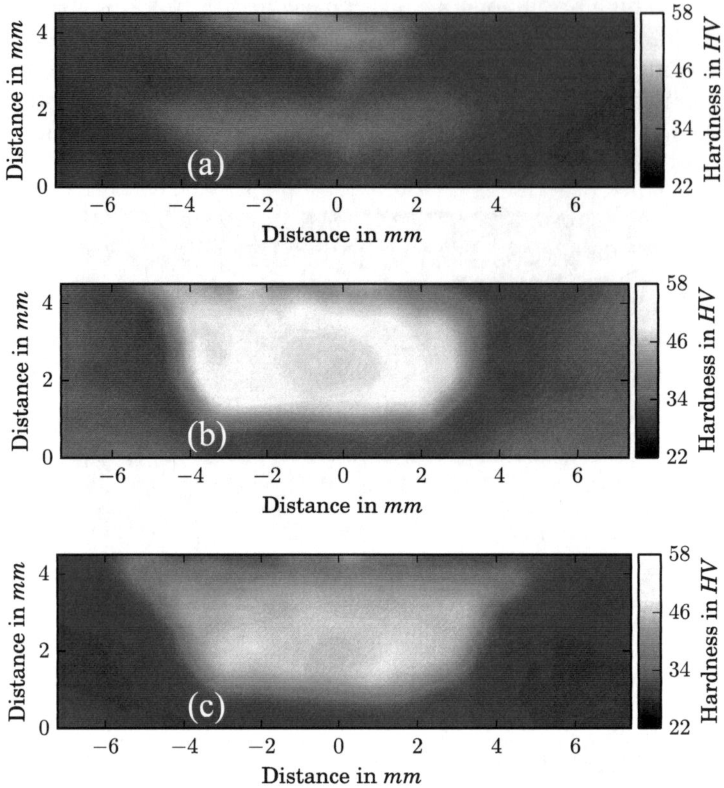


**Figure 7.** Hardness mappings recorded with a Berkovich indenter, 500 *mN* indentation force and 300 *µm* spacing between the indents. Sample 1 (a), sample 2 (b) and sample 3 (c)

## 4. Temperature measurements and modelling

To understand the different behaviour regarding the mechanically assisted dissolution depending on rotation speed and pass number, temperature measurements shown in Figure 8 were taken into account. The process temperature in direct proximity of the probe deviates from the measured one by up to 100 °*C*, which was proven in literature by measurements inside the probe [20]. Nevertheless, the relation between the temperature data of sample 2 and 3 can be evaluated to get insights into the reasons for the different outcome after FSP.

Beyond that, the diffusion length $\boldsymbol{d}$ depending on time $\boldsymbol{t}$ was incrementally calculated by Equation 1, assuming, that the Diffusion is thermally activated:

$$\boldsymbol{d(t)} = \int_0^{\boldsymbol{t}} \sqrt{\boldsymbol{t}\, \boldsymbol{D_0}\, e^{-\frac{\boldsymbol{Q}}{\boldsymbol{k_B T(t)}}}}\, \boldsymbol{dt} \quad (1)$$

With $\boldsymbol{T}$ being the temperature, which changes with time $\boldsymbol{t}$, $\boldsymbol{D_0}$ is the prefactor of diffusion, $\boldsymbol{Q}$ is the activation energy and $\boldsymbol{k_B}$ is the Boltzmann constant. Both $\boldsymbol{D_0}$ and $\boldsymbol{Q}$ where taken from [20]. The result for sample 2 and 3 is also plotted in Figure 8. Because sample 1 was produced with the same process parameters as sample 3 and thus yielded almost identical results, it was not taken into account.

It is obvious, that the reachable diffusion length of up to 100 *nm* does not contribute effectively to the distribution of Cu in the stir zone. While thermally activated, the diffusion is only occurring during the tens of seconds when the temperature reaches its peak values during to the passage of the probe. Although this is sufficient to bring a small amount of Cu into solid solution.

With the temperatures of sample 2 and 3 being very similar and also constant across all passes (pass 4 of sample 3 was not recorded, due to software failure), no explanation is obvious. Even the noticeable higher rotation speed, with which sample 2 was produced, does not seem to have significant influence on the temperature.

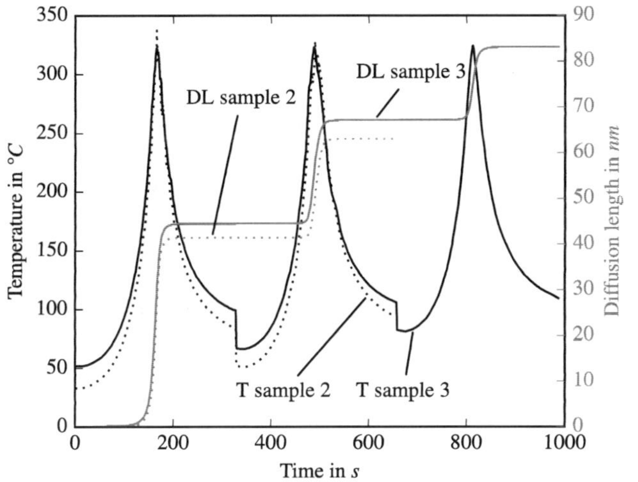


**Figure 8.** Attainable diffusion length in sample 2 and 3 over process time in pass 1 and 2 (sample 2) and passes 1, 2 and 3 (sample 3).

Comparing sample 1 and 3, there has to be a starting point of the Cu dissolution reaction even at constant rotation speed, just by increasing the pass number. This starting point is therefore assumed to be caused by a chain reaction, where enough Cu dissolved to harden the material, which in return leads to an increase of necessary deformation energy at constant rotation rate to dissolve more Cu. On the atomic length scale the effect could be caused by a change of dislocation movement. Once the coarse particles are parted by shear forces to a size at which it is energetically favorable to cut a particle instead of bypassing it, this chain reaction of mechanical dissolution can begin. Why this process does not have any effect on the temperature and why some of the Cu particles after the process still have their initial size is part of ongoing investigation.

## 5. Conclusions

- An effective mechanical alloying process starts with >1500 *RPM* and >2 passes at 50 *mm/min* feed rate for Cu in AA1050, but not all copper particles can be dissolved completely.
- Cu rich aluminium is apparently distributed in complex laminar and stable motion along the processing direction, which leads to significant variation in Cu content in stir zone.
- An extreme hardening effect is accomplished by solid solution hardening, the formation of intermetallic phases and the microstructure characteristic for FSP.
- Temperature measurements give no explanation for a starting point of the dissolution process.
- Complex interplay of interdiffusion at the particle matrix interface with subsequent hardening is suggested, which leads to chain reaction after the starting point of dissolution.
- Further investigations will include temperature measurements inside the probe and a raise of the Cu content to test, if a surpassing of the maximum solution limit is possible.

**Acknowledgements**
The research project HA 1549/6-1 is funded by the German Research Foundation (DFG). The authors would like to thank for the funding and the support. Also, David Stein and Florian Fleischmann shall be thanked for the support by practical help and fruitful discussions.

IOP Conf. Series: Materials Science and Engineering **480** (2019) 012020 doi:10.1088/1757-899X/480/1/012020

## References


[1] Ogino Y, Yamasaki T, Murayama S and Sakai R 1990 Non-equilibrium phases formed by mechanical alloying of Cr · Cu alloys *J. Non. Cryst. Solids* **117–118** 737–40

[2] Gaffet E, Louison C, Harmelin M and Faudot F 1991 Metastable phase transformations induced by ball-milling in the Cu · W system *Mater. Sci. Eng. A* **134** 1380–4

[3] Gente C, Oehring M and Bormann R 1993 Formation of thermodynamically unstable solid solutions in the Cu-Co system by mechanical alloying *Phys. Rev. B* **48** 13244–52

[4] Gonzalez G, Sagarzazu A, Bonyuet D, D'Angelo L and Villalba R 2009 Solid state amorphisation in binary systems prepared by mechanical alloying *J. Alloys Compd.* **483** 289–97

[5] Karoline B, Kormout S, Pippan R and Bachmaier A 2016 Deformation-Induced Supersaturation in Immiscible Material Systems during High-Pressure Torsion ** 1–19

[6] Bachmaier A, Schmauch J, Aboulfadl H, Verch A and Motz C 2016 On the process of co-deformation and phase dissolution in a hard-soft immiscible Cu-Co alloy system during high-pressure torsion deformation *Acta Mater.* **115** 333–46

[7] Min G, Lee J-M, Kang S-B and Kim H-W 2006 Evolution of microstructure for multilayered Al/Ni composites by accumulative roll bonding process *Mater. Lett.* **60** 3255–9

[8] Rahmatabadi D, Mohammadi B, Hashemi R and Shojaee T 2018 An Experimental Study of Fracture Toughness for Nano/Ultrafine Grained Al5052/Cu Multilayered Composite Processed by Accumulative Roll Bonding *J. Manuf. Sci. Eng.* **140** 101001–11

[9] Hsieh C C, Shi M S and Wu W 2012 Growth of intermetallic phases in Al/Cu composites at various annealing temperatures during the ARB process *Met. Mater. Int.* **18** 1–6

[10] Zens A, Gnedel M, Zaeh M F and Haider F 2018 The effect of additive geometry on the integration of secondary elements during Friction Stir Processing *IOP Conf. Ser. Mater. Sci. Eng.* **373**

[11] Xu W F, Liu J H and Chen D L 2011 Material flow and core/multi-shell structures in a friction stir welded aluminum alloy with embedded copper markers *J. Alloys Compd.* **509** 8449–54

[12] Khorrami M S, Samadi S, Janghorban Z and Movahedi M 2015 In-situ aluminum matrix composite produced by friction stir processing using FE particles *Mater. Sci. Eng. A* **641** 380–90

[13] Lee I S, Kao P W and Ho N J 2008 Microstructure and mechanical properties of Al--Fe in situ nanocomposite produced by friction stir processing *Intermetallics* **16** 1104–8

[14] Hassan K A A, Norman A F, Price D A and Prangnell P B 2003 Stability of nugget zone grain structures in high strength Al-alloy friction stir welds during solution treatment *Acta Mater.* **51** 1923–36

[15] Zohoor M, Besharati Givi M K and Salami P 2012 Effect of processing parameters on fabrication of Al-Mg/Cu composites via friction stir processing *Mater. Des.* **39** 358–65

[16] Abnar B, Kazeminezhad M and Kokabi A H 2014 The Effect of Cu Powder During Friction Stir Welding on Microstructure and Mechanical Properties of AA3003-H18 *Metall. Mater. Trans. A* **45** 3882–91

[17] Hsu C J, Kao P W and Ho N J 2007 Intermetallic-reinforced aluminum matrix composites produced in situ by friction stir processing *Mater. Lett.* **61** 1315–8

[18] Hsu C J, Kao P W and Ho N J 2005 Ultrafine-grained Al--Al 2 Cu composite produced in situ by friction stir processing *Scr. Mater.* **53** 341–5

[19] Karthik G M, Ram G D J and Kottada R S 2017 Friction stir selective alloying *Mater. Sci. Eng. A* **684** 186–90

[20] Marstatt R, Krutzlinger M, Luderschmid J, Constanzi G, Mueller J F J, Haider F and Zaeh M F 2018 Intermetallic layers in temperature controlled Friction Stir Welding of dissimilar Al-Cu-joints *IOP Conf. Ser. Mater. Sci. Eng.* **373**

IOP Conf. Series: Materials Science and Engineering **480** (2019) 012021 doi:10.1088/1757-899X/480/1/012021

# On the correlation of hammer-peened surfaces and process, material and geometry parameters

**S Pfeiffer[1,*], M Fiedler[2], T Bergelt[1], M Kolouch[2], M Putz[2] and M F-X Wagner[1]**

[1] Technische Universität Chemnitz, Chair of Materials Science, Chemnitz, Germany
[2] Fraunhofer Institute for Machine Tools and Forming Technology IWU, Chemnitz, Germany

[*] e-mail: steffen.pfeiffer@mb.tu-chemnitz.de

**Abstract.** Hammer-peening can replace extensive hand-polishing during the processing of freeform surfaces, but it is rarely applied in practice due to the interplay of numerous parameters, including the process and tool parameters as well as the geometry and the material of the workpiece. In the present study we systematically investigate the effect of tool radii, tool angles, pressure, path and impact distance and the path definition (with respect to the direction of previous surface milling) on the resulting surface roughness and strain hardening of a cast iron, a low- and a high-alloyed steel using a pneumatic tool setup and complementary Finite Element (FE) simulations. Our experimental results show that an impact and path distance of 0.18 mm, a pressure of 7.3 bar and a tool radius of 8 mm reduce the surface roughness of a face-milled surface with an average roughness of $R_a$ = 0.8 μm to an average roughness of $R_a$ = 0.5 μm for all materials under consideration. The FE simulations indicate that, for a ball-milled surface, the surface roughness $R_a$ decreases (and the arithmetical mean height $S_a$ increases) the most for a tool angle of 45° and for a hammer-peening path perpendicular to the previous ball-milling path. These results contribute to the development of a mathematical model that allows to identify suitable parameter combinations prior to hammer-peening processing for a given workpiece surface.

## 1. Introduction

Machine Hammer Peening (MHP) is a finishing technique that is manually applied on weld toes to increase fatigue strength [1] and recently automatized technologies are now also used on die and mould freeform surfaces [2]. An oscillating spherical hammer head is moved over the surface of the workpiece, producing incremental deformations in the surface layer. Typical frequencies are of the order of magnitude of 100 Hz according to the operating principle of the actuator. There are piezo-electric [3], electro-magnetic [4] and pneumatic tools [5] that can be attached to a tooling machine or to a robot. The surface is deformed by numerous impacts that are characterized by the mass of the hammer head, the impact angle and the feed rate. The impact paths and distances between the impacts and the paths themselves are well defined and result in several advantages compared to other surface treatments: a good accessibility of freeform surfaces compared to techniques with continuous contact of tool and workpiece (e.g., deep rolling) as well as a defined amount of deformation compared to treatments with non-uniform impact intensities (e.g., shot peening). MHP decreases the surface roughness, induces work hardening and compressive residual stresses and results in microstructural changes of the surface layer if appropriate process parameters are used [6]. These surface modifications can further result in an improvement of fatigue and wear resistance and beneficial tribological properties. An overview of MHP in comparison to other burnishing technologies is given in [2].

Mannens et al. [7] investigated the influence of impact force, impact angle, stroke length and lubricant in an experimental study on the residual stresses and the roughness of stainless steel X3CrNiMo13-4 (DIN no. 1.4313). They achieved a roughness reduction of up to 80% for $R_a$ and 63% for $R_z$ and maximum residual stresses of 1100 MPa orthogonal and 800 MPa parallel to the impact path located in a depth of 0.15 mm below the surface. These residual stresses are induced from plastic deformation that can also result in a grain refinement up to ultra-fine grained microstructures, as demonstrated in a study of Revilla-Gomez et al. [8] on welded and hammer-peened samples of a low alloyed steel S690. Using electron backscatter diffraction, they demonstrated that MHP results in an increase of the dislocation density especially near the surface and up to 250 µm below the surface of the specimen. Bleicher et al. [9] studied the influence of the impact angle on the surface roughness of a high alloyed steel X155CrVMo12 (DIN no. 1.2379) and a plain carbon steel C45E (DIN no. 1.1191). They showed that a decreasing impact angle leads to a decrease of the normal component and to an increase of the transverse component of the impact force in a range of 80° to 60°. In addition, the surface roughness is decreasing for the softer plain steel. For larger angles the force components remain almost the same and in addition they increase if a higher peening distance is used for machining. Furthermore, they [9] presented experiments on embedding tungsten-carbide particles (in a powder suspension) in near surface layers by MHP. An almost homogeneous layer of embedded particles can be generated, significantly increasing the surface roughness and also the wear resistance. Baptista et al. [10] presented a fully dynamic FE simulation calculating multiple impacts on weld toes of the duplex stainless steel type S31803 and of an austenitic stainless steel type 304L. Their focus was on different weld toe radii, impact load and number of hammering passages of MHP and their effect on fatigue life improvement. Klocke et al. [11] used predefined fields in FE simulations to transfer strain hardening information and residual stresses from a first mechanical simulation to a second fluid-structure interaction simulation to study the potential of structured tool surfaces by MHP for foil-free forming of stainless steel sheets of cold work steel X155CrMoV12. They showed that MHP results in a reduction of the contact area to only 6 % and in low friction coefficients indicating that a total separation of the contact partners might mostly be achieved.

Despite the numerous studies on tools with different working principles and a large number of materials, there is no general approach for defining process parameters that result in the required surface properties. Due to the interaction of tool and process parameters with the geometry and material of the workpiece, suitable parameter combinations can differ for every machining task and even for different areas of the same workpiece. In this study we investigate these dependencies for three different materials experimentally using a pneumatic experimental setup and in complementary FE simulations. We determine roughness and hardness values of the initial face-milled and ball-cut surfaces and for the peened surfaces. These data contribute to a mathematical process model of MHP that offers useful starting values for a given machining task.

## 2. Methods

### *2.1. Materials*

In the present study three typical materials used for the production of dies and moulds were considered. A cast iron EN-JS 2070 (GGG70), a low-alloyed steel 40CrMnMo8-6 (DIN no. 1.2312) and a high-alloyed steel X155CrVMo12-1 (DIN no. 1.2379) were tested in the as-received conditions with no further heat treatment. EN-JS 2070 was in an annealed condition (hardness: 224±5 HBW 2.5/187.5) and both steels 40CrMnMo8-6 (287±2 HBW 2.5/187.5) and X155CrVMo12-1 (223±3 HBW 2.5/187.5) were in a normalized condition.

*2.2. Uniaxial compression tests*

During MHP, elevated strain rates occur and it is important to consider this effect in the FE simulations. Compression tests on cylindrical specimens with a diameter and a height of 6 mm were therefore performed at room temperature for all three materials at three different strain rates to obtain the true plastic strain - true stress relations to describe the material behavior. Static ($\dot{\varepsilon} = 10^{-3}s^{-1}$) and quasi-static ($\dot{\varepsilon} = 10^{-2}s^{-1}$) compression tests were performed in a conventional testing machine Zwick/Roell Allround-Line Z020 with a 20 kN load cell. Furthermore dynamic compression tests ($\dot{\varepsilon} = 10^{2}s^{-1}$) were carried out using a drop tower with a drop weight of 600 kg. Strains were measured in all experiments by digital image correlation (GOM GmbH, Germany) at the surface of the compression specimens. Stresses were calculated from the reaction forces measured by the load cell of the conventional testing machine (static tests) and from data obtained from a strain gauge applied at the punch of the drop weight (dynamic tests), respectively.

*2.3. MHP experiments*

All experiments were performed on a machine tool Heckert HEC 630 by Starrag AG equipped with the interfaces for the pneumatic MHP tool (figure 1a). On the workpiece side, a flat plate (one for each of the three materials of this study) was mounted on a force measuring platform (Kistler 9255B with a range of 10 to 40 kN and a rigidity of 3 µm/kN). The three plates were face-milled with a roughness of $R_z = 10$ µm in order to ensure the same conditions for all experiments.

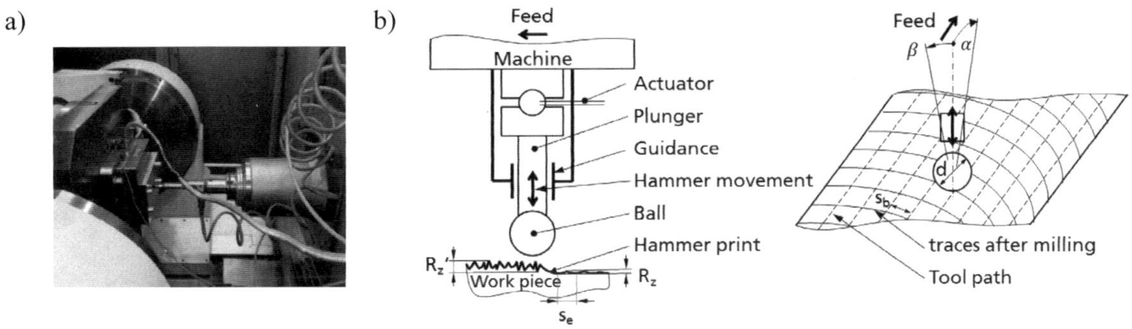


**Figure 1.** Experimental arrangement and process parameters of the MHP setup. a) Machine tool (Starrag AG HEC 630) with the pneumatic MHP tool on the right hand side and the face-milled plate on the left hand side. b) Schematic representation of the process parameter of MHP.

The following parameter ranges (see figure 1b) were selected as the scope of observation for the experimental investigations (table 1). They are based on the technical boundary conditions of the MHP tool and on results of first simple FE simulations. With the help of the predicted penetration depth of one indention, the investigation range of path distance as well as the feed rate could be narrowed. To reduce the number of trials, a design of experiments (DoE) using the software tool Cornerstone 7 was performed. A d-optimal design with a square model approach was used. In order to achieve the highest possible significances, the functions are adapted to a Gaussian distribution using a Box-Cox transformation plot. A qualitative review was done with the residuals probability plots. The software is also used to derive a mathematical process model that describes the relations of test and result parameters based on the different MHP tests (experiment and simulation). In addition to the already described input variables, essential output variables that characterize the surface of the workpiece after MHP were defined (table 1).

To evaluate the manufactured sample plates and to determine the parameters of the resulting surface, both mobile and stationary measuring devices were used. The stationary confocal microscope (Walter Uhl, technische Mikroskopie GmbH & Co. KG; z-resolution 5 to 20 nm and x/y-resolution of 1 to 3 µm) and the mobile roughness measuring device (Mahr Gmbh - MarSurf PS10; profile resolution 8 nm) were used to determine the surface roughness. In addition, the mobile hardness measuring device (PCE Instruments GmbH - PCE-3500; range from 20 to 70 HRC) served to evaluate surface hardening.

**Table 1:** Overview of the selected input and output variables of the experimental investigations.

| **test parameters** | | | **result parameters** | |
|---|---|---|---|---|
| *description* | *setting* | *unit* | *description* | *unit* |
| feed rate | 1.0, 3.0, 5.0 | m/min | roughness along the feed direction | µm |
| path distance | 0.1, 0.2, 0.3 | mm | roughness across the feed direction | µm |
| tool pressure | 2.5, 4.9, 7.3 | bar | surface hardness | HRC |
| us | 4.0, 6.0, 8.0 | mm | reaction force at the work piece | N |

*2.4. FE simulations*

Accompanying FE simulations were carried out using the commercial software ABAQUS 6.12. Figure 2a shows the initial geometry of a representative workpiece volume. In contrast to the experiments, the surface is ball-cut with a radius of 8 mm and a path distance of 0.7 mm. Two parameters were investigated using the FE method. First we studied the orientation of the MHP path with respect to the ball-cut paths as described by the angle φ. Figure 2b shows the four different MHP paths that were considered. The full circles correspond to the locations of impacts. In a second study, different impact or tool angles α (see also figure 1) were investigated for a MHP path of φ = 0°.

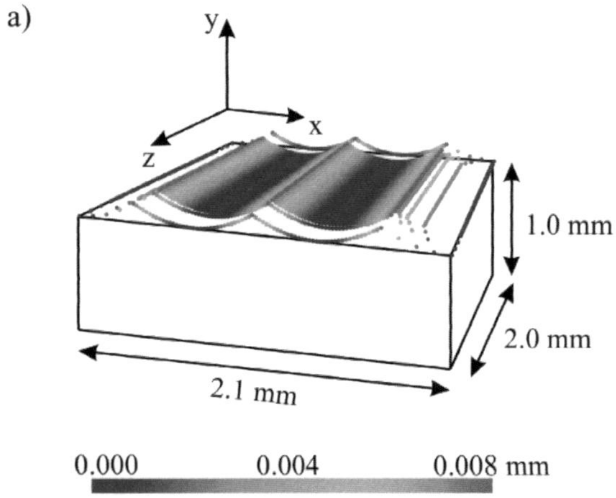


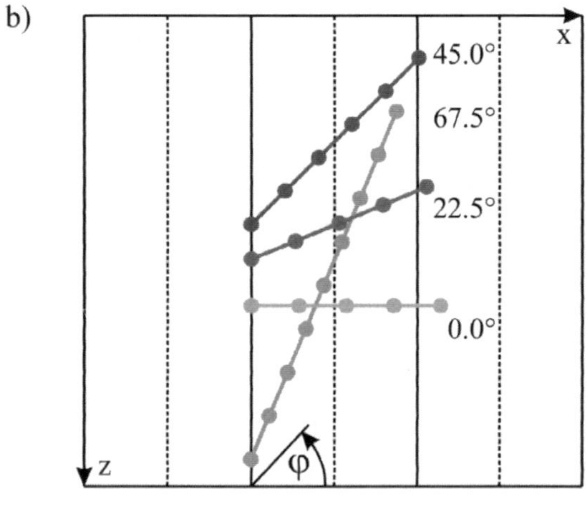


**Figure 2.** Specimen geometry and MHP paths used in the FE simulation. a) Schematic representation of the specimen geometry and b) the definition of four different MHP paths with respect to the x-direction of the coordinate system of the FE-model. The FE mesh consists of about 100,000 finite elements. In a) only the colored nodes of the surface are shown (color corresponds to the y-coordinate). The specimen covers two full ball-cut paths to investigate subsequent MHP processing of the surface.

The FE mesh consisted of 100,000 brick elements of type C3D8R. True stress-true strain data, measured experimentally in static, quasi-static and dynamic uniaxial compression tests, was used as input to describe the elasto-plastic material behavior. To describe the elastic properties a Young's modulus of E = 210 GPa and a Poisson's ratio of ν = 0.3 were used for all three materials. A MHP path consists of several single impacts. Each impact is calculated in one simulation that consists of three steps: First, a displacement-controlled step to ensure the contact of sphere and workpiece. Second, loading and unloading using a gravity load boundary condition, and third a displacement-

IOP Conf. Series: Materials Science and Engineering **480** (2019) 012021 doi:10.1088/1757-899X/480/1/012021

controlled step to separate tool and workpiece. Sphere (master) and work piece (slave) contact is modeled using the penalty method in normal direction. Tangential friction is considered using a friction coefficient of $\mu = 0.1$. A gravity load is used to apply the load to the system consisting of the sphere with unity inertia and the massless workpiece. The gravity was chosen according to a validation with the experiment where a distance of 0.2 mm of two impacts results in an overlap of 50 % of the areas of remaining indents for an impact angle of $\alpha = 90°$. A displacement boundary with $u_x = u_y = u_z = 0$ is set at the bottom surface of the workpiece. The boundary conditions of the rigid sphere (with a radius of 6 mm) were defined at a reference point. A rotation boundary with $\theta_x = \theta_y = \theta_z = 0$ was used in all steps. The displacement boundary conditions (to ensure the contact of sphere and workpiece in the first step and to separate both parts in the last step) are applied in a local coordinate system according to the impact angle $\alpha$. The load boundary condition (gravity) is also defined in that local coordinate system. At the end of each simulation a so-called restart file was written that contains information about the displacements of all nodes and the strain hardening. This information is transferred in the following simulation using a predefined field.

## 3. Results and Discussion

### *3.1. Uniaxial compression tests*

The three materials EN-JS 2070, 40CrMnMo8-6 and X155CrVMo12-1 were tested in uniaxial compression tests at static ($\dot{\varepsilon} = 10^{-3}s^{-1}$), quasi-static ($\dot{\varepsilon} = 10^{-2}s^{-1}$) and dynamic ($\dot{\varepsilon} = 10^{2}s^{-1}$) loading conditions. Figure 3 shows the true stress – true plastic strain curves. All three materials exhibit a positive strain rate sensitivity: An increasing strain rate results in higher yield stresses. The initial yield strengths for static loading are 364.2 MPa (EN-JS 2070), 703.4 MPa (40CrMnMo8-6) and 378.6 MPa (X155CrVMo12-1), respectively. These yield strengths increase with increasing strain rate up to 560.2 MPa (EN-JS 2070), 719.0 MPa (40CrMnMo8-6) and 526.9 MPa (X155CrVMo12-1) for dynamic loading conditions. This data was used as input to describe the plastic material behavior in the subsequent FE simulations.

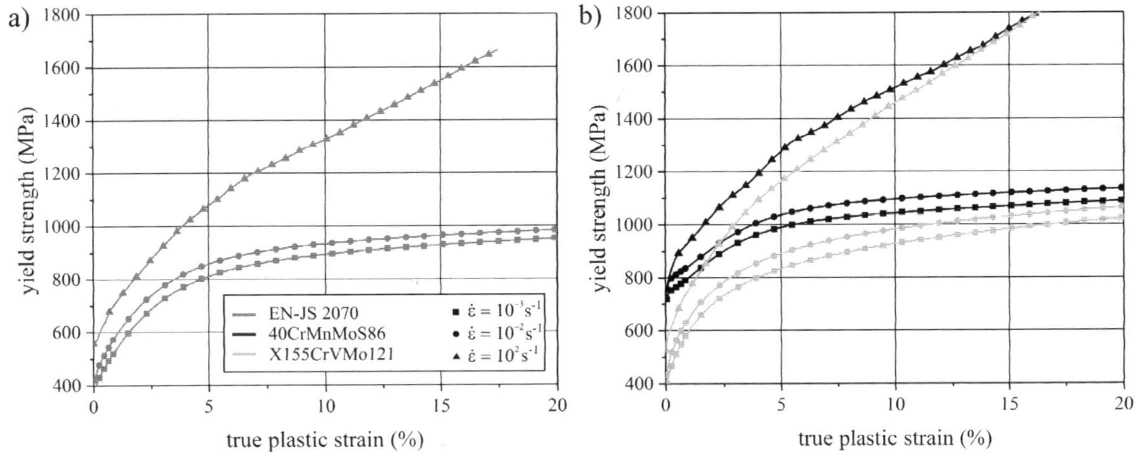


**Figure 3.** Strain hardening behavior of a) EN-JS 2070, b) 40CrMnMoS8-6 and X155CrVMo12-1 under uniaxial compressive loading for three different strain rates. The yield stress – true plastic stain curves indicate a positive strain rate sensitivity for all three materials.

### *3.2. MHP experiments*

Figure 4a shows a photograph of five machining test sections on an experimental plate of 40CrMnMoS8-6. Experimental results using the other two materials are not shown here, but are included in the relationships of roughness and process parameters (determined in the evaluation of the

IOP Conf. Series: Materials Science and Engineering **480** (2019) 012021 doi:10.1088/1757-899X/480/1/012021

DoE) presented later in this paper (see also figure 5). The evaluation of the test processing was performed in the marked middle section in order to minimize the influence of the acceleration behavior of the machine tool on the machining result. The roughness was measured along and across the feed direction for each test area (N1 to N35) in order to be able to independently evaluate the influence of individual parameters. Even very good optical results are possible, for example, with parameter set N5 (40CrMnMoS8-6, 2.5 bar, tool radius 4 mm, feed rate 1.0 m/min, path distance 0.1 mm). In contrast, parameter set N4 (40CrMnMoS8-6, 2.5 bar, tool radius 4 mm, feed rate 5.0 m/min, path distance 0.3 mm) does not offer a satisfactory surface quality. The only differences in these parameter sets are the feed rate and the path distance. The overlap of the individual ball impacts in connection with the plastic deformation in the workpiece is not sufficient to be able to produce a homogeneous surface. However, the different parameter sets are important in order to be able to evaluate the influence of individual parameters separately. As a result, all parameter sets in the test range (table 1) were evaluated and incorporated into the process model. For this purpose, the roughness values recorded for each parameter set with the mobile roughness measuring device were used.

Since the comparatively large test plates cannot be evaluated with the confocal microscope, parameter sets with a particularly large difference before and after peening were repeated with small samples. A corresponding, detailed evaluation is shown in figure 4b with parameter set N8 (40CrMnMoS8-6, 7.3 bar, tool radius 8 mm, feed rate 1.0 m/min, path distance 0.1 mm). Basically, the measuring methods of tactile and optical roughness measurement are fundamentally different and so are their numerical results (for milled surfaces they differ by a factor of 2 and for peened surfaces by a factor of 4). The roughness values of the optical method are generally larger than those of the tactile method, since the used probe tip (radii 2 μm) has a mechanically integrating effect. In contrast, the optical process captures the valleys in much greater detail and depth. The tactile, mobile measurement method can still be used for evaluation due to its flexibility, handling and the comparability of the measured values among themselves.

Using a confocal microscope, the real surface profiles before (figure 4b, left, $R_a = 2.2$ μm) and after hammer peening (figure 4b, right, $R_a = 0.9$ μm) can be clearly displayed. The three specimens were pre-machined by face milling. The individual milling grooves are clearly visible in both representations and remain in their basic structure even after hammer peening. However, the roughness peaks are largely leveled by the hammering process and pressed into the valleys.

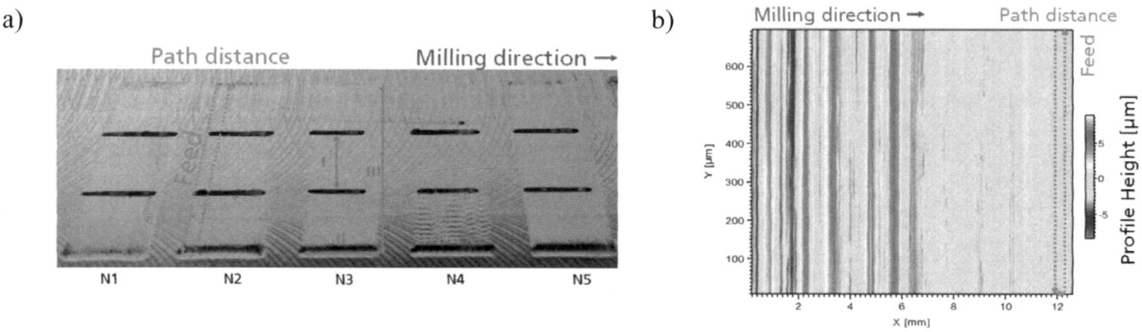


**Figure 4.** Surface topology before and after peening. a) Width (7 mm, II) and length (21 mm, III) of each machining test (N1-N5 pictured). Due to the axle acceleration, only the middle range (7 mm, I) was used for the evaluation. b) Example of the evaluation by means of a confocal microscope for one parameter set (40CrMnMoS8-6, 7.3 bar, tool radius 8 mm, feed rate 1.0 m/min, path distance 0.1 mm) across the peening direction.

As a result of the DoE, where all experiments for all three materials as well as parameter sets were considered, the following relationships can be inferred: It can be seen that larger tool radii result in

IOP Conf. Series: Materials Science and Engineering **480** (2019) 012021 doi:10.1088/1757-899X/480/1/012021

finer surface roughness due to the tool geometry, despite greater mass that needs to be accelerated with each impact, see figure 5. The roughness in the feed direction decreases with increasing tool radius from approx. $R_a$ = 0.29 µm at 4 mm radius (confidence interval ±0.05 µm) to almost $R_a$ = 0.1 µm at 8 mm radius (confidence interval ±0.01 µm). The roughness across the feed direction decreases with increasing tool radius from approx. $R_a$ = 0.66 µm at 4 mm radius (confidence interval ±0.11 µm) to almost $R_a$ = 0.18 µm at 8 mm radius (confidence interval ±0.02 µm). Since the roughness values for the complete evaluation of all measuring tracks were determined by the tactile measuring method, these values cannot be compared directly with the measured values of the confocal measurement. The difference between the two measuring directions results from the previous milling process. The milling grooves of the face cutter are significantly more marked transversely to the feed direction and cannot be completely smoothed by the peening process. Basically, the larger the tool radius, the lower the surface roughness. In addition, the confidence level narrows, which lead one to expect a smaller dispersion of the hammer peening results.

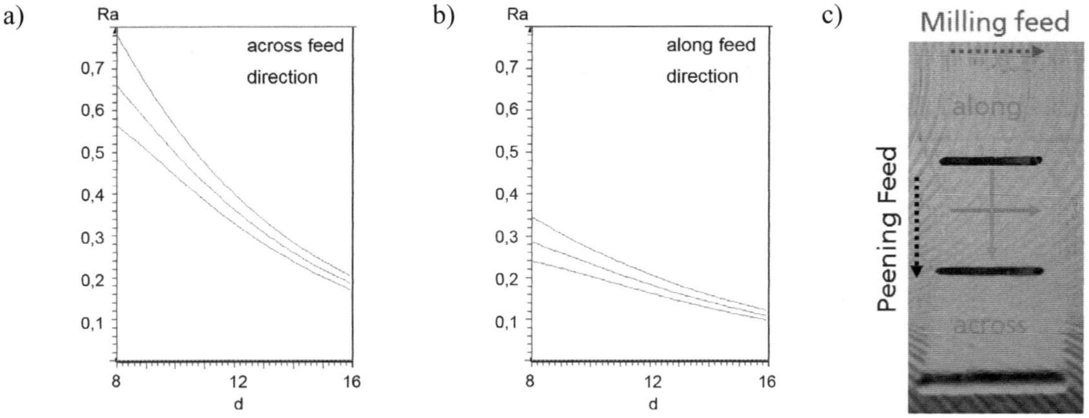


**Figure 5.** Evaluation of experimental data obtained for the three materials used in this study: Surface roughness a) along and b) across the feed direction as a function of the ball diameter. For small ball diameters, the achievable surface roughness varies considerably between the respective measuring directions as shown in the surface image in c). The larger the ball diameter, the closer the results approach.

A higher operating pressure of the pneumatic tool leads to a higher beat frequency. As a result, higher feed rates can be used. The surface roughness in the feed direction directly depends on this tool frequency. In contrast, the surface roughness across the feed direction depends directly on the selected path distance. That means that the surface roughness shows significant correlations in the evaluation. By contrast, the values of surface hardness and force signal vary strongly. The high variation in hardness can be attributed to the locally very different deformation. The hardness cannot be measured repeatedly at comparable locations using the mobile hardness tester. In addition, the force signal cannot be accurately evaluated due to strong noise effects and different beats (interference pattern between the impacts). In order to be able to make a statement about the relationships with the force acting on the workpiece, the first peak of each impact was evaluated and averaged over several successive impacts. Based on the DoE and under consideration of all MHP experiments so far the best parameter set for the investigated initial surfaces (face-milled with a roughness of $R_z$ 10 µm) of the three materials in this study consists of a pressure of 7.3 bar, a tool radius of 8 mm and a path distance of 0.18 mm. Specifically, MHP processing with these parameters results in an average reduction of the initial roughness of $R_a$ = 0.8 µm to $R_a$ = 0.5 µm in the case of 40CrMnMoS8-6 steel.

*3.3. FE simulations*

In contrast to the experiments, the initial surface of the work-piece was considered to be ball-cut (with a radius of 8 mm and a path distance of 0.7 – a technologically relevant initial surface condition) in the FE simulations in order to provide well-defined initial conditions. Two parameters, the orientation of MHP paths relative to the ball-cut paths and the impact angle, were investigated. All results presented below refer to the high-alloyed steel X155CrVMo12-1. In a first simulation (not shown here) a single impact on a flat surface was considered to validate the model. The gravity parameter was modified such that an impact distance of 0.2 mm results in an overlap of 50 % of the remaining indents of two adjacent impacts. This amount of gravity was then used in all simulations with ball-cut work-piece surfaces. Figure 6 shows the results of the first study with different MHP paths relative to the ball-cutting paths defined in figure 2. In these simulations the impact angle was always $\alpha = 90°$ (tool radius 6 mm; path distance 0.2 mm). The initial surface is characterized by a roughness of $R_a = 1.9$ (dashed lines in figure 6). With increasing angle of the MHP path, the path distance from one minimum to the next minimum of the ball milling grooves increases. Due to the MHP processing the surface is deformed (full lines) and the roughness decreases. MHP along a path with $\varphi = 0°$ results in $R_a = 0.74$. With increasing angle $\varphi$ a rougher surface can be expected, e.g., for $\varphi = 67.5°$ the roughness increases to $R_a = 1.43$. In contrast, the maximum plastic strain is independent of the MHP path with approx. 6 % and is located at a depth of 50 µm (the location of that maximum is not on the plotted path in figure 6, but in the remaining work-piece volume). Minor changes are observed for the average equivalent plastic strain (arithmetic average) along the MHP path that ranges from 1.4 % ($\varphi = 0°$) to 1.8 % ($\varphi = 67.5°$). Figure 6 shows that the deformed surface is always located below the initial surface. This observation is only valid for the considered path in this analysis. The volume of the work-piece is constant, and therefore, regions with displacements relative to the indentation direction must exist. The largest plastic deformations along the MHP path occur at the maxima of the initial milling grooves. The global maxima of the equivalent plastic strain are located beside the considered path.

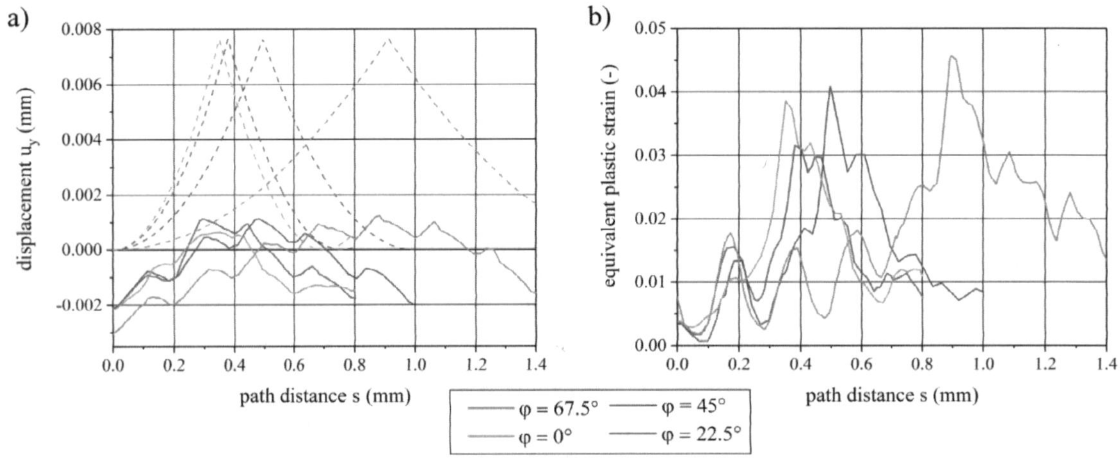


**Figure 6.** Results of the parameter study on the orientation of the MHP path relative to the previous ball-cutting path. a) Displacement of the surface in impact direction along the MHP path. The deformed surface (full lines) is always below the undeformed surface (dashed lines). b) Corresponding equivalent plastic strain along the MHP path. Most pronounced deformation occurs at the former ridges of the milling grooves.

A second study was focused on the impact angle α. Four different angles (90°, 67.5°, 45° and 22.5°) were investigated for the same MHP path with $\varphi = 0°$. The results for the impact perpendicular to the workpiece surface ($\alpha = 90°$) are already presented in figure 6 (green lines). In this study the same amount of gravity as before was used, but the working direction was inclined relative to the surface.

Figure 7 shows the initial and deformed surface along the MHP path for a tool angle of α = 45° as well as the equivalent plastic strain. Furthermore the displacements in all three directions are shown for a large part of the work-piece surface. The highest amount of deformation occurs again at the former maxima of the ball milling grooves. An average plastic strain of 2.6 % was evaluated, while the maximum overall plastic equivalent strain is 6.8 % (located beside path s). The location of this maximum is independent of the tool angle with a depth below the surface of 50 μm. The roughness is decreased due to the MHP to $R_a = 0.77$ (α = 90°), $R_a = 0.73$ (α = 75°), $R_a = 0.66$ (α = 60°) and $R_a = 0.49$ (α = 45°). Clearly, a decreasing tool angle (investigated here between 90° and 45°) results in a lower roughness. This result seems to disagree with previously reported experimental results, e.g. in [7], where a decreasing tool angle was associated with an increased roughness. Most likely this discrepancy is caused by the method of evaluation used in our FE simulation, which only considers the line path right below the MHP tool. Using, instead, the arithmetical mean height $S_a$ to assess the roughness of the peened surface the reverse relation (in qualitative agreement with [7]) is found: A decreasing impact angle results in an increase of $S_a$. For this two-dimensional parameter all nodes of the surface are used in contrast to the line parameter $R_a$. Hence the different impact numbers may have an effect on the determined roughness parameter. Obviously, different roughness measures affect the physical interpretation of MHP simulations; further work is required to validate the effect of different theoretical approaches.

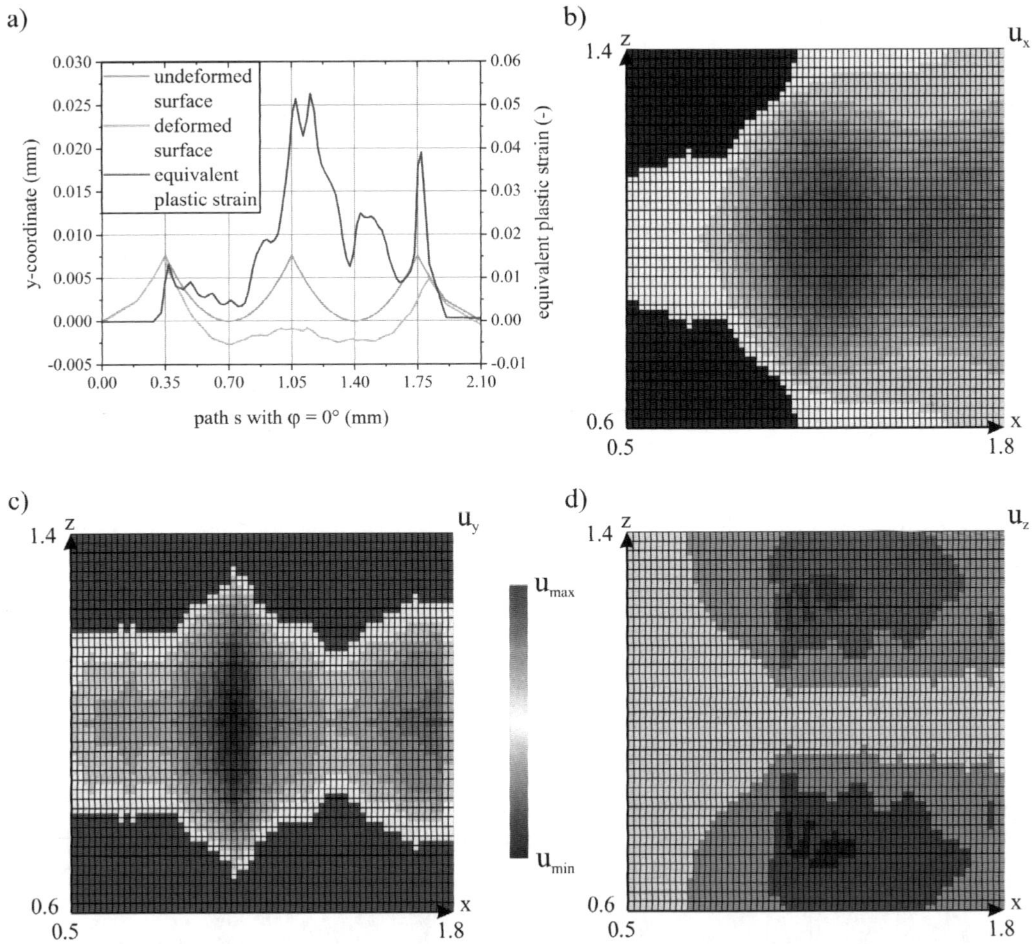


**Figure 7.** Results for a MHP path with φ = 0° and an impact angle α = 45°. a) Initial and deformed surface after MHP processing as well as the equivalent plastic strain along path s. b) - d) displacements at the workpiece surface $u_x$, $u_y$ and $u_z$.

Figure 7 also shows the displacement field after five impacts with a tool angle α = 45° and φ = 0° according to figure 2. The displacement fields exhibit a mirror symmetry relative to the MHP path (parallel to the x-direction at z = 1.0). This evaluation with $u_{xx}$ ($u_{max}$ = 4.75 µm, $u_{min}$ = -0.22 µm), $u_{zz}$ ($u_{max}$ = 1.65 µm, $u_{min}$ = -1.65 µm) and $u_{yy}$ ($u_{max}$ = 0.98 µm, $u_{min}$ = -8.57 µm) shows that the material is deformed towards the x-direction (according to the tool angle) and also in positive and negative z-direction away from the MHP path. The displacements in y-direction also show that some material at greater distance from the MHP path is raised upwards with displacements of a smaller order of magnitude compared to the other components of u.

## 4. Summary and conclusions

In this study, experiments and accompanying FE simulations were performed considering the machine hammer peening (MHP) of three materials with face-milled and ball-cut surfaces. The results contribute to the early development of a mathematical model that captures the correlations of process, material and geometry parameters on the resulting surfaces. The experimental results presented in this paper refer to 40CrMnMoS8-6 while the numerical results correspond to X155CrVMo12-1 and represent only a part of the overall testing program. The main results and conclusions of this study can be summarized as follows:

- Uniaxial compression tests under static ($\dot{\varepsilon} = 10^{-3}s^{-1}$), quasi-static ($\dot{\varepsilon} = 10^{-2}s^{-1}$) and dynamic ($\dot{\varepsilon} = 10^{2}s^{-1}$) loading conditions provided true stress-true strain data indicating a positive strain rate sensitivity. These data sets were used to describe the material behavior in the FE simulations.
- On the basis of a statistical experimental design, detailed investigations of the peening process were carried out investigating various parameters (e.g., tool pressure, tool radius, path distance and feed rate) and their effect on the resulting surface geometry. Based on all experimental results included in the mathematical model, the lowest roughness can be achieved using a path distance of 0.18 mm, a pressure of 7.3 bar and a tool radius of 8 mm.
- Using FE simulations, the MHP of a ball-cut surface with different MHP paths and tool angles was investigated. The results demonstrate that the roughness $R_a$ is most effectively reduced for a MHP path perpendicular to the previous milling path and for a tool angle of 45°. Furthermore, the depth of the maximum plastic strain is independent of the MHP path or of the tool angle and is located at a depth of 50 µm below the work-piece surface.

**Acknowledgements**

The authors gratefully acknowledge funding by the Federal Ministry for Economic Affairs and Energy (BMWi), Germany, through projects ZF4131902P07 (SP) and ZF4016221P07 (MF).

## References

[1] Mangerig I, Kroyer R and Koller M 2017 Experimental and numerical analyses of the effectiveness of high-frequency peening processes *Steel Construction* **10** 54-66
[2] Schulze V, Bleicher F, Groche P, Guo Y B and Pyun Y S 2016 Surface modification by machine hammer peening and burnishing *CIRP Annals* **65** 809-832
[3] Kolouch M, Wabner M and Kurth R 2017 Hammer peening with a piezo-actuated tool for surface finishing *MM Science journal* **10** 2055-2061
[4] Bleicher F, Lechner C, Habersohn C, Kozeschnik E, Adjassoho B and Kaminski H 2012 Mechanism of surface modification using machine hammer peening technology *CIRP Annals* **61** 375-378
[5] Wied J, Scheil J, Klamser M and Berger C 2011 Impact experiments and finite element simulation of surface roughness reduction by machine hammer peening *Mat.-wiss u. Werkstofftech.* **42** 827-832

[6] Habschied M, de Graff B, Klumpp A and Schulze V 2015 Fertigung und Eigenspannungen *Journal of Heat Treatment and Materials* **70** 111-121

[7] Mannens R, Trauth D, Mattfeld P and Klocke F 2018 Influence of Impact Force, Impact Angle, and Stroke Length in Machine Hammer Peening on the Surface Integrity of the Stainless Steel X3CrNiMo13-4 *Procedia CIRP* **71** 166-171

[8] Revilla-Gomez C, Buffiere J-Y, Verdu C, Peyrac C, Daflon L and Lefebvre F 2013 Assessment of the Surface Hardening Effects from Hammer Peening on High Strength Steel *Procedia Engineering* **66** 150-160

[9] Bleicher F, Lechner C, Habersohn C, Obermair M, Heindl F, Ripoll M R 2013 Improving the tribological characteristics of tool and mould surfaces by machine hammer peening *CIRP Annals* **62** 239-242

[10] Baptista R, Infante V and Branco C 2011 Fully Dynamic Numerical Simulation of the Hammer Peening Fatigue Life Improvement Technique *Procedia Engineering* **10** 1943-1948

[11] Klocke F, Trauth D, Schongen F and Shirobokov A 2014 Analysis of friction between stainless steel sheets and machine hammer peened structured tool surfaces: experimental and numerical investigation of the lubricated interaction gap *Prod. Eng. Res. Devel.* **8** 263-272

IOP Conf. Series: Materials Science and Engineering **480** (2019) 012022 doi:10.1088/1757-899X/480/1/012022

# Finite element simulations on the relation of microstructural characteristics and the formation of different types of adiabatic shear bands in a β-titanium alloy

**S Winter[1,2,*], S Pfeiffer[1], T Bergelt[1] and M F-X Wagner[1]**

[1] Institute of Materials Science and Engineering, Materials Science Group, Chemnitz University of Technology, 09125 Chemnitz, Germany
[2] now at Fraunhofer Institute for Machine Tools and Forming Technology, IWU, 09126 Chemnitz, Germany

[*] e-mail: sven.winter@iwu.fraunhofer.de

**Abstract.** Deformation at high strain rates in combination with local thermal softening can lead to the formation of so-called adiabatic shear bands. These shear bands occur when the macroscopic deformation is concentrated in microscopic areas. The fundamental thermo-mechanical processes and micromechanical mechanisms acting during adiabatic shear band formation are not fully understood yet. In particular, the influence of different initial microstructures on shear band formation has not sufficiently clarified. In an earlier investigation, adiabatic shear bands formed in a β-titanium alloy with different initial microstructures were characterized in detail. It was shown that different volume fractions of the α-phase result in the formation of different types of shear bands. In the present study, a simple modeling approach based on the finite element (FE) method is used to investigate in a systematic parameter study (with a special focus on mechanical and thermal variables) how these different types of adiabatic shear bands can occur in the same material. The FE simulations clearly show that microstructural characteristics – such as size and volume fraction of the α-phase in the β-titanium alloy – have a strong influence on shear band formation. This study contributes to a more detailed understanding of the microstructural and thermo-mechanical mechanisms leading to the formation of different types of adiabatic shear bands in titanium alloys.

## 1. Introduction and experimental background

Because of their low thermal conductivity, titanium alloys are susceptible to the formation of adiabatic shear bands (ASB) [1–3]. Generally, ASB occur at high strain rates ($<10^2\ s^{-1}$) when the macroscopic deformation is concentrated in microscopic areas [4,5]. This leads to a local temperature increase because the heat cannot be dissipated quickly enough into the surrounding material. This effect results in local thermal softening which in turns leads to a further concentration of deformation in this region of the sample.

The mechanical behavior of β-titanium alloys is strongly affected by microstructural parameters like β-grain size, morphology and volume fraction of primary and secondary α-phase precipitates [6–9], but the effect of these microstructural features on shear band morphology is not yet well understood. The present study is part of an ongoing experimental and theoretical investigation on the evolution of adiabatic shear bands in the β-titanium alloy Ti-10V-2Fe-3Al for different types of initial microstructures: β-titanium with primary α-phase and β-titanium with primary and secondary α-phase

IOP Conf. Series: Materials Science and Engineering **480** (2019) 012022 doi:10.1088/1757-899X/480/1/012022

precipitates. We carried out extensive compression-shear tests in a previous experimental study [10] to specifically characterize the influence of the different initial microstructural characteristics on the formation of ASB. Compression-shear specimens, proposed by Meyer et al. [11], are geometrically derived from conventional cylindrical compression samples, but the top and lower surfaces are cut at a small angle of inclination with respect to the central axis. When loaded in compression, this special geometry leads to a superposition of shear stresses. While the resulting stress state is inhomogeneous [12], compression-shear specimens allow to test a material for its tendency to exhibit localized (shear) deformation. Compression-shear testing has been successfully used to characterize localized deformation in functional materials [13,14], where complex stress-states can considerably affect phase transitions [15]; this technique is, moreover, specifically well-suited to analyze the formation of ASB and subsequent material failure in structural materials at high strain rates [16,17]. All experiments considered here [10] were performed on compression-shear specimens (width a height and width of 6 mm and an angle of inclination of 6°) in a Split-Hopkinson pressure bar (initial strain rate: $10^3$ $s^{-1}$), [18,19]. Deformation was stopped at a nominal (uniaxial) compressive strain of 10 % for each microstructural condition, facilitating an investigation of the different stages of nucleation and propagation of the shear bands. Our results confirmed that the formation mechanisms of adiabatic shear bands strongly depend on the initial microstructure [20]. Two different types of adiabatic shear bands can occur in the deformed compression-shear specimens, as exemplarily shown in Figure 1.

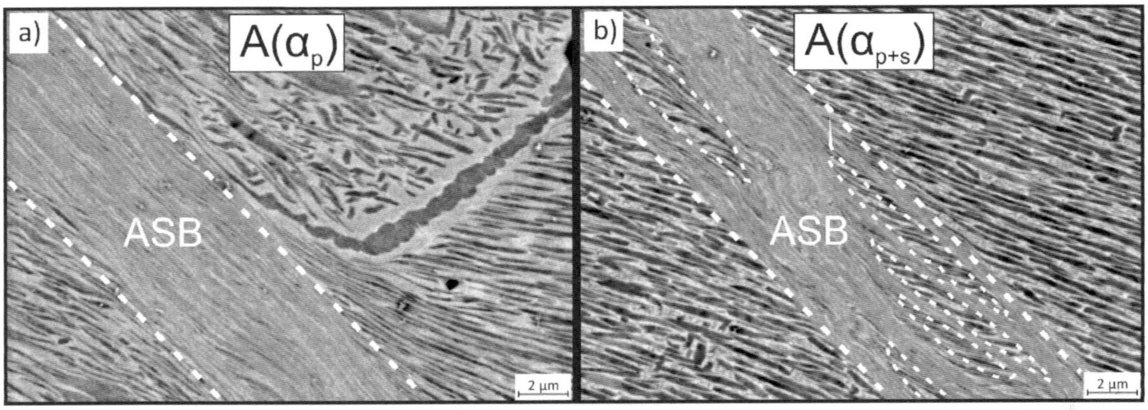


**Figure 1.** Scanning electron micrographs of two differently heat-treated states of the β-titanium alloy Ti-10V-2Fe-3Al, after 10 % nominal axial deformation of a compression-shear specimen (strain rate $10^3$ $s^{-1}$). a) The single-step heat-treated state A($\alpha_p$) exhibits the primary α-phase (dark zones) with a volume fraction of about 40 %. This state forms a single adiabatic shear band (ASB) with a width of 5 µm. b) The two-step heat-treated state A($\alpha_{p+s}$) contains the same content of primary α-phase and an additional volume fraction of about 15 % of the secondary α-phase. This state forms ASBs that are characterized by many fine branches.

Figure 1a) shows a typical ASB in the single-step heat-treated state **A($\alpha_p$)**, which was aged at a temperature of 700 °C for 1h. This state exhibits about 40 vol.-% of the lamellar primary α-phase ($\alpha_p$), which is precipitated according to the Burgers relationship [21]. When shear deformation occurs, a major part of the plastic deformation is transmitted to the lamellar α-precipitates (black lamellar structures in the scanning electron microscopy (SEM) images). In addition, a rotation of the α-precipitates towards the propagation direction of the adiabatic shear band is observed. Inside the shear band, there is a strong deformation of the α-precipitates in the shearing direction. The thick shear band completely deforms a volume width a characteristic thickness of about 5 µm. Figure 1b) shows a similar micrograph of the material heat-treated in two steps (step 1: similar to the condition A($\alpha_p$); step 2: additional aging at 500 °C for 8h). This material condition differs from A($\alpha_p$) only by an additional volume fraction (about 15 %) of the secondary α-phase ($\alpha_s$), which is also precipitated according to the

Burgers relationship. The content and size of the primary α-phase are very similar to the one-step heat-treated condition. Moreover, the macroscopic stress-strain behavior of the two material conditions is also very similar [20]. Interestingly, however, the stopped compression-shear experiments demonstrate that the secondary α-phase content leads to the formation of a quite different, branching type of ASB [20]. The additional lattice strain in the material, introduced by the smaller secondary α-phase precipitates, prevents rotation of the large primary α-precipitates during shearing. It is assumed that, due to the smaller distances between the different α-precipitates, the shear band is deflected by the very small $\alpha_s$-precipitates (average size below 200 nm, [20]) when passing through the β-matrix. Because of the lower strength of the α-phase [22] the shear band propagates predominantly along an α-rich path.

The stopped compression-shear experiments and in particular the subsequent microstructural investigations of the specimens [20] summarized briefly above have clearly shown that an increased volume fraction of the α-phase (by means of finely precipitating secondary α-phase) strongly influences the evolution of ASB - in particular in the nominal deformation range of 10 %, where the shear bands still propagate in the compression-shear specimens and the material has not yet failed. The present study is motivated by these experimental findings and aims at identifying microstructural parameters that affect the formation and propagation of the different types of ASB by simple numerical simulations. We present first results obtained from a simple Finite Element (FE) modeling approach, considering a model system that shows how different types of ASB can be formed in the same base material, but with different initial microstructures. The systematic parameter study identifies simple but relevant relations between mechanical, microstructural and thermal parameters and ASB propagation.

## 2. Finite element simulations

For this study, 2D FE simulations were performed with the commercial FE package *Abaqus*. Because of the strong effects of both strain rate and temperature on ASB formation, the FE simulation was mechanically-thermally coupled and dynamically explicit calculations were performed. A 2D compression-shear specimen was modeled with an inclination angle of 6° similar to the geometry of the samples used in our previous experiments (Figure 1). The inclination of the sample is necessary to locally create shear stresses that can result in shear localization during loading. For the numerical compression-shear tests, various (much simplified, but based on the real microstructure of the single-step heat-treated $A(\alpha_p)$ state) microstructures were modeled. In the simulations, the microstructure had to be represented by two different material definitions. The matrix approximated the behavior of β-titanium and the precipitates that of α-titanium.

The parameters used in the simulations are summarized in Table 1. To model the α-phase, material parameters of pure α-titanium from literature [23] were assumed for the modulus of elasticity, density, thermal conductivity and specific heat capacity. The parameters for β-titanium correspond to experimentally determined values of the solution-annealed condition of the alloy Ti-10V-2Fe-3Al [24]. To model the elasto-plastic behavior, a flow curve of the state $A(\alpha_p)$ (determined experimentally in a compression test under quasi-static conditions, [20]) was used for the β-titanium matrix. This state represents the macroscopic mechanical behavior of a hardened matrix with lamellar α-precipitates the most. According to [22], the α-phase has a lower strength than the surrounding β-titanium microstructure; therefore the flow curve of the matrix was multiplied by 0.5 in a first order approximation to account for the properties of the α-precipitates (Figure 2). This corresponds to the strength of pure grade 3 α-titanium [25] and can be considered as a realistic value for the material definition.

The area below the flow curves corresponds to the energy introduced into the material in the simulated compression-shear test. The Taylor-Quinney factor, determining the fraction of the total energy that is converted to heat and thus increases the temperature in the material [26], was assumed to be 0.9 for the entire simulation, which is a common value for the simulation of ASB [27–29], although recent work [30] shows that the Taylor-Quinney factor can be strongly material-dependent.

IOP Conf. Series: Materials Science and Engineering **480** (2019) 012022 doi:10.1088/1757-899X/480/1/012022

While both thermal softening and the strain-rate dependence of the material (which strongly affect ASB formation in a macroscopic mechanical test) were not considered in our simulations, the boundary conditions and the special geometry of the 2D sample ensured that localized deformation occurred. While highly simplified, our simulations therefore still allowed to qualitatively analyze the effect of different microstructural parameters on ASB formation.

**Table 1.** Parameters for the material definitions of the β-titanium matrix and the α-titanium precipitates.

| | **α-titanium – Ti(α)** (precipitates) | **β-titanium – Ti(β)** (matrix) |
|---|---|---|
| Young's modulus in *MPa* | 105000 | 110000 |
| Poisson's ratio | 0.3 | |
| flow curve | **Ti(α)** = 0.5 · Ti(β) | **Ti(β)** = [flow curve A($\alpha_p$)] |
| density in *g/cm³* | 4.5 | 4.6 |
| thermal conductivity in *W/m·K* | 22 | 8 |
| specific heat capacity in *J/Kg·K* | 523 | 560 |
| Taylor-Quinney factor | 0.9 | |

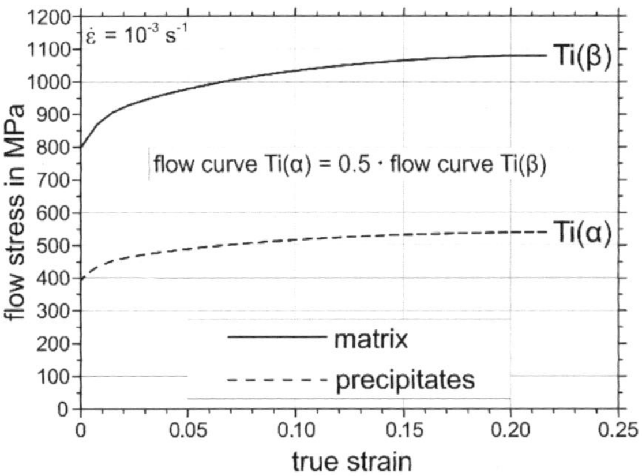


**Figure 2.** Flow curves of the matrix (Ti(β)) and the precipitates (Ti(α)) that were used for the FE simulations. The flow curve for Ti(β) was determined experimentally under quasi-static conditions (state A($\alpha_p$)) in a conventional compression test. The flow curve of the softer α-phase was approximated by multiplication of the curve of Ti(β) by 0.5.

The mesh of the compression shear sample consisted of 4-node elements (CPE4RT); the mesh at the precipitates was finer than that in the adjacent sample volume. In a sample with, for instance, 90 precipitates, we used approx. 14300 elements. We did not include grain boundaries, i.e., the simulated sample in a first approximation represents a single grain in which localized shear occurs due to the geometric/ mechanical instability. To reduce computation times for this first parameter study, a plane-strain state was assumed, and friction effects (e.g., between the sample and an experimental setup) were neglected. The specimen was instead deformed by simply applying a strain rate of $10^3$ $s^{-1}$ directly at the top and bottom (displacement boundary conditions). Below, the simulation results are presented at 10 % nominal plastic deformation, to allow for a qualitative comparison with the experimental results from our previous studies.

## 3. Results and discussion

Figure 3 shows results of the modeled compression-shear test in terms of the undeformed specimen (labeled "structure" in Figure 3) the distributions of equivalent (von Mises) stresses and strains, respectively. In order to be able to represent an initial condition as a reference configuration, a specimen consisting only of pure β-titanium structure (without precipitates) is considered first. The calculated stress distribution at 10 % nominal axial compression shows the expected stress concentration along the short diagonal of the specimen. The deformation is therefore also concentrated almost entirely in a small band along the shortest diagonal. We do not analyze the deformation of the sample near its corners in detail because it is primarily related to the simplified (displacement) boundary conditions used in the present study. The calculated strain in the shear plane is significantly higher at the corners than in the middle of the specimen; this agrees well with more realistic simulations of the three-dimensional compression-shear samples, [12]. The specimen shape used for the simulation is therefore well suited to produce a stress concentration together with a strongly localized deformation in a narrow area, which can be regarded as a shear band in the context of our simple analysis. A sample without α-precipitates as shown in Figure 3 leads to a strongly localized and closed structure of the shear band. A branching or deflection of this shear band is not recognizable, as is expected due to a lack of additional microstructural elements.

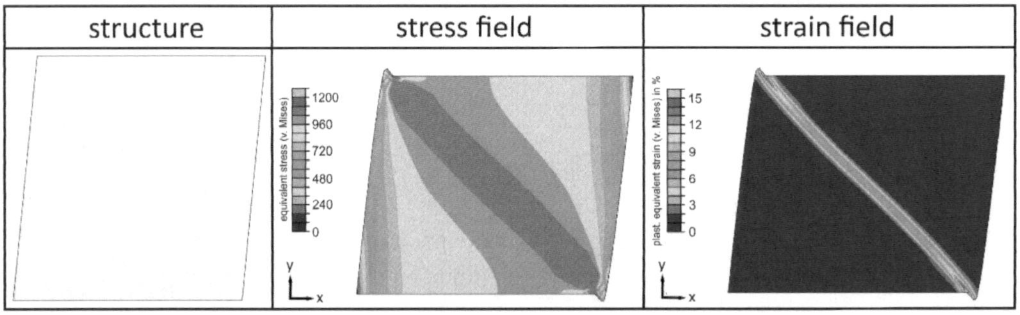


**Figure 3.** Simulation of the compression-shear deformation of pure β-titanium (without α-precipitates) at a nominal axial strain of 10 %. A pronounced concentration along the shear plane can be observed in the calculated stress and strain fields. Deformation of the precipitation-free matrix material (used as a simple reference condition) leads to a closed structure of the shear band.

In the following, the influence of microstructural parameters on shear band formation is examined in greater detail. For this purpose, different amounts of precipitates (approximated to have an oval shape) with the material definition of α-titanium were added to the β-titanium matrix. The precipitates represent the lamellar primary α-phase and are thus oriented similar to the (2D) Burgers relationship: the angle between individual precipitates is always 60°, resulting in three different orientations. To define a regular distribution of these precipitates, the first precipitate was oriented with an angle of 0° and placed in the lower left corner of the compression-shear sample. The next one was moved at a defined distance (depending on the total number of precipitates to be considered in the different simulations) in X-direction and rotated by 60°. This process was repeated (in X- and Y-direction, using further rotations by 60°) until the sample was filled with a regular pattern of precipitates.

Figure 4 (presenting information in a layout similar to Figure 3) shows three microstructures with different amounts of precipitates (the labels in the top row correspond to the number of precipitates in X-direction multiplied by the number in Y-direction) and the corresponding stress and strain fields. Changing the number of precipitates (and, to preserve the volume fraction of the α-phase, changing their size) clearly affects shear band formation. In the 5 x 4 sample with relatively large precipitates, the calculated stress field shows that there is not a single monotonous stress concentration (as in the pure β-microstructure): Instead, in the shear plane between the precipitates, areas with significantly lower equivalent stress values are observed. This is directly related to the lower strength of the precipitates that consequently carry much of the deformation (and thus partially unload the surrounding matrix). The strain field shows a deflection of the main region of deformation out of the shear plane; a slight splitting into two shear bands can be observed, where the deformation is primarily localized on an α-phase rich path in (and in proximity to) the shear plane. The deformation is somewhat more pronounced in the upper shear band, which is most likely related to the location and orientation of the precipitates next to the upper left corner.

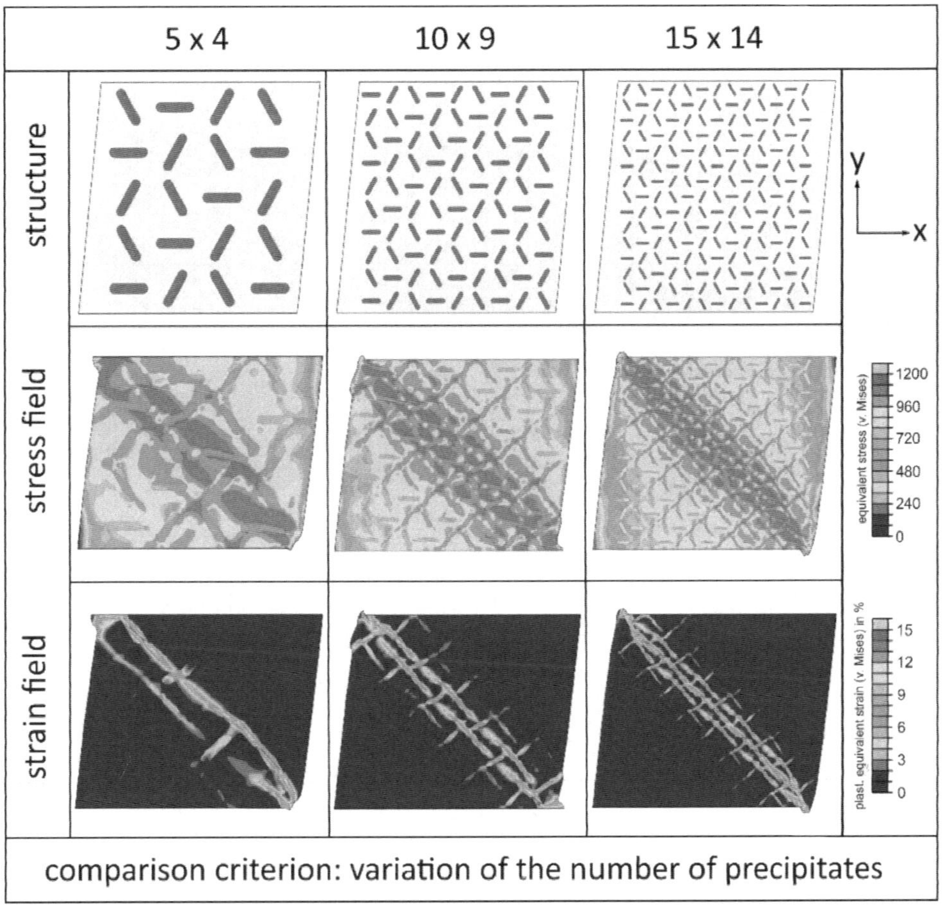


**Figure 4.** Simulated microstructures and the corresponding stress and strain fields at 10 % axial compression for different sizes of α-precipitates (at a constant volume fraction of 17.5 %). With increasing number of (soft) precipitates in the shear plane, the tendency to deflect or split the shear bands increases significantly.

When the number of precipitates is increased to 90 (10 x 9) or 210 (15 x 14), the calculated strain fields show a more pronounced tendency for shear band splitting. As more (and finer) precipitates lie in or directly next to the shear plane, growth of the initial shear band is more strongly affected by the α-phase, and individual precipitates can deflect the growth direction of the shear band by offering a path of low resistance to mechanical deformation. Because of the presence of many smaller α-lamellae, shear band splitting can occur more easily. Depending on the orientation of the particles, either a deflection or splitting of the shear bands occurs at different points in the simulated samples. This leads to the formation several approximately parallel shear bands and a more widely spread region of increased stresses compared to the pure β-titanium reference simulation. To summarize, the presence of the α-phase, and the sizes, orientation and distribution of the corresponding precipitates, clearly affect shear band formation. Despite the many simplifications used in our 2D simulations, this behavior agrees well with the experimental observations on shear band formation in the two aged states (Figure 1): The state A($\alpha_{p+s}$) is characterized by a significantly higher amount of small $\alpha$-precipitates and also tends quite strongly to deform by the formation of branched ASB.

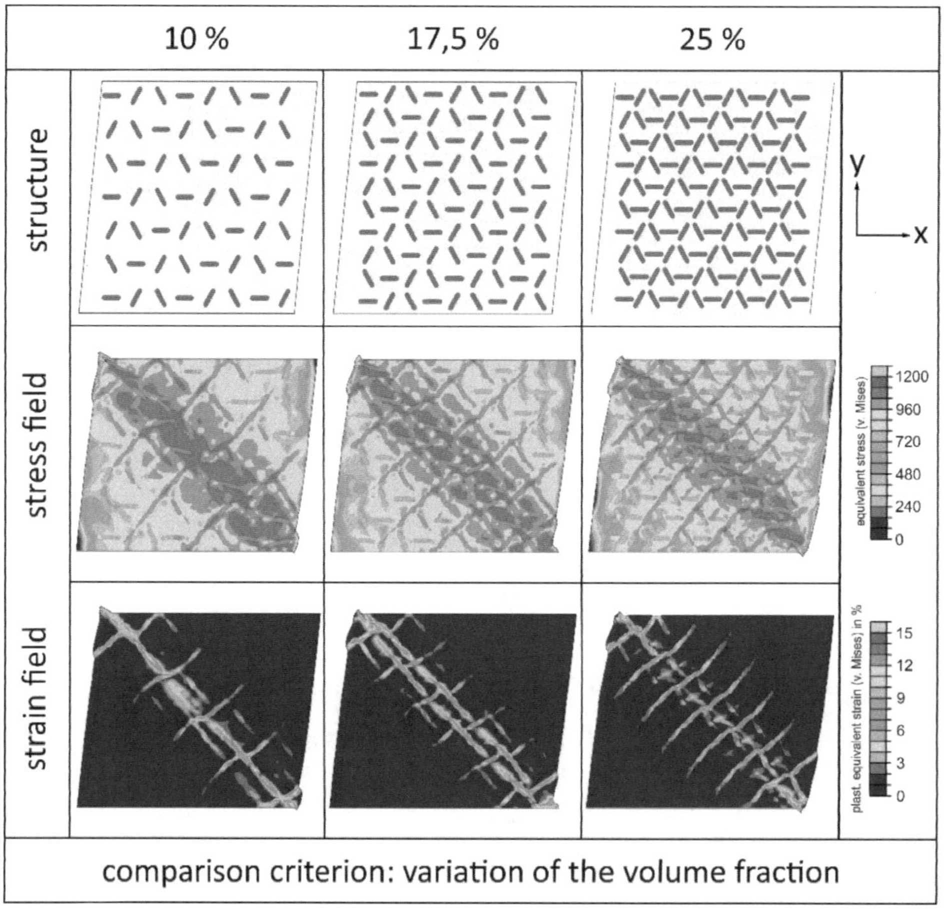


**Figure 5.** Simulations of representative microstructures with different volume fractions (but constant particle size) of the $\alpha$-phase at 10 % axial compression. A certain critical volume fraction of precipitates in the microstructure (17.5 %) is necessary in order to observe shear band branching into two sub-bands running parallel to the shear plane. If the content (and thus also the particle spacing) is too low (10 %), a closed band is formed that passes primarily through the $\beta$-matrix. If particle distances are too short and/or the volume fraction is high (25 %), deformation is redistributed into the soft $\alpha$-lamellae.

The simulation results presented in Figure 5 show the influence of different volume fractions (at a constant particle size) of the $\alpha$-precipitates on ASB formation. The compression-shear test with a volume fraction of 17.5 % corresponds to the previous 10 x 9 sample (Figure 4). As already discussed, this condition is characterized by a deformation in two shear bands that are formed parallel to the shear plane. When the volume fraction is reduced to 10 %, the mean distance between the precipitates is obviously too large for deflection or branching of the shear band. Only one main shear band is formed, which runs predominantly through the $\beta$-matrix. Only in areas where the precipitates are aligned perpendicular to (and across) the shear plane, the deformation partially occurs in the precipitates. Interestingly, when the volume fraction is instead increased to 25 %, there is also no distinct branching of the shear band (as observed in the corresponding strain field plot). Due to the significantly lower strength of the precipitates, most of the deformation in the shear plane can occur in the closely packed precipitates. The main path for the propagating shear band then is located inside the α-lamellae, which are positioned conveniently and in sufficient number in the main shear plane of the sample. Because the distance between the precipitates is small, strain is concentrated in the closely spaced softer precipitates and can also be deflected orthogonally (even away from the shear plane, but only inside additional precipitates). Clearly, this very high content of the α-phase leads to a widely spread strain field into the proximity of the shear plane even though branching of the ASB cannot be observed.

In summary, the simulation results show that shear band splitting and deflection is directly affected by the volume fraction of α-precipitates. When the volume fraction is too low, deformation occurs primarily in the β-matrix in a closed shear band. When, in contrast, the volume fraction is too high (and hence the distances between precipitates are very small), the energy introduced into the sample during deformation is consumed and distributed mainly into the softer precipitates. Consequently, there is no well-defined, localized deformation in the shear plane. Considering real microstructures as observed in [20], this means that a critical volume fraction of α-phase must be present in β-titanium in order for ASB to form distinct branches. In our simple and qualitative 2D FE simulations, the critical volume fraction is approx. 17.5 % in the 10 x 9 compression shear specimens. In the actual experiments, ASB branching was observed in the state A($\alpha_{p+s}$) with approx. 55 % α-phase ($\alpha_p + \alpha_s$), whereas in the material state A($\alpha_p$) (with approx. 40 % $\alpha_p$-phase and larger distances between the individual lamellae) only a homogeneous shear band occurred, which corresponds closely to the simulated sample 10 x 9 with 10 % of the α-phase.

## 4. Summary and conclusions

In the Ti-10V-2Fe-3Al material conditions tested experimentally in our previous studies, the formation and propagation of ASB is characterized by the complex interaction of, for example, microstructure, temperature, multi-axial stress states (as well as local stress gradients), physical properties of matrix and precipitates, and strain rate. While the 2D FE simulations performed in the present study cannot, as expected, provide quantitative information on critical microstructural parameters for ASB type selection, the simple model system investigated numerically does allow for an evaluation of relevant microstructural trends that are in good agreement with our experimental observations. Particularly the information on a critical volume fraction of the α-phase, and on the effect of soft phases providing favorable pathways for shear band growth, provides a basis for further investigations into the interesting phenomenon of ASB branching in Ti-10V-2Fe-3Al.

In summary, the following main results on the influence of microstructure on shear band formation in Ti-10V-2Fe-3Al can be directly applied to qualitatively interpret experimental observations:

(1) The tendency to split a shear band increases significantly with increasing number and decreasing size of soft α-lamellae in the shear plane.

(2) A critical volume fractions of the α-phase in the β-titanium matrix must be reached before branching of shear bands can occur.

(3) Increasing free distances between α-precipitates, in contrast, lead to the formation of conventional shear bands without branching.

## References

[1] Meyers M A and Pak H-R 1986 Observation of an adiabatic shear band in titanium by high-voltage transmission electron microscopy *Acta Metall.* **34** 2493–9
[2] Meyers M A, Subhash G, Kad B K and Prasad L 1994 Evolution of microstructure and shear-band formation in α-hcp titanium *Mech. Mater.* **17** 175–93
[3] Yang Y and Wang B F 2006 Dynamic recrystallization in adiabatic shear band in α-titanium *Mater. Lett.* **60** 2198–202

[4] Bai Y and Dodd B 2012 *Adiabatic Shear Localization* (Pergamon Press)
[5] Zener C and Hollomon J H 1944 Effect of Strain Rate Upon Plastic Flow of Steel *J. Appl. Phys.* **15** 22–32
[6] Jackson M, Jones N G, Dye D and Dashwood R J 2009 Effect of initial microstructure on plastic flow behaviour during isothermal forging of Ti-10V-2Fe-3Al *Mater. Sci. Eng. A* **501** 248–54
[7] He L, Dehghan-Manshadi A and Dippenaar R J 2012 The evolution of microstructure of Ti-6Al-4V alloy during concurrent hot deformation and phase transformation *Mater. Sci. Eng. A* **549** 163–7
[8] Chen W, Sun Q, Xiao L and Sun J 2012 Thermal stability of bulk nanocrystalline Ti-10V-2Fe-3Al alloy *Mater. Sci. Eng. A* **536** 223–30
[9] Winter S and Wagner M F-X 2016 On the effect of incremental forming on alpha phase precipitation and mechanical behavior of beta-Ti-10V-2Fe-3Al *IOP Conf. Ser. Mater. Sci. Eng.* **118** 012026
[10] Winter S, Schmitz F, Clausmeyer T, Tekkaya A E and Wagner M F-X 2017 High temperature and dynamic testing of AHSS for an analytical description of the adiabatic cutting process *IOP Conf. Ser. Mater. Sci. Eng.* **181** 012026
[11] Meyer L W, Staskewitsch E and Burblies A 1994 Adiabatic shear failure under biaxial dynamic compression/shear loading *Mech. Mater.* **17** 203-14
[12] Pfeiffer S, Frint P and Wagner M F-X 2017 Analysis of the complex stress state during early loading in cylindrical compression-shear specimens *IOP Conf. Ser. Mater. Sci. Eng.* **181** 012025
[13] Elibol C and Wagner M F-X 2015 Investigation of the stress-induced martensitic transformation in pseudoelastic NiTi under uniaxial tension, compression and compression-shear *Mat. Sci. Eng.* A **621** 76-81
[14] Elibol C and Wagner M F-X 2015 Strain rate effects on the localization of the stress-induced martensitic transformation in pseudoelastic NiTi under uniaxial tension, compression and compression–shear *Mat. Sci. Eng.* A **643** 194-202
[15] Wagner M F-X and Eggeler G 2006 Stress and strain states in a pseudoelastic wire subjected to bending rotation *Mech. Mater.* **38** 1012-25
[16] Meyer L W and Krüger L 2007 *ASM Handbook vol* **8** (Ohio: ASM International) pp 452–4
[17] Pursche F and Meyer L W 2011 *Engineering Transactions* **59** 67-84
[18] Kolsky H 1949 An investigation of the mechanical properties of materials at very high rates of loading. *Proc. Phys. Soc. Sect. B* **62** 676–700
[19] Pouya M, Winter S, Fritsch S and Wagner M F-X 2017 A numerical and experimental study of temperature effects on deformation behavior of carbon steels at high strain rates *IOP Conf. Ser. Mater. Sci. Eng.* **181** 012022
[20] Winter S 2017 *Mikrostrukturelle Einflussparameter auf die adiabatische Scherbandbildung in einer metastabilen β-Titanlegierung, PhD Thesis* (TU Chemnitz)
[21] Burgers W G 1934 On the process of transition of the cubic-body-centered modification into the hexagonal-close-packed modification of zirconium *Physica* **1** 561–86
[22] Hamajima T, Lütjering G and Weissmann S 1973 Importance of slip mode for dispersion-hardened β-titanium alloys *Metall. Trans.* **4** 847–56
[23] Nemat-Nasser S, Guo W G and Cheng J Y 1999 Mechanical properties and deformation mechanisms of a commercially pure titanium *Acta Mater.* **47** 3705–20
[24] MacHai C and Biermann D 2011 Machining of β-titanium-alloy Ti-10V-2Fe-3Al under cryogenic conditions: Cooling with carbon dioxide snow *J. Mater. Process. Technol.* **211** 1175–83
[25] ASTM B381-13: 2013 *Standard Specification for Titanium and Titanium Alloy Forgings* (West Conshohocken: ASTM International)

[26] Taylor G I and Quinney H 1934 The Latent Energy Remaining in a Metal after Cold Working. *Proc. R. Soc. London* **143** 307–26
[27] Bejjani R, Balazinski M, Attia H, Plamondon P and L'Éspérance G 2016 Chip formation and microstructure evolution in the adiabatic shear band when machining titanium metal matrix composites *Int. J. Mach. Tools Manuf.* **109** 137–46
[28] Polyzois I and Bassim N 2015 Microstructural simulation of adiabatic shear band formation in AISI 4340 steel using Voronoi Tessellation *Comput. Mater. Sci.* **109** 157–71
[29] Bäker M, Rösler J and Siemers C 2002 A finite element model of high speed metal cutting with adiabatic shearing *Comput. Struct.* **80** 495–513
[30] Rittel D, Zhang L H and Osovski S 2017 The dependence of the Taylor–Quinney coefficient on the dynamic loading mode *J. Mech. Phys. Solids* **107** 96–114

IOP Conf. Series: Materials Science and Engineering **480** (2019) 012023 doi:10.1088/1757-899X/480/1/012023

# Study on determination of flow behaviour of 6060-aluminium and AZ31-magnesium thin sheet by means of stacked compression test

**M Graf[1,*], T Henseler[2], M Ullmann[2], R Kawalla[2], U Prahl[2] and B Awiszus[1]**

[1] Chemnitz University of Technology, Professorship Virtual Production Engineering, Reichenhainer Str. 70, 09126 Chemnitz, Germany
[2] TU Bergakademie Freiberg, Institute of Metal Forming, Bernhard-von-Cotta-Str. 4, 09599 Freiberg, Germany

[*] e-mail: marcel.graf@mb.tu-chemnitz.de

**Abstract**. The characterization of materials is essential for process development and process optimization. As the virtual techniques are becoming increasingly common, the material data must be identified and modelled in consideration of the process-relevant parameters (forming temperature, strain rate, sample position, main stress state and orientation). In order to ensure the model functionality, two light metal alloys – aluminium EN AW-6060 (AlMgSi0.5) and magnesium alloy AZ31 – were studied, which differ in terms of their forming behaviour and sample surface state. In order to establish a reliable test method for the determination of flow curves for thin sheet, the materials were upset in the cylindrical and stacked compression test. These were subsequently modelled by the semi-empirical Hensel-Spittel approach, for use in commercial Finite Element (FE) software.

## 1. Introduction

The development and optimisation of process technologies without the use of mathematical formulations and FE-methods is impossible to illustrate nowadays. In the area of metal forming, especially non-linear constitutive equations for the flow behaviour of materials with high accuracy are of great importance. In addition to the calculation of material flow and machine loads during forming, flow curves also provide information regarding hardening and softening processes in the material. As the flow behaviour is greatly influenced by material state as well, it is important to investigate the material properties for various initial states and to take these into account during each step of the process simulations [1]. For bulk materials, the calculation of flow curves from the cylindrical compression test has proven to deliver accurate material models. Alternate approaches with plane-strain compression tests, torsion tests as well as simple tension tests have been applied. Nevertheless, when considering the thin dimensions of the sheet material, commonly available mechanical testing methods have limited maximum strains.

As one of the pioneers, PAWELSKI [2] proposed to stack multiple thin specimens in order to evaluate the flow behaviour of sheet material. Here, the author's investigations primarily focused on the influence of varying friction scenarios, during hollow cylinder compression tests. Based on that, the authors BERNRATH *et al.* [3] investigated layered compression tests of steel without any guides as additional stabilising methods. Conical compression tests without any lubrication have been conducted successfully. Currently named as stacked compression test or layer compression test, accurate flow behaviour determination for steel, aluminium and magnesium sheet have been reported [4–7]. Although defined and constant friction between tool and specimen are understood to be self-evident during mechanical testing, there is no in-depth literature that gives insight regarding the friction conditions in between stacked specimens. Now, this still represents an uncertainty, because contradictory statements are found in the cited literature. As presented in [2] and [3], in order to achieve equal results for stacked versus full material compression specimens it is essential that relative movement between layers is avoided. The maximum transferable shear stress between adjacent layers must be equal to that of the full sample; otherwise, the measured flow stresses will be too low. With sufficiently high friction between adjacent layers or permanent attachment through diffusion welding processes, the determined flow stress as well as material flow remains similar to that of the full material specimen. Under these circumstances, the number of layers has shown no influence. Therefore, lubrication between the layers should be avoided. ALVES *et al.* [4] confirm that only sticking friction conditions at the layer interface of aluminium specimens ensures stable deformation, although the authors also tested lubrication in between layers, as suggested by MERKLEIN and KUPPERT [5]. Slipping of layers due to insufficient interface friction can be identified by a deviation in the flow curve. Concerning the magnesium alloy AZ31, flow curves from stacked compression tests by BEHRENS *et al.* [6] have been compared to simple tension tests. Although the authors have applied lubricant to all layer interfaces, they also state that only the flow curves have been evaluated, where the slipping of the layers to each other has not taken place. While the authors of [7] only lubricate the tool side with $MoS_2$ and not the layer interfaces, their results deliver good accordance with the flow curves from stacked compression tests and simple tension tests. Nevertheless, a very low friction coefficient between tool and specimen is known to lead to instability in stacked compression tests (specimens begin to tilt and slip) [2]. [5] and [7] both align the rolling direction in stacked compression samples in order to characterise anisotropic material behaviour. The flow curves are calculated by accounting for strains in both horizontal directions.

The friction and the surface expansion correlate directly with the surface roughness of the contact partners. However, it is significant to note which contact partner has the coarser topography. It should be that for all forming processes, the forming tool has the higher hardness. Hence, its surface structure should also be finer, so that the roughness does not dent the work piece surface and inhibit the material flow. Therefore, the roughness profile of the work piece can be levelled due to the increasing normal stresses during forming with compression stresses. With continuous forming the work piece takes over the topography of the tool. Certainly, rough surfaces can provide the lubricant with additional lubricant reservoirs as well as low contact points at the beginning of the forming process. However, if too much lubricant is trapped in the roughness cavities, it cannot escape and the roughness cannot be eliminated. In contrast, the roughness of surfaces without tool contact in the forming process, for instance during stretching, can increase. This depends not only on the initial surface but also on the near-surface microstructure. Investigations by KIENZLE *et al.* [8] and GRÄFENS [9] showed that the microstructure is directly proportional to the roughness change. This is due to the activation of slip and twinning systems as deformation relevant mechanisms on the microscale.

The present study addresses the fundamentals in preparation for stacked compression tests using the example of 6060-aluminium and AZ31-magnesium thin strips. The methods, with respect to initial states and material specific deformation properties, which have led to flow models are explained in detail.

## 2. Materials and experimental procedure

### *2.1. Materials*

Because of the discontinuities in the literature concerning the boundary conditions to carry out the stacked compression test, firstly, an experimental study with different setups of thin discs of a commercially available aluminium alloy was conducted. These tests were carried out using extruded rods of the aluminium alloy EN AW-6060 (see Table 1). The average grain size of this material is 5 μm, measured by the linear intercept method.

**Table 1.** Nominal chemical composition (wt.-%) of the investigated extruded EN AW-6060 determined via optical emission spectrometry (OES).

| Mg | Zn | Mn | Zr | Cu | Si | Fe | Ca | Sb | Al |
|---|---|---|---|---|---|---|---|---|---|
| 0.46 | 0.017 | 0.05 | <0.001 | 0.01 | 0.43 | 0.21 | 0.0047 | 0.0127 | *bal.* |

Furthermore, sheets of an AZ31 magnesium alloy with a thickness of 1.0 mm and the chemical composition shown in **Table 2** have been upset with stacked specimens. These sheets were produced via the twin-roll casting process (TRC) with subsequent hot rolling and a final heat treatment in the pilot plant of the Institute of Metal Forming (TU Bergakademie Freiberg). The microstructure is characterised by fine recrystallised grains with an average size of 7 μm and even distribution measured by the linear intercept method.

**Table 2.** Nominal chemical composition (wt.-%) of the investigated AZ31 sheet determined via OES.

| Al | Zn | Mn | Zr | Cu | Si | Fe | Ca | Sn | Mg |
|---|---|---|---|---|---|---|---|---|---|
| 2.99 | 0.969 | 0.377 | 0.002 | 0.001 | 0.017 | 0.003 | <0.001 | <0.005 | *bal.* |

### *2.2. Setup and experimental procedure*

The baseline studies to determine the boundary conditions and fitting testing parameters were performed on the extruded aluminium alloy at room temperature and a strain rate range from 0.01 to 0.5 $s^{-1}$. The samples were taken from the same batch. In principle, the stacked compression tests were carried out on three geometries, some of which were stacked with three, eight and nine discs. Compression tests with full cylinder aluminium samples (D10 mm x H18 mm) and the calculation of their flow curves serve as reference. In order to correlate the stacked compression test with the full sample, discs with D10 mm x H6 mm as well as D10 mm x H2 mm were machined. They were stacked in different height-diameter ratios. The thinner disc dimensions were used for the comparability of sheet metal dimensions to full samples. Special care was taken during the exact positioning of the samples. Furthermore, polished carbide plates with an average roughness of $R_a$ = 0.13 μm were utilized. The aluminium sample surfaces had an average roughness of $R_a$ = 0.55 μm, as a result of the machining process. To increase the roughness up to $R_a$ = 1.1 μm in the contact area between two discs, selected specimen surfaces were ground with sandpaper (240 grit). The roughness was measured by the portable surface roughness tester Surftest 201 by Mitutoyo. The contact zone between the specimens and the tool surface was lubricated with a graphite solution. All surfaces were cleaned with acetone before stacking and testing.

In order to carry out isothermal compression tests, the stacked specimens as well as full cylinders were placed in upsetting cups, as shown in Figure 1. The tests were carried out in the hot forming simulator (WUMSI) at constant strain rates ranging from 0.01 $s^{-1}$ to 1 $s^{-1}$ (see [10]). In the case of hot compression tests with the magnesium alloy the upsetting cups were heated at approx. 5 K/min to the desired test temperature (250 °C to 350 °C) in an air circulation oven and held for 30 minutes. In this way, a homogeneous temperature distribution in the upsetting cups is achieved and the stacked specimens have all passed a similar temperature profile.

For the study of AZ31 magnesium thin sheets, round blanks with a diameter of D10 mm were punched out of the 1.0 mm sheet ensuring that the blanks get a straight cut edge with no bend and are subsequently stacked to a height of 15 mm (see **Figure 1**). The height-to-diameter ratio of 1.5 ensures the least possible influence due to friction at the tool surface, without causing the samples to buckle. Further, to ensure constant friction conditions, a graphite suspension was applied to the tool surface (carbide plates) prior to each test. As suggested by the references in literature, additional lubrication between specimen layers was omitted in order to avoid relative movement of the layers during testing. For selected compression tests, the rolling direction of the individual sheet metal blanks was marked and aligned during stacking.

For all compression tests, the flow stress (true stress) was calculated depending in the actual cross section according to the VON MISES theory and was plotted against the logarithmic strain computed from the crosshead displacement. The bulging of the compressed specimens due to friction at the tool surface was taken into account by applying the correction function by SIEBEL [11]. The coefficients of friction assumed here can be taken from the following discussion. Softening effects due to an increase in temperature from dissipated forming energy were eliminated numerically by use of the thermodynamic temperature factor $K_\vartheta = \exp(-m_1 * \vartheta)$ with $m_1 = -0.00148$ °C$^{-1}$ (for Al-6060) and $m_1 = -0.00427$ °C$^{-1}$ (for AZ31). At least three parallel tests were carried out to ensure a sufficient degree of statistical certainty. The flow curves presented in the following represent the average of multiple tests.

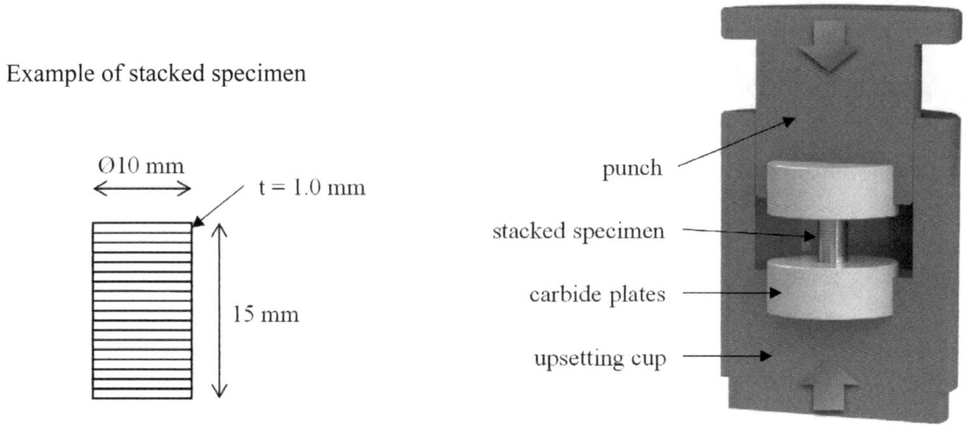


**Figure 1.** Setup of stacked specimens and upsetting cup to ensure straight compression as well as an even temperature distribution during hot deformation tests.

## 3. Results and discussion

### *3.1. Modelling of flow behaviour of EN AW-6060 at room temperature*

The study on the aluminium alloy was done at constant strain rates in the range of 0.01 to 0.5 s$^{-1}$. For the subsequent calculation of flow curves a friction coefficient of $\mu = 0.05$ was assumed. In Figure 2 (left) comparison between flow curves of the full cylinder specimens versus the three-layered, eight-layered and nine-layered samples is shown.

In comparison to the flow curves of the full samples, the stacked samples show a deviation of less than 5 %. It can therefore be assumed that the calculation of the flow curve including numeric corrections for friction as well as dissipated energy are valid for stacked compression tests under these conditions. In the case of shear fraction of single layers or relative movement of adjacent layers, changes in the force-displacement curves are well noticeable. At this point, the termination criterion of the experiments has been reached. However, if the compression is continued until a maximum height reduction of 10 mm, the centred discs would be most often damaged (Figure 2, right). During the presented study, stacked compression tests with a diameter-height-ratio of 10:18 with 2 mm thick discs could not be realized. Very likely, the large number of slip planes has led to instability.

The modelling of the flow behaviour was done with the semi-empirical flow curve model by HENSEL and SPITTEL [12] for cold forming processes according to:

$$k_f = A \cdot e^{m_1 \cdot \vartheta} \cdot \varphi^{m_2} \cdot e^{m_4/\varphi} \cdot \dot{\varphi}^{m_3} \tag{1}$$

IOP Conf. Series: Materials Science and Engineering **480** (2019) 012023 doi:10.1088/1757-899X/480/1/012023

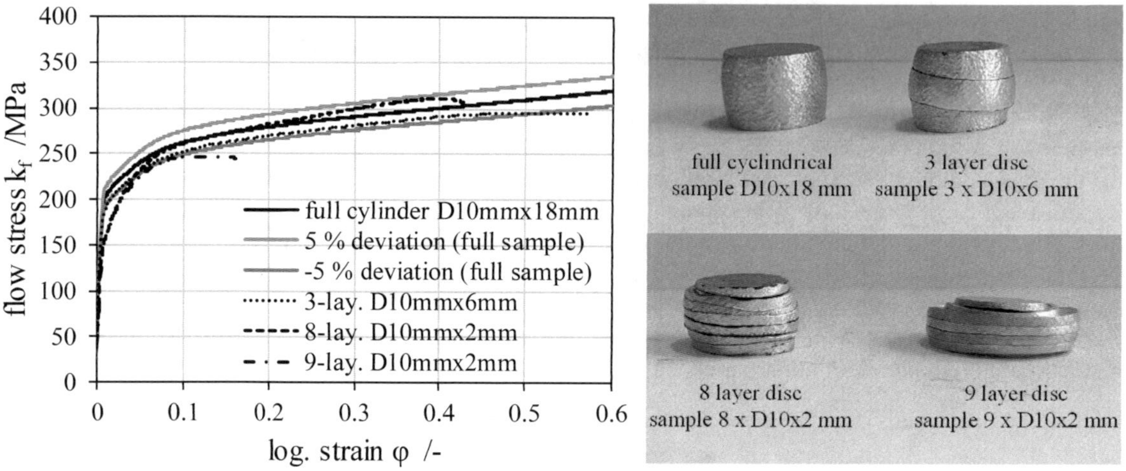


**Figure 2.** Comparison of the flow curve and the influence of the test geometry at 20 °C and a strain rate of $\dot{\varphi} = 0.01\ s^{-1}$ (left) for the aluminium alloy; Geometry of the compressed sample with height reduction of 10 mm at room temperature (right).

The determined model coefficients A and $m_1$-$m_4$ for the investigated aluminium alloy are shown in Table 3, resulting in a mean coefficient of determination $r^2 = 0.9785$. Therefore, a very good agreement with the experimental results can be assumed (Figure 3).

**Table 3.** Material specific coefficients of EN AW-6060 for the cold flow curve term.

| | A | $m_1$ | $m_2$ | $m_3$ | $m_4$ |
|---|---|---|---|---|---|
| $k_f$ | 145.734 | 0.0435 | 0.1087 | 0.0093 | -0.000079 |

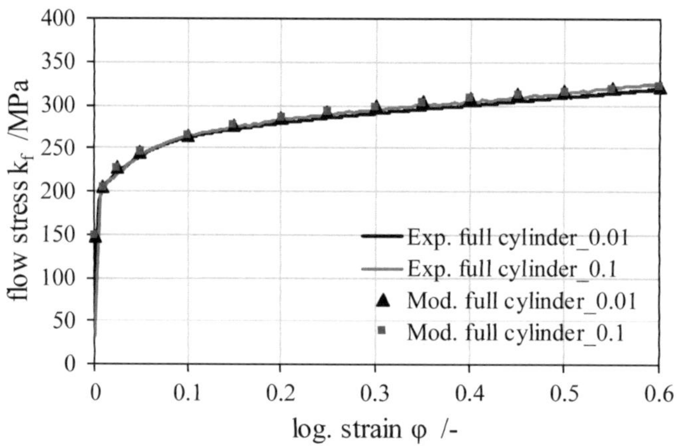


**Figure 3.** Comparison of the experimental and modelled flow curve at room temperature for the investigated aluminium alloy at strain rates $0.1\ s^{-1}$ and $0.01\ s^{-1}$.

It should be noted, that the here developed flow model for the aluminium alloy is only valid under the investigated forming parameters temperature and strain rate. Further compression tests and investigations are necessary for conditions beyond these boundaries. In general, the chosen model can be implemented in any FE-simulation program, which allows the integration of user-defined flow stress calculation.

By applying a 2D Finite Element Method (FEM) simulation in *Simufact.Forming V15*, the above mentioned flow model has been used to confirm a theory that the friction coefficient between adjacent layers in the stacked compression test must be higher than the friction coefficient between the tool and the specimen (see Table 4). In the shown simulation a comparison was done by varying the friction coefficient between adjacent layers from $\mu_{disc} = 0.05$ to $\mu_{disc} = 0.2$ at a height reduction of $\Delta h = 10$ mm. The friction coefficient between the samples and the upsetting tool was set to $\mu = 0.05$ for all calculations.

**Table 4.** Comparison of FEM simulations by varying the friction coefficient between adjacent discs, but maintaining a constant friction coefficient $\mu_{tool} = 0.05$ of to the tool surface.

| | $\mu_{disc} = 0.05$ | $\mu_{disc} = 0.12$ | $\mu_{disc} = 0.20$ |
|---|---|---|---|
| 9 layers ($\Delta h = 10$ mm) | | | |
| 8 layers ($\Delta h = 10$ mm) | | | |

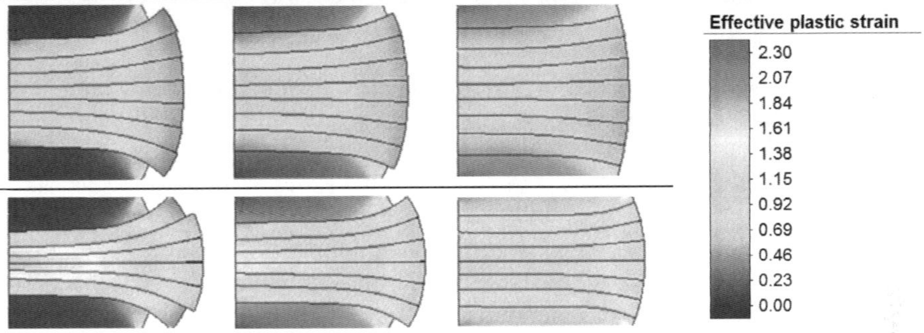


The results shown in Table 4 visualise precisely that lower friction coefficients between adjacent discs caused by redundant lubrication provide slip planes and thus relative movement of the layers. Therefore, these stacked samples are not valid for the calculation of flow curves as the theories for stable deformation during compression tests do not apply anymore. In practice this means, that the roughness of the discs surface (high roughness to adjacent layers) as well as of the tool surface (low roughness) must be machined adequately to ensure stable compression conditions during stacked compression tests. Lubrication is only needed at the tool-specimen-interface.

### *3.2. Modelling of the flow behaviour of AZ31 under hot forming conditions*

The flow behaviour of the AZ31 sheet, as shown in Figure 4, was characterised at the temperatures 250 °C, 300 °C and 350 °C and at the strain rates $0.01\ s^{-1}$, $0.1\ s^{-1}$ and $1\ s^{-1}$ for logarithmic strains $\varphi < 1$. Strain hardening is followed by softening due to dynamic recrystallisation until a steady state is reached for all testing parameters. Strain hardening effects increase with strain rate and decrease with higher temperatures. The diffusion-controlled process of dynamic recrystallisation shows the opposite dependency, leading to a later steady-state with increasing strain rate and decreasing temperatures. Although the rolling direction of each layer has been aligned, the compressed sample remained circular, without any effects on in-plane anisotropy.

The external appearance and cross sections of compressed specimens have been analysed to evaluate the validity of the flow curves. Indications of horizontal or diagonal sliding of individual layers meant the exclusion of these individual experimental data from further investigation. A cross-section of a valid compression specimen is shown in Figure 5, as well. In order to correctly estimate the true friction coefficient between the stacked specimen and the tool surface, 2D-FEM simulation in *Simufact.Forming V15* have been conducted here as well. As expected, due to the rough surface of the hot-rolled and heat treated AZ31 sheets the friction between adjacent layers is very high, as well as the friction at the specimen-tool-interface. Here, these conditions have led to successful stacked compression tests, as individual layers were barely exposed to relative movement.

IOP Conf. Series: Materials Science and Engineering **480** (2019) 012023 doi:10.1088/1757-899X/480/1/012023

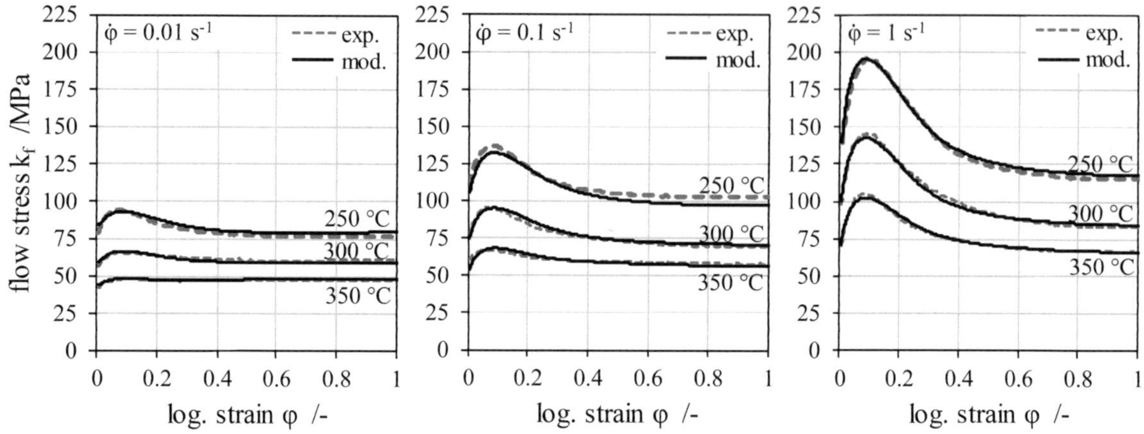


**Figure 4.** Comparison of experimental results and flow curve modelling developed from stacked compression tests with AZ31 sheets at various strain rates and temperatures.

In the FEM simulation shown here, the later described flow model for AZ31 has been adapted, as well as a constant friction coefficient of $\mu_{disc} = 0.5$ between adjacent layers. As seen in Figure 5, the simulation shows good agreement with the specimen cross-section for a high friction coefficient $\mu_{tool} = 0.2$ at the specimen-tool-interface. This conceivably confirms the theory by [2], that higher friction at the specimen-tool-interface leads to more stable upsetting of stacked specimens.

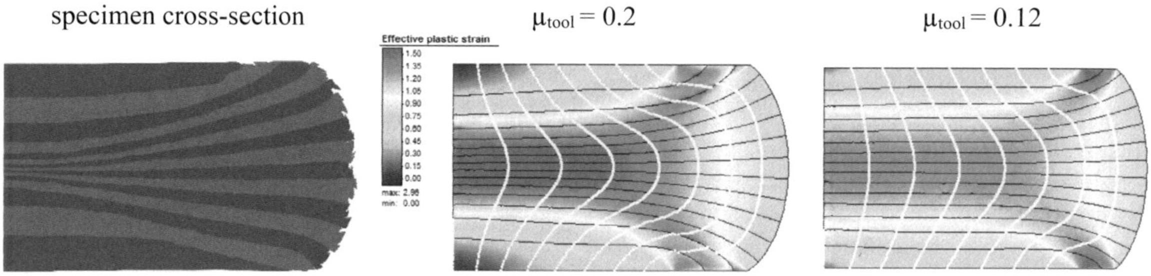


**Figure 5.** Coloured layers from the cross-section of a tested specimen at 300 °C and 0.1 s$^{-1}$ (determined in optical micrograph) in comparison to FEM-simulations ($\mu_{disc} = 0.5$) and varying friction coefficients $\mu_{tool}$ at the specimen-tool-interface.

For the modelling of the flow stresses of the AZ31 stacked compression tests a previously utilised approach by REICHELT *et al.* [13] for TRC-AZ31 was adopted. Here, a split approach was used for the description of the flow stress in the region of logarithmic strain $\varphi \leq 0.2$ and $\varphi > 0.2$, in order to represent the steep strain hardening followed by rapid dynamic recrystallisation and long-range steady state. In the present study, this approach was modified by changing the discrete transition to a smooth transition via a weight factor as a function of logarithmic strain, strain rate and temperature:

$$k_{f_{AZ31}} = (1-\delta_W)k_{f_{hard.}} + \delta_W k_{f_{soft.}} \qquad (2)$$

$$\delta_W = 0,5 + {}^1/_\pi \tan^{-1}\left[w_1 \vartheta^{w_2} \dot\varphi^{w_3} \left(\varphi - \varphi_k \vartheta^{w_4} \dot\varphi^{w_5}\right)\right] \qquad (3)$$

The coefficients $m_1$-$m_9$ for the semi-empirical flow curve model by HENSEL and SPITTEL [12] have been determined by regression of the experimental data. This approach takes into account the change in flow stress due to temperature $\vartheta$, logarithmic strain $\varphi$ and strain rate $\dot\varphi$:

$$k_f = A \cdot e^{m_1 \cdot \vartheta} \cdot \vartheta^{m_9} \cdot \varphi^{m_2} \cdot e^{m_4/\varphi} \cdot (1+\varphi)^{m_5 \cdot \vartheta + m_6} \cdot e^{m_7 \cdot \varphi} \cdot \dot{\varphi}^{m_3 + m_8 \cdot \vartheta} \quad (4)$$

The parameters $w_1$, $w_2$ and $w_3$ define the slope of the weight-function and the parameters $\varphi_k$, $w_4$ and $w_5$ define the location of the transition point and these were determined by regression of the experimental data (see **Fehler! Verweisquelle konnte nicht gefunden werden.**), as well.

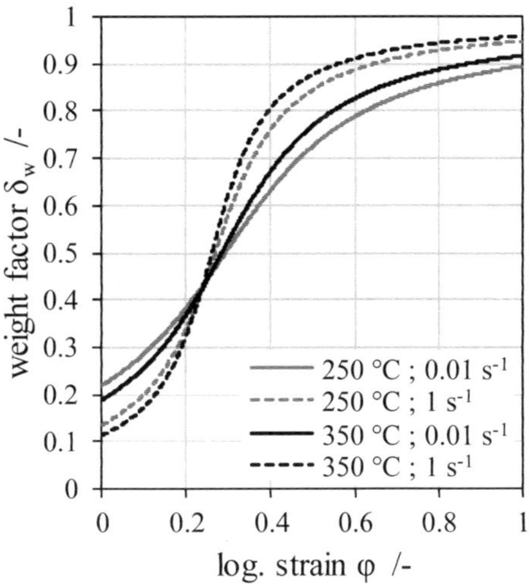


**Figure 6.** Representation of the weight factor as a function of the logarithmic strain for two temperatures and two strain rates respectively.

As seen in Figure 4, the model approach with the determined coefficients presented in **Table 5** is in good agreement with the experimental data. The mean coefficient of determination $r^2$ for all presented flow curves show a conformity > 0.9 matched to the experimental results. Hence, the material model can be applied in numeric simulations in the temperature range of 250 °C to 350 °C, at strain rates from 0.01 $s^{-1}$ to 1 $s^{-1}$ and logarithmic strains from 0.005 to 1.

**Table 5.** Material specific coefficients for the flow curve terms $k_{f_{hard.}}$ and $k_{f_{soft.}}$ (*a*), as well as the coefficients for the weight function (*b*) determined by regression of the experimental data.

| (*a*) | A | $m_1$ | $m_2$ | $m_3$ | $m_4$ | $m_5$ | $m_6$ | $m_7$ | $m_8$ | $m_9$ |
|---|---|---|---|---|---|---|---|---|---|---|
| $k_{f_{hard.}}$ | 1.203563 | -0.01248 | 0.303596 | 0.206172 | - | - | - | -2.91774 | - | 1.671411 |
| $k_{f_{soft.}}$ | $2968 \cdot 10^6$ | 0.005883 | -0.22306 | 0.064209 | - | - | - | 0.159763 | - | -3.37845 |

| (*b*) | $\varphi_k$ | $w_1$ | $w_2$ | $w_3$ | $w_4$ | $w_5$ |
|---|---|---|---|---|---|---|
| $\delta_w$ | 0.434768 | 0.177448 | 0.693086 | 0.147218 | -0.087559 | -0.018842 |

## 4. Conclusion

In the presented study stacked compression tests for 6060-aluminium and AZ31-magnesium alloys were realised successfully, in order to provide material-state-specific flow models for numeric simulations. The following conclusions and handling recommendations can be derived from the presented stacked compression test results:

- The authors have concluded from the literature research, that high friction between adjacent layers in combination with low friction to the upsetting tool prohibits relative layer movement and ensured stable upsetting.
- A flow curve model for the hot deformation behaviour of 1.0 mm AZ31 sheet was presented for logarithmic strains 0.005 to 1, as well as a flow curve model for 6060-aluminium extruded rod at room temperature for logarithmic strains 0.005 to 0.6.
- A study on aluminium disks with different height-diameter-ratios, as well as parallel FEM-simulations have led to the conclusion that the friction coefficient between adjacent layers needs to be greater than the friction coefficient at the specimen-tool-interface.
- As a result of the rough surface quality of the hot rolled AZ31 sheets, relative layer movement did not occur and stable upsetting of stacked specimens was achieved.

**Acknowledgments**
The authors thank the German Research Foundation (DFG) for the financial support of the project "Characterisation and modelling of the biaxial material behaviour of twin-roll-cast, hot rolled and annealed AZ31 sheets" (AW6/34-1; UL471/3-1). This research has the DFG grant number 396576920.

**References**
[1] Graf M, Ullmann M, Korpala G and Kawalla R 2013 Materialkennwerte als Basis für die numerische Simulation von Warmumformprozessen *Proceedings 22. Verformungskundliches Kolloquium (Planeralm)* ed B. Buchmayr pp 49–55
[2] Pawelski O 1967 Über das Stauchen von Hohlzylindern und seine Eignung zur Bestimmung der Formänderungsfestigkeit dünner Bleche *Archiv für das Eisenhüttenwesen* **38** pp 437–442
[3] Bernrath G, Volles R and Kopp R 2006 Multi-Layer Compression Tests under Hot Forming Conditions *steel research international* Issue **77** pp 265–270
[4] Alves L M, Nielsen C V and Martins P A F 2011 Revisiting the Fundamentals and Capabilities of the Stack Compression Test *Exp Mech* **51** pp 1565–1572
[5] Merklein M and Kuppert A 2009 A method for the layer compression test considering the anisotropic material behavior *Int. J Mater Form* **2** pp 483–486
[6] Behrens B-A, Bouguecha A, Bonk C and Dykiert M 2017 Experimental Characterization and Material Modelling of an AZ31 Magnesium Sheet Alloy at Elevated Temperatures under Consideration of the Tension-Compression Asymmetry *J. Phys.: Conf. Ser* **896** 12019.
[7] Herzig N, Abdel-Malek S and Meyer L W 2010 Experimentelle Ermittlung und Modellierung dynamischer Fließortkurven an Blechwerkstoffen *Proc. 9. LS-DYNA Forum* (Bamberg) pp 29-40
[8] Kienzle O and Mietzner K 1967 *Atlas umgeformter metallischer Oberflächen* (Berlin, Heidelberg: Springer Verlag)
[9] Gräfen H 1993 *Lexikon Werkstofftechnik: Berichtigter Nachdruck* ed H Gräfen (Berlin, Heidelberg: Springer Verlag)
[10] Henseler T, Ullmann M, Korpala G, Klimaszewska K, Kawalla R and Berge F 2017 Influence of Deformation Controlled Strain Rate on Tensile and Compression Behaviour of Magnesium and Steel Wire *Mater. Sci. Forum* **892** pp 89–96
[11] Siebel E 1932 *Die Formgebung im bildsamen Zustande, theoretische Grundlagen der technischen Formgebungsverfahren* (Düsseldorf: Stahleisen)
[12] Hensel A and Spittel T 1978 *Kraft- und Arbeitsbedarf bildsamer Formgebungsverfahren* (Leipzig: Deutscher Verlag für Grundstoffindustrie)
[13] Reichelt S, Neh K, Helbig M and Ullmann M 2016 On the Importance of Accurate Material Data for Modeling and Simulation of Forming Processes *Proceedings of the 19th International ESAFORM Conference on Material Forming* (Nantes, France) ed F Chinesta, E Cueto, E Abisset-Chavanne (Melville, N. Y.: American Institute of Physics)

IOP Conf. Series: Materials Science and Engineering **480** (2019) 012024 doi:10.1088/1757-899X/480/1/012024

# Characterisation of the activation and reaction behaviour and determination of the emissivity of reactive nickel-aluminium particles with regenerated fibre Bragg gratings

**S Grohmann[1,*], M Lindner[2], G Langhans[1], J Roths[2] and M F Zaeh[1]**

[1] Institute for Machine Tools and Industrial Management (*iwb*), Technical University of Munich, Boltzmannstrasse 15, 85748 Garching, Germany
[2] Photonics Laboratory, Munich University of Applied Sciences, Lothstrasse 34, 80335 Munich, Germany

[*] e-mail: sandra.grohmann@iwb.mw.tum.de

**Abstract.** Reactive particles represent a customisable heat source for joining applications as each reactive particle is able to undergo an exothermic, self-sustaining reaction. Microwaves are used to homogeneously initiate this reaction and allow the resulting temperature-time profiles to be influenced. The possibility to derive cause-effect relationships between the different particle structures, the activation energy, and the resulting maximum temperatures is essential for developing a resource-efficient, tailored heat source in production engineering. The reaction of reactive nickel and aluminium particles generates heat and thus temperatures up to 1500 K within milliseconds. A promising temperature measurement method despite an unknown emissivity are optical fibre sensors. These feature high frame rates, a compact design, and resistance to high temperatures as well as to electromagnetic waves. In this paper, the investigation of the use of regenerated fibre Bragg gratings (RFBG) as a new, innovative approach to characterise reactive particles is described. Experimental studies with an infrared camera and RFBG demonstrated that RFBG retain their functionality while being exposed to microwaves and high temperatures. The good accordance of the recorded RFBG-based temperature evolutions with infrared thermography data confirmed the suitability of RFBG as a sophisticated characterisation method and for determining the emissivity of reactive metal particles.

## 1. Introduction

Industrially established joining technologies like welding, soldering, and adhesive bonding, require a sophisticated heat source, which guarantees an efficient and controllable energy input. Reactive particles meet these requirements, since they are able to provide energy by undergoing a self-sustaining, exothermic reaction. Particularly, when joining materials with different thermo-physical properties, a customisable energy input is essential to avoid thermal damage of the components in the heat-affected zones and to promote the bond formation. Measuring temperature profiles is a standard procedure used in academia and industry to quantify the impact of the heat source on the material. However, the characterisation of reactive particles is a demanding challenge. High-temperature measurement with frame rates in the range of kilohertz and an unknown emissivity is not yet state of the art. Optical fibre sensors, such as regenerated fibre Bragg gratings (RFBG), represent a promising solution for investigating reactive particles. They enable recording high-resolution temperature-time evolutions and thereby can contribute to a profound understanding of the activation and reaction behaviour, which is necessary for the use of reactive particles in mechanical engineering.

### *1.1. Activation and characterisation of reactive particles*

In the field of combustion synthesis, which is also known as self-propagating high temperature synthesis (SHS), powder materials or green compacts consisting of homogeneously mixed educts are used for the industrial manufacturing of advanced materials [1]. Moreover, reactive systems in the form of foils, wires, and bimetallic particles are investigated in scientific research. Figure 1 illustrates the different types of reactive particles and various intrinsic structures of particles with two reactants.

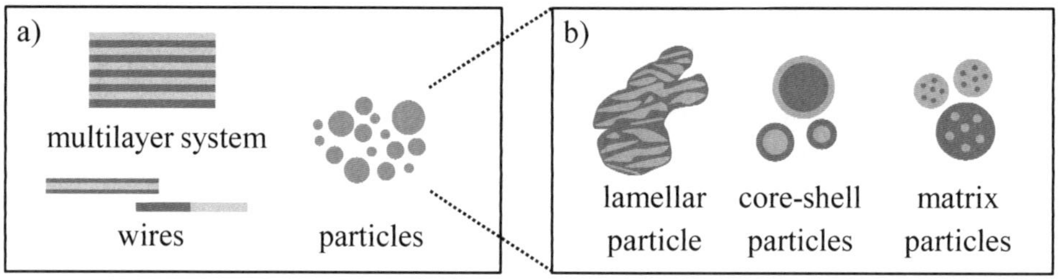


**Figure 1.** Scheme of a) different types of reactive systems and b) intrinsic structures of bimetallic reactive particles according to [2].

Bimetallic reactive particles can be divided into three categories according to their intrinsic structure: *core-shell particles* are obtained through electroless [3] or galvanic [4] plating processes, *lamellar particles* are commonly synthesised via high energy ball milling [5], and *matrix particles* are produced by either ultrasonic powder consolidation [2] or by planetary ball milling. Reactive particles represent an innovative heat source for thermal joining applications as each particle is able to release a defined amount of energy depending on the stoichiometric ratio of the contained reactants. Furthermore, they are a stable and compact alternative to brittle or fragile multilayer systems and wires, which are not or only to a limited extent applicable onto freeform surfaces.

A prominent example of a reactive system, which has induced increased interest in academia and industry, is nickel and aluminium. Nickel aluminides are high performance materials that form five intermetallic phases: NiAl, $NiAl_3$, $Ni_2Al_3$, $Ni_3Al$, and $Ni_5Al_3$. Due to their advantageous characteristics like excellent mechanical properties, low densities, high melting points, and good oxidation as well as chemical resistance, NiAl and $Ni_3Al$ have received the most attention [6; 7]. The combustion reaction of reactive systems with an equimolar ratio of nickel and aluminium exhibits the highest enthalpy of reaction of $-62 \pm 2$ kJ/mol at 298 K [8]. Furthermore, these nickel aluminides feature the highest melting points of 1911 K [7]. Sophisticated investigations of the cause-effect relationships between the intrinsic structure and the resulting activation and reaction behaviour provided important findings about the reaction mechanisms in nickel and aluminium systems and demonstrated the importance of a liquid aluminium phase for the start of the chemical reaction [9–11].

A further, significant option to influence the characteristics of the combustion reaction is the ignition technique, which directly correlates with the heating rate. Reactive systems can generally be activated through a locally confined or a homogeneous energy input, which triggers either the propagating or the simultaneous combustion mode [12]. It is important to note when studying reactive particles, that the available contact area between the single particles may promote, limit, or inhibit the formation of a self-propagating reaction front. A widely used approach in combustion synthesis to ensure sufficient contact is to manufacture compacted pellets. However, varying densities of the green compacts affect thermal conduction processes and lead to deviations of the maximum combustion temperature $T_{max}$ as well as to differences in the reaction kinetics.

Analysing loose reactive particles in order to avoid these unintended effects on the activation and reaction behaviour requires a homogeneous energy input. Typical activation methods to initiate the simultaneous combustion mode include induction heating [13], a modified hot isostatic pressing (HIP) process [14], and heating in an oven, which might simultaneously be used for dynamic scanning calorimetry (DSC) [15] or time-resolved X-ray diffraction (TRXRD) [16] experiments. As heat transfer is often accompanied by temperature gradients that can affect the uniformity of the reaction, energy transfer into the reactive particles is a valuable alternative.

In this context, electromagnetic waves represent an efficient energy transfer method and are especially suitable for the volumetric heating of particles [17; 18]. Microwaves at a frequency of 2.45 GHz have been successfully used to initiate the combustion reaction of nickel and aluminium particles [19; 20]. Major advantages of utilising microwaves include the possibility to realise high heating rates, adjust them in a relatively wide range, and influence exothermic phase transformations. Furthermore, microwaves allow to manipulate the reaction even after the initiation, which is a unique option considering the high reaction rates in combustion reactions of nickel and aluminium. However, discrepancies in the experimental data in scientific publications of the ignition temperature $T_{ig}$ and the maximum combustion temperature $T_{max}$ can often be attributed to different experimental set-ups and only partially comparable temperature measurement techniques.

The ability of microwave-activated reactive particles to reach high temperatures within milliseconds to seconds is quite advantageous for reducing cycle times, but extremely challenging for measuring temperature profiles. Measurement methods must guarantee high frame rates to capture changes in temperatures, which might indicate intermetallic phase transformations, and should not interact with the electromagnetic field. Even though thermocouples are an elaborated and established measurement method, they are not suitable in this case as they reflect a certain amount of microwave radiation and might also act as transmission line or antenna.

Contactless temperature measurement methods, which detect emitted electromagnetic radiation, such as infrared thermography, are compatible with microwaves, but require profound knowledge of the emissivity of the material. In contrast to emissivity values for bulk materials, which are often documented in table books, reference values for particles are not reliable since powder beds feature non-reproducible cavities. These, in turn, depend on the particle shape, the size distribution, and the degree of compaction. Sophisticated mathematical equations [21; 22] are valuable approaches, but are not able to consider the phase transitions during the combustion reaction.

A promising technique for the desired temperature measurement are optical fibres, which have already been successfully applied in the form of sapphire fibres in combustion synthesis [20]. Fibre Bragg gratings represent an innovative high-temperature measurement method for reactive particles and enable the determination of the emissivity of powder beds when combined with an infrared camera system.

### *1.2. Regenerated fibre Bragg gratings*

Optical fibre sensors based on Bragg gratings have been successfully used in a wide range of applications, like measuring the temperature gradient during the casting process of an aluminium alloy [23], in gas turbines [24], or in chemical and nuclear reactors [25]. Particularly noteworthy are their immunity to electromagnetic fields [26] as well as their small size [27]. Typical optical fibre diameters of 125 µm and sensor lengths from 0.5 mm up to a few millimetres enable their integration into even small-sized structures and thus a compact design of the experimental set-up. In addition, fibre Bragg gratings offer the possibility of wavelength multiplexing, which means that multiple measurement locations can be included in a single optical fibre [28; 29].

Fibre Bragg gratings (FBG) consist of refractive index modulations with a period length $\Lambda$ of about 0.5 µm inside the fibre's core. If broadband waves propagate through the optical fibre and experience the refractive index modulation, the waves will be partially reflected at each refractive index step. The reflected waves interfere constructively if their wavelength is equal to the Bragg wavelength and destructively for all other wavelengths. The Bragg wavelength $\lambda_B$ [30] is given by

$$\lambda_B = 2\, n_{eff} \Lambda \,, \tag{1}$$

where $\eta_{eff}$ is the effective refractive index. If heat is applied, the Bragg wavelength increases due to the thermo-optic effect and the thermal expansion of the silica fibre. For large temperature differences, a nonlinear relationship between the temperature und the Bragg wavelength occurs [23; 31; 32]. Fibre Bragg gratings, which are inscribed in optical fibres by using the photosensitivity of the germanium doped core, are referred to as Typ*I*-FBG [33]. If a Typ*I*-FBG is exposed to temperatures above approximately 600 K, the grating will degrade until it vanishes completely [34]. Temperature stable Bragg gratings, which have encountered a specific temperature treatment, like regenerated fibre Bragg gratings (RFBG), are resistant up to 1550 K [35]. With these RFBG, high-temperature measurements are possible.

## 2. Experimental set-up and materials

In order to evaluate the potential of RFBG sensors as an innovative temperature measurement method, specific RFBG and suitable sensor packages have been developed. Furthermore, an elaborated experimental set-up is required which allows both the integration of the fabricated sensors and the simultaneous observation with an infrared camera as a reference and additional source of information for the activation and reaction behaviour.

### *2.1. Fabrication of RFBG and sensors*

The RFBG were produced employing a custom-built inscription set-up and a thermal annealing process. The FBG had a length of 3 mm and a Bragg wavelength of 1550 nm. The optical fibre sensors were mounted inside of glass capillaries with outer diameters of 360 µm as depicted in Figure 2.

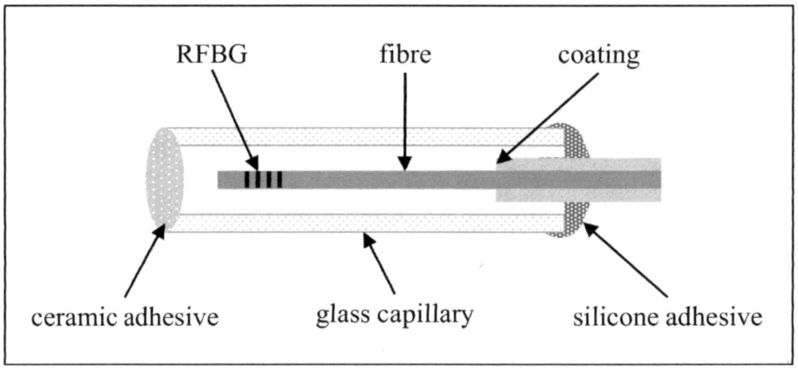


**Figure 2.** Scheme of the RFBG sensor package.

The end of each capillary was sealed with a ceramic adhesive to avoid unintended contamination and direct contact with the reactive particles. The Bragg wavelength was measured with a FBG Interrogator (I4, FAZ Technology Ltd., Dublin, Ireland). Its data sampling rate was 1 kHz.

The FBG were calibrated from room temperature up to 1073 K and the temperatures were determined on the basis of a fifth order polynomial according to [23]. Above 1073 K, the polynomial was linearly extrapolated.

### *2.2. Experimental set-up and temperature measurement*

The investigated reactive systems consisted of reactive particles with fine lamellar structures of nickel and aluminium (Indium Corporation, Clinton, USA). Circular disk-shaped powder samples with identical masses were prepared with the help of a template with a diameter of 18 mm and a height of 1 mm. A groove within the template ensured a reproducible positioning of the fibre sensor within the sample. Boron nitride sheets (HeBoSint, Henze Boron Nitride Products AG, Lauben, Germany) served as temperature resistant base material. Furthermore, boron nitride interacts with microwaves only to a limited extent, which is of major importance for the experimental set-up shown in Figure 3.

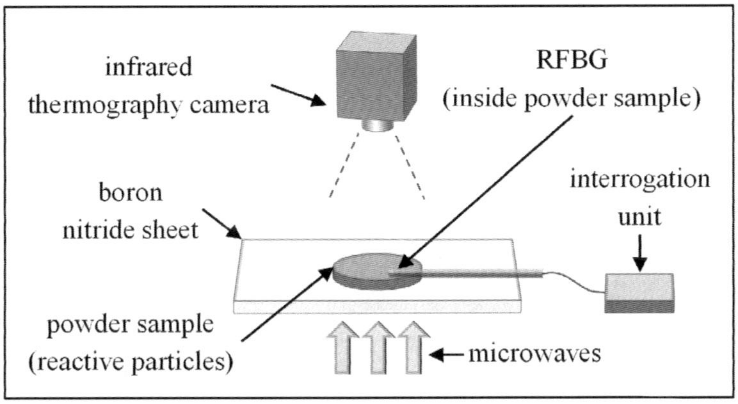


**Figure 3.** Scheme of the experimental and measurement set-up for the microwave activation of reactive particles.

A magnetron operating at a frequency of 2.45 GHz (MH2000S-215BB, Muegge GmbH, Reichelsheim, Germany) was used for the microwave activation of the prepared samples. An infrared (IR) camera (ImageIR 8300, InfraTec GmbH, Dresden, Germany) captured two-dimensional information of the reaction process at a frame rate of 1 kHz. As the emissivity $\varepsilon$ of the reactive particles is not known, a value of $\varepsilon = 1$ was assumed, which allowed an initial estimation of the observed temperature distribution.

### 3. Results and discussion

Initial experiments aimed at investigating the influence of the microwaves on the optical fibre sensors and the RFBG. Therefore, a fabricated sensor was placed on the boron nitride sheet without any reactive particles and the microwave power was increased to 450 W at 50 W intervals. Neither the infrared camera nor the interrogation unit detected changes of the sensor's behaviour.

For the following experiments, the microwave power was set to 200 W. Three samples of reactive particles were activated as depicted in Figure 3. Scanning electron micrograph images of unreacted and microwave-activated lamellar particles are shown in Figure 4.

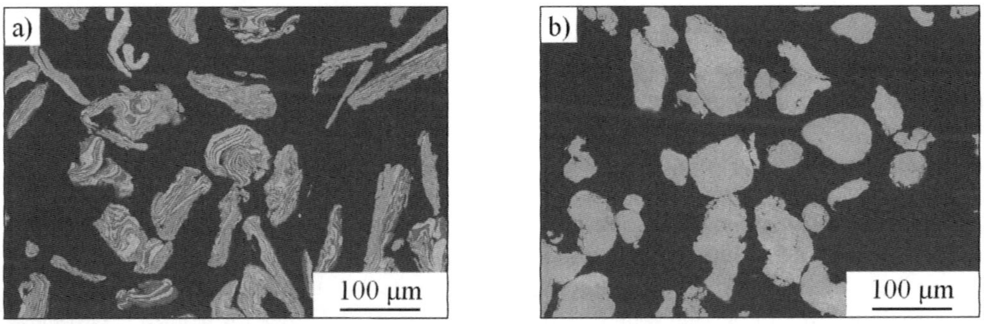


**Figure 4.** Scanning electron micrograph images of a) unreacted particles consisting of nickel (light grey) and aluminium (dark grey) and b) reacted particles.

The scanning electron micrograph images illustrate that both the shape and the size distribution of the unreacted particles deviated. Spherically and needle-shaped particles with fine lamellar structures consisting of nickel and aluminium were found. Microwave-activated particles featured nickel aluminides, which was also indicated by an energy dispersive X-ray spectroscopy analysis.

During the activation, which was performed three times, reactive particles were monitored with an RFBG-based sensor and the infrared camera. Figure 5 shows selected results.

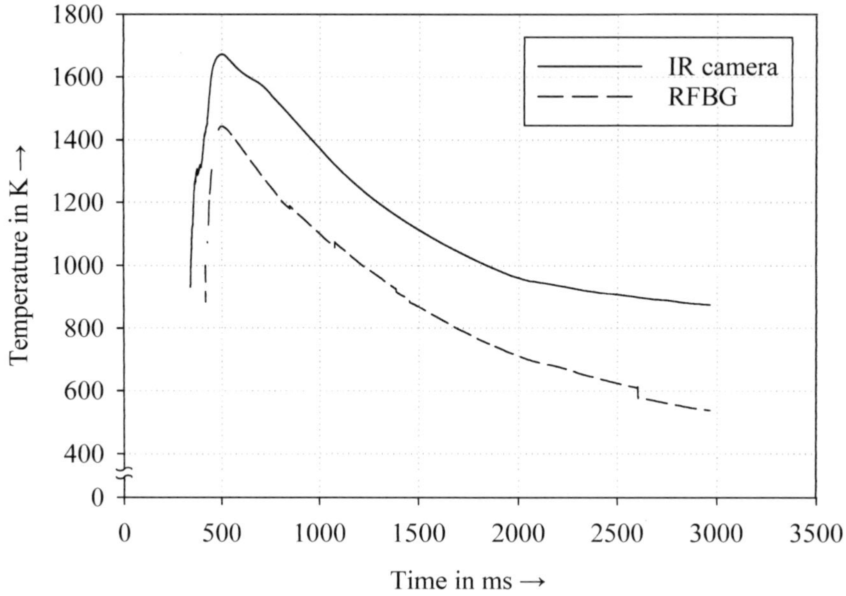


**Figure 5.** Temperatures as a function of time measured by a fibre sensor (RFBG) and with the infrared camera (IR camera).

Important conclusions can be drawn from the measured temperature-time profiles. The qualitative course of the recorded RFBG-based temperature evolutions show good accordance with the infrared thermography data. Expectedly, due to the overestimated emissivity, higher temperature values were measured with the infrared camera. Maximum temperatures were generated through the exothermic reaction of the reactive particles within 500 ms after the activation with the microwaves. A comparison of the measured values of the three different RFBG sensors (not shown for reasons of clear arrangement) indicated that the qualitative courses of the curves were identical, while maximum temperatures differed (1500 K ± 51 K). Similar observations for the data of the infrared camera suggested that deviations of the particle shapes and the size distribution of the prepared samples might have influenced the uniformity of the analysed combustion reactions and thus led to minor variations of the measured temperatures.

The great advantage of measuring with two independent systems is that the results can be used to derive the emissivity of the reactive particles during the reaction. For the example illustrated in Figure 5, an emissivity of 0.86 was calculated for the maximum temperature. This possibility represents a highly valuable approach not only for the field of combustions synthesis, but also for other metal-powder-based processes, like for example laser beam melting.

Moreover, the measured data formed the basis to evaluate the activation and reaction behaviour of the reactive particles. It can be stated that the obtained temperature-time profiles are comparable with other findings in literature. According to [36], it is assumed that interactions between a liquid aluminium and a solid nickel phase took place as the temperature-time profiles featured single peaks. As mentioned before, microwaves enable an efficient energy transfer and high heating rates, which promote solid-liquid or liquid-liquid interactions instead of solid state diffusion processes.

It also becomes evident that there are missing values for the RFBG-based fibre sensor. This probably resulted from predominant temperature gradients along the optical fibre sensor, which distorted the spectrum of the RFBG and prevented the calculation of the Bragg wavelength. However, this evaluation is essential for deriving temperature values. Nevertheless, informative temperature-time profiles were obtained.

A further important outcome is the suitability of the fibre sensors for continuous recording regardless of the current temperature. In contrast to the fibre sensors, the infrared camera required a pre-selection of a specific temperature range. Consequently, several samples had to be activated in order to obtain a complete temperature-time profile starting from room temperature. As mentioned before, even similarly prepared powder beds feature deviating emissivity values which affects the comparability and complicates matching of the data. Therefore, the results of the infrared camera in Figure 5 are limited to temperatures ranging from 873 to 1773 K. Complete temperature-time profiles without missing values or limited temperature ranges allow the evaluation of the ignition temperature $T_{ig}$ and the enthalpy of reaction, which in turn is essential for analyses of the reaction characteristics.

Finally, in order to evaluate a possible sensor drift of the RFBG during the exposure to microwaves and high temperatures, additional RFBG spectra were determined before and after the reaction in a temperature-stabilised room temperature furnace. The shape of the spectra before and after the use of the RFBG were almost identical. The differences of the Bragg wavelengths were 10.3 pm ± 8.7 pm, which demonstrated the resistance to electromagnetic waves as well as to the reaction conditions of reactive particles. Furthermore, this highlights the potential of using optical fibre sensors with RFBG as an innovative characterisation method in combustion synthesis.

## 4. Conclusions

Reactive particles, which were originally used in the field of combustion synthesis, represent a promising heat source for joining applications, such as welding, soldering, and adhesive bonding. Depending on the stoichiometric ratio of the reactants, a customisable amount of energy can be released by each particle through an exothermic reaction. As heat transfer is often accompanied by temperature gradients, which influence the resulting activation and reaction behaviour, the use of electromagnetic waves at a frequency of 2.45 GHz as an efficient energy transfer is a valuable ignition method. However, high temperatures, electromagnetic waves, high reaction rates, and unknown emissivity values of powder beds represent demanding challenges for conventional temperature measurement methods. A sophisticated approach are optical fibre sensors based on regenerated fibre Bragg gratings (RFBG). Their successful deployment for the characterisation of reactive nickel and aluminium particles with a lamellar structure has been demonstrated in this paper. Temperature-time profiles, recorded with RFBG-based sensors and an infrared camera, showed good accordance. The additional possibility to determine the emissivity of metallic particles demonstrated the potential of RFBG for a profound evaluation of reactive particles in mechanical engineering.

## Acknowledgements

This research is kindly funded by the Deutsche Forschungsgemeinschaft (DFG, German Research Foundation) – 3376550608. The authors would also like to thank A Reindl and G Marchi for their friendly support.

## References

[1] Rogachev A S and Mukasyan A S 2014 *Combustion for Material Synthesis* (Hoboken: Taylor and Francis)

[2] Gu Z, Cui Q, Chen J, Buckley J, Ando T, Erdeniz D, Wong P, Hadjiafxenti A, Epaminonda P, Gunduz I E, Rebholz C and Doumanidis C C 2013 Fabrication, characterization and applications of novel nanoheater structures *Surf. Coat. Technol.* **215** 493–502

[3] Schreiber S, Theodossiadis G D and Zaeh M F 2017 Combustion synthesis of reactive nickel-aluminum particles as an innovative approach for thermal joining applications *IOP Conf. Ser.: Mater. Sci. Eng.* **181** 12008

[4] Schreiber S and Zaeh M F 2018 Electroplating of aluminium microparticles with nickel to synthesise reactive core-shell structures for thermal joining applications *IOP Conf. Ser.: Mater. Sci. Eng.* **373** 12008
[5] Rogachev A S, Shkodich N F, Vadchenko S G, Baras F, Kovalev D Y, Rouvimov S, Nepapushev A A and Mukasyan A S 2013 Influence of the high energy ball milling on structure and reactivity of the Ni+Al powder mixture *J. Alloys Compd.* **577** 600–605
[6] Deevi S C and Sikka V K 1996 Nickel and iron aluminides: an overview on properties, processing, and application *Intermetallics* **4** 351–315
[7] Morsi K 2001 Review: reaction synthesis processing of Ni–Al intermetallic materials *Mat. Sci. Eng.* A **299** 1–15
[8] Hu R and Nash P 2005 The enthalpy of formation of NiAl *J. Mater. Sci.* **40** 1067–1069
[9] Mukasyan A S, Lau C and Varma A 2001 Gasless combustion of aluminium particles clad by nickel *Combust. Sci. Technol.* **170** 67–85
[10] Andrzejak T, Shafirovich E and Varma A 2007 Ignition mechanism of nickel-coated aluminum particles *Combust. Flame* **150** 60–70
[11] Shteinberg A S, Lin Y-C, Son S F and Mukasyan A S 2010 Kinetics of high temperature reaction in Ni-Al system: influence of mechanical activation *J. Phys. Chem.* A **114** 6111–6116
[12] Moore J J and Feng H J 1995 Combustion synthesis of advanced materials: Part I. Reaction parameters *Prog. Mater Sci.* **39** 243–273
[13] Lee H-Y, Ikenaga A, Kim S-H and Kim K-B 2007 The effects of induction heating rate on properties of Ni–Al based intermetallic compound layer coated on ductile cast iron by combustion synthesis *Intermetallics* **15** 1050–1056
[14] Matsuura K, Kitamura T and Kudoh M 1996 Combustion synthesis of Al-Ni alloys by a pseudo-HIP process *J. Japan Inst. Light Metals* **46** 383–388
[15] Biswas A and Roy S K 2004 Comparison between the microstructural evolutions of two modes of SHS of NiAl *Acta Mater.* **52** 257–270
[16] Mukasyan A S, White J D E, Kovalev D Y, Kochetov N A, Ponomarev V I and Son S F 2010 Dynamics of phase transformation during thermal explosion in the Al–Ni system *Phys. Rev. B: Condens. Matter* **405** 778–784
[17] Buchelnikov V D, Louzguine-Luzgin D V, Xie G, Li S, Yoshikawa N, Sato M, Anzulevich A P, Bychkov I V and Inoue A 2008 Heating of metallic powders by microwaves: Experiment and theory *J. Appl. Phys.* **104** 113505
[18] Crane C A, Pantoya M L, Weeks B L and Saed M 2014 The effects of particle size on microwave heating of metal and metal oxide powders *Powder Technol.* **256** 113–117
[19] Karimbeigi A, Zakeri A, Aboutalebi M R and Sharifi P 2013 Study on microwave synthesis and characterisation of bulk nickel aluminides *Mater. Sci. Technol.* **28** 86–91
[20] Rosa R, Veronesi P, Casagrande A and Leonelli C 2016 Microwave ignition of the combustion synthesis of aluminides and field-related effects *J. Alloys Cmpd.* **657** 59–67
[21] Bernhard F 2014 *Handbuch der technischen Temperaturmessung* vol. 2, ed. F Bernhard (Berlin Heidelberg: Springer-Verlag)
[22] Boley C D, Mitchell S C, Rubenchik A M and Wu S S 2016 Metal powder absorptivity: modeling and experiment *Appl. Opt.* **55** 6496–6500
[23] Lindner M, Tunc E, Weraneck K, Heilmeier F, Volk W, Jakobi M, Koch A W and Roths J 2018 Regenerated Bragg grating sensor array for temperature measurements during an aluminum casting process *IEEE Sensors J.* **18** 5352–5360
[24] Dutz F J, Lindner M, Heinrich A, Seydel G C, Bosselmann T, Koch A W and Roths J 2018 Multipoint high temperature sensing with regenerated fiber Bragg gratings *Proceedings of SPIE (Fiber Optic Sensors and Applications XV)* **10654** 1065407-1–1065407-11
[25] Laffont G, Cotillard R, Roussel N, Desmarchelier R and Rougeault S 2018 Temperature resistant fiber Bragg gratings for on-line and structural health monitoring of the next-generation of nuclear reactors *Sensors (Basel)* **18** 1–20

[26] Lee B 2003 Review of the present status of optical fiber sensors *Opt. Fiber Technol.* **9** 57–79
[27] Udd E 1995 An overview of fiber-optic sensors *Rev. Sci. Instrum.* **66** 4015–4030
[28] Lindner E, Chojetztki C, Brueckner S, Becker M, Rothhardt M, Vlekken J and Bartelt H 2009 Arrays of regenerated fiber Bragg gratings in non-hydrogen-loaded photosensitive fibers for high-temperature sensor networks *Sensors (Basel)* **9** 8377–8381
[29] Askins C G, Putnam M A and Friebele E J 1995 Instrumentation for interrogating many-element fiber Bragg grating arrays *Proceedings of SPIE (Smart Sensing, Processing, and Instrumentation)* **2444** 257–266
[30] Erdogan T 1997 Fiber grating spectra *J. Lightwave Technol.* **15** 1277–1294
[31] Pal S, Sun T, Grattan K T V, Wade S A, Collins S F, Baxter G W, Dussardier B and Monnom G 2004 Non-linear temperature dependence of Bragg gratings written in different fibres, optimised for sensor applications over a wide range of temperatures *Sens. Actuators* A: *Physical* **112** 211–219
[32] Laffont G, Cotillard R and Ferdinand P 2013 Multiplexed regenerated fiber Bragg gratings for high-temperature measurement *Meas. Sci. Technol.* **24** 1–5
[33] Hill K O and Meltz G 1997 Fiber Bragg grating technology fundamentals and overview *J. Lightwave Technol.* **15** 1263–1276
[34] Erdogan T, Mizrahi V, Lemaire P J and Monroe D 1994 Decay of ultraviolet-induced fiber Bragg gratings *J. Appl. Phys.* **76** 73–80
[35] Canning J, Stevenson M, Bandyopadhyay S and Cook K 2008 Extreme Silica Optical Fibre Gratings *Sensors (Basel)* **8** 6448–6452
[36] Thiers L, Mukasyan A S and Varma A 2002 Thermal explosion in Ni-Al system *Combust. Flame* **131** 198–209

IOP Conf. Series: Materials Science and Engineering **480** (2019) 012025 doi:10.1088/1757-899X/480/1/012025

# Method comparison for the determination of stretch zone parameters

**U Hähnel, M Markus and P Hübner**

Hochschule Mittweida - University of Applied Sciences, Faculty Engineering Sciences

**Abstract.** The fracture behaviour of ductile failing materials is evaluated by parameters of yielding fracture mechanics. According to ISO 12135 the determination of fracture mechanical characteristic values involves, among other things, analysing the stretch zone width (SZW) in the scanning electron microscope (SEM). This procedure is very time-consuming and requires great experience in handling SEM images of fracture surfaces. An alternative method for determining the stretch zone parameters works with 3D elevation profiles of the fracture surface in the stretch zone area taken with a digital microscope. The fracture mechanical parameters determined using the above methods were finally compared. The $J_i$ and $\delta_i$ values of scrap steels resulting from the digital determined SZW show a good agreement with analytically determined crack initiation values ($J_{i/BL}$, $\delta_{i/BL}$).

## 1. Introduction

Cases of failure characterized by larger plastic deformations before fracture are evaluated by means of yielding fracture mechanics, also known as elastic-plastic fracture mechanics. The J-integral and CTOD (Crack Tip Opening Displacement) concept are the most important tools in yielding fracture mechanics. The relationship between the fracture mechanical stress parameter $J$ or $\delta$ and the stable crack propagation $\Delta a$ is shown in the crack resistance curve (R curve). The course of the J-integral- or CTOD-R curves contain the consecutive stages of blunting, crack initiation and stable crack propagation. Blunting is the rounding and bulging of a sharp fatigue crack of ductile materials under mode I loading. During the bulging process, pores are already formed in the non-cracked material in front of the crack tip. These pores are caused by breaking of inclusions or detachment of matrix material from inclusions. Newly formed and already existing pores grow. Bars of matrix material remain between the pores. The initiation of the crack is regarded as the moment when the first bars between the rounded crack tip and the pores formed in front of the crack tip tear.

Blunting creates a stretch zone in front of the crack tip, whose geometry is defined by the stretch zone width (SZW) in crack propagation direction and by the stretch zone height (SZH) perpendicular to it (Figure 1). The stretch zone is a microscopically detectable feature of crack initiation. If the SZW reaches a critical value, the crack starts. The corresponding physical crack initiation values $J_i$ and $\delta_i$ are regarded as failure criteria for ductile components, provided that they are geometry independent [1, 2].

Two main standards – ISO 12135 and ASTM E 1820 – are used to determine fracture mechanical characteristic values under quasi-static loading. In the method according to ISO 12135, the crack initiation point as well as the crack tip parameters $J_i$ and $\delta_i$ are defined as the intersection of the R curve with the blunting line. Additionally, it is necessary to determine the critical stretch zone width $\Delta a_{SZW}$ by means of analysing the fracture surface in the scanning electron microscope (SEM). The shape of the J-integral / CTOD-R curve corresponds to the function

$$J\ or\ \delta = \alpha + \beta\Delta a^{\gamma}\ [3] \quad (1)$$

with the constants $\alpha, \beta \geq 0$ and $0 < \gamma < 1$. The description of the blunting line results from

$$J = 3{,}75 \times R_m \times \Delta a \text{ and} \quad (2)$$

$$\delta = 1{,}87 \times \frac{R_m}{R_{p0,2}} \times \Delta a\ [3] \quad (3)$$

with $R_m$ and $R_{p0,2}$ as material parameters from corresponding tension tests. However, a determination of the physical crack initiation values $J_i$ and $\delta_i$ according to ISO 12135 is only permissible if a clear distinction between the stretch zone and the stable crack propagation on the fracture surface is possible. In this case, $J_i$ and $\delta_i$ also result from the orthogonal intersection of $\Delta a_{SZW}$ with the R curve. Alternatively, the technical crack initiation value can be determined with a stable crack propagation of $\Delta a = 0.2$ mm.

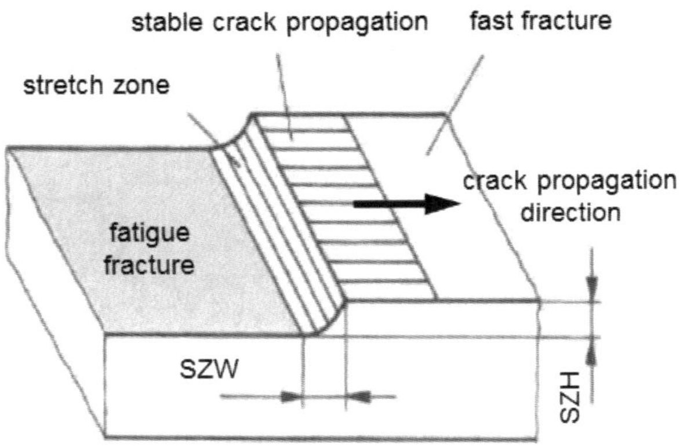


**Figure 1.** Stretch zone schematic.

## 2. Experimental and materials

### *2.1. Methods for determining the stretch zone parameters*

According to ISO 12135 the critical SZW measured in SEM represents the mean of nine local measurements along the entire width of the fracture surface. At the local measuring sites, at least five measurements should be carried out [3]. The detection of SZH in the SEM can't take place at the same time as SZW. Therefor a tilting of the specimen is necessary.

The determination of the stretch zone parameters in the SEM can be subject to large scattering, thus it requires extensive experiences in the analysis of fracture surfaces according to ISO 12135 [3]. However, round robin tests of 14 participating institutions on the same set of SEM images in [4] showed approximately the same scatter of SZW regardless of the existing experience.

Alternative methods for determining the stretch zone parameters include various SEM techniques [5, 6], roughness profiles [7], optical systems such as white light interferometry [5] or the principle of confocality [8] and FEM simulation [9]. In [5], the position and dimension of the stretch zone were defined by combining a 2D SEM image with the 3D reconstruction of the fracture surface. For the 3D reconstruction, two images were taken by tilting the primary electron beam instead of tilting the specimen. Using the generated anaglyph image and the 3D profile of the fracture surface created by special software, SZW and SZH could be determined simultaneously. The authors of [6] have carried out comparative SEM investigations with the common secondary electron technique and backscatter electron contrast to determine the critical SZW. Using the backscattering technology, it should be possible to image the fracture surface even more precisely. Results from roughness measurements could only be used in combination with additional SEM images to determine the stretch zone

IOP Conf. Series: Materials Science and Engineering **480** (2019) 012025 doi:10.1088/1757-899X/480/1/012025

parameters due to the existing scatter. In order to identify the stretch zone more precisely and thus to minimize the scattering of the $J_i$ values, the 2D SEM images in [8] were supplemented with images from a laser scanning microscope. The SZW based on 3D surface profiles displayed lower values than the SEM results. Numerical determination of the SZW is also possible. [9] presents a FEM model based on a strongly deformed stretch zone.

For our own fracture mechanical investigations on scrap steels (St38b-2, St52-3), in addition to the standardized SEM method, a procedure was applied in which topology images of the fracture surface in the area of the stretch zone were taken with a digital microscope. This approach combines the advantages of relatively simple handling with a low expenditure of time compared to scanning electron microscopy. Furthermore, it offers the possibility to evaluate both stretch zone parameters, SZW and SZH, simultaneously over a larger area of the specimen by means of 3D elevation profile per measuring site.

### *2.2. Standard method according to ISO 12135*

In order to determine the crack initiation values of scrap steel (St38b-2, St52-3), fracture mechanics tests at -30°C were performed. After fatigue precracking (crack length 2 mm), J-Δa and CTOD-Δa curves were plotted with the partial unloading compliance method using three-point bend specimens with a thickness of 10 mm. The crack lengths achieved (fatigue precrack, stable crack propagation, fast fracture) were measured using a stereo microscope.

At first the crack initiation values were determined analytically via the intersection of the R curve with the blunting line. The $J_{i/BL}$ or $\delta_{i/BL}$ values are valid and as a result transferrable to components if the thickness criterion according to ISO 12135

$$J_{i/BL} \; or \; \delta_{i/BL} < J_{max} \; or \; \delta_{max} \; [3] \tag{4}$$

is complied.

The fracture surface analysis is performed with the scanning electron microscope EVO 40 from Zeiss to measure the local SZW. For this purpose, the individual SZW were marked in the SEM images and evaluated using the image processing software Adobe Photoshop CS6 Extended.

### *2.3. Digital microscopy*

For the fracture mechanical evaluation of the scrap steels St38b-2 and St52-3 topology images of the fracture surface were analysed at nine measuring sites in the area of the stretch zone with the digital microscope VHX-900F from Keyence. In addition to the time advantage over the SEM method, this procedure offers the possibility of evaluating both stretch zone parameters, SZW and SZH, per local measuring site on the basis of a 3D elevation profile.

The nine local measuring sites along the specimen width were set by shifting the microscope stage perpendicular to the crack propagation direction by 0.5 mm each. The 3D elevation profiles were captured at 700x magnification. For this purpose, the VHX works with the D.F.D. method ("Depth from Defocus"). To calculate the 3D depth information, the blur extent of 2D images is analysed. A stepwise shift of the lens relative to the measuring object along the height axis, which takes place automatically here, enables image acquisition from all contour levels. In the top position of the lens, the highest point of the considered partial fracture surface is sharply imaged, in the lowest position the lowest point. In order to generate the 3D overall image of the surface profile, the sharp image areas for all contour levels are determined and joined together. Combined with glare removal, the D.F.D process and the Keyence image processing software can be used to precisely image metallic surfaces with complex geometries. The entire procedure, from the image acquisition to the generation of the 3D fracture surface profile per measuring site takes about 30 s. A height-colour map, which can be superimposed on the raw image, shows the surface texture and structure. The height scaling can be adjusted by means of a tilt correction.

Instead of the five individual measurements in the SEM, the Keyence software offers a measuring grid with which it is possible to average the local stretch zone parameters over a larger area.

## 3. Results and discussion

*3.1. Standard method according to ISO 12135*

For all tested specimens MP1 to MP6 ductile failure due to crack resistance behaviour was found (Figure 2). The optically measured crack lengths correspond well with the values determined during the test using the compliance method. Some specimens show very high $J_{i/BL}$ values that exceed the $J_{max}$ criterion (equation 4). Thus, there is still a geometry dependence. In this case a limitation to $J_{max}$ is recommended.

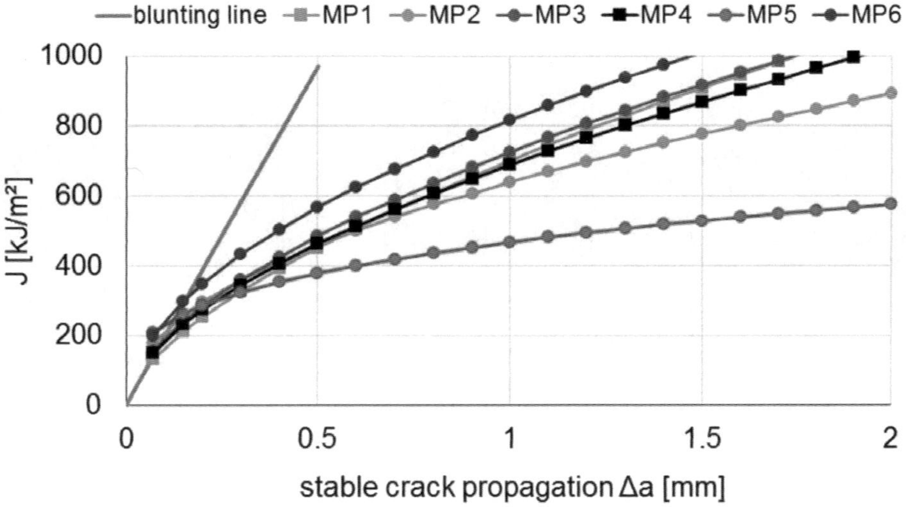


**Figure 2.** J-R curves of scrap steel at -30°C (St38b-2: MP1, MP4 / St52-3: others).

On the basis of SEM overview images, it was possible to identify the different fracture surface areas and to set the measuring sites for determining the local SZW (Figure 3a). A typical ductile fracture surface without cleavage fracture parts is shown in the area of stable crack propagation (Figure 3b). The local SZW $\Delta a_{SZW,L}$ represents the mean value of five measurements per measuring site (Figure 3c). The mean value of all $\Delta a_{SZW,L}$ defines the critical SZW per specimen whose intersection point with the J- or CTOD-R curve represents the crack initiation values $J_{i/SZW}$ and $\delta_{i/SZW}$.

The determination of the SZW in the SEM requires a great deal of time, especially if the SZH is also to be recorded for the validation of the CTOD values. Furthermore, with five individual measurements at the nine local measuring sites, only a very small area of the stretch zone is used for averaging over the entire specimen. The scattering of the measured values, which is often mentioned in literature, as well as the necessary experience to carry out this procedure additionally motivate the development of alternative methods.

*3.2. Digital microscopy*

Figure 4 shows the generated 3D elevation profile of a local measuring site of St38b-2. The 3D profile can be rotated in all directions so that a suitable measuring position can be easily found. For the fracture surface analyses of scrap steel, a grid of 30 measuring lines at a distance of 5 µm each was selected, which is oriented perpendicular to the crack front and corresponds to a measuring range of 145 µm. The measurement of SZW and SZH is carried out via a horizontal or vertical point to point measurement using an average profile displayed in 2D. The position of this profile section can be freely selected within the defined measuring range. The local SZW and SZH result from the mean value of the 30 individual measurements. Figure 5 shows an example measurement on the St38b-2.
With this optical method, the position and dimensions of the stretch zone can be recorded simultaneously. Another advantage over the SEM procedure is the larger local measuring range. Using

the selected measuring grid, 16 % of the net thickness of the specimen ($B_N$ = 8 mm) is used to evaluate the stretch zone parameters. Thus, individual over- or underestimates of SZW and SZH have less influence on the overall result.

Comparable 3D profiles of the fracture surface could also be recorded with the 3D laser scanning microscope VK-X200 from Keyence. Laser scanning microscopes use the principle of confocality to measure height data at the surface. In further investigations it has to be resolved whether the higher resolution up to 0.5 µm compared to the digital microscope (up to 2 µm) provides an information gain for the stretch zone analysis. According to the authors, however, there are doubts because the expected error in the determination of SZW and SZH is much greater.

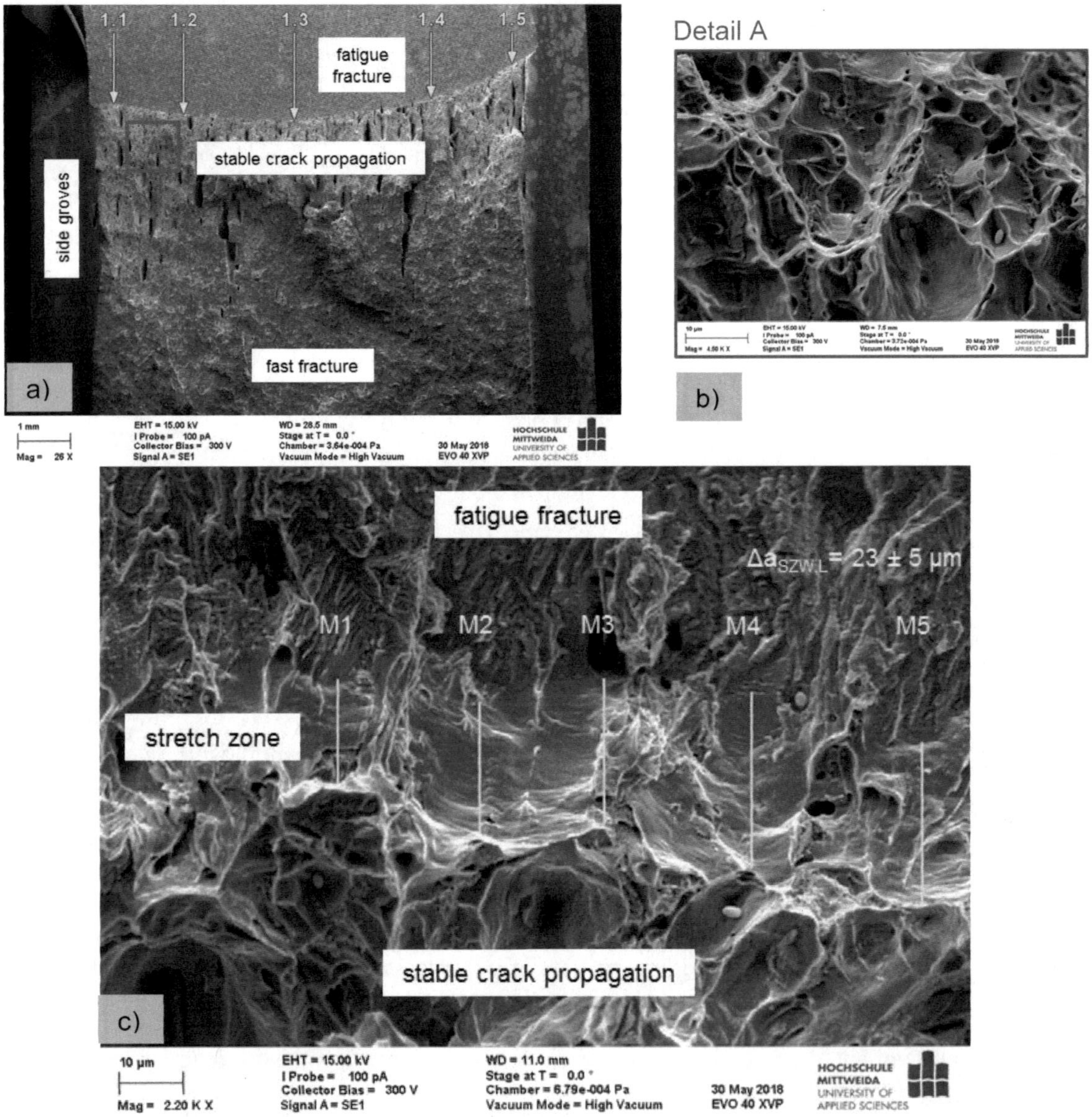


**Figure 3.** Measurement of SZW in SEM for St38b-2: overview of the fracture surface specimen MP1 with the measuring sites 1.1 to 1.5 (a), fracture surface in the area of $\Delta a$ (b), local stretch zone of measuring site 1.2 with measurements M1 to M5(c).

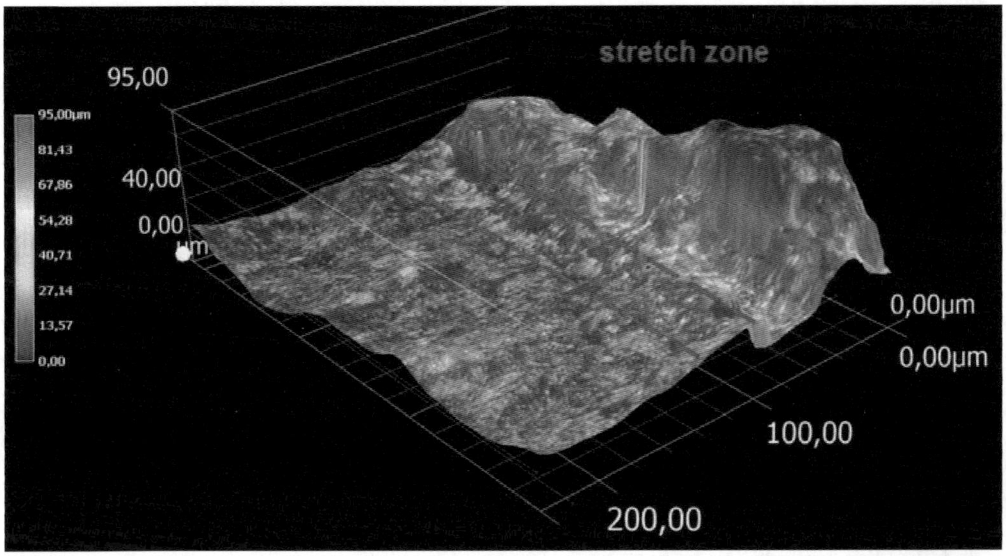


**Figure 4.** 3D elevation profile of a partial fracture surface of St38b-2 / stretch zone marked between dashed lines.

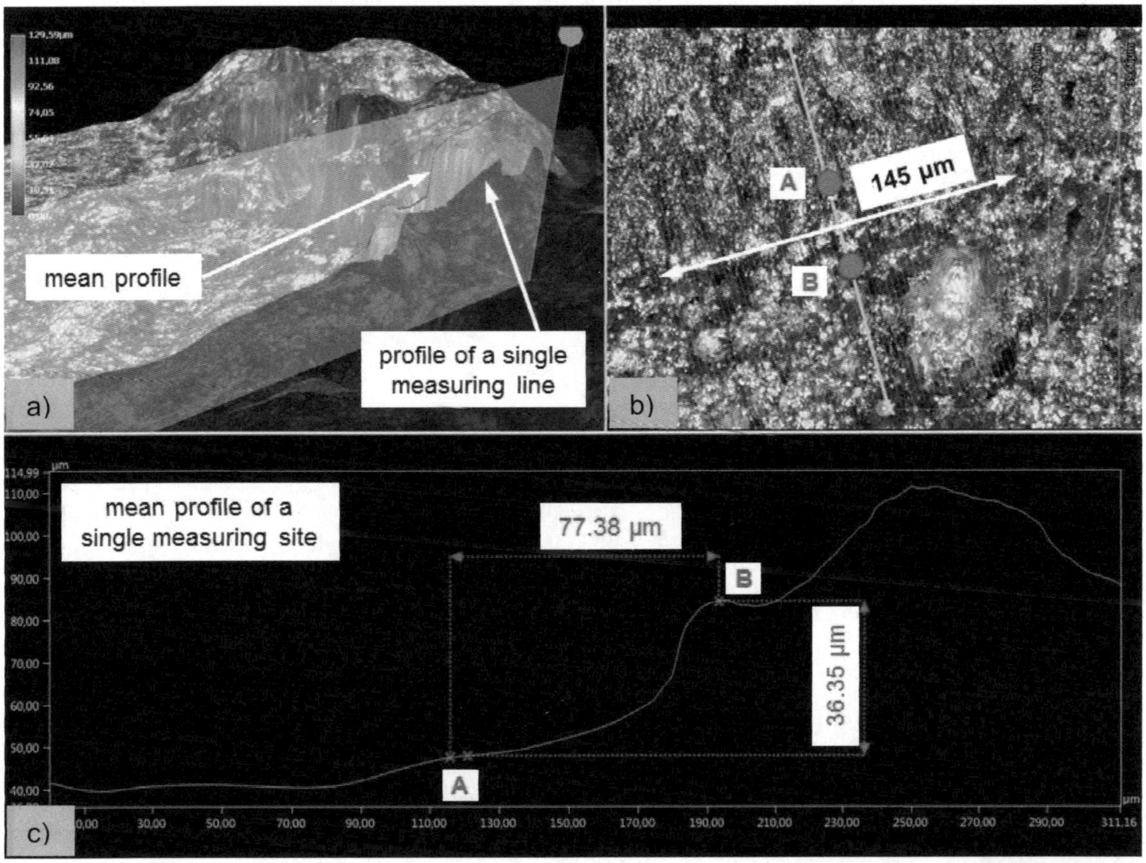


**Figure 5.** Determination of the stretch zone parameters with VHX at a single measuring site / St38b-2: finding measuring position (a), measuring grid (b), mean profile for measuring SZW and SZH.

IOP Conf. Series: Materials Science and Engineering **480** (2019) 012025 doi:10.1088/1757-899X/480/1/012025

### *3.3. Comparison of results*

The stretch zone parameters and the resulting crack initiation values of the scrap steels St38b-2 and St52-3 are compared in Table 1 with the analytically determined $J_{i/BL}$ / $\delta_{i/BL}$ values.

**Table 1.** Average stretch zone parameters and resulting crack initiation values.

| Scrap steel | $\Delta a_{SZW}$ SEM (µm) | $\Delta a_{SZW}$ VHX (µm) | SZH (µm) | $J_{i/BL}$ (kJ/m²) | $J_{i/SZW}$ SEM (kJ/m²) | $J_{i/SZW}$ VHX (kJ/m²) | $\delta_{i/BL}$ (µm) | $\delta_{i/SZW}$ SEM (µm) | $\delta_{i/SZW}$ VHX (µm) |
|---|---|---|---|---|---|---|---|---|---|
| St38b-2 | 29±1 | 77±3 | 25±3 | 148±38 | 85±12 | 150±9 | 115±3 | 108±5 | 146±12 |
| St52-3 | 34±2 | 76±4 | 26±4 | 204±11 | 133±27 | 193±18 | 181±10 | 157±12 | 199±14 |

A comparison of $J_{i/BL}$ and $J_{i/SZW}$ (VHX) shows a good agreement. The $J_{i/SZW}$ values determined with the SEM method are significantly lower. This statement can be illustrated using the example of the J-R mean value curve of St38b-2 (Figure 6). The deviation is attributed the lower SZW from the SEM. In contrast to literature, e.g. [5] and [8], the SZW was underestimated compared to the optical method. The analytically determined $\delta_{i/BL}$ values lie between the results of $\delta_{i/SZW}$ (SEM) and $\delta_{i/SZW}$ (VHX). However, a validation of the CTOD values via the double measured SZH is currently not possible. The corresponding measured values in Table 1 were systematically underestimated. The stretch zone parameters recorded with the VHX method show a low scatter and, with the measuring range that can be used for evaluation, provide reliable values for determining $J_i$ and $\delta_i$.

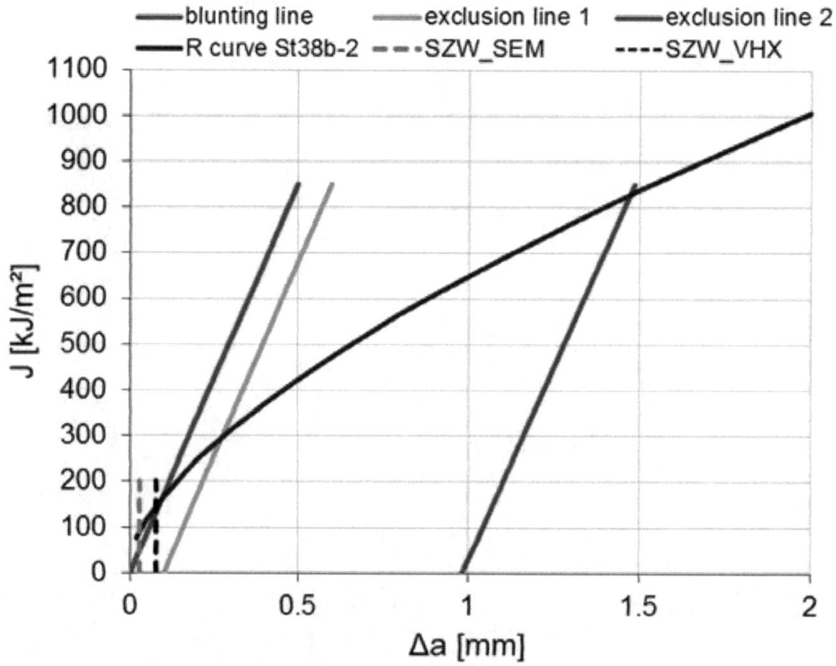


**Figure 6.** Average J-R curve of St38b-2 at -30°C.

## 4. Conclusions

An alternative method for determining the stretch zone parameters is presented. The SZW and SZH are analysed with the digital microscope VHX-900F from Keyence using a 3D elevation profile of the fracture surface. For comparison with the standard method according to ISO 12135, in which the SZW is determined in the SEM, the results from static fracture mechanics tests on scrap steel at -30°C were used. The crack initiation values $J_{i/SZW}$ and $\delta_{i/SZW}$ derived from the digital microscopically determined SZW show good agreement with the analytical $J_{i/BL}$ / $\delta_{i/BL}$ values.

However, a repeatable, precise determination of the stretch zone parameters with the digital microscope requires the validation of the process by comparison measurements on different materials. If a general match with $J_{i/BL}$ / $\delta_{i/BL}$ can be demonstrated, the exclusive use of the VHX method with its advantages

- low time expenditure
- user-friendly approach
- detection of position and dimensions of the stretch zone
- simultaneous determination of SZW and SZH
- averaging over a larger measuring range

instead of the standard procedure according to ISO 12135 is conceivable.

**References**

[1] Blumenauer H and Pusch G 1993 *Technische Bruchmechanik* 3., stark überarbeitete Ausgabe (Leipzig – Stuttgart: Deutscher Verlag für Grundstoffindustrie) pp 67-73

[2] Schödel M 2005 *Bruchmechanische Untersuchungen der Rissöffnung bei stabilem Risswachstum in dünnem Blech aus Al 5083* GKSS vol 2006/6, ed GKSS Forschungszentrum in der Helmholtz-Gemeinschaft pp 44-45

[3] ISO 12135 2002 *Metallic materials – Unified method of test for the determination of quasistatic fracture toughness*

[4] Heerens J, Cornec A and Schwalbe K-H 1988 Results of round robin on stretch zone width determination *Fatigue Fract. Eng. Mater. Struct.* **11** 19-29

[5] Weidner A, Mottitschka T, Biermann H and Henkel S 2013 Determination of stretch zone width and height by powerful 3D SEM imaging technology *Engineering Fracture Mechanics* **108** 294-304

[6] Tarpani J R, Bose W W and Spinelli D 2003 Backscattered electron microscopy technique enhancing stretch zone width imaging for initiation fracture toughness measurements *Mater. Charact.* **51** 159-170

[7] Pluvinage G and Lanvin A 1993 Stretch zone geometrical measurement, a particular way to measure fracture toughness *Fatigue Fract. Eng. Mater. Struct.* **16** 955-972

[8] Yu W, Fan M, Gao H, Yu D, Xue F and Chen X 2018 Effect of long-term aging on the fracture toughness of primary coolant piping material Z3CN20.9M *Nucl. Eng. Des.* **327** 150-160

[9] Saxena S, Ramakrishnan N and Dutta B 2009 Determination of stretch zone width using fem *Engineering Fracture Mechanics* **76** 911-920

IOP Conf. Series: Materials Science and Engineering **480** (2019) 012026 doi:10.1088/1757-899X/480/1/012026

# Automated hole expansion test with pneumatic crack detection

**A Leonhardt[1,*], V Kräusel[1] and U Paar[2]**

[1] Chemnitz University of Technology / Institute for Machine Tools and Production Processes / Professorship for Forming and Joining / Reichenhainer Straße 70 / 09126 Chemnitz / Germany
[2] Volkswagen Sachsen GmbH / Serienplanung Presswerk / Glauchauer Straße 40 / 08058 Zwickau / Germany

[*] e-mail: andre.leonhardt@mb.tu-chemnitz.de

**Abstract.** The hole expansion test according to ISO 16630:2009 is a widely used method for determining the edge crack sensitivity, e. g. for advanced high strength steels (AHSS). However, the optical crack detection causes a subjective stop of the test, which is also difficult to automate. These existing disadvantages are eliminated by a newly developed method to stop the hole expansion test using pressure drop at an edge crack through the entire sheet thickness. The pneumatic crack detection ensures a high degree of objectivity and automation. The functionality of the new stopping method was experimentally proven by the comparison with the optical method, the investigation of the influence of different parameters and by testing of good part and reject batches from series production. The pneumatic method stops the test when through-thickness cracks are less pronounced up to a hole expansion ratio $\lambda$ of approx. 35%, i. e. for materials sensitive to edge cracks. Thus, this variant provides more objective results than the optical method. The test results between the two methods are comparable in the range of $\lambda \leq 100\%$.

## 1. Introduction

Advanced high strength steels (AHSS) are increasingly being used in the automotive industry due to increasing efficiency and safety standards, but their formability is severely limited compared with that of conventional steels. They include dual-phase steels (DPS) which consist of a soft, ferritic matrix in which martensitic phases are embedded in an island shape to increase strength. A typical failure of these steels is edge cracking through sheet thickness, which is caused by forming of shear-cut component edges. However, new investigation methods are developed with great effort to determine the sensitivity to edge cracking since the failure due to edge cracking cannot be reliably predicted using conventional test methods such as the forming limit curve. **Table 1** summarizes a selection of the variants for determining the edge crack sensitivity for open and closed cutting lines described in literature.

**Table 1.** Testing procedures for determining the edge crack sensitivity.

| **Testing procedures for open cutting line** | **Testing procedures for closed cutting line** |
|---|---|
| • Hole expansion test (ISO 16630) [1]<br>• Collar forming test [2, 3]<br>• KOBE tensile specimen with hole [2, 4] | • Tensile test with shear-cut edges [3]<br>• NAKAZIMA tensile specimen / BMW edge cracking test [2, 3]<br>• Half-specimen dome test [5]<br>• DIABOLO test [6, 7]<br>• Side bending test [8]<br>• In-plane bending test [2, 7]<br>• Double bending test [2]<br>• Simulative flange test [9]<br>• YOSHIDA test [2, 8] |

The hole expansion test is a common procedure and standardized in ISO 16630:2009 [1]. A punched sheet sample is expanded by a conical punch until at least one crack runs through the entire sheet thickness. The hole expansion ratio λ (%) is then calculated as the characteristic value for the edge crack sensitivity using the following formula:

$$\lambda = \frac{D_h - D_0}{D_0} \cdot 100 \qquad (1)$$

$D_0$ is the hole diameter before expanding, and $D_h$ is the expanded diameter at the stop of the test.

A major problem of the hole expansion test lies in the distinct detection of a continuous crack, which is the stop criterion of the test. This point of time is currently determined by a metrologist and highly subjectively influenced. In addition, this optical crack detection method is difficult to automate. Various methods for crack detection have already been investigated in technical literature. These include, among others:

- Digital image correlation (DIC) [10, 11]
- Optical forming analysis [12]
- Active and passive thermography
- Detection of a defined force drop [10]

However, these solutions are very complex to implement, require cost-intensive measurement technology and can only be used to a limited extent. In addition, some methods react sensitively to external influences.

## 2. Hole expansion test with pneumatic crack detection

The development of a method for an automated procedure to determine the edge crack sensitivity of car body sheet materials eliminates the currently predominant subjective assessment of the inspector. The basic structure corresponds to the conventional hole expansion test in which the open die is replaced by a closed pressure cylinder (**Figure 1**). In the first stage, the conical punch is inserted into the hole of the clamped sheet until a seal is reached between punch and sheet. The pressure cylinder is then filled with compressed air to create an overpressure $p_0$. The hole is expanded using the punch in stage 3. As soon as a crack occurs that runs through the entire sheet thickness, the compressed air escapes to the outside and the pressure drops. This is detected with a manometer and used as a criterion for stopping the testing procedure.

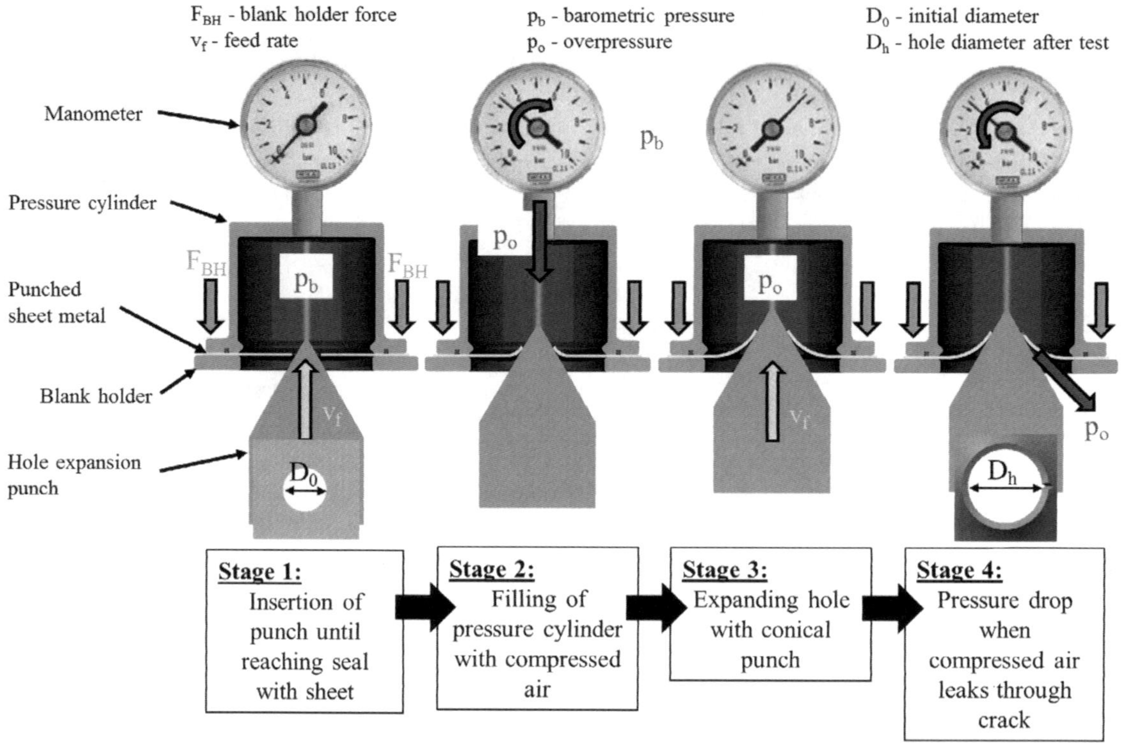


**Figure 1.** Schematic structure and sequence of the hole expansion test with pneumatic crack detection.

The method of pneumatic crack detection enables an objective stop of the test, which is independent of the tester. In addition, automation is possible as the pressure drop can be used as a control signal. Furthermore, the test can be implemented with simple and cost-effective tool and testing technology, and can be used on standard testing machines.

## 3. Experimental evidence of functional capability

### *3.1. Experimental plan and testing procedure*

A comprehensive test program was carried out to validate the functionality of the hole expansion test with pneumatic crack detection (**Table 2**).

The tests were carried out on a servo spindle press, whereby the tool was used both for punching and for hole expansion with pneumatic crack detection. An initial pressure of 6 bar existed inside the pressure cylinder. The measurement of the pressure was executed by a pressure sensor. The hole expansion test was stopped as soon as a pressure drop of 0.03 bar occurred. The tests with optical crack detection were carried out by trained personnel from various steel manufacturers. The cold-formable dual-phase steel HC450XD was used for the majority of the tests as this material is particularly sensitive to edge cracking. The reason for this property is the structure of a soft, ferritic matrix in which hard martensite particles are embedded, and which contains the previous damage caused by shear cutting. Further investigated materials included an austenitic stainless steel, representing a possible alternative to dual-phase steel, and a 6000 aluminum alloy in outer skin quality. First, a comparison was made between the pneumatic and the optical crack detection. Scattering was determined in two test series, on the one hand by using an increased number of samples, and on the other hand with different materials having various hole expansion ratios. Furthermore, investigations were carried out regarding the influence of different parameters such as the hole insertion method, the cutting clearance and the pre-hole diameter, which influence the hole expansion capability. These investigations were performed in order to test the sensitivity of the new crack detection variant. In the end, batches of dual-phase steel were taken from series production, which led to reject components due to occurring edge cracks. These batches were also tested using the pneumatic stopping method in the hole expansion test. The maximum crack depths were also determined in addition to measuring the expanded diameters.

**Table 2.** Test program for the validation of the hole expansion test with pneumatic crack detection.

| Name | Criterion | Material | Sheet thickness [mm] | Remarks | No. |
|---|---|---|---|---|---|
| **A.1** | **Comparison** of optical and pneumatic method | Dual-phase steel HC450XD | 1.50 | • Pneumatic method<br>• Optical method by two steel manufacturers | 15 |
| **A.2** | **Comparison** of optical and pneumatic method | Dual-phase steel HC450XD | 1.50 | • Pneumatic method<br>• Optical method by one steel manufacturer | 5 |
| | | Dual-phase steel HC450XD | 2.00 | | |
| | | Austenitic stainless steel H500 | 1.50 | | |
| | | 6xxx aluminum EN AW-6016 | 1.04 | | |
| **B.1** | Influence of **hole insertion method** | Dual-phase steel HC450XD | 1.50 | • Shear cutting<br>• Drilling<br>• Drilling and reaming<br>• Abrasive waterjet cutting<br>• Laser cutting<br>• Wire eroding | 5 |
| **B.2** | Influence of related **cutting clearance** | Dual-phase steel HC450XD | 1.50 | $c_s$ = 6.67–33.33% | 5 |
| **B.3** | Influence of **initial hole diameter** | Dual-phase steel HC450XD | 1.50 | Ø = 6.65–22.00 mm | 5 |
| **C** | Investigation of **good and bad batches** | Dual-phase steel HC450XD | 2.00 | Different reasons for edge cracking in bad parts from series production | 5 |

*3.2. Results*

Test series A.1 (**Figure 2a**) illustrates the subjective evaluation of a continuous crack when applying the optical method. On the one hand, this is demonstrated by the high fluctuations within a test series where the test is partly stopped too late. Thus the expansion of the hole is too high, the edge cracks are more pronounced and the maximum crack depths are larger. On the other hand, subjectivity is also reflected in the differences between the two results of the optical method to stop the test. Earlier stopping of the test can be detected for the highly edge crack-sensitive dual-phase steel by applying the pneumatic method so that the values for the hole expansion ratio are lower than those obtained by the optical method. Continuous edge cracks were found in all specimens, although they were less pronounced than the cracks determined with the optical method to stop the test. In addition, the fluctuations of the pneumatic method were the same or lower than those of the optical method. These results were confirmed in the second investigation comparing the two methods (A.2) for materials with a hole expansion ratio of up to approx. 35% (**Figure 2b**). However, it is evident that the cracks in the pneumatic variant were more pronounced and the maximum crack depths were larger with decreasing sensitivity to edge cracking and with increasing hole expansion ratio. The reason for this phenomenon is that the cracks must continue to grow until the pressure drop occurs as the material contacts the conical punch due to the higher expansion. Earlier or comparable stopping of the test can be detected with the pneumatic method up to a hole expansion ratio of approx. 100%. Above this value, the edge cracks are more likely to be detected with the optical variant.

a) A.1 (15 samples each) b) A.2 (5 samples each)

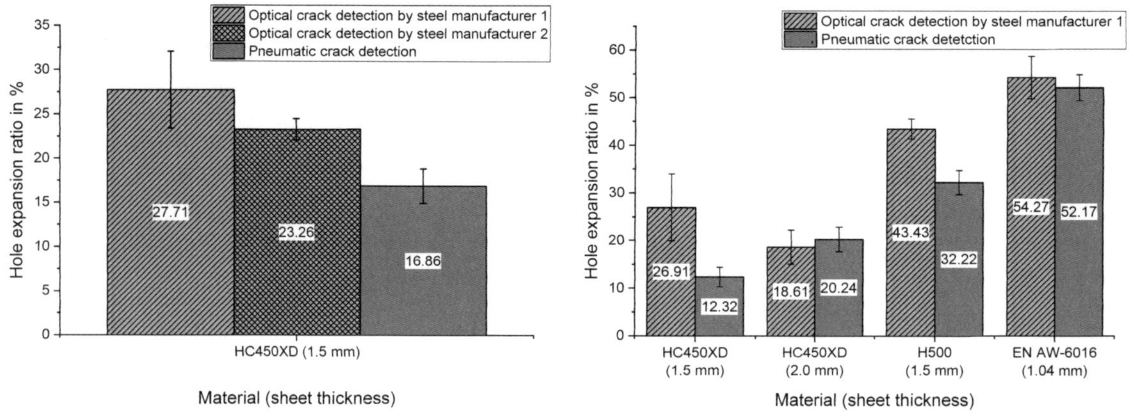


**Figure 2.** Comparison of hole expansion ratios during stop of the test, using optical and pneumatic method for a dual-phase steel sensitive to edge cracks (a) and for other car body materials (b).

The investigations of dual-phase steel with regard to the influence of various process parameters confirm the suitability of the hole expansion test with pneumatic crack detection (**Figure 3a-c**). **Figure 3a** shows the strong influence of the type of hole insertion (B.1) on the hole expansion ratio of HC450XD. The values for shear cutting are lowest since the cutting edge is pre-damaged by strain hardening and the occurring microcracks. This is particularly critical as this cutting method is generally used for high-volume sheet metal production. The hole expansion ratio can be increased by inserting the holes with a laser beam. Laser cutting is often used for tool and component testing in the early stages of product development processes. The higher hole expansion ratio must be taken into account since edge cracking can be caused by later application of shear cutting in series production. The other variants investigated for inserting the pre-hole also increase the expanded diameters, but are rather unsuitable for large series production. Shear cutting also shows a decrease in the hole expansion ratio with increasing cutting clearance in test series B.2 (**Figure 3b**). Only minor differences can be determined up to a related cutting clearance of 20.00%. This is followed by a sharp drop in the hole expansion ratio since a cutting clearance of 33.33% forms a large burr that is the starting point for crack initiation. Furthermore, an influence of the initial hole diameter (B.3) can be determined whereby the hole expansion ratio decreases with increasing diameter (**Figure 3c**). The number of continuous cracks along the circumference is higher for larger diameters, which means that more cracks cause a pressure drop. These cracks must therefore be less pronounced, but continuous, so that an earlier stop of the test occurs.

HC450XD with a sheet thickness of 2.0 mm was tested within a batch test (C) which is used for a component of the Volkswagen Group MQB platform. Three batches of the steel, which did not lead to reject production, provided the reference values. These expansion ratios were above 20% (**Figure 3d**). Furthermore, three batches were investigated in which edge cracks occurred in the outer edge of the component during the drawing stage, and one batch with edge cracks in the collar hole. These two failure types are typical for material batches with insufficient hole expansion capability. This was proven using the hole expansion test with pneumatic crack detection on the basis of the lower hole expansion ratios. A further batch with cracks in the flange was tested. This error, however, was caused by a forming tool that was not incorporated, leading to damage in the cutting edge in this area. As a result, the hole expansion ratio here lay above 20%, which was similar to the reference batches.

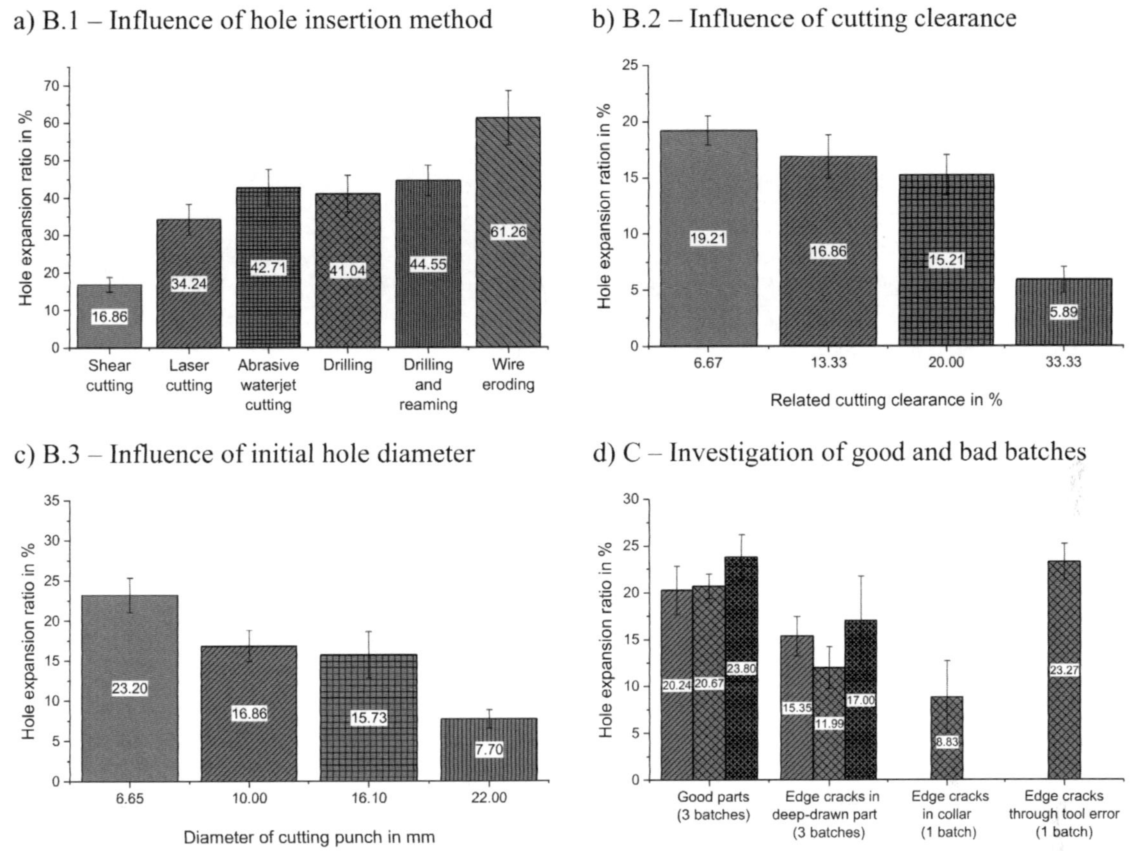


**Figure 3.** Investigation of the influence of various parameters on the hole expansion with pneumatic stop method for the dual-phase steel HC450 XD 1.5 mm (a-c), investigation of the edge crack sensitivity of good and bad batches of the dual-phase steel HC450XD 2.0 mm with pneumatic crack detection (d).

## 4. Conclusion and outlook

The hole expansion test for determining the sensitivity to edge cracking can be made more objective and automatable by using the introduced and investigated method of pneumatic crack detection. This method can be easily implemented. Compared to the optical method, it enables an earlier stop of the test as well as lower fluctuations for metallic car body materials up to a sheet thickness of 2.0 mm, which are highly sensitive to edge cracks ($\lambda$ < 35%). Using the pneumatic variant, better or comparable results can be achieved up to a hole expansion ratio of approx. 100% compared to the optical method for stopping the test. The optical method is to be preferred for values of $\lambda$ above 100% due to the increase in crack formation for increasing hole expansion ratios since the crack depths required for the pressure drop are too large. The hole expansion test with pneumatic crack detection exhibits sufficient sensitivity to determine the influence of various process parameters and to identify material-related reject batches of a dual-phase steel. Subsequent research will deal with further validation of the pneumatic crack detection in other materials and by using other test centers. This will

IOP Conf. Series: Materials Science and Engineering **480** (2019) 012026 doi:10.1088/1757-899X/480/1/012026

lead to further optimization and, if successful, standardization. The method will also be implemented in an experimental rig or in a universal testing machine. The application of the new testing method is conceivable for material users and material manufacturers for standard testing of selected material batches in the series process and for targeted optimization of new materials with regard to edge crack sensitivity. On the one hand, the characteristic value of the hole expansion can be implemented in the simulation model when designing new components. On the other hand, this test provides an opportunity to qualify pressed parts and tools more quickly, to carry out inspections of incoming material for quality control and to avoid rejects, and to clarify the causes of rejects by cross-checking the material in ongoing series production.

## References

[1] N N 2009 Metallic materials - Sheet and strip - Hole expanding test *International Organization for Standardization*

[2] Westhäuser S 2014 Experimentelle Prüfmethoden zur quantitativen Bewertung der Kantenrissempfindlichkeit *Salzgitter Mannesmann Forschung*

[3] Larour P, Pauli H, Freudenthaler J and Grünsteidl 2011 Alternative stretch flangeability characterisation methods for AHSS steel grades *Proceeding IDDRG 2011 International Coference Towards sustainable sheet forming processes in Bilbao (Spain)*

[4] Gläsner T, Sunderkötter C, Rommel S, Volk W, Hoffmann H and Golle R 2015 Evaluation eines Verfahrens zur Bestimmung der Kantenrisssensitivität *Whitepaper www.umformtechnik.net Verlag Meisenbach*

[5] Shih H C, Hsiung C K and Wendt B 2014 Optimal Production Trimming Process for AHSS Sheared Edge Stretchability Improvement *SAE Technical Paper* **2014-01-0994**

[6] Beier T, Gula G, Woestmann S and Keßler L 2015 Eine Bewertung des Umformpotenzials von Schnittkanten zur Auslegung von Formgebungsprozessen mit höherfesten Stahlwerkstoffen *Proceedings TB-040 35. EFB-Kolloquium Blechverarbeitung 2015 in Bad Boll* 139–150

[7] Held C, Liewald M and Sindel M 2010 Erweiterte Werkstoffprüfverfahren zur Charakterisierung von Leichtbaublechwerkstoffen im Hinblick auf die Kantenrisssensitivität *Materials Testing-Materials and Components Technology and Application* **52/9** 596–602

[8] Yoshida H, Sato K, Matsuno T, Yoshida T, Takahashi Y and Nitta J 2013 Evaluation and Improving Methods of Stretch Flangeability *Nipon Steel Technical Report* **103** 18–24

[9] Huang M, Zhou D and Siekirk J 2008 Stretch Flangeability of DP490 *Great Designs in Steel Seminar 2008 www.autosteel.org*

[10] Dünckelmeyer M, Karelová A, Krempaszky C and Werner E 2009 Instrumented hole expansion test *Proceedings International Doctoral Seminar 2009 in Smolenice Castle (Slovakia)*

[11] Chiriac C and Chen G 2008 Local Formability Characterization of AHSS - Digital Camera Based Hole Expansion Test Development *Proceedings IDDRG 2008 International Conference Best in class stamping in Olofstrom (Sweden)* 81–92

[12] Chen X, Yang L, Chen X, Chirac C, Du C and Zhou D 2011 Measurement of Strain Distribution for Hole Expansion with Digital Image Correlation (DIC) System *SAE Technical Paper* **2011-01-0993**

# Calculation of stress near the cracks in welded plates with taking into account the residual deformation

**H Pokhmurska[1,*], O Maksymovych[2], A Dzyubyk[1] and L Dzyubyk[1]**

[1] National University "Lviv Polytechnic", Lviv, 79013, Ukraine
[2] University of Science and Technology, Bydgoszcz, 85-796, Poland

[*] e-mail: hanna.pokhmurska@mb.tu-chemnitz.de

**Abstract.** An algorithm for determining the stress state of plates with cracks caused by residual deformations, including taking into account the contact of their edges, is proposed. The residual stresses in the plates are determined by calculation in the experimental method. The study of residual stresses and stresses at cracks in plates of various shapes is performed. It was established that ignoring contact with the edges of cracks can lead to significant errors in the calculation of stresses.

There are stresses due to residual deformations in the elements of the structures. These stresses may be compatible with stresses due to operational loads, and therefore they must be taken into account when calculating strength and durability. To study the stresses at cracks in welded structures, which often occur in the vicinity of welds, it is necessary to pre-determine the residual stresses. The question of determining the residual stresses in the welded elements of structures was considered in [1-5]. The distribution of residual stresses is influenced by the selected welding modes, the thermomechanical characteristics of the materials that need to be determined for a wide range of temperatures, structural transformations, conditions of fixation, etc. In this regard, the determination of residual stresses by theoretical methods can be carried out only for certain classes of problems.

In this work, the calculation-experimental method was used to determine the residual stresses in the plates [6]. The algorithm for determining the stresses at the cracks in the plates due to residual deformations is based on the method of integral equations. It is taken into account that residual stresses can be compressible. For such cases, the research was done taking into account the contact of the edges of the cracks. Other approaches to the study of stresses around cracks in welded structures considered in [7-15].

**Determination of residual stresses in continuous plates.** Consider the problem of determining the state of the plate, caused by the residual (initial) deformations $\varepsilon_x^0, \varepsilon_y^0, \gamma_{xy}^0$ with the condition that its edge traction-free. To do this we use Hooke's law.

$$\varepsilon_x = \varepsilon_x^0 + \frac{1}{E}\left(\sigma_x - \nu\sigma_y\right),\ \varepsilon_y = \varepsilon_y^0 + \frac{1}{E}\left(\sigma_y - \nu\sigma_x\right),\ \gamma_{xy} = \gamma_{xy}^0 + \frac{1}{G}\tau_{xy}, \qquad (1)$$

where $\varepsilon_x, \varepsilon_y, \gamma_{xy}$ – complete deformations that satisfy the conditions of compatibility of deformations [16]

IOP Conf. Series: Materials Science and Engineering **480** (2019) 012027 doi:10.1088/1757-899X/480/1/012027

$$\frac{\partial^2 \varepsilon_x}{\partial y^2} + \frac{\partial^2 \varepsilon_y}{\partial x^2} - \frac{\partial^2 \gamma_{xy}}{\partial x \partial y} = 0 . \tag{2}$$

Tension in the plate $\sigma_x, \sigma_y, \gamma_{xy}$ satisfy the equilibrium equation

$$\frac{\partial \sigma_x}{\partial x} + \frac{\partial \tau_{xy}}{\partial y} = 0 , \frac{\partial \tau_{xy}}{\partial x} + \frac{\partial \sigma_y}{\partial y} = 0 . \tag{3}$$

Consider a more detailed case involving plates with a straight weld seam placed along the axis *Oy*. Remaining deformations in this case in the literature are accepted as non-zero at $|x| < d$ as

$$\overset{0}{\varepsilon}_x = C_x f(x), \overset{0}{\varepsilon}_y = C_y f(x), \overset{0}{\gamma}_{xy} = 0 , \tag{4}$$

where $f(x)$ – known function, $C_x = -\frac{S_x}{E}, C_y = -\frac{S_y}{E}$, $S_x = k_x \sigma_T, S_y = k_y \sigma_T$, $k_x, k_y$ – dimensionless parameters, $\sigma_T$ – yield point, $d$ – width of the plastic zone.

Here is a typical welding case, when the residual deformations are negative. At [6]

$$f(x) = 1 + a_0 \left( \frac{x}{d} \right)^2 - (3 + 2a_0) \left( \frac{x}{d} \right)^4 + (2 + a_0) \left( \frac{x}{d} \right)^6 , \tag{5}$$

where $a_0$ – constant located on the interval *[0,2]*. In this case, the description of initial deformations includes four parameters $d, a_0, k_x, k_y$, which must be determined so that the calculated and experimental methods of stress found differ little from one another.

The solution of the problem is represented as the sum of the partial and correction components.

**Partial solution.** For a partial solution we will accept: $\sigma_x = 0, \varepsilon_y = 0$, $\tau_{xy} = 0, \gamma_{xy} = 0$. Then from the second formula (1) we get $\sigma_y = -E\overset{0}{\varepsilon}_y$, and then we determined from the first formula $\varepsilon_x = \overset{0}{\varepsilon}_x + \nu \overset{0}{\varepsilon}_y$. Thus, we obtain a partial solution for stresses and deformations in the form

$$\sigma_x = 0, \ \ \sigma_y = -EC_y f(x) = S_y f(x), \tau_{xy} = 0 , \tag{6}$$

$$\varepsilon_x = (C_x + \nu C_y) f(x), \ \ \varepsilon_y = 0, \ \ \gamma_{xy} = 0 . \tag{7}$$

Inspection makes sure that tension (6) satisfying the equilibrium equation (3), and deformation (7) – compatibility equation (2).

Stress vector $(X_0, Y_0)$ on an arbitrary plane, which corresponds to a partial solution, is determined by the formulas

$$X_0 = 0, Y_0 = \sigma_y \sin\gamma , \tag{8}$$

where $\gamma$ – the angle between the axis *Ox* and perpendicular to the plane.

**Correction solution.** To determined it it is necessary to consider the problem of the theory of elasticity for a given plate, to the boundary of which the above-identified forces with the opposite sign are applied $X = 0, Y = -S_y f(x) \sin\gamma$.

Let's accept, that the well-known expressions for Muskhelishvili's complex potentials for plates with a free limit loaded by a system of forces $(X_k, Y_k)$, applied at points $z_k$, $k = 1,...,K$ (For a limited plate of force, the forces must be balanced themselves). We will write down these potentials in the form

$$\Phi(z) = \sum_{k=1}^{K} \Phi_0(z, z_k, X_k, Y_k), \Psi(z) = \sum_{k=1}^{K} \Psi_0(z, z_k, X_k, Y_k), \tag{9}$$

where $\Phi_0(z, z_k, X_k, Y_k)$, $\Psi_0(z, z_k, X_k, Y_k)$ – known functions.

Consider the case when the forces applied to the boundary are distributed on the arc $AB$ and are equal to $X(t), Y(t)$. Then the complex potentials corresponding to this load will be

$$\Phi(z) = \int_{AB} \Phi_0\left(z, t, X(t), Y(t)\right) ds,\ \Psi(z) = \int_{AB} \Psi_0\left(z, t, X(t), Y(t)\right) ds, \tag{10}$$

Taking into account that $X = 0, Y = -S_y f(x)\sin\gamma$, we get

$$\Phi(z) = S_y \int_{AB} \Phi_0\left(z, t, 0, 1\right) f(\xi) d\xi,\ \Psi(z) = S_y \int_{AB} \Psi_0\left(z, t, 0, 1\right) f(\xi) d\xi, \tag{11}$$

where $\xi = \mathrm{Re}\,t$. Functions $\Phi_0(z,t,X,Y)$, $\Psi_0(z,t,X,Y)$ are linear relatively to $X, Y$ and $\sin\gamma = -dx/ds$.

These potentials in the inner points of the plate for which the sub-integral functions are smooth can be calculated using known quadrature formulas. In particular, we assume that the boundary equation is written in a parametric form $z = \omega(\theta),\ 0 < \theta \le 2\pi$. Using the quadrature formula of rectangles, which, for periodic smooth functions, with increased accuracy, we have

$$\Phi(z) = hS_y \sum_{n=1}^{M} \Phi_0\left(z, t_n, 0, 1\right) f(x_n) x'(\theta_n), \tag{12}$$

$$\Psi(z) = hS_y \sum_{n=1}^{M} \Psi_0\left(z, t_n, 0, 1\right) f(x_n) x'(\theta_n), \tag{13}$$

where $h = 2\pi / M, \theta_n = nh$, $t_n = \omega(\theta_n)$, $x_n = \mathrm{Re}\,t_n, x'(\theta) = \mathrm{Re}\left(\omega'(\theta)\right)$.

When determining hoop stress $\sigma_\theta$ on the edge of the plate just enough to calculate the function $\Phi$, as here $\sigma_\theta + N = 4\,\mathrm{Re}\,\Phi(z)$, where $N + iT = \left(X + iY\right)\dfrac{ds}{dt}$. Integral function at the point $t$ at limit of concentrated force $(X, Y)$ is

$$\Phi_0(z,t,X,Y) \sim \frac{(X+iY)}{2\pi(t-z)}, \tag{14}$$

Therefore, we shall represent $\Phi_0(z,t,0,1) = -\dfrac{1}{2\pi i(t-z)} + F(t,z)$, where $F$ – regular function.

Then after the limit crossing, using the Pleemelj-Sokhotski formula, we obtain

$$\sigma_\theta = N + 4\,\mathrm{Re}\,\tilde{\Phi}(z),\ N = -S_y f(x)\left(dx/ds\right)^2, \tag{15}$$

where $\tilde{\Phi}(z)$ - is equal to the function $\Phi$, in which the Cauchy integral is considered in the sense of the main value. Well-known quadrature formulas can be used to calculate this integral.

*Results of calculations.* Consider a rectangular plate with half-edges $a$ and $b$ with residual deformations (4), (5) at $b/a = 1{,}4$, $d/a = 0{,}3$, $a_0 = 0$. Complex potentials $\Phi_0, \Psi_0$ were determined in a closed form on the basis of the Muschelishvili method using the conformal mapping of the area occupied by a plate on a circle of the unit radius [16]. The display function was expanded into a Taylor series, which then held up to one hundred variables. At fig. 1 the distribution of found residual relative stresses is given $\sigma_y / S_y$ in cross-sections $y = 0,\ 0{,}5a,\ a,\ 1{,}3a$, and the value is given near the curves $y/a$.

Results of calculations of relative stresses $\sigma_y / S_y$ in cross-sections $x = ma$ at $m = 0,\ 0{,}2,\ 0{,}5$ depending on the coordinate $y$ see Fig. 1 on the right.

The given graphs practically coincide with the data of calculations, which are obtained in [6] for this task with another method.

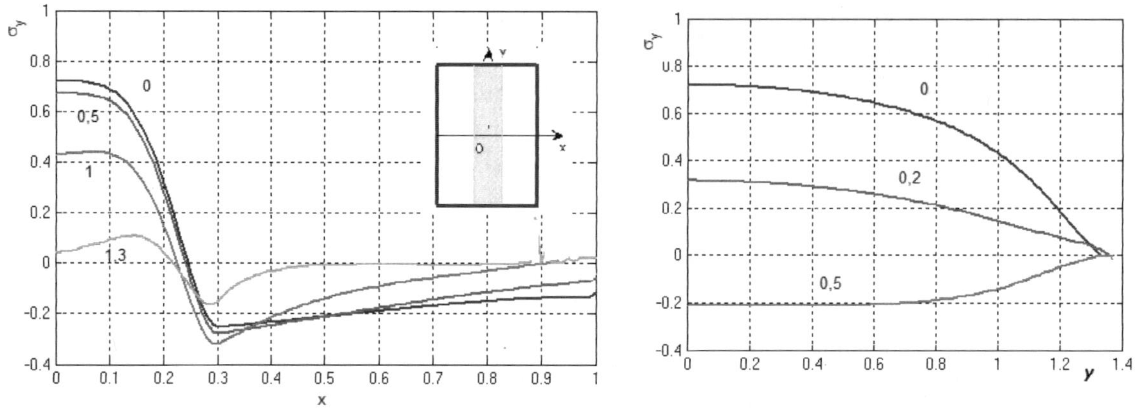


**Figure 1.** The distribution of residual relative stresses $\sigma_y / S_y$ is given in cross-sections $y = 0,\ 0{,}5a,\ a,\ 1{,}3a$, and the value is given near the curves $y / a$.

Calculation of residual stresses in a rectangular plate at $b / a = 1{,}5$ when the distribution of residual deformations has the form $f(x) = \sqrt{1 - x^2 / d^2}$ also showed complete convergence of calculation results with data [6], which are obtained by other methods.

**Residual stresses in the half-plane and the lane.** Consider the half-plane $y<0$ load-free edge. On the basis of the halfplane, it is expedient to simulate the distribution of residual stresses in a plate of large size in the vicinity of the seam near the limits. Determining a correction solution reduces to the consideration of the half-plane, which is under the influence of the load $p(x) = \sigma_y(x,0) - i\tau_{xy}(x,0) = -S_y f(x)$, with $p \neq 0$ at $-d < x < d$. Complex potentials for this case are determined by the formulas [16]

$$\Phi(z) = \frac{1}{2\pi i}\int_{-d}^{d}\frac{p(t)dt}{z-t},\quad \Omega(z) = \frac{1}{2\pi i}\int_{-d}^{d}\frac{\overline{p(t)}dt}{z-t}, \tag{16}$$

and $\Psi(z) = \Omega(z) - z\Phi'(z) - \Phi(z)$.

Let $f(x) = 1 - x^2 / d^2$. In this case

$$\Phi(z) = \frac{C}{2\pi i d^2}\left[\left(z^2 - d^2\right)L(z,d) + 2dz\right],\ \Phi'(z) = \frac{C}{2\pi i d^2}\left[2zL(z,d) + 4d\right],$$
$$\Omega(z) = \frac{\overline{C}}{2\pi i d^2}\left[\left(z^2 - d^2\right)L(z,d) + 2dz\right], \tag{17, 18}$$

where $L(z,d) = \ln\frac{z-d}{z+d}$, $C = -S_y$.

Adding partial and correction components, we obtain the solution of the problem of residual stresses in the half-plane.

Consider the lane $-H < y < 0$. The complex potentials for a strip are represented in the form

$$\Phi(z) = \Phi_p(z) + \Phi_\Delta(z),\ \Omega(z) = \Omega_p(z) + \Omega_\Delta(z), \tag{19}$$

where $\Phi_p(z), \Omega_p(z)$ – potentials for the half-plane $y < 0$, loaded with forces applied to the band at $y = 0$. Expressions for potentials $\Phi_\Delta(z), \Omega_\Delta(z)$, which when $y > -H$ defined by integrals of smooth functions, is given in [5].

For example, calculations of stresses in the band due to residual deformations at $\overset{0}{\varepsilon}_y = -S_y(1 - x^2 / d^2) / E$. Calculated stresses $\sigma_y, \sigma_x, \tau_{xy}$ devided by $S_y$ depending on the relative coordinate $x/d$ at *d/H=0.05* shown in fig. 2. With numbers *0, 1, 2, 3, 4, 5* near the curves correspond to the cross sections *y=-0.1jd* at *j=0, 1, 2, 3, 4, 5.*

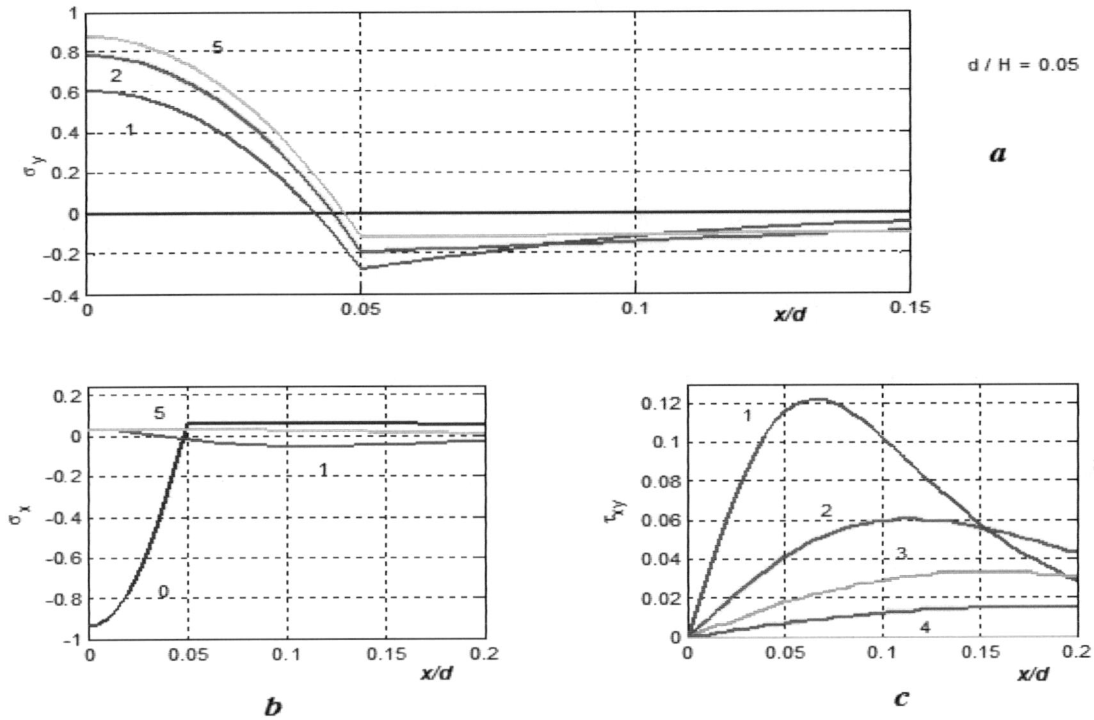


**Figure 2**. Calculated stresses $\sigma_y, \sigma_x, \tau_{xy}$, which are attributed to the magnitude $S_y$ depending on the relative coordinate $x/d$ at *d/H=0.05.*

Obtained stresses with $d/H = 0{,}1$ are well consistent with the data for a rectangular plate of fig. 1.

In fig. 3 shows the distribution of relative stresses $\sigma_x / S_y$ (a) and $\sigma_y / S_y$ (b) at *x=0,* $-H/2 < y < 0$ depending on the width of the distribution zone of the initial deformations. Here curves 0,...,5 correspond to the values $d/H = 0.05, 0.1, 0.25, 0.5, 0.75, 1$.

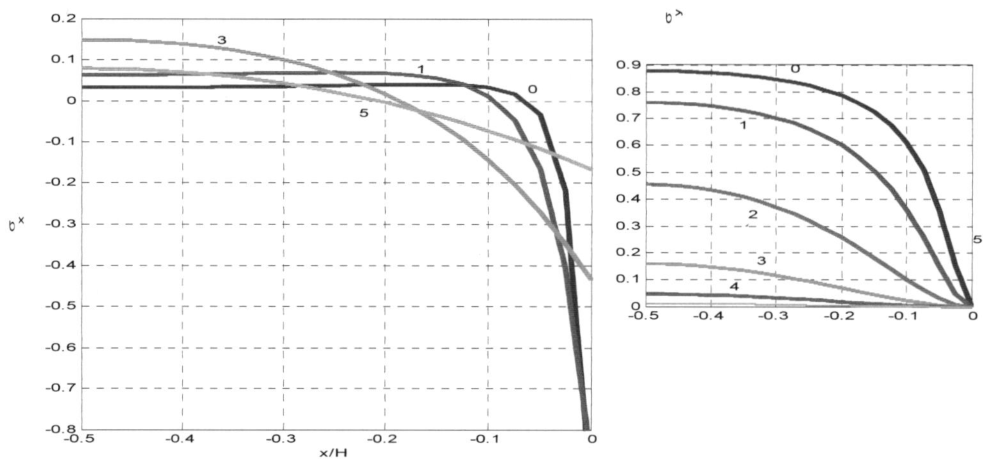


**Figure 3.** The distribution of relative stresses $\sigma_x / S_y$ (a) and $\sigma_y / S_y$ (b) at $x=0$, $-H/2<y<0$.

**Determining in a rectangular plate with residual stresses.** Consider now the case when in a plate with residual stresses a rectilinear cracks a half length $d$ centered at the point $(x_c, y_c)$, which is tilted at an angle $\alpha$ to axis $Ox$. The stresses in such a plate are represented as the sum of stresses in a solid plate with residual deformations and stresses corresponding to the correction solution. Denote the stress vector in the solid plate on the line on which the crack is as $(X_L, Y_L)$. Then, in order to determined the correction solution, we consider the problem of the elasticity theory for a plate with a crack, to the edges of which the effort $(-X_L, -Y_L)$ is applied. We will also consider the case, when, the edges can contact with each other. Then the above-mentioned efforts will be applied only in sections of the crack, where there is no contact, and in the contact area we assume that there are no tangential stresses (smooth contact) and normal edge displacements are the same. To solve this problem we use an algorithm of work [17].

Calculated relative values $F_{I,II} = K_{I,II} / (S_y \sqrt{\pi d})$ taking into account the contact of the edges for the case $d/a = 0,2$ at $y_c = 0$, $\alpha = 0$, depending on the location of the center $x_c$ for the left *(A)* and right tips *(B)* the crack is shown in Fig. 4. on the left.

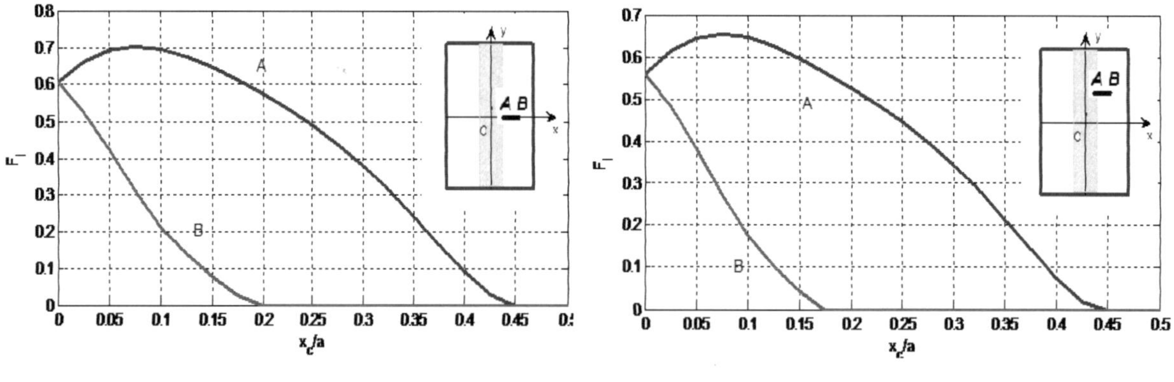


**Figure 4.** Calculated relative values $F_{I,II} = K_{I,II} / (S_y \sqrt{\pi d})$ for the left *(A)* and right tips *(B)* the crack.

Similar results are parallel to the axis *Ox* cracks at $y_c / a = 0{,}5$ shown in fig. 4 on the right.

Figure 5 shows the distribution for an inclined crack depending on the angle $\alpha$ when placing the center of the crack at the points $(0,0)$ i $(0, 0, 2a)$.

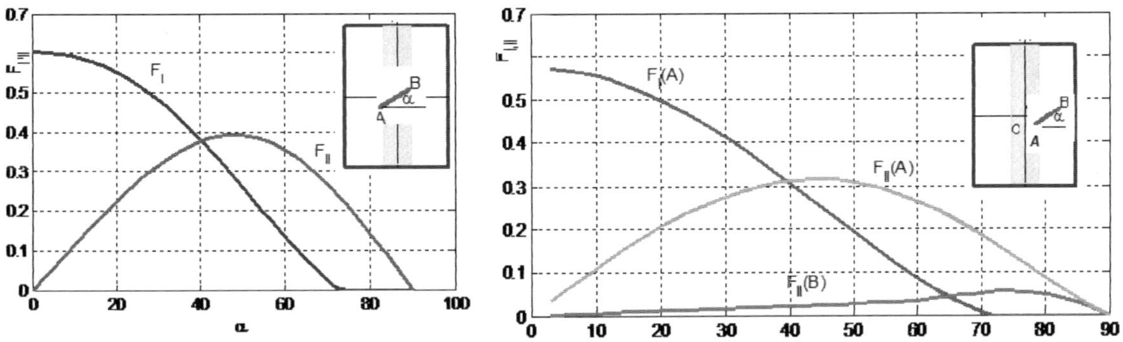


**Figure 5.** Calculated relative values $F_{I,II} = K_{I,II} / (S_y \sqrt{\pi d})$ for the left *(A)* and right tips (*B)* the crack.

Based on the calculations made, the following conclusions can be made: when determining SIF for cracks, which are in the field of residual stresses, it is necessary to take into account the contact of the edges of the cracks; the maximum value for cracks perpendicular to the weld seam SIF $K_I$ is achieved if the crack is shifted; the contact of the shores, as a rule, takes place near the peak of the seam.

Conclusion. An algorithm for determining the stress state of plates with cracks, caused by residual deformations is proposed. It has been established that ignoring contact with the edges of the cracks can lead to significant errors in determining SIF under the action of residual stresses. SIF can reach the maximum values for the shifted relative to the center of seam cracks.

## References

[1] Lu J 1996 *Handbook of measurement of residual stresses* (Fairmont Press)
[2] Masubuchi, K 2013 *Analysis of welded structures: residual stresses, distortion, and their consequences* (Pergamon)
[3] Radaj D 2003 *Welding residual stresses and distortion: Calculation and measurement* (Düsseldorf: DVS-Verlag)
[4] Fricke S, Keim E and Schmidt J 2001 Numerical weld modeling—a method for calculating weld-induced residual stresses *Nuclear Engineering and Design* **206** 139-150
[5] Vrancken B et al. 2014 Residual stress via the contour method in compact tension specimens produced via selective laser melting *Scripta Materialia* **87** 29-32
[6] Osadchuk V 2005 Vyznachennya zalyshkovykh napruzhen′ u zvarnomu stykovomu zyednanni dvokh pryamokutnykh plastyn / V. Osadchuk, YU. Tokovyy // Mashynoznavstvo **9/99** 3-9
[7] He M Y, Evans A G and Hutchinson J W 1994 Crack deflection at an interface between dissimilar elastic materials: role of residual stresses *International Journal of Solids and Structures* **31** 3443-3455
[8] Beghini M and Bertini L 1990 Fatigue crack propagation through residual stress fields with closure phenomena *Engineering Fracture Mechanics* 36 379-387
[9] Nejad R M, Farhangdoost K and Shariati M 2015 Numerical study on fatigue crack growth in railway wheels under the influence of residual stresses *Engineering Failure Analysis* **52** 75-89
[10] Prime M B 1999 Residual stress measurement by successive extension of a slot: the crack compliance method *Applied Mechanics Reviews* **52** 75-96

[11] He M Y, Evans A G and Hutchinson J W 1994 Crack deflection at an interface between dissimilar elastic materials: role of residual stresses *International Journal of Solids and Structures* **31** 3443-3455

[12] Galatolo R and Lanciotti A 1997 Fatigue crack propagation in residual stress fields of welded plates *International Journal of Fatigue* **19** 43-49

[13] Chang K-H, and Lee C-H 2007 Residual stresses and fracture mechanics analysis of a crack in welds of high strength steels *Engineering Fracture Mechanics* **74** 980-994

[14] Dzyubyk A R, Nykolyshyn T M and Porokhovs'kyi Yu V 2016 Influence of Residual Stresses on the Limit Equilibrium of a Pipeline with Internal Crack of Arbitrary Configuration *Materials Science* **52** 89-98

[15] Osadchuk V A, Tsymbalyuk L I and Dzyubyk A R 2012 Determination of the triaxial distribution of residual stresses in welded joints of structural elements with rectilinear seams and estimation of their influence on joint strength in the presence of crack-type defects *Journal of Mathematical Sciences* **183** 150-161

[16] Muskhelishvili N I 2013 Some basic problems of the mathematical theory of elasticity (Springer Science & Business Media)

[17] Maksymovych O, Pasternak I, Sulym H and Kutsyk S 2014 Doubly periodic cracks in the anisotropic medium with the account of contact of their faces *Acta mechanica et automatica* **8** 160-164

IOP Conf. Series: Materials Science and Engineering **480** (2019) 012028 doi:10.1088/1757-899X/480/1/012028

# Operating principle of volatile corrosion inhibitors in the jar test

**L-M Rymer, M Sieber*, S Lautner and F Faßbender**

EXCOR Korrosionsforschung GmbH, Magdeburger Straße 58, D-01067 Dresden, Germany

* e-mail: maximilian.sieber@excor.de

**Abstract.** An appropriate corrosion protection is required for transport and storage of high-quality goods and semi-finished parts made from metals. The application of volatile corrosion inhibitors (VCI) as a means of temporary corrosion protection is based on the release of corrosion inhibitors from films, papers, diffusers or oils to the gaseous phase. By adsorption on the metal surface, the inhibitors prevent the degradation of the primary oxide layer and hence impede corrosion. In principle, the anticorrosive effect is independent of the part geometry, because the protected surfaces do not need direct contact with the VCI material. Also, the VCI method does not require costly cleaning of parts prior to further use after transport and storage. The EXCOR jar test is a common method to examine the efficiency of VCI. A metallic sample and a VCI material are exposed in a jar. After conditioning in dry and humid atmosphere, corrosive stress is applied to the sample. To clarify the VCI´s mechanism of action, the test steps, their duration and the type of inhibitors were varied. Thus, the operating principle of VCI during the test was identified. The acquired knowledge enables to understand and tailor the corrosion protection by VCI in service.

## 1. Introduction

Corrosion is an extensive problem for the worldwide industry. The cost caused by corrosion amounts to hundreds of billions of dollars every year. Especially, the manufacturers and suppliers of the metal-processing industry are striving to improve the corrosion resistance of their metallic products during transportation and storage. Bare metals are prone to condensed humidity on their surface, which already leads to corrosive attack, even if no further corrosion-promoting substances are existent in the considered environment. To temporarily prevent corrosion of metallic products, in particular volatile corrosion inhibitors (VCI) present a convenient method. The VCI sublimate from films, papers, diffusers and/or oils – which serve as VCI donors – to the gaseous phase within a packaging. Finally, the inhibitors adsorb on the metal surface to stabilize the primary oxide layer and thus reduce corrosion attack. The corrosion protection with VCI possesses considerable advantages. They offer protection for several years and on any part geometry. Furthermore, except for the above-mentioned common carriers, VCI can be incorporated in manifold packaging products, such as polymer-, paper- or cardboard-based trays, carriers etc. Moreover, no expensive cleaning after transport and storage is necessary. And at last, the VCI ingredients exhibit an eco-friendly behavior. [1–7]

To investigate the corrosion protection effect of the VCI materials, there are different established test methods. The tests are generally based on the exposure of metallic samples in VCI packaging materials, especially films or papers, to alternating climates with varying temperature and humidity so that a corrosive impact occurs on the sample surface. The samples are mostly bare metal sheets made from iron-, copper- or aluminum-based alloys or galvanized steel. In one common test layout

(Figure 1), a metallic sample is mounted in an auxiliary frame, which completely avoids contact between the samples′ surfaces and the VCI packaging (so-called distance test). In another test layout (Figure 2), the metallic sample′s surface is in direct contact with the VCI packaging to simulate contact of metal surface and packaging material, which leads to crevice-like corrosion conditions when humidity accumulates between sample surface and packaging (so-called contact test). In each case, the efficiency of VCI is assessed after a defined number of climate cycles with regard to a reference packaging material, which does not contain any corrosion inhibitors. In the most cases, the duration of these tests amounts to 20 to 30 d to receive significant results. The criterium for the assessment of the VCI products is the number of cycles until significant corrosion occurs on the metallic sample, which is related to the number of cycles until corrosion occurrence on a metallic sample in a non-VCI packaging. In contrast, the so-called jar test is a rapid testing method due to the reduction of the test duration to 24 h instead of 30 d. The jar test procedure conduces to the determination of the efficiency of VCI for steel samples. The metallic sample is exposed in a jar containing VCI films or papers and a corrosive stress is applied by a simple climate regime. Because of its reduced test duration, the jar test is suitable for receiving rapid and efficient test results. The test procedure will be explained in detail in section 2.

**Figure 1.** Experimental setup of the distance test. The metallic specimen (metal sheet) is mounted in an auxiliary frame, packed in a film and exposed to a climatic stress test (no contact between metallic specimen and film).

**Figure 2.** Experimental setup of the contact test. The metallic specimen′s surface is in direct contact with the VCI packaging during the climatic stress test.

To derive comprehensive indications for industrial applications, the inhibition effect of VCI during corrosion attack on different metals requires a more detailed study. With the focus on steel, the test stages of the jar test and their duration are varied in the present examination to determine the functionality and mode of action of two different types of VCI-films.

## 2. Experimental and materials

The jar test procedure constitutes a rapid test method to examine the VCI effect on steel, which is used for quality control of mainly VCI films and papers. For the present investigation, cylindrical steel samples with a diameter of 16 mm were used. The chemical composition is given in Table. 1.

**Table 1.** Nominal composition of the sample material S235JRG.

| Element | C | Mn | P | S | Fe |
|---|---|---|---|---|---|
| Concentration in wt-% | <0.2 | <1.4 | <0.045 | <0.045 | balance |

The steel samples were ground with grade 320 abrasive paper on a grinder Saphir 520 (ATM, Germany) for 90 s each in co- and counterrotation mode followed by rinsing under deionized water and acetone p. a. To place the metallic sample in an Erlenmayer flask with a volumetric capacity of 1 l, a rubber plug with a diameter of 54 mm and an axial bore was used. The axial bore was rimmed by two slots, where the VCI material were fixed during the test. These stripes of VCI-film had the dimension 25 mm x 150 mm. The metallic sample and the VCI stripes are fixed in the rubber plug and put into the flask. After 20 h storage of the flask at room temperature, 10 ml of a water glycerol mixture (density 1.076 g/cm³ at 23±2 °C, mass fraction glycerol to water approx. 1:3) prepared from deionized water and glycerol (p.a. grade) are added to the flask, which is then stored at room temperature for another 2 h. Afterwards, the flask is put into an incubator (ULP500, Memmert, Germany) at 40 °C (pre-heated) for 2 h. Figure 3 shows the complete experimental arrangement of the jar test. The actual prevalent temperatures and humidity in the jar are depicted in Figure 4. The measurement originates from a data logger (temperature, humidity logger RHT10, Extech, U.S.), which was inside the jar during the whole test. Water condenses on the steel sample during the first 30 min due to thermal inertia of the steel sample (temperature below the dew point of the already heated atmosphere). Then, the condensed water remains on the steel surface because of the high humidity of the atmosphere within the jar.

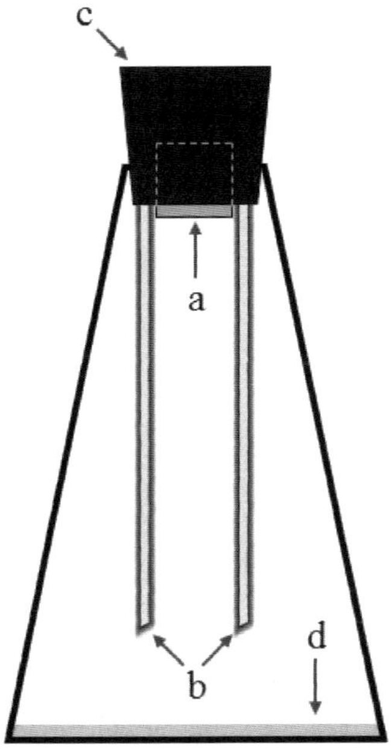


**Figure 3.** Experimental arrangement of the jar test procedure. A ground steel sample (a) and the VCI-containing material (b) are mounted in a rubber plug (c) and in the course of the test the water glycerol mixture (d) is added to the Erlenmeyer flask.

In a variation of the described test, the corrosive load is imposed on the sample differently. Instead of putting the flask into an incubator, pipettes with frozen water are positioned in the rubber plug on the top of the metallic sample. Thereby, the sample is cooled below the dew point, which leads to condensation of humidity on the sample surface and thus to the corrosive attack. The pipettes with frozen water are replaced every 15 min for an overall time of the corrosive load of 2 h. During that time, the flask is positioned in a thermostatically controlled water bath (40 °C). The flask is immersed in the water bath approx. 1 cm, so that the water glycerol mixture on the bottom of the flask partially evaporates and thus increases the humidity in the flask, but the temperature of the atmosphere in the flask is not significantly affected (during the 2 h, the temperature of the atmosphere in the flask does not significantly exceed room temperature).

The total test duration thus includes the dry conditioning for 20 h, the humid conditioning for 2 h and the corrosive stress at elevated temperature (40 °C in incubator) and/or at room temperature (ice-water-filled pipettes) for 2 h, each. Afterwards, the metallic sample is removed from the flask, is dried in an incubator at 40 °C and is then evaluated by visual inspection. The corrosive impact is described using a generalized evaluation scale. The rating definition is divided into four grades as can be seen in Figure 5. If a steel sample shows an overall corrosive attack, it will receive the evaluation 0 (no corrosion protection). The grade 3 is assigned to samples without any corrosion spot (good corrosion protection). In between the corroded area decreases notably from value 1 (low corrosion protection) to 2 (moderate corrosion protection). [2]

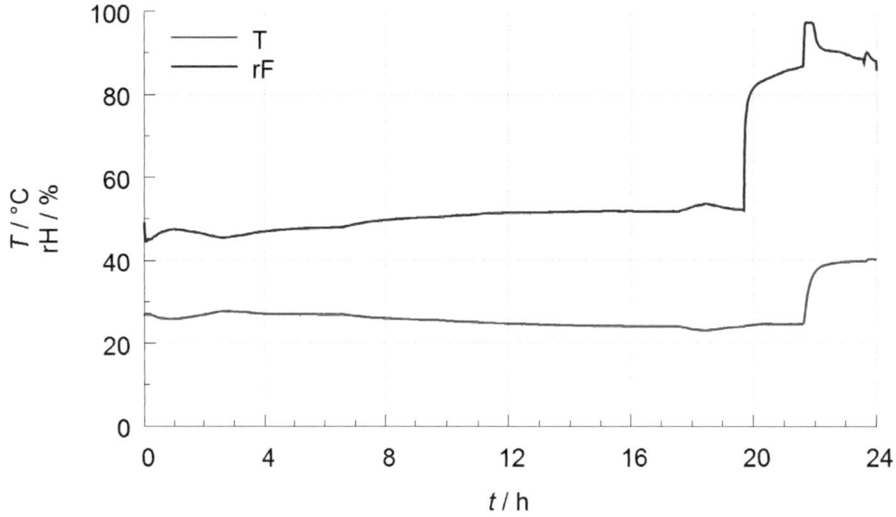


**Figure 4.** Temperature and relative humidity of the atmosphere in the flask during the jar test procedure.

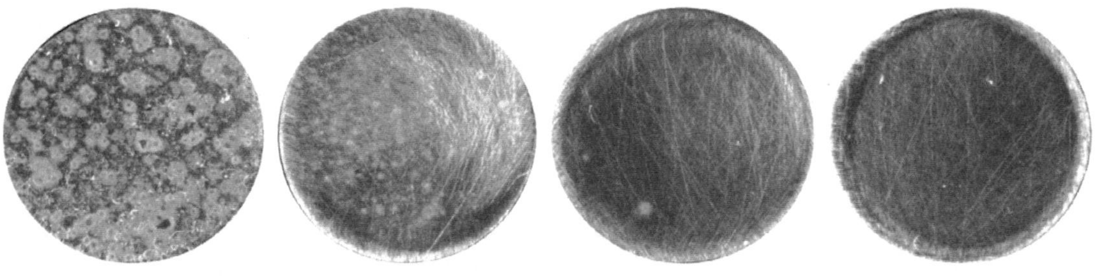

**Figure 5.** Representative visual occurrence of the steel sample surface for the 4 grades of the evaluation scale with regard to VCI efficiency; 0 – no corrosion protection, 1 – low corrosion protection, 2 – moderate corrosion protection, 3 – good corrosion protection (from left to right).

To identify the mechanism of action of VCI during the jar test, the test stages, their duration and the type of inhibitor were varied. The investigated VCI types in this work are denoted as Type I and Type II. The films consisted of polyethylene (PE). Their thickness amounted to 80 µm. VCI Type I and Type II had fundamentally different composition with regard to the contained corrosion inhibitors. Generally, this includes both organic and inorganic substances, for instance volatile organic acids, cyclic compounds or salts, which act as inhibitors after dissociation as well as non-volatile stabilizers which moderate the inhibitor release and prevent the VCI product from premature decay of its technical properties (both in terms of corrosion protection as well as ageing, discoloration etc.). However, the detailed composition cannot be disclosed here for reasons of confidentiality, since both

Type I and Type II are research VCI compositions similar to commercially available products. The variation of the duration of the three jar test phases are shown in Figure 6. The duration of the dry conditioning was modified between 20 h, 1 h and 0 h. The duration of the humid conditioning was varied between 2 h and 0 h. The duration the corrosive attack was not modified. While normally, the VCI material remains in the flask during the whole test, it was removed from the flask after different stages of the test to investigate into the adsorption/desorption behavior of the VCI during dry and humid conditioning as well as during corrosive stress. Each variation of the test stages and their duration included control test runs without any VCI material. Thus, the appropriateness of the corrosive stress applied in the test should be proven.

For the evaluation of the data, at least four samples per test variation were produced and visually assessed utilizing the scale described above. Moreover, the average value and the minimum and maximum of the obtained data were established for the discussion.

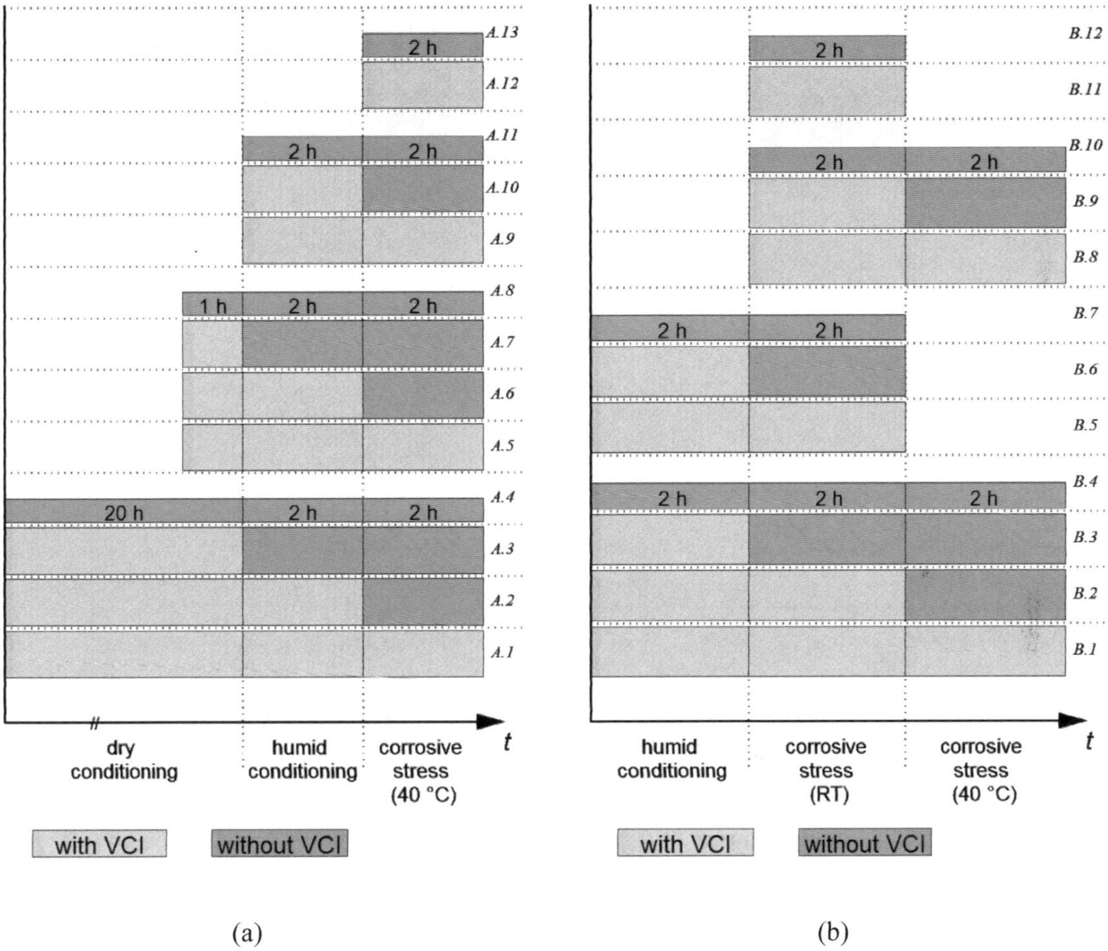


(a) (b)

**Figure 6.** Experimental matrix of the investigated jar test – the corrosive stress applied in the incubator at 40 °C is referenced as "corrosive stress (40 °C)" and the corrosive stress applied by placing ice-water-filled pipettes on the sample is referenced as "corrosive stress (RT)", since the temperature of the atmosphere in the flask does not significantly exceed room temperature in this case. The variation of the test includes the removal of the VCI material after different stages and the variation of the duration of single stages of the test (a – corrosive stress only at elevated temperature, b – corrosive stress at room temperature and elevated temperature).

## 3. Results and discussion

The obtained results of the tests including only corrosive stress at elevated temperature (40 °C) (tests A.x) or at room temperature and at elevated temperature (tests B.x) are given in Figures 7 and 8, respectively.

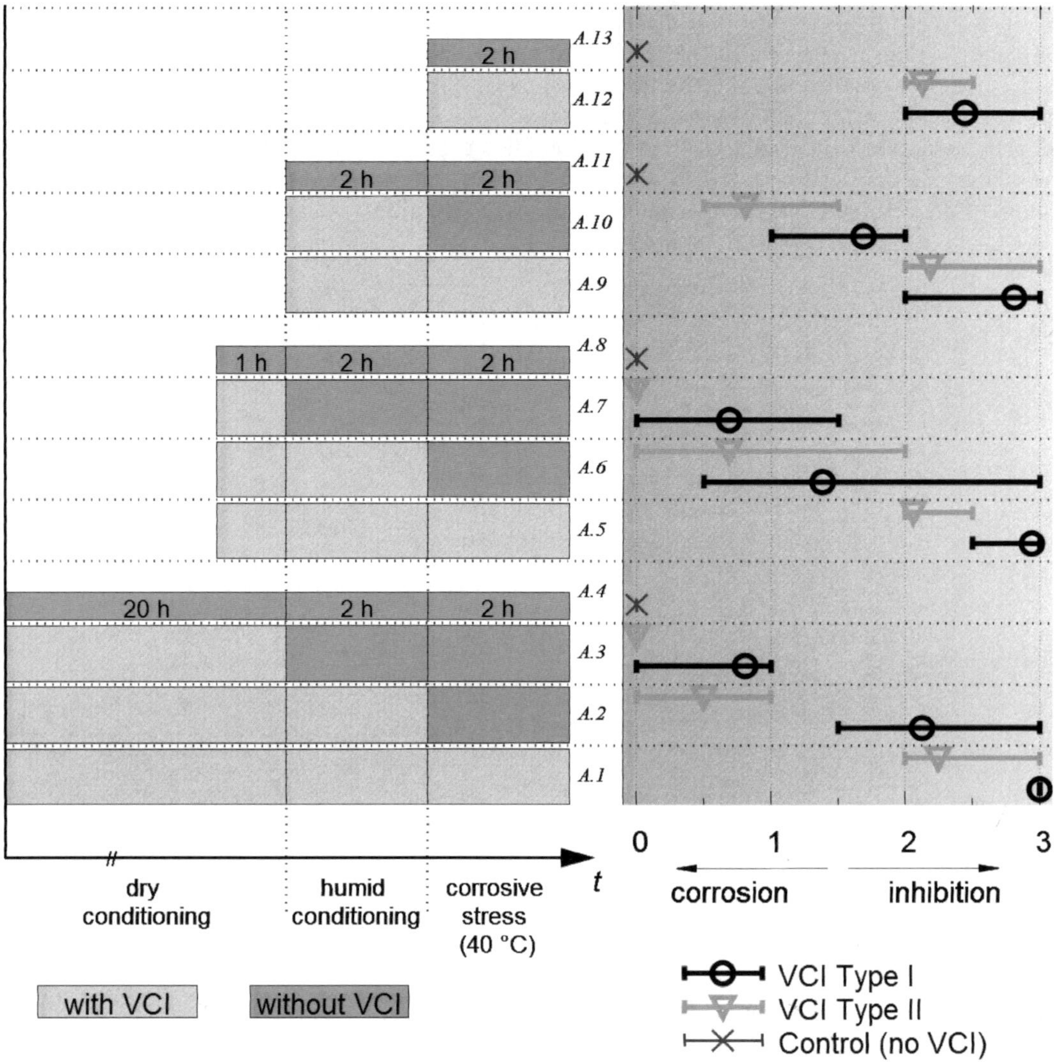


**Figure 7.** Obtained average results and min/max values (8 samples, each) according to the evaluation scale for the corrosion inhibition effect (0 = no inhibition, i.e. heavily corroded sample; 3 = good inhibition, i.e. non-corroded sample) of VCI Type I and Type II (right) in dependence of the variation of the jar test for corrosive stress at elevated temperature (40 °C, left). The effect of the 20 h (test no. A.1–A.4) or 1 h dry conditioning (test no. A.5–A.8) on the inhibition is low compared to the 2 h humid conditioning (test no. A.9–A.11) for both types. When the VCI material is present during the corrosive stress, the inhibition effect is good for both Type I and Type II (test no. A.1, A.5, A.9, A.12). When the VCI is present only during humid conditioning, but is removed from the flask before the corrosive load (test no. A.2, A.6, A.10), the inhibition effect of Type I is more significant than for Type II. When the VCI is removed from the flask after dry conditioning (test no. A.3, A.7), the inhibition is rather low for both VCI.

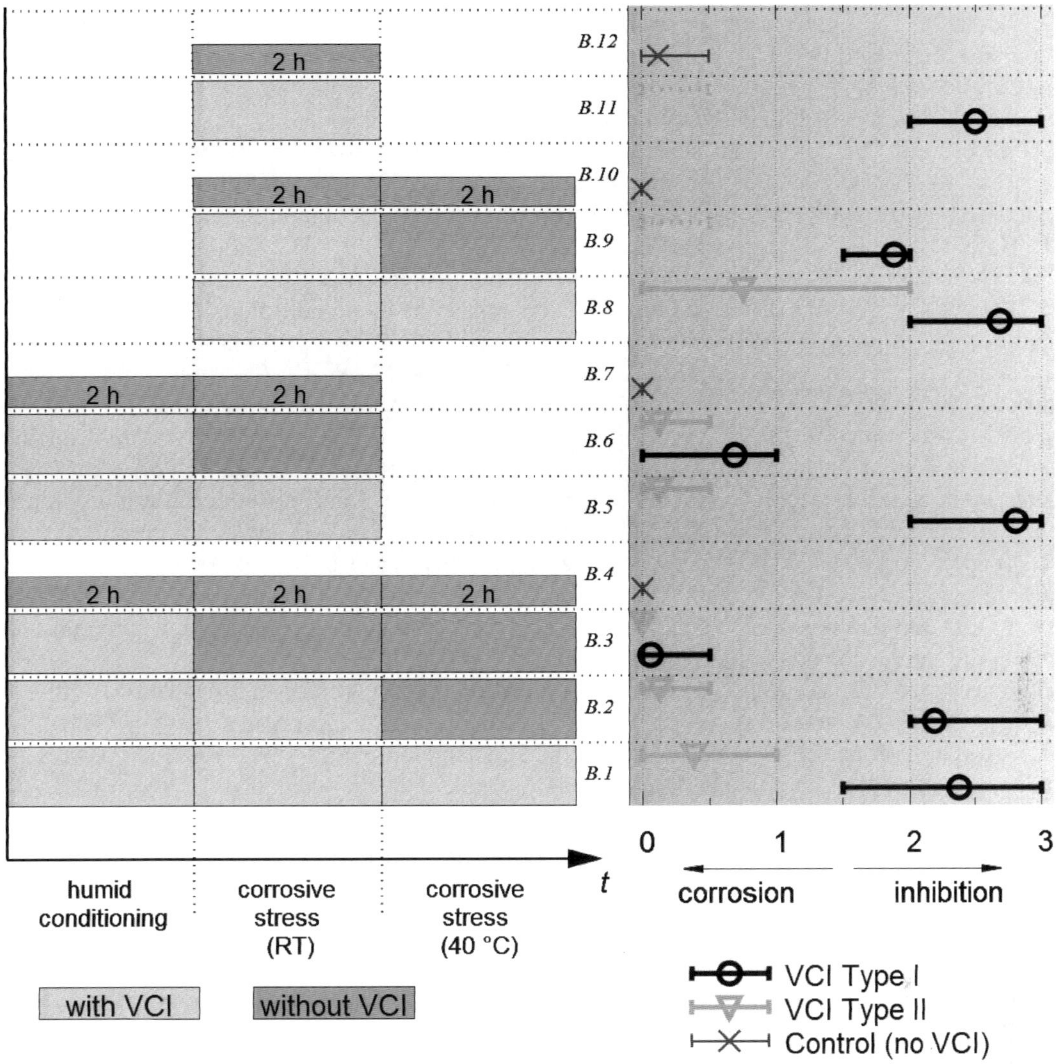


**Figure 8.** Obtained average results and min/max values (4 samples, each) according to the evaluation scale for the corrosion inhibition effect (0 = no inhibition, heavily corroded steel sample surface; 3 = good inhibition, non-corroded steel sample surface) of VCI Type I and Type II (right) in dependence of the variation of the jar test for sequential corrosive stress at room temperature (RT) and elevated temperature (40 °C) (left). VCI Type I shows good inhibition when present during the corrosive stress at both room temperature and 40 °C (test no. B.1, B.8). In the test, VCI Type II fails for corrosive stress at room temperature (evaluation 0–1, test no. B.5). The humid conditioning before the corrosive load has a rather low effect on the inhibition effect for both Type I and Type II (compare test no. B.1 and B.8; B.2 and B.9; B.3 and B.10). Test no. B.8 contains a single outlier with moderate inhibition effect for VCI Type II. Dotted lines (Type II, test no. B.9, B.11) refer to tests, which have not been carried out because it was obvious from test no. B.5 that Type II fails under corrosive load at room temperature.

Overall, the jar test displays average grades of 0.8 and 3.0 for VCI Type I, whereas the average grades of Type II are between 0.0 and 2.4. The presence of VCI during the corrosive stress is essential for a good corrosion protection effect in the test for both types. The appropriate grades to confirm this statement vary between 2.5 and 3.0 for Type I and from 2.1 to 2.4 for Type II in Fig. 7 (test no. A.1,

A.5, A.9, A.12). The same statement applies to VCI Type I by considering the results shown in Fig. 8 (test no. B.1, B.5, B.8, B.11). For corrosive stress at room temperature, however, Type II fails in the test even when the VCI is present during said corrosive stress (Fig. 8, test no. B.5). Test no. B.5 was independently repeated with another four samples and yielded identical results, so that the statement was consolidated. Therefore, tests no. B.9 and B.11 were not carried out with VCI Type II (dotted lines in Fig. 8). For both VCI Type I and Type II, when the VCI is present during the corrosive load, the conditioning generally only has a minor effect on the corrosion inhibition ability of the respective products in the considered time scale (Fig. 7, test no. A.1, A. 5, A.9, A.12 and Fig. 8, test no. B.1, B.5, B.8, B.11). When the VCI is removed from the flask before corrosive stress, however, VCI Type I and Type II show significantly different behavior with regard to the conditioning. When VCI Type I is present during humid conditioning (2 h) but is removed before the corrosive stress at 40 °C, there is still a significant inhibition effect (Fig. 7, test A.2, A.6, A.10). For VCI Type II under the same conditions, the inhibition effect is significantly less pronounced. The conditioning at low relative humidity – independently of the presence of VCI products in time of corrosive stress at 40 °C – contributes to the corrosion inhibition only to a small amount. Removal of the VCI after 1 h or 20 h of dry conditioning leads to only weak inhibition effect for Type I and no noticeable inhibition effect anymore for Type II (Fig. 7, test A.3, A.7). Also, in case of corrosive stress at room temperature, neither Type I nor Type II show relevant inhibition ability when only present before during humid conditioning (Fig. 8, test B.6). Interestingly, for Type I, corrosive stress at 40 °C after the corrosive stress at room temperature does not lead to enhanced corrosion of the steel sample, whether or not VCI is present during the corrosive stress at 40 °C (Fig. 8, test B.2, B.9).

The performed examination provides some indications of the dominating mechanisms of action of the VCI materials. A graphical representation of the findings is given in Figure 9. Generally, the release of inhibitors and their effect on the sample surface is obviously enhanced in the presence of humidity. Whether this is mainly due to better release and transportation of the VCI in humid atmosphere and/or due to stronger bonding of the inhibitors to the metal surface in the presence of humidity will have to be addressed in future investigation. However, the required amount of VCI for corrosion protection in the atmosphere and especially on the metallic surface is achieved more comprehensively for Type I than for Type II. Further, compared to VCI of Type II, the VCI of Type I are affected to a lesser extent by the removal of the VCI product and by the variation of the temperature of the atmosphere during corrosive load. There is a conception of a "retentive memory" of the metallic sample, implying the remain of the inhibitor on the surface after the removal of the donating product. This memory could be ruled by the different adsorption strength and desorption rates of the inhibitors or it could be caused by a different mode of action regarding the corrosion inhibition, especially with regard to the reaction mechanism. Since VCI Type II obviously shows higher activity in case of higher temperature, the corrosion protection effect for this type could be dominated by a chemical or electrochemical reaction with either a relatively high temperature dependence of the reaction rate and/or a temperature-dependent activation energy. Additional tests equivalent to experiment B.6, but with prolonged conditioning of 20 h and 200 h showed low and moderate corrosion protection for VCI Type II, respectively. According to the reaction model, this indicates a temperature-dependent reaction rate as the dominating factor instead of the activation energy. For the latter, there would have been no effect of the prolonged conditioning time at lower temperature, while a lower reaction rate at lower temperature can be compensated by longer conditioning time. Therefore, in the test, this reaction would take place only during the corrosive load at elevated temperature in presence of the VCI or during a longer conditioning phase and could therefore explain the test results for Type II. In contrast, the behavior of VCI Type I could thus be dominated by an adsorption mechanism and therefore its protection effect would show less temperature-dependence considering that both adsorption and desorption activity increase at elevated temperature. Additionally, the adsorption of inhibitors from Type I is obviously rather strong so that after removal of the VCI from the flask and therefore the shifting of the adsorption/desorption equilibrium in favor of desorption, the corrosion protection effect is still significant.

IOP Conf. Series: Materials Science and Engineering **480** (2019) 012028 doi:10.1088/1757-899X/480/1/012028

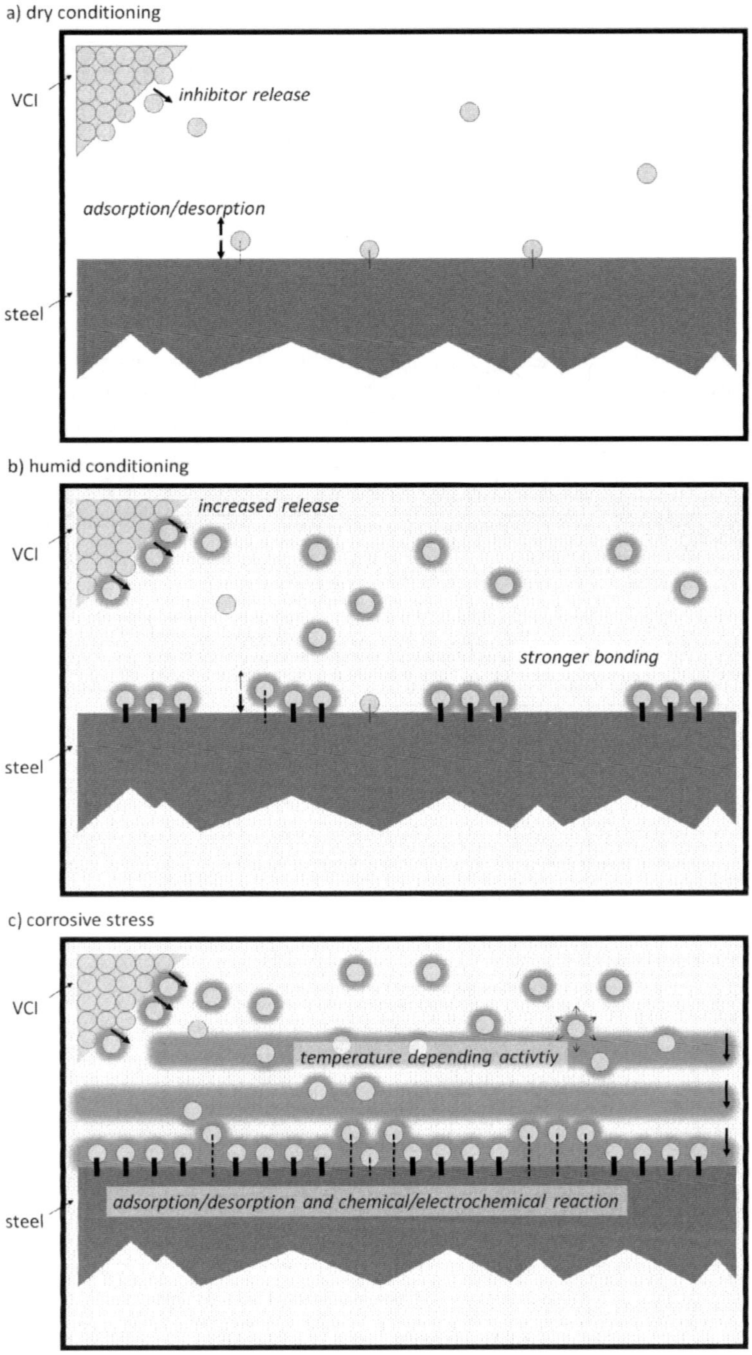


**Figure 9.** Mechanism of inhibitor-substrate interaction during the stages of the test. Obtained results indicate an increased release of inhibitors and/or stronger bonding of the inhibitors on the steel surface during humid conditioning (b) compared to dry conditioning (a). Additionally, the activity of the inhibitors during corrosive load under condensation of humidity on the sample surface is temperature-dependent and dominated by a chemical/electrochemical reaction or adsorption/desorption behavior depending on the type of VCI (c).

Further experiments are necessary for clarification and validation of the proclaimed mechanism. These will include the variation of the conditioning time before corrosive stress at room temperature as well as a variation of the amount of VCI in the flask via the surface of the stripes of VCI-film mounted in the rubber plug. Also, the effect of humidity on the release of VCI will have to be addressed by setting a defined humidity during conditioning via the variation of the water/glycerol mixture used for the test. To support the results, methods to quantify sublimation and adsorption of VCI based on chemical analysis of VCI-donating and VCI-adsorbing materials are under development.

## 4. Conclusions

In the present study, the corrosion protection effect of VCI was examined. Therefore, a jar test procedure was used, which is a rapid testing method to obtain results on corrosion inhibition within a short time period. To identify the operating principle and the VCI mechanism of action leading to the anti-corrosion effect, the test stages of the jar test and their duration were varied and two different types of VCI-films with fundamentally different composition were considered. The results can be summarized as follows:

(1) The presence of the VCI-containing films during the corrosive stress decreases corrosion damage at the sample surface significantly. Only in few of the varied test regimes, a corrosion protection effect is still noticeable when the VCI film is removed after conditioning and before the corrosive stress. VCI Type I shows an overall superior protection effect compared to Type II.

(2) In the test, humid conditioning contributes to the inhibition effect more than dry conditioning. This can be explained with the enhanced release/transport of VCI and/or a stronger bonding of the adsorbents on the sample surface in presence of high humidity.

(3) In the test, the temperature during the phase of corrosive stress plays an important role regarding the protection effect, especially for VCI Type II, which failed in the test when the temperature during corrosive stress was not increased significantly above room temperature. Additional experiments have shown that VCI Type II is generally able to provide corrosion protection also at non-elevated temperatures when the humid conditioning times are increased. This is attributed to fundamentally different mechanisms of action of the investigated types. The results indicate that the dominating effect of the protection by Type I is its adsorption behavior on the metal surface, while the dominating effect for Type II is a chemical or electrochemical reaction with temperature-dependent reaction rate.

**Acknowledgments**
The authors gratefully acknowledge technical assistance by Katrin Schäfer and Daniela Wegner (both EXCOR Korrosionsforschung GmbH, Dresden, Germany).

## References

[1] Ansari F A, Verma C, Siddiqui Y S, Ebenso E E and Quraishi M A 2018 Volatile corrosion inhibitors for ferrous and non-ferrous metals and alloys: A review *Int. J. Corros. Scale Inhib.* **7** 126–150
[2] Valdez B, Schorr M, Cheng N, Beltran E and Salinas R 2018 Technological applications of volatile corrosion inhibitors *Corros. Rev.* **36** 227–238
[3] Subramanian A, Natesan M, Muralidharan V S, Balakrishnan K and Vasudevan T 2000 An Overview: Vapor Phase Corrosion Inhibitors *Corrosion* **56** 144–155
[4] Koehler S and Reinhard G 2014 VCI containing package material – mode of functioning *Int. J. Corros. Scale Inhib.* **3** 286–306
[5] Kuznetsov Y I 2015 Progress in science of corrosion inhibitiors *Int. J. Corros. Scale Inhib.* **4** 15–34
[6] Andreev N N and Kuznetsov Y I 2012 Volatile Inhibitors of Metal Corrosion. I. Vaporization *Int. J. Corros. Scale Inhib.* **1** 16–25

[7] Andreev N N and Kuznetsov Y I 2012 Volatile inhibitors of metal corrosion. II. Interaction of systems being protected with the environment and corrosion prevention conditions *Int. J. Corros. Scale Inhib.* **1** 146–153

IOP Conf. Series: Materials Science and Engineering **480** (2019) 012029 doi:10.1088/1757-899X/480/1/012029

# Evaluation of volatile corrosion inhibitors in the presence of condensation water by electrochemical methods

**M Sieber*, S Lautner and F Faßbender**

EXCOR Korrosionsforschung GmbH, Magdeburger Straße 58, D-01067 Dresden, Germany

* e-mail: maximilian.sieber@excor.de

**Abstract.** Volatile corrosion inhibitors (VCI) are a means of temporary corrosion protection based on the release of inhibitors from films, papers, diffusers or oil to the gaseous phase. By adsorption on a metal surface, the inhibitors prevent degradation of the primary oxide layer and impede corrosion. Thus, VCI provide corrosion protection for high-quality goods and semi-finished parts made from metals during transportation and storage. The application of oil on the metal surface for corrosion protection and its removal can be avoided. Due to rapid volatilization of the VCI after unpacking, no costly cleaning of parts prior to further use is required. In a new approach for evaluation of the protective effect of VCI, a metallic sample is exposed in a small chamber filled with air alongside a VCI-containing product. The air is humidified and the sample is cooled down under dewpoint, which leads to condensation of humidity and represents corrosive stress. The state of the metallic surface under the condensed water film is characterized by electrochemical methods. The present work demonstrates the evaluation of the anticorrosive effect of various VCI-films on mild steel. In summary, the application-oriented test allows for quick and quantified evaluation of VCI-containing products with a tabletop device.

## 1. Introduction

The globalized, specialized and liberalized economy of modern times comprises a global flow of commodities. Semi-finished products and high-quality goods are not only transported from manufacturers and customers, but the transport chains also include subcontractors, intermediaries and distributers. The immense flow of goods is reflected by a value of exported goods of 1.4 trillion $ for the German economy in 2017 [1]. According to the commodity index, approx. 57 % of these goods, still equaling a worth of 820 billion $, are typically made from or at least include metallic materials. During transportation by land, sea and air, even appropriately packaged commodities may meet harsh conditions in terms of temperature and humidity. Especially for products, which do not face intensive corrosive stress during service and are therefore neither made from corrosion-resistant materials nor adequately protected by permanent coatings, the corrosive stress imposed during transportation may cause severe damage. Also, with regard to long-term storage of such parts temporary protection is required to avoid corrosion damage of the bare metallic surfaces. Temporary corrosion protection is classified in the following categories: (1) application of a temporary coating based on organic solvent, water, oil or wax; (2) use of a desiccant within a packaging, which absorbs water vapor and therefore prevents corrosive stress imposed by humidity; (3) use of volatile corrosion inhibitors (VCI), which are set free from the packaging material itself (films, paper) or from a diffuser placed within a packaging.

The major drawback of temporary coatings is the expense for application of these coatings before shipment/storage and the subsequent removal prior to use. The VCI method does not require any operations on the metallic surfaces, but is able to deliver corrosion protection especially under climatic stress. The mode of action includes:

1 the release of the volatile corrosion inhibitors from the packaging material or from a VCI-containing material placed within the packaging,
2 the diffusion of the inhibitors in the atmosphere within the packaging,
3 the adsorption of the inhibitors on the metallic surface and the interaction of the inhibitors with the primary oxide layer of the metal by:
 3.1 formation of a passivating electrolyte film on the surface,
 3.2 formation of an electrolyte film in which the primary oxide layer is hardly soluble,
 3.3 formation of solid compounds or complexes by a reaction of the inhibitor and the base metal and/or
 3.4 formation of a conversion layer.

When corrosive stress occurs, the degradation of the primary oxide layer is thus impeded and the corrosive attack is slowed down significantly or inhibited completely. The corrosion protection is provided independently of the parts´ geometry, since contact between the part and the VCI material is not mandatory. The costly application and removal of temporary coatings and the disposal of any additionally required cleaner is avoided. After unpacking of the parts, the inhibitors volatilize rapidly. Therefore, no cleaning is required prior to further use. Further information on the VCI method and current developments is given in [2-8].

Appropriate evaluation of VCI-containing products is required both for quality control and for research and development. The testing of VCI-containing products with regard to inhibitor content and the corrosion protection effect is classified in three major categories:

1 chemical analysis
 - of a VCI-containing product
 - of an atmosphere, in which a VCI-containing product was exposed
 - of a metallic surface, which was exposed to an atmosphere alongside a VCI-containing product
2 climatic stress test of a metallic sample in a packaging consisting of or enclosing a VCI-containing material
3 electrochemical measurement of a metal surface in an aqueous solution including VCI or in a solution obtained from elution of a VCI-containing product

The chemical analyses give information on the quality and quantity of inhibitors, which are contained in the product or which are set free to the atmosphere, depending on the test setup. However, the actual corrosion protection ability of the product is not considered and the analyses are rather costly in terms of time, required equipment and expertise. Climatic stress tests focus on the actual corrosion protection effect of VCI under the respective conditions. However, they are prone to errors and deviation when carried out in different institutions or by different workers, since subjective visual assessment of the metallic surface after the test almost always provides the basis for the evaluation of the VCI containing product. Further, alternating climate tests are only fully comparable for metallic samples of equal mass, since the thermodynamic inertia of the sample is the ruling factor for condensation of water on the sample and thus for the corrosive load. While electrochemical measurements may provide quantitative results on inhibition effects, the information obtained with this approach is very often misleading. As a matter of fact, the mode of action of the corrosion inhibition under atmospheric conditions, which apply during service, is substantially different to the mode of action in aqueous solution in a conventional electrochemical cell setup.

The present study focusses on a new test method, which uses electrochemical measurements to characterize the surface state of a metallic sample in a test setup, which replicates the VCI mode of action in a more accurate way. A metallic sample is exposed in a small test chamber. A VCI-containing is also placed in the chamber. After a defined amount of time, during which inhibitors volatilize from the product, reach the metallic sample and there interact with the primary oxide layer, a corrosive stress is applied by humidifying the atmosphere in the chamber and by actively cooling down the sample below the dew point. Electrochemical measurements are carried out under the film of condensed water. The phenomenological test consequently reflects the VCI mode of action, since it covers the release of the inhibitors from the product, the transport to the metallic surface by diffusion, the interaction with the surface and the ability to impede corrosion under a reproducible and defined climatic stress. With regard to the electrochemical measurements, numerous studies have established correlations between values obtained from direct current (d.c.) or alternating current (a.c.) electrochemical methods and the state of a passive or corroding surface. Polarization of a sample with d.c. voltage or application of a voltage sweep will result in high current densities in case of a high surface activity. In an equilibrium state, this almost certainly indicates corrosion under the considered test conditions (only in non-equilibrium state, high activity may shortly occur during passive film formation). In contrast, current densities will be low in case of a passive (non-corroding) surface state. A.c. methods like electrochemical impedance spectroscopy provide further insights, since they include information on the amplitude ratios of potential and current signals as well as on the phase shift between said signals. By varying the excitation frequency, the behavior of the system can be characterized with regard to different time domains. For corrosion of metals in electrolytes with rather low ion concentration, the system behavior is often capacitive due to the dominating double layer of charge carriers at the metal/electrolyte interface. It has been established that a high resistance against charge transfer across this interface and a low capacitance correlate with a passive sample state, while in case of ongoing corrosion, the resistance against charge transfer across the interface is rather low and the capacitance is rather high [9-12]. Only when considerable amounts of corrosion products are formed and block the surface, the resistance against charge transfer will take rather high values although corrosion occurs.

Hereinafter, the use of the developed test method and apparatus to evaluate a VCI-containing product by an electrochemical measurement under a film of condensed water will be demonstrated. A VCI film is assessed with regard to its ability to protect a mild steel surface against corrosion under climatic stress to prove the suitability of the test for quality inspection as well as research and development of new VCI-containing products.

## 2. Experimental and materials

### *2.1. Sample material and test arrangement*

Metallic samples with a diameter of 16 mm and a height of approx. 10 mm were made from steel 235JRG. The nominal composition of the material is given in Table 1. Sample preparation included grinding the upper surface of the cylindric samples with SiC-paper (600 grit), rinsing with deionized water, rinsing with ethanol and drying under warm air. Afterwards, the samples were stored in a desiccator for at least 3 days. As a VCI-containing material, a commercially available film was used. The film was made from polyethylene and had a thickness of 100 μm. The composition of the film is not disclosed for reasons of confidentiality.

**Table 1.** Nominal composition of the sample material S235JRG.

| Element | C | Mn | P | S | Fe |
|---|---|---|---|---|---|
| Concentration in wt-% | <0.2 | <1.4 | <0.045 | <0.045 | balance |

The test arrangement is schematically shown in Figure 1 and Figure 2. The test apparatus is shown in Figure 3. The test setup consisted of a cylindrical test chamber. The chamber was made from poly(methyl methacrylate) (PMMA-tube) and polyoxymethylene (POM-lid). The volume of the chamber was approx. 4.5 l. The VCI film was brought into the chamber so that both the inside of the cylindrical tube and of the lid was covered. The bottom of the chamber consisted of 2 aluminum

IOP Conf. Series: Materials Science and Engineering **480** (2019) 012029 doi:10.1088/1757-899X/480/1/012029

plates, which were insulated thermally and electrically by POM. A separate peltier element was mounted to each of the aluminum plates. One of the plates and the POM formed a reservoir, in which water was injected during the course of the test by a peristaltic pump. By heating of the aluminum plate under the water reservoir via the attached peltier element, the atmosphere in the chamber was humidified. Also, an extruded fin-shaped heatsink was mounted on the aluminum plate. Thus, by cooling of the plate via the peltier element, humidity condensed on the fin-shaped heatsink and the atmosphere in the chamber was dehumidified. On the other aluminum plate, a sample holder was placed, which contained the prepared steel sample. The sample holder consisted of polyamide PA12 and was epoxy-coated. The sample holder had an aperture, under which the grinded surface of the metallic sample was positioned. A ring seal was inserted between said surface and the sample holder and thus the active sample surface was defined. Above the aperture, a platinized titanium mesh was positioned in a distance of approx. 2 mm to the sample surface. The mesh served as counter electrode for the electrochemical measurement. A platinum wire was also positioned above the aperture so that its tip had a distance of approx. 1 mm to the sample surface. The wire served as a pseudo-reference electrode during the measurement. The pseudo-reference electrode was chosen to avoid contamination of the system by the use of conventional reference electrodes containing corrosion-promoting ions like chloride, sulfate or the like. The sample holder was such that the sample was thermally connected to the aluminum plate and thus to the peltier element. By cooling via the peltier element, the sample temperature could be set below the dew point of the atmosphere in the chamber so that humidity condensed on the sample surface. This simulates the condensation of water, which typically occurs during transportation when cold parts are brought into a heated room from the outside or during night/day cycle. Additionally, an auxiliary profile made from extruded, anodized aluminum was placed above the aperture in a distance of approx. 10 mm. The profile was thermally connected to the same aluminum plate as the sample and thus to the peltier element. While humidity condensed on the sample, it would therefore also condense on the auxiliary profile. From there, a small amount of condensed humidity would drop on to the sample. This simulates so-called container sweat, which typically forms during transportation when the whole packaging and its included atmosphere is cooled down from higher to lower temperatures and moisture precipitates. In that case, the contained humidity condenses inside the packaging and drops onto the packaged goods. After the end of the test, the metallic sample was dried outside the chamber under a stream of warm air.

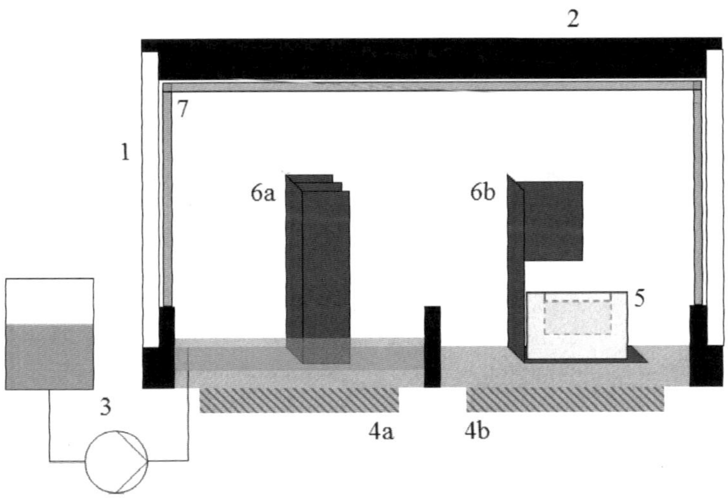


**Figure 1.** Principal test arrangement with PMMA cylindrical tube (1), POM lid (2), water reservoir with pump (3), two separate peltier elements mounted below aluminum plates (4), sample holder with sample (5), two fin-shaped aluminum heatsinks (6) and VCI-film (7).

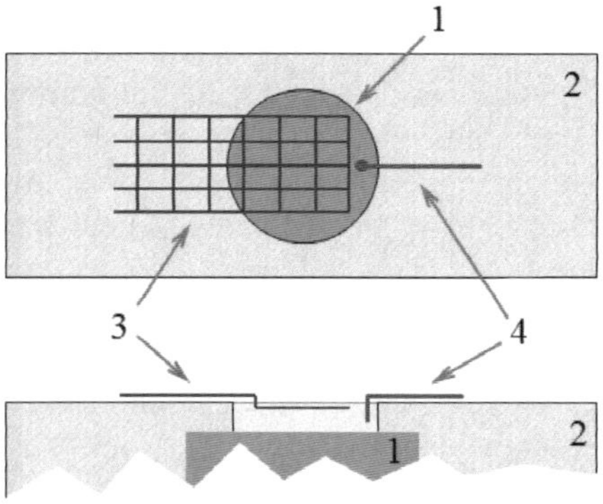


**Figure 2.** Principal layout of the sample holder (2) with sample (1) and auxiliary electrodes, namely a platinized titanium mesh (3) and a platinum wire (4) acting as counter and reference electrode, respectively.

**Figure 3.** Overview image of the tabletop device, which contains the test arrangement and a user interface.

*2.2. Test regime and electrochemical measurements*

The requirements with regard to the test regime were as follows: (1) simulate critical climatic stress imposed on metallic goods during transportation, (2) ensure wetting of sample surface and auxiliary electrodes to allow for electrochemical measurement during some state of the test, (3) impose sufficient stress to significantly corrode the sample in absence of any appropriate VCI material, (4) allow for differentiation of VCI-containing products with regard to their corrosion protection effect, (5) allow for short-term examination of the corrosion protection effect of VCI-containing products (preferably within 1 working day). The test regime as well as the principal layout of the apparatus to realize the test regime, which is disclosed in section 2.1, have been the subject of extensive preliminary investigation.

The principal course of the test with regard to the conditions in the test chamber is shown in Figure 4. At the beginning of the test, the chamber as well as the aluminum plates are at room temperature. In the first stage of the test (2 h), both the peltier elements are kept in passive state, i.e. their temperature is not regulated, water is pumped into the chamber and humidifies the atmosphere. Meanwhile, inhibitors are set free from the VCI-containing product and may diffuse in the chamber and reach the metallic sample surface. In Stage 2, a short corrosive load is applied. The sample in the sample holder is cooled down 3 K below the dew point for 15 min. A small amount of humidity condenses on the surface. The short duration of the cooling phase is neither sufficient to wet all electrodes nor does additional condensed humidity drop on the sample from the auxiliary aluminum profile, which is thermally connected to the sample. The metallic sample is afterwards heated 5 K above the dew point for 15 min. Meanwhile, the water reservoir is emptied and the corresponding aluminum plate with the fin-shaped heat sink is cooled down to 10 °C so that the air in the chamber is dehumidified to ensure that the small amount of condensed humidity on the metallic sample evaporates quickly and completely. Stage 2 thus simulates a mild corrosive stress by short-time condensation of humidity and subsequent drying and is carried out thrice. In Stage 3 of the test, the peltier elements are in passive state again. Again, water is injected into the reservoir in the chamber and humidifies the atmosphere. Afterwards, stage 4 is entered, during which the main corrosive stress is imposed. Therefore, the aluminum plate below the filled water reservoir is heated to 60 °C, which leads to excessive evaporation and thus oversaturation of the atmosphere in the chamber with humidity. In the same time, the metallic sample and the attached auxiliary aluminum profile are cooled down 7 K below the dew point. This leads to severe condensation both on the metallic sample and the auxiliary profile. After approximately 30 min, condensed humidity on the auxiliary profile starts to drop on to the metallic sample. After approximately 45 min in stage 4, the metallic sample surface as well as the auxiliary electrodes are sufficiently wetted to allow for electrochemical measurements. Because of the heating of the water reservoir, the atmosphere in the chamber is also heated and reaches a temperature of approx. 35 °C. The corrosive stress in this stage originates from the condensation of humidity directly on the sample surface, which is below the dew point, and the condensed water dropping on the surface from the auxiliary fin-shaped profile (cf. container sweat).

In the standard procedure, the test is terminated after approx. 60 min in stage 4. This equals a total test duration of approx. 7 h. The corrosion protection ability of the VCI-containing product is assessed on the basis of the measured polarization resistance and/or impedance at this point of time. The described procedure is used for quality assessment of VCI-containing products. In a variation of the test for research and development purposes, stage 4 is prolonged until the electrochemical measurements indicate corrosion initiation (see section 3.2). The point of time in the test when corrosion initiation is indicated serves as another measure to quantify the corrosion protection ability of VCI-containing products.

Electrochemical measurements were carried out with a workstation IM6 (Zahner, Germany). The use of the platinum wire pseudo-reference electrode did not allow for a simple determination of the electrochemical potential of the sample. However, electrochemical impedance spectroscopy (EIS) and the determination of the polarization resistance of the system by a small-range potential-sweep do not require knowledge of the absolute potential, but are based on the evaluation of the change of current density relative to the change of the potential. EIS was carried out with an amplitude of 5 mV in a 10 Hz – 25 mHz frequency range. 4 frequencies were measured per decade. The impedance value at a particular frequency was averaged from 2 full sine-wave periods for each frequency. For reasons of stability and to avoid sample corrosion during the measurement, the absolute d.c. current flow between sample and counter electrode was kept below 150 nA at any time. The impedance spectra were fitted using Microsoft Excel VBA scripts. Polarization resistance measurements were carried out by applying a polarization of 0–9 mV (scan rate: 1 mV/s) relative to the equilibrium potential against the pseudo-reference electrode (open circuit). The small polarization avoids sample corrosion due to the measurement. In the mentioned potential range, the current-density-potential-behavior is approximately linear, so that the polarization resistance is obtained by determination of the mean inverse slope of the recorded curve in the measured range. By sequentially carrying out EIS and/or polarization resistance measurements, time-resolved information on the sample´s surface state was obtained.

IOP Conf. Series: Materials Science and Engineering **480** (2019) 012029 doi:10.1088/1757-899X/480/1/012029

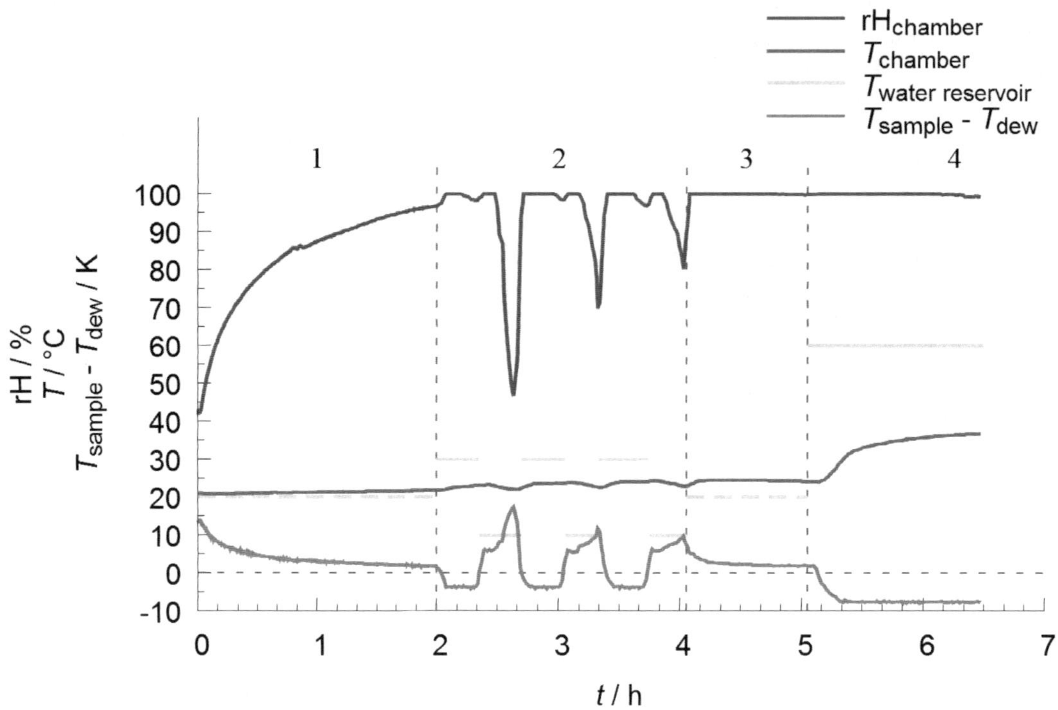


**Figure 4.** Evolution of relative humidity and temperature in the test chamber and of the sample temperature with respect to the dew point over the course of the test. Additionally, the set temperature for the water reservoir in the chamber is plotted (dashed line: peltier element below the water reservoir is in passive state, i.e. no temperature regulation). Corrosive stress is applied on the sample in test phases 2 and 4, when the sample is cooled below the dew point of the atmosphere in the chamber.

## 3. Results and discussion

### *3.1. Electrochemical behavior of the metallic sample in the presence of condensation water*

Electrochemical measurements under a film of condensation water are peculiar because of the very low electrical conductivity of the water film, which acts as electrolyte. Insight into the system behavior is gained by EIS. The system behavior is capacitive (negative phase angles) and is ruled by two distinguishable processes, dominating in the higher frequency range (greater than approx. 10 Hz) or in the lower frequency range (lower than approx. 10 Hz). By deliberate variation of the electrolyte conductivity and the corrosivity of the electrolyte (addition of small amounts of inhibiting or corrosion-promoting salts to the water film) and of the distances of the auxiliary electrodes to the sample surface, it was established that the high-frequency capacitive behavior is due to the low-conductive electrolyte, which acts as an imperfect dielectric between the sample surface and the auxiliary electrodes. Meanwhile, the low-frequency capacitive behavior is ruled by the charge transfer at the interface between the water film and the metallic sample´s surface. It therefore represents the metallic sample´s surface state with low capacitance and high charge transfer resistance in case of passivity and high capacitance and low charge transfer resistance in case of activity, i.e. corrosion.

Only in prolonged measurements with the occurrence of greater amounts of corrosion products, the charge transfer resistance will potentially increase again after some time due to blocking of the surface by corrosion products. With the formation of the corrosion products and their solution in the electrolyte, the conductivity of the electrolyte is also significantly increased up to the point where the high-frequency capacitive behavior vanishes completely, since the electrolyte no longer acts as an imperfect dielectric, but provides higher electrical conductivity. In comparison, the absorption of any VCI has a rather small effect on the electrolyte conductivity so that the high-frequency capacitive behavior can still be observed in presence of VCI-containing products in the test chamber. The validity of the d.c. measurement despite of the contribution of the high electrolyte resistance will be addressed in the discussion of the results in section 3.2. The assessment of the corrosion protection ability of VCI-containing products in d.c. (sweep) mode saves effort in terms of instrumentation of the test apparatus and analysis of the measured data. However, EIS measurements will potentially provide additional information depth for research and development of both the test method and VCI-containing products and therefore are carried out, too. Since only the low-frequency behavior reflects the sample´s surface state and allows to distinguish between the passive and the active, i.e. corroding, state, the impedance measurement is limited to this frequency range. Different electrolyte conductivities due to solution of corrosion products and/or small deviations of the distance between the electrodes are accounted for by normalizing the spectrum with regard to the obtained impedance at the highest measured frequency.

The resulting equivalent circuit for fitting of the impedance data consists of a normalized resistance (electrolyte resistance) in line with a parallel connection of another resistance (charge transfer at the interface between electrolyte and metallic sample surface) and an imperfect capacitance (constant phase element reflecting capacitive behavior at the interface between electrolyte and metallic sample surface with a pseudo-capacitance $Q$ and an exponent $n$). The exponent $n$, which is a measure for the imperfectness of the electrolyte/metal interface in terms of heterogeneous distribution of time constants (e.g. influenced by surface roughness, homogeneity of primary oxide layer etc.), is mostly in a range of 0.8–0.95 for the system under investigation. The transition from the passive state to the active, corroding state is denoted by values of the pseudo-capacitance of approx. $10^{-4}$ $F{\cdot}s^{n-1}{\cdot}cm^{-2}$. The value has preliminarily been established by stopping tests immediately after impedance measurements and examining the sample for traces of corrosion. For a pseudo-capacitance exceeding the mentioned value, corrosion spots are visible on the sample. For values just slightly exceeding $10^{-4}$ $F{\cdot}s^{n-1}{\cdot}cm^{-2}$, the corrosion spots are invisible to the bare eye, but can be identified using a light microscope with at least 5x magnification. In the same way, a decrease of the polarization resistance obtained from the DC (sweep) electrochemical measurement to values below approx. 30 $k\Omega{\cdot}cm^2$ indicates corrosion initiation, while values below approx. 5 $k\Omega{\cdot}cm^2$ indicate severe corrosion (either pitting or corrosion of large fractions of the surface). For support, Table 2 shows representative images of the sample surface and the obtained values from the EIS and polarization resistance measurements from preliminary experiments.

**Table 2.** Preliminary tests – correlation of the visual appearance of the active sample surface (diameter: 0.95 cm, contrast and color intensity increased by image manipulation), the pseudo-capacitance $Q$ and the polarization resistance $R_{pol}$ obtained from electrochemical measurements immediately before the end of the tests.

| 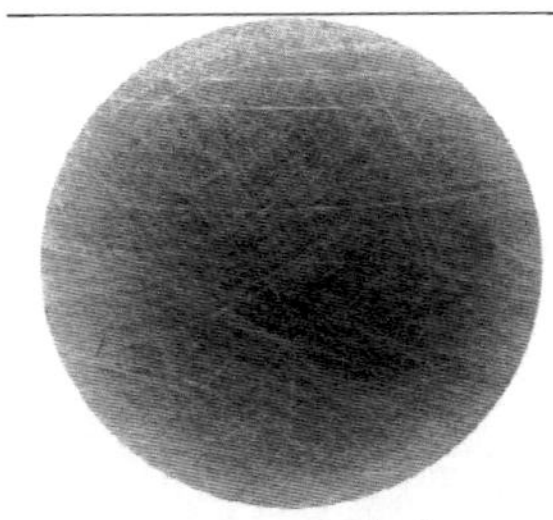 | 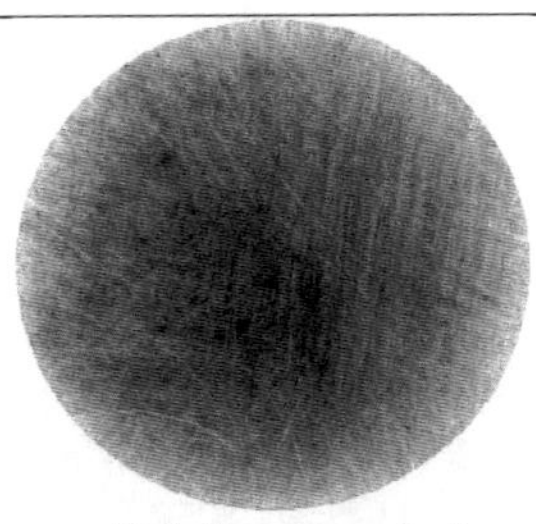 |  |  |
|---|---|---|---|
| not corroded | slightly corroded | moderately corroded | heavily corroded |
| $Q = 7{\cdot}10^{-5}$ $F{\cdot}s^{n-1}{\cdot}cm^{-2}$ | $Q = 1{,}2{\cdot}10^{-4}$ $F{\cdot}s^{n-1}{\cdot}cm^{-2}$ | $Q = 5{,}3{\cdot}10^{-4}$ $F{\cdot}s^{n-1}{\cdot}cm^{-2}$ | $Q = 1{,}8{\cdot}10^{-3}$ $F{\cdot}s^{n-1}{\cdot}cm^{-2}$ |
| $R_{pol} = 57$ $k\Omega{\cdot}cm^2$ | $R_{pol} = 23$ $k\Omega{\cdot}cm^2$ | $R_{pol} = 7$ $k\Omega{\cdot}cm^2$ | $R_{pol} = 0{,}1$ $k\Omega{\cdot}cm^2$ |

IOP Conf. Series: Materials Science and Engineering **480** (2019) 012029 doi:10.1088/1757-899X/480/1/012029

### *3.2. Assessment of the corrosion protection effect of a VCI-containing polyethylene film*

Figure 5 shows the evolution of the charge transfer resistance and the pseudo-capacitance obtained from EIS measurement during a climatic stress, which was prolonged compared to the original cycle (cf. Figure 4) by cyclically applying an increase and decrease of the sample temperature with respect to the dew point. While this is accompanied by evaporation and condensation of humidity and consequently increased corrosive stress in the respective cycles, the water film, which wets the sample surface and the auxiliary electrodes, stays intact during the rapid temperature variation and thus allows for continuous assessment of the sample′s surface state by electrochemical measurement. In the present case, both the polarization resistance $R_{pol}$ and the pseudo-capacitance $Q$ at the interface between the metallic sample and the film of condensed water indicate corrosion initiation during the first two of the additional temperature cycles. Additionally, the charge transfer resistance $R_{ct}$ obtained by fitting of the obtained impedance spectra with the described equivalent circuit (section 3.1) is shown alongside the polarization resistance in Figure 5. $R_{ct}$ shows larger fluctuation over time compared to $Q$, which is mainly due to the fact that $Q$ is of higher significance for the fit of the spectra. However, the values of $R_{ct}$ and $R_{pol}$ are in the same order of magnitude, which proves the validity of the d.c. measurement to obtain $R_{pol}$. The comparably large electrical resistance of the film of condensed water, which serves as the electrolyte, contributes to the overall polarization resistance $R_{pol}$, but not to the value of $R_{ct}$. The latter one is obtained only from the low-frequency range and thus exclusively represents the charge transfer behavior at the interface between the metallic sample and the film of condensed water. Since both are in the same order of magnitude, the polarization resistance is obviously not greatly affected by the low electrical conductivity of the water film under the considered circumstances.

Two major test modes are derived for the assessment of VCI-containing products, regardless of the carrier material (e.g. films, paper, diffuser, oils etc.):

1 The polarization resistance is measured once after 6 h in the presented climatic regime (Figure 4). The obtained value is used to determine, whether the sample is in a corrosion-active or in a passive (inhibited) state. Thus, the corrosion protection effect of the VCI-containing product is assessed. This method is especially suitable for quality inspection purposes, since the assessment can be carried out automatically with respect to the established limit value of the polarization resistance So, no electrochemical expert knowledge is required to evaluate a VCI-containing product as ok/not ok.

2 The test is prolonged by a number of cycles, which impose further corrosive stress on the metallic sample by variation of its temperature with respect to the dew point. The polarization resistance is measured repeatedly during the cycles and the moment of corrosion initiation as well as the severity of the corrosive attack can be evaluated. This allows for discrimination between different VCI-containing products, which pass test mode (1) and is especially suitable for research and development purposes.

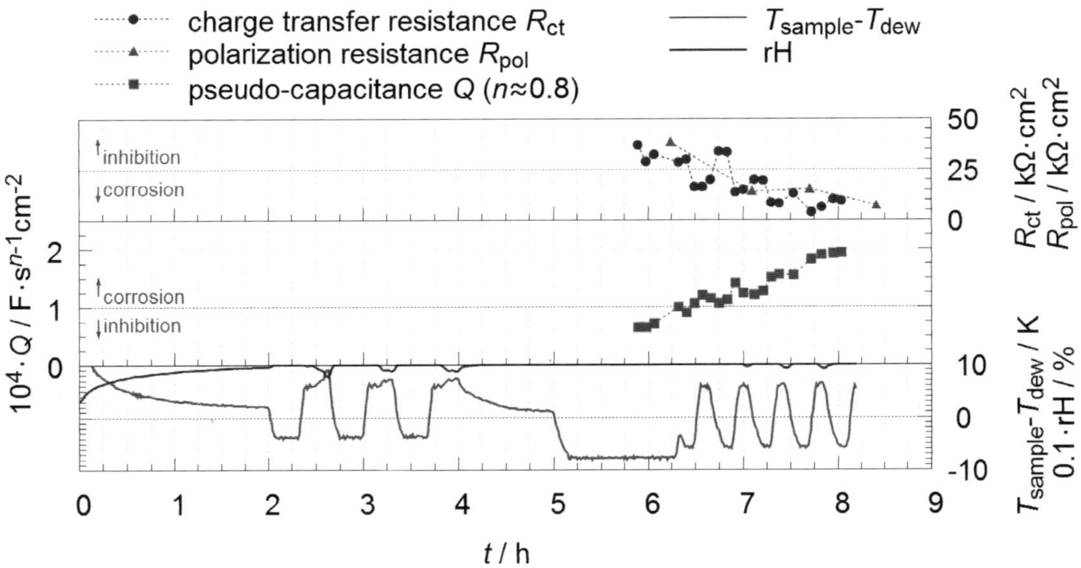


**Figure 5.** Evolution of the electrochemical parameters, namely the pseudo-capacitance $Q$, the charge transfer resistance $R_{ct}$ (both obtained from EIS) and the polarization resistance $R_{pol}$ (obtained from d.c. potential sweep) over the course of a climatic stress, which adds a cycled temperature variation to the regime shown in Figure 4. Comparison of both the obtained values for pseudo-capacitance and polarization resistance with the established limits for corrosion initiation (dotted lines) shows that corrosion is initiated during the first two additional temperature cycles. The increased activity at the interface between the metallic sample and the film of condensed water is reflected by decreasing resistance values and an increasing value of the pseudo-capacitance.

## 4. Conclusions

The VCI-method provides temporary corrosion protection during transport and storage of goods with metallic surfaces. A new test method and test apparatus for the assessment of the corrosion protection effect of VCI-containing products have been presented. The test includes exposition of a bare metallic sample to a defined volume of atmosphere within a test chamber alongside a VCI-containing product. During the course of the test, the metallic sample is cooled down below the dew point of the atmosphere. Consequentially, humidity condenses on the sample surface and imposes a corrosive stress. The sample´s surface state under the film of condensation water is evaluated objectively and quantitatively by electrochemical measurements. The values obtained from a.c. as well as d.c. electrochemical measurements correlate with the occurrence of corrosion on the metallic sample surface. For the evaluation of VCI-containing products for quality inspection or R&D purposes, a test regime and limit values of the polarization resistance and the capacitance at the metal/electrolyte interface have been established. The test arrangement is integrated in a tabletop device and allows for the assessment of the corrosion protection effect of VCI-containing products within a working day.

## Acknowledgments

The authors gratefully acknowledge the financial support of the project by the German Federal Ministry for Economic Affairs and Energy within the framework of the ZIM funding program.

## References

[1] German Federal Statistical Office: Exports and imports in 2017 (foreign trade), last accessed on 2018-08-09

[2] Subramanian A, Natesan M, Muralidharan V S, Baiakrishnan K and Vasudevan T 2000 An Overview: Vapor Phase Corrosion Inhibitors *Corrosion* **56** 144–155

[3] Kuznetsov Y I 2015 Progress in the science of corrosion inhibitors *Int. J. Corros. Scale Inhib.* **4** 15–34
[4] Andreev N N and Kuznetsov Y I 2012 Volatile Inhibitors of Metal Corrosion. I. Vaporization *Int. J. Corros. Scale Inhib.* **1** 16–25
[5] Andreev N N and Kuznetsov Y I 2012 Volatile Inhibitors of Metal Corrosion. II. Interaction of systems being protected with the environment and corrosion prevention conditions *Int. J. Corros. Scale Inhib.* **1** 146–153
[6] Koehler S and Reinhard G 2014 VCI containing package material – mode of functioning *Int. J. Corros. Scale Inhib.* **3** 286–306
[7] Ansari F A, Verma C, Siddiqui Y S, Ebenso E E and Quraishi M A 2018 Volatile corrosion inhibitors for ferrous and non-ferrous metals and alloys: A review *Int. J. Corros. Scale Inhib.* **7** 126–150
[8] Valdez B, Schorr M, Cheng N, Beltran E and Salinas R 2018 Technological applications of volatile corrosion inhibitors *Corros. Rev.* **36** 227–238
[9] Kim K T, Kim H W, Chang H Y, Lim B T, Park H B and Kim YS 2015 Corrosion Inhibiting Mechanism of Nitrite Ion on the Passivation of Carbon Steel and Ductile Cast Iron for Nuclear Power Plants *Adv. Mater. Sci. Eng.* **2015** 408138-1–408138-16
[10] Tan Y J, Bailey S and Kinsella B 1996 An investigation of the formation and destruction of corrosion inhibitor films using electrochemical impedance spectroscopy (EIS) *Corros. Sci.* **38** 1545–1561
[11] Khaled K F and Abdel-Rehim S S 2011 Electrochemical investigation of corrosion and corrosion inhibition of iron in hydrochloric acid solutions *Arabian J. Chem.* **4** 397–402
[12] Popova A, Sokolova E, Raicheva S and Christov M 2003 AC and DC study of the temperature effect on mild steel corrosion in acid media in the presence of benzimidazole derivatives *Corros. Sci.* **45** 33–58

IOP Conf. Series: Materials Science and Engineering **480** (2019) 012030 doi:10.1088/1757-899X/480/1/012030

# Hydroabrasive wear on high carbide infiltration materials

**V Wesling*, R Reiter and T Müller**

Clausthal University of Technology / Institute for Welding and Machining

*e-mail: office@isaf.tu-clausthal.de

**Abstract.** Hydroabrasive wear, a subtype of erosion, is a frequently occurring wear phenomenon. The type and extent of the damage depend on a large number of material characteristics and process conditions. In this research project, high carbide infiltration materials used in deep drilling technology were tested for their resistance to this erosive stress. The flow angle was varied stepwise between 15° and 90°. The resulting damage was quantified by volumetric material loss. Furthermore, the wear mechanisms were determined by surface examinations using SEM. The result of this project is an improved understanding of the effect of different process and material parameters on the tribological system. An improved knowledge of the critical quantities will allow the estimation of the effects of changes in tribological systems under erosive load and finally a reduction of material removal in process plants by adaptation of single variables.

## 1. Introduction

Wear is a complex system property that is influenced by many parameters. Erosion wear, caused by a particle-laden flow, is a frequently occurring wear phenomenon. A subtype of erosion is hydroabrasion, in which the particles are carried in a water flow. The damaging effect is caused by the impacting particles on the surface, whereby the damaging effect is decisively dependent on the angle of impact and the properties of the sample material. Hydroabrasion occurs in a large number of technical applications, for example in pipeline construction, hydropower turbines or in deep drilling technology. The type and extent of the damaging effect depend on a large number of material characteristics and process conditions.

One of the most important parameters of erosive wear is the angle at which the particle-laden flow hits the surface of the material. The impulse acting on impact can be divided into two components, one perpendicular and one parallel to the surface. The different wear mechanisms act differently depending on the amounts of the respective components. At a low angle of impact, and thus a very high parallel impulse component, the affected material is mainly removed by micro-chipping or microploughing. At very high angles of impact, the vertical impulse component predominates, whereby the surface is removed by fatigue wear. The result is subsurface cracking at the location of maxium hertzian pressure parallel to the surface that leads to plate-shaped breakouts. Between these extreme cases, the mechanisms mentioned above have a superimposing effect and micro-chipping and surface breakdown can be observed at the same time. In addition, ductile materials are abraded more quickly by micro-chipping, whereas hard and brittle materials tend to be abraded by breaking or shattering [1].

## 2. Experimental and materials

### *2.1. Materials*

Within the frame of these investigations, two rolled steel materials and four infiltration materials used in deep drilling technology with a high carbide content were investigated. The steel materials are S355 and the highly wear-resistant steel Creusabro 8000 [2]. Its structure consists of martensit, bainite and retained austenite, which promises an increased wear resistance.

The infiltration materials, Materials 01 - 04, are pseudo alloys of a high copper content binder with embedded tungsten carbides. The resistance to abrasion is produced by the carbides, while the ductile matrix has to withstand shock loads. In the production of these materials [3], a powder mixture of the carbides is given in a graphite form. The remaining matrix material, the low-melting solder, is placed on top. When this arrangement is heated in an inert gas atmosphere, the liquid solder infiltrates into the capillaries of the fixed bed, where it bonds with the powder contained therein. By the superposition of gravity, adhesion and cohesion of the liquid metal, low pore densities can be achieved. In the case of these infiltration materials, a binder of copper, approximately 30 wt.% nickel, 22 wt.% manganese and a small percentage of zinc was used.

### *2.2. Hydroabrasive wear testing*

The testing is carried out in a test rig (Figure 1) specially developed by the Institute of Welding and Machining (ISAF). This test bench enables the testing of components and material samples at different flow angles. The abrasive material is introduced into the highly accelerated water jet in front of the jet nozzle, which accelerates it in the direction of the sample. The tests were conducted with the set of parameters listed in Table 1.

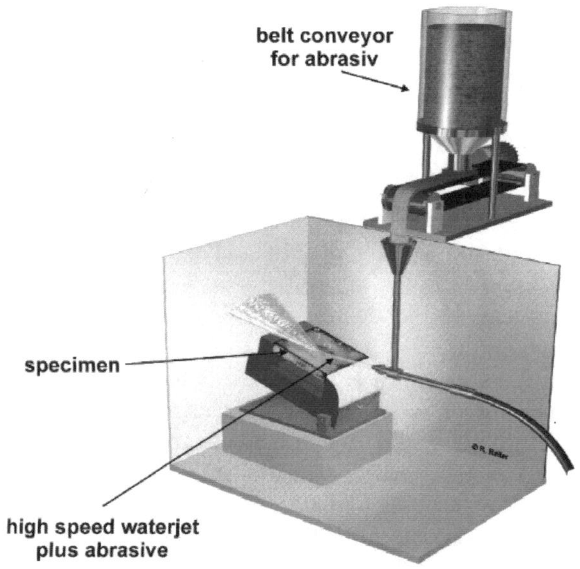


**Figure 1.** Principle of the test rig.

IOP Conf. Series: Materials Science and Engineering **480** (2019) 012030 doi:10.1088/1757-899X/480/1/012030

**Table 1.** Testing parameters.

| Testing parameter | Variations |
|---|---|
| Abrasive | Broken quartz sands |
| Particle sizes | 0 µm – 1000 µm |
| Mass flow | 150 g/min |
| Angle of impingement | 15°, 30°, 45°, 60°, 75°, 90° |
| Flow velocity | 95 m/s |

### *2.3. Metallographic examination methods*

Specimens of the infiltration materials were embedded in a warm embedding agent, grinded and polished in a Mecatech polishing maschine with diamond abrasive past (6 µm, 3 µm and 1 µm) and finally polished with OP-S suspension (0,04 µm). The light microscopic examinations were conducted with a Leica DM6 M. Further information about the material behaviour is provided by the investigations of the surfaces in the SEM (CamScan Serie 4, acceleration voltage 10 kV, working distance 35 mm). The specimens were cleaned after the experiment and vapor-plated with gold to prevent charging of the embedded quartz particles.

## 3. Results and discussion

In this chapter, the results of the tests of S355, Creusabro 8000, and Material 01 - 04 are presented with angles of impingement of 30°, 60° and 90°. Testing with these angles allows an assessment of the resistance of the materials to the different wear mechanisms. In order to ensure the comparability of the results, the volumetric material loss is considered.

### *3.1. Microstructural characterization of the infiltration material*

Material 01 (Figure 2) and Material 02 (Figure 3) contain block-shaped fused tungsten carbide (FTC) and tungsten monocarbide (WC). These materials differentiate in the size of the FTC and WC, and in the mixing ratio of the two carbide types. Material 02 has a higher proportion of WC. Material 01 has 35% WC and 25% FTC, while Material 02 has 40% WC and 15% FTC.

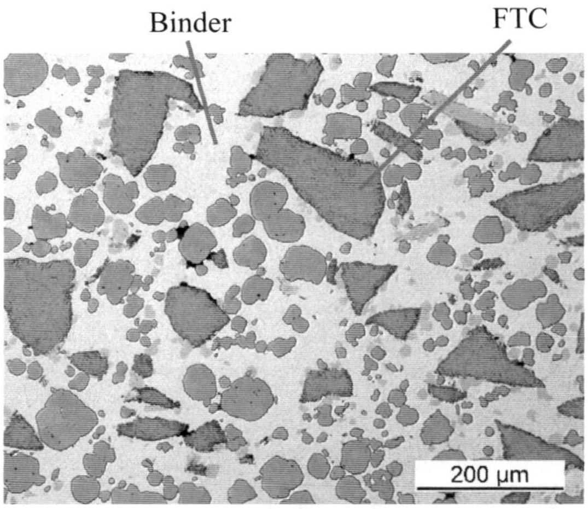


**Figure 2.** Material 01.

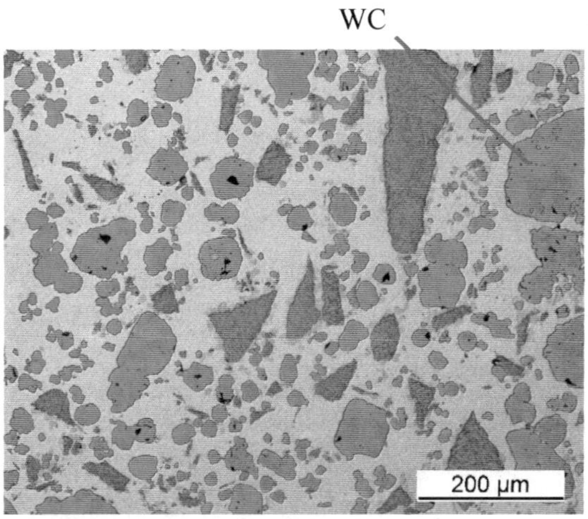


**Figure 3.** Material 02.

IOP Conf. Series: Materials Science and Engineering **480** (2019) 012030 doi:10.1088/1757-899X/480/1/012030

Due to its inner structure, which consists of fine WC and $W_2C$ lamellas [4], the FTC has a higher material hardness and fracture toughness than the WC. The reinforcement in Material 03 (Figure 4) only consists of block-shaped FTC, with a degree of filling of 45%.

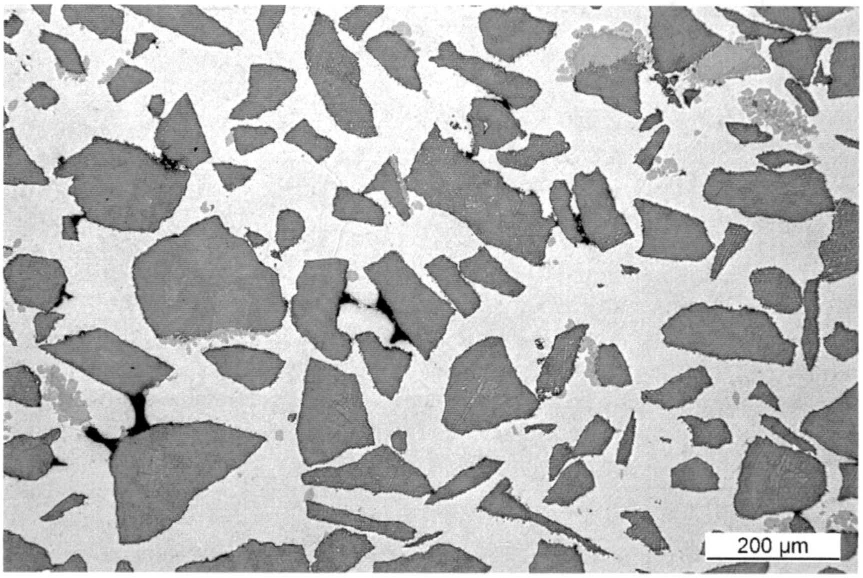


**Figure 4.** Material 03.

The Material 04 (Figure 5) also uses a multimodal reinforcement consisting of FTC and WC. There are fine-grained WCs in the binder between the large FTCs. The large FTCs should withstand the abrasive attack of the larger particles, while the fine-grained WCs should prevent the binder from being washed out. The reinforcement contains 25% WC and 35% FTC.

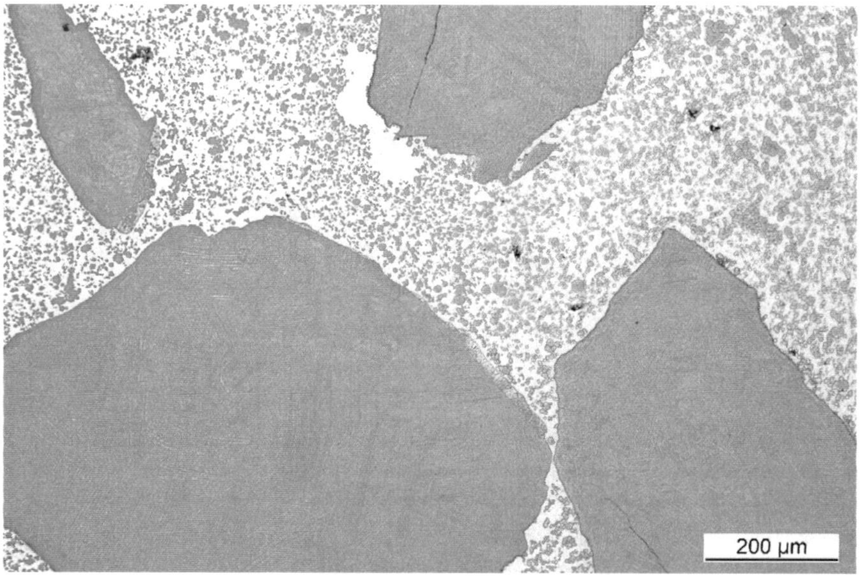


**Figure 5.** Material 04.

### *3.2. Angle of impingement 30°*

At this angle of impingement (Figure 6), Material 03 and 04 have the lowest volumetric loss. The highest material loss can be found at S355 and Material 02. The graphs of the materials Creusabro 8000 and Material 01 are close to each other.

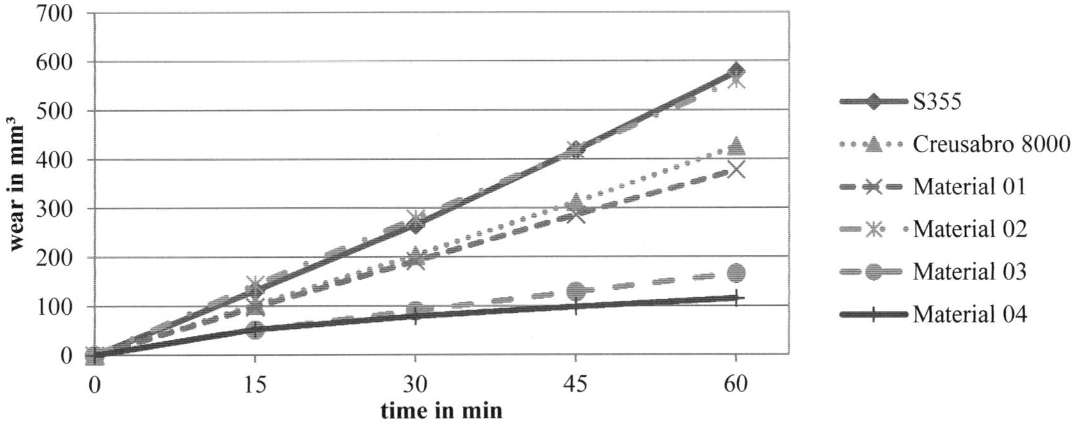


**Figure 6.** Volumetric wear, angle of impingement 30°.

### *3.3. Angle of impingement 60°*

As the angle of impingement increases (Figure 7), the wear of infiltration materials increases. Material 02 has such a high wear rate that the test had to be stopped after 45 min due to the failure of the specimens. The volumetric wear of Material 04 is also the lowest here. The material loss of Material 03 is the second lowest, whereby the difference to the wear of Creusabro 8000 is considerably smaller than at the angle of impingement of 30°.

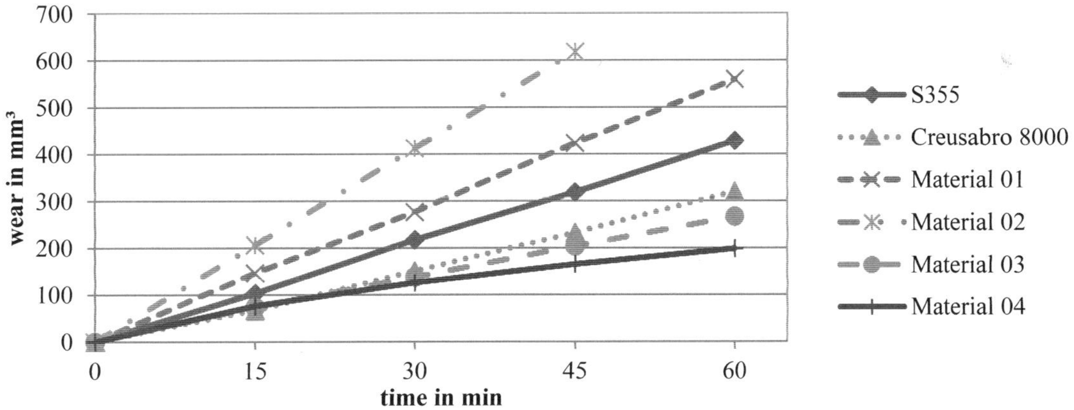


**Figure 7.** Volumetric wear, angle of impingement 60°.

### *3.4. Angle of impingement 90°*

In the case of impact wear (Figure 8), the materials S355, Creusabro 8000 and Material 03 are close to each other. Material 02 shows the worst wear behaviour in this case. The lowest material loss was determined here at Material 04. Additionally, the Material 04 shows a higher wear rate in the first testing interval. The wear rate drops from 4.3 $mm^3\ min^{-1}$ in the first interval to 2.0 $mm^3\ min^{-1}$ in the last interval.

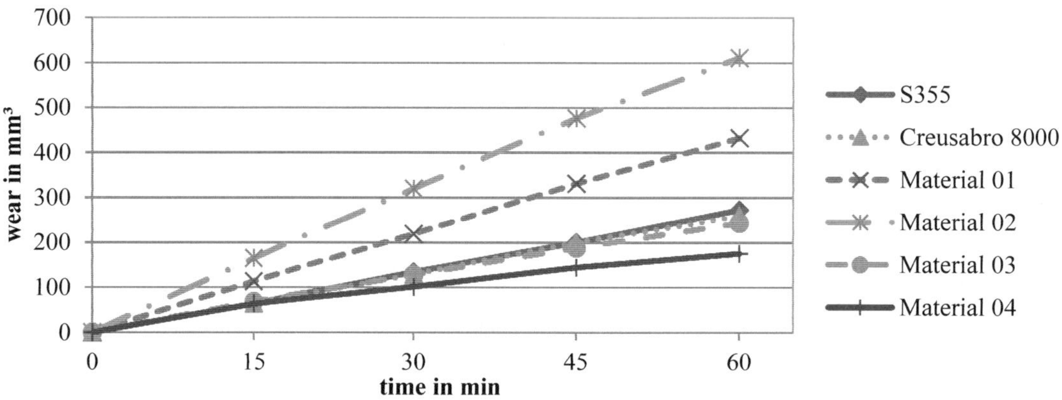


**Figure 8.** Volumetric wear, angle of impingement 90°.

### *3.5. Surface characterization after abrasive wear*

The surface of a sample made of S355 (Figure 9) shows clear traces of the micro-chipping after the wear test. Occasional deformations of the surface and embedded quartz particles can be observed.

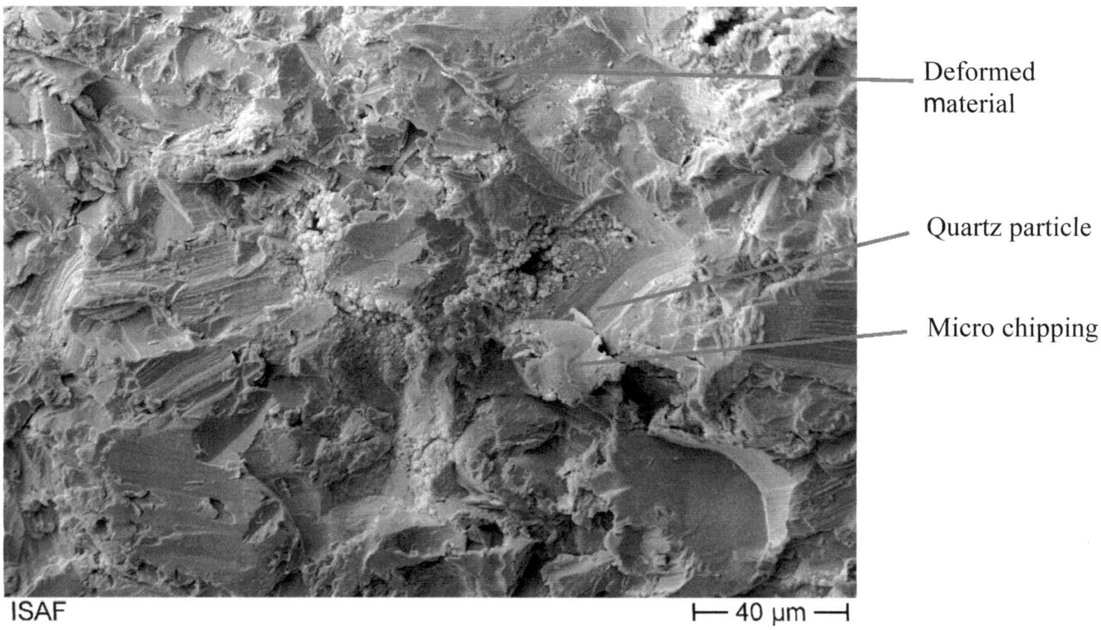


**Figure 9.** S355, SEM, angle of impingement 60°.

The examination of Material 01 (Figure 10) shows more signs of reshaping and shattering, but also isolated traces of micro-chipping. Furthermore, glass-fractured edges, surfaces and even splinters can be detected. This mainly concerns the brittle WC, which breaks very quickly on impact of quartz particles. The larger hard phases at the left edge of the image are large FTC particles. These have no visible cracks.

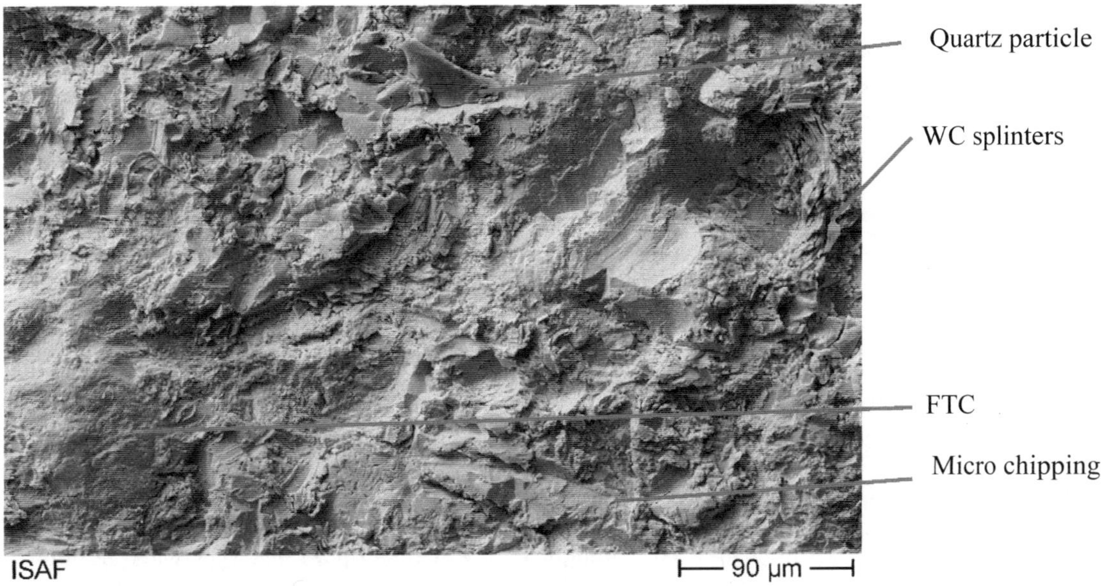


**Figure 10.** Material 01, SEM, angle of impingement 60°.

*3.6. Comparison of wear rates*

Since not all specimen showed linear wear behaviour over time and premature specimen failure occurred, the wear rates are considered in the following. These wear rates were calculated over the last intervals of the respective samples, which have a comparable gradient. The materials S355 and Creusabro 8000 show a significantly higher wear rate at a low angle of impingement. It can therefore be deduced that these materials are more susceptible to micro-chipping and ploughing, but can withstand surface shattering or deformation well. The high carbide materials have the highest wear rates at a medium angle of impingement. This proves a lower susceptibility to micro-chipping. However, the different behaviour of these materials in the case of pure impact wear indicates different resistance to surface shattering.

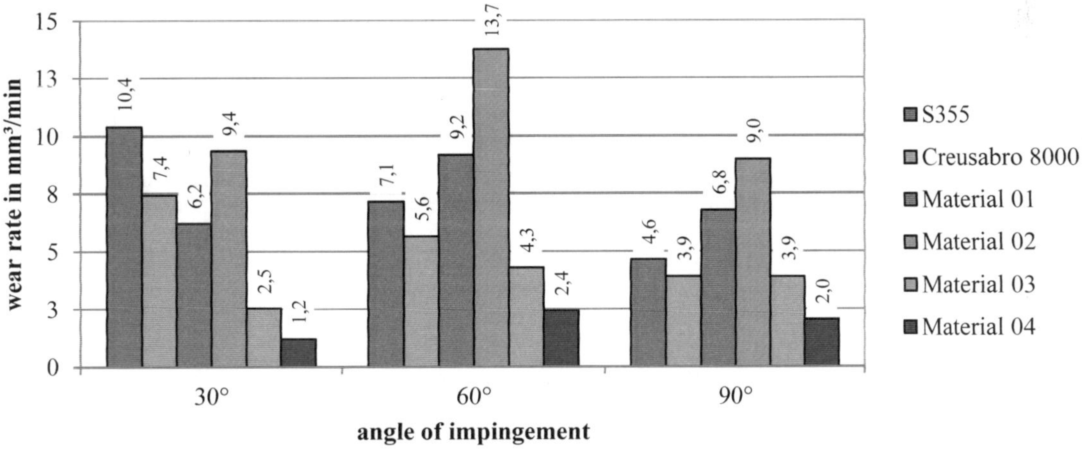


**Figure 11.** Wear rates at different angles of impingement.

Material 01 and 02 show higher wear rates than Material 03 and 04. This can be explained by the different fracture susceptibility of the carbides. The large WC particles in the materials tend to a brittle

IOP Conf. Series: Materials Science and Engineering **480** (2019) 012030 doi:10.1088/1757-899X/480/1/012030

fracture behaviour, while the FTC particles have a significantly higher fracture toughness. Therefore, the WC particles of the materials Material 01 and Material 02 do not offer any relevant resistance to the impacting quartz particles, which means that the binder can be removed more quickly by the erosive stress. On the other hand, the small WC particles in Material 04 protect the binder from being washed out by the small quartz particles in the water jet. Due to the lower particle size, their tendency to break decreases. At the same time, the impacts of the larger quartz particles are absorbed by the FTC particles, which leads to lower wear rates of the Material 03 and Material 04.

Overall, the Material 04 shows the highest resistance at all flow angles.

## 4. Conclusions

At low angles of impingement, highly carbide-containing infiltration materials such as Material 03 and Material 04 show a higher resistance to hydroabrasive stress than steels such as Creusabro 8000. However, the precondition for this is that sufficiently fracture-resistant hard particles such as FTC are used as reinforcement. When more brittle hard particles such as WC are used, the breakage of the carbides by impacting particles reduces the wear resistance significantly. The exception are fine particles, as in Material 04, which lie between larger carbides, as these do not have to endure high impact loads. Only Material 04 was able to show a significantly lower wear rate than S355 in the high angle of impingement range. In this case, the strengths of the different hard materials were combined in a multimodal reinforcement, which increased the resistance to the predominant wear mechanisms at all investigated angles.

## References

[1] Wesling V, Reiter R and Oligmüller J 2008 Untersuchungen zum Erosionsverschleiß an schweißtechnisch hergestellten Beschichtungen *Materialwissenschaft und Werkstofftechnik* WILEY-VCH Verlag GmbH & Co. KGaA 83-87

[2] Industeel ArcelorMittal 2016 Creusabro® 8000: a high performance wear and impact resistant steel datasheet

[3] Gerk C, Yan-Gerk L, Wesling V and Reiter R 2016 Highly carbide filled composite materials for the mining and drilling industry *Materials Science and Engineering* **118**

[4] Kiefer R and Benesovsky F 1963 Hartstoffe *Springer Verlag*

[5] Wesling V, Reiter R, Hamje J and Müller T 2018 Schweißtechnische Generierung von Verschleißschutzschichten *Abschlussbericht WE 2846/19-1* Deutsche Forschungsgesellschaft

[6] Padhy M K and Saini R P 2008 A review on silt erosion in hydro turbines *Renewable and Sustainable Energy Reviews* **12** 1974-1987

[7] Chauhan A K, Goel D B and Prakash S 2008 Erosion behaviour of hydro turbine steels *Bull. Mater. Sci.* **31** 115–120

[8] Momber A W and Kovacevic R 2003 Hydro-abrasive erosion of refractory ceramics, *Journal of Materials Science* **38** 2861–2874

[9] Neopane H P 2015 Sediment Erosion in Hydro Turbines *Renewable and Sustainable Energy Reviews* **49** 1100-1113

IOP Conf. Series: Materials Science and Engineering **480** (2019) 012031 doi:10.1088/1757-899X/480/1/012031

# Influence of the heat-treatment prior to plastic deformation on the aging behavior and the hardness of the aluminum alloy 6056

**L Winter[1,*], F Einer[1], C Geisler[2,3], R J Hellmig[2,3], K Hockauf[1] and T Lampke[1]**

[1] Materials and Surface Engineering Group, Institute of Materials Science and Engineering, Chemnitz University of Technology, D-09107, Germany
[2] Institute of Materials Engineering, University of Siegen, D-57068, Germany
[3] EJOT GmbH and Co. KG., Industrial Fasteners Division, Bad Berleburg, D-57319 Germany

[*] e-mail: lisa.winter@mb.tu-chemnitz.de

**Abstract.** Motivated by the improvement of the cost and resource efficiency of the manufacturing process of age-hardening aluminum components, the influence of an initial heat-treatment, prior to plastic deformation, on the artificial aging behavior and the hardness is investigated. Systematic work is done by comparing three initial heat-treatment conditions of the age-hardening aluminum alloy 6056 prior to low-temperature plastic deformation: a solid-solution heat-treated, a naturally aged and a stabilized one. Two plastic deformation processes, linear extrusion and compression, each with two different strains, respectively, were performed and followed by artificial aging. The hardness was measured prior to plastic deformation as a function of the natural aging time, and after plastic deformation as a function of the artificial aging time. Stabilization annealing of the aluminum alloy inhibited natural aging. After plastic deformation and artificial aging, the stabilized conditions showed an increased hardness, as compared to the solid-solution heat-treated, respectively naturally aged, condition. The artificial aging behavior of all tested conditions was only slightly influenced by the respective plastic deformation process. However, the hardness was similar for the extruded, respectively compressed, conditions in their respective initial heat-treatment conditions.

## 1. Introduction

Age-hardening aluminum alloys stand out for lightweight-applications, not only due to their high specific strength and good machinability, but especially due to their capability to adjust required mechanical properties by aging treatments. Thermomechanical treatment of these aluminum alloys allows to directly affect the precipitation kinetics of the formation of precipitates.

*Kim et al.* [1–3] developed an effective strategy for age-hardening alloys to achieve high strength and hardness, while preserving a good ductility by a combination of heat treatment and severe plastic deformation. Thereby, equal-channel angular pressing (ECAP) of the alloy leads to a significant increase in hardness and strength due to the introduced strain hardening and the high dislocation density. To compensate for the accompanying loss in ductility, the low-temperature deformation was realized subsequently to a solid-solution heat-treatment of the alloy. During the following artificial aging, the accelerated precipitation kinetics, which is caused by the high dislocation density, results in the formation of fine and finely dispersed precipitations which preserve the high strength of the deformed alloy. Simultaneously, thermal recovery leads to a regain of ductility (which has been decreased by work hardening). Nevertheless, the mechanical properties are strongly dependent on the introduced strain, the aging time and the artificial aging temperature [4,5]. *Hockauf et al.* [4] showed, that the strengthening effect of the precipitation hardening is overcompensated by the recovery process during artificial aging when a critical introduced strain is exceeded. Further, low temperatures for artificial aging and short aging times lead to the best resulting combination of strength and ductility [5,6].

The aspired main objective is a transfer of this method into industrial application to improve the cost and resource efficiency of the manufacturing process of components, which require a proper heat-treatment to realize precipitation hardening. The aim is a reduction of heat-treatment times and therefore reduced manufacturing costs by utilization of the accelerated precipitation kinetics due to the low-temperature plastic deformation. As "precipitation engineering" was also successfully performed with cryo-rolling as a severe plastic deformation process [7,8], we want to evaluate the adaptability of this method for the more conventional processes of linear extrusion and compression. To investigate the influence of the heat-treatment prior to deformation on the hardening behavior, a solid-solution heat-treated condition is compared with a naturally aged condition, which represents an industrial process with an interim storage period at room temperature. As a third option for heat-treatment, a stabilized condition is chosen. Stabilization annealing at temperatures below 100 °C for several hours after solid-solution heat-treatment leads to an increased room temperature stability and inhibits natural aging of the aluminum alloy [9].

The purpose of the present study is to systematically investigate the effect of the initial heat-treatment prior to deformation, the plastic deformation process and the artificial aging time on the hardness of an Al-Si-Mg-alloy containing copper. For our investigations, the aluminum alloy 6056 was chosen due to the possibility of natural and artificial aging and its pronounced hardening effect, the good workability and the wide usage in industrial applications.

## 2. Experimental and materials

For this study, the age-hardening aluminum alloy 6056 (AlSi1MgCuMn) was used. The wire-casted material with a diameter of 5.26 mm was provided by EJOT GmbH & Co. KG (Bad Berleburg, Germany). The chemical composition and the mechanical properties of the tested alloy are given in table 1 and table 2.

**Table 1.** Chemical composition of the aluminum alloy 6056 (AlSi1MgCuMn).

| Element | Si | Mg | Cu | Mn | Fe | Cr | Zn | Ti | Zr | Al |
|---|---|---|---|---|---|---|---|---|---|---|
| wt.-% | 1.06 | 0.89 | 0.78 | 0.59 | 0.16 | 0.01 | 0.19 | 0.04 | 0.13 | balance |

**Table 2.** Mechanical properties of the wire-casted aluminum alloy 6056. The standard deviation is given in absolute values.

| Wire diameter in mm | Yield strength $R_{p0.2}$ in MPa | Ultimate tensile strength $R_m$ in MPa | Elongation $A_{100mm}$ in % |
|---|---|---|---|
| 5.26 | 173 ± 1 | 188 ± 2 | 6.3 ± 1.3 |

For analyzing the natural aging behavior of the tested alloy, a solid-solution heat-treated and a stabilized condition were compared. The solid-solution heat-treatment was done at 530 °C for 1 h and quenching in water to room temperature (RT) afterwards. For stabilization annealing, the solid-solution heat-treated condition was heat-treated at 80 °C for 5 h, subsequently.

To examine the influence of the deformation process and the introduced strain on the artificial aging behavior, the tested aluminum alloy was extruded for $\varphi_E = 1.2$ and $\varphi_E = 0.8$, respectively compressed for $\varphi_C = 0.55$ and $\varphi_C = 0.7$. The deformation processes were carried out on three different heat-treated conditions: a solid-solution heat-treated, a condition which has been naturally aged for

IOP Conf. Series: Materials Science and Engineering **480** (2019) 012031 doi:10.1088/1757-899X/480/1/012031

7 days, and a stabilized condition. The parameters for the respective heat-treated and deformed conditions are given in table 3.

The extrusion process was carried out at EJOT GmbH & Co. KG (Bad Berleburg, Germany) at room temperature. The introduced strain by the extrusion process $\varphi_E$ is defined by the diameter reduction of the wire-casted material and is calculated from the ratio of the wire cross-section area before ($A_0$) and after ($A_1$) extrusion:

$$\varphi_E = \ln\frac{A_0}{A_1} \quad (1)$$

The compression process was performed at room temperature and at a strain rate of $10^{-3}\cdot s^{-1}$ in a Zwick Roell servohydraulic testing machine (Zwick Roell, Ulm, Germany). To minimize friction during the compression process, the cylindrical specimens (5 x 5 mm) were lubricated with molybdenum disulphide. The introduced strain by the compression process $\varphi_C$ is specified by the height reduction of the wire specimens and is calculated from the ratio of the specimen height before ($h_0$) and after ($h_1$) compression:

$$\varphi_C = \ln\frac{h_1}{h_0} \quad (2)$$

To prevent natural aging of the aluminum alloy, the extrusion, respectively the compression, and the following artificial aging treatment were done immediately after the respective initial heat-treatment. Otherwise, the specimens were stored at -7 °C, if a continuous route was unrealizable.

All plastically deformed conditions were artificially aged at 120 °C, with annealing times ranging from 1 to 600 min. Vickers hardness was measured on the cross section of the wire specimens with 20 indentations from edge to core. For the compressed conditions, an automatic hardness tester KB250BVRZ (KB Prüftechnik GmbH, Hochdorf-Assenheim, Germany) with Vickers hardness HV0.5 was used. The extruded conditions were measured at EJOT GmbH & Co. KG (Bad Berleburg, Germany) using an automatic hardness tester DuraScan 70 G5 (EMCO-TEST Prüfmaschinen GmbH, Kuchl, Österreich) with Vickers hardness HV0.1.

**Table 3.** Heat-treatment parameters for the initial material conditions of the wire-casted aluminum alloy 6056 used for the respective plastic deformation process.

| Deformation process | Heat-treatment parameters | | |
|---|---|---|---|
| | Solid-solution heat-treatment | Natural aging | Stabilization annealing |
| extrusion | 530 °C for 50 min | 7 days | 80 °C for 7 h |
| compression | 530 °C for 1 h | 7 days | 80 °C for 5 h |

## 3. Results

To illustrate the natural aging behavior as a function of the heat-treatment condition of the aluminum alloy 6056, figure 1 shows the development of hardness as a function of the natural aging time. The solid-solution treated condition exhibits an almost linear increase in hardness with increasing aging time. Starting at 68 HV0.5 after 90 min of natural aging time, the hardness increases to 91 HV0.5 after 4 weeks. In contrast, the stabilization annealing leads to a more stable, yet higher initial hardness of 80 HV0.5. However, after 3 weeks of natural aging time, the hardness increases as well (up to 86 HV0.5), but remains below the solid-solution treated condition.

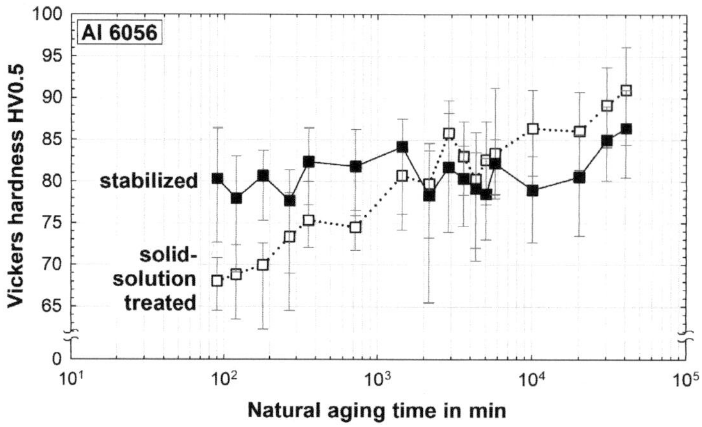


**Figure 1.** Vickers hardness (mean values) of the aluminum alloy 6056 in solid-solution heat-treated and in stabilized condition as a function of the natural aging-time. Scatter bars depict the maximum and minimum values. The stabilized condition shows a more stable hardness within the first weeks of natural aging time, as compared to the solid-solution treated condition.

The artificial aging behavior of the solid-solution treated, the naturally aged and the stabilized condition after extrusion, respectively compression, for the introduced strains is shown in figure 2. In general, the plastic deformation process and the introduced strain have a minor influence on the achieved peak-hardness for the respective heat-treatment conditions.

For all heat-treatment options, stabilization annealing of the aluminum alloy before plastic deformation results in the highest peak-hardness. Artificial aging for 300 min after stabilization annealing and extrusion leads to a peak-hardness of 162 HV0.1 for both introduced strains (see figures 2 a, b). For this aging time, the stabilized and compressed conditions also show a local hardness peak of 152 HV0.5, but the global peak-hardness of 156 HV0.5 is reached later, after 540 min of artificial aging time (see figures 2 c, d). However, the hardness profile of the stabilized and compressed conditions exhibits significant drops: for $\varphi_C = 0.55$ at 420 min of artificial aging time and for $\varphi_C = 0.7$ at 120 min and 360 min (see figures 2 b, d).

As compared to the stabilized conditions, the hardness of the naturally aged, respectively solid-solution heat-treated, conditions remain generally lower.

An acceleration of the artificial aging behavior by an increase in the introduced strain was only noticeable for the solid-solution treated and subsequently extruded, respectively compressed, conditions. Extrusion with $\varphi = 0.8$ after solid-solution treatment does not result in a peak-hardness within the chosen time for artificial aging of 360 min. In contrast, an increase in the introduced strain up to $\varphi_E = 1.2$ leads to an accelerated hardening and to a pronounced peak already after 180 min of artificial aging time.

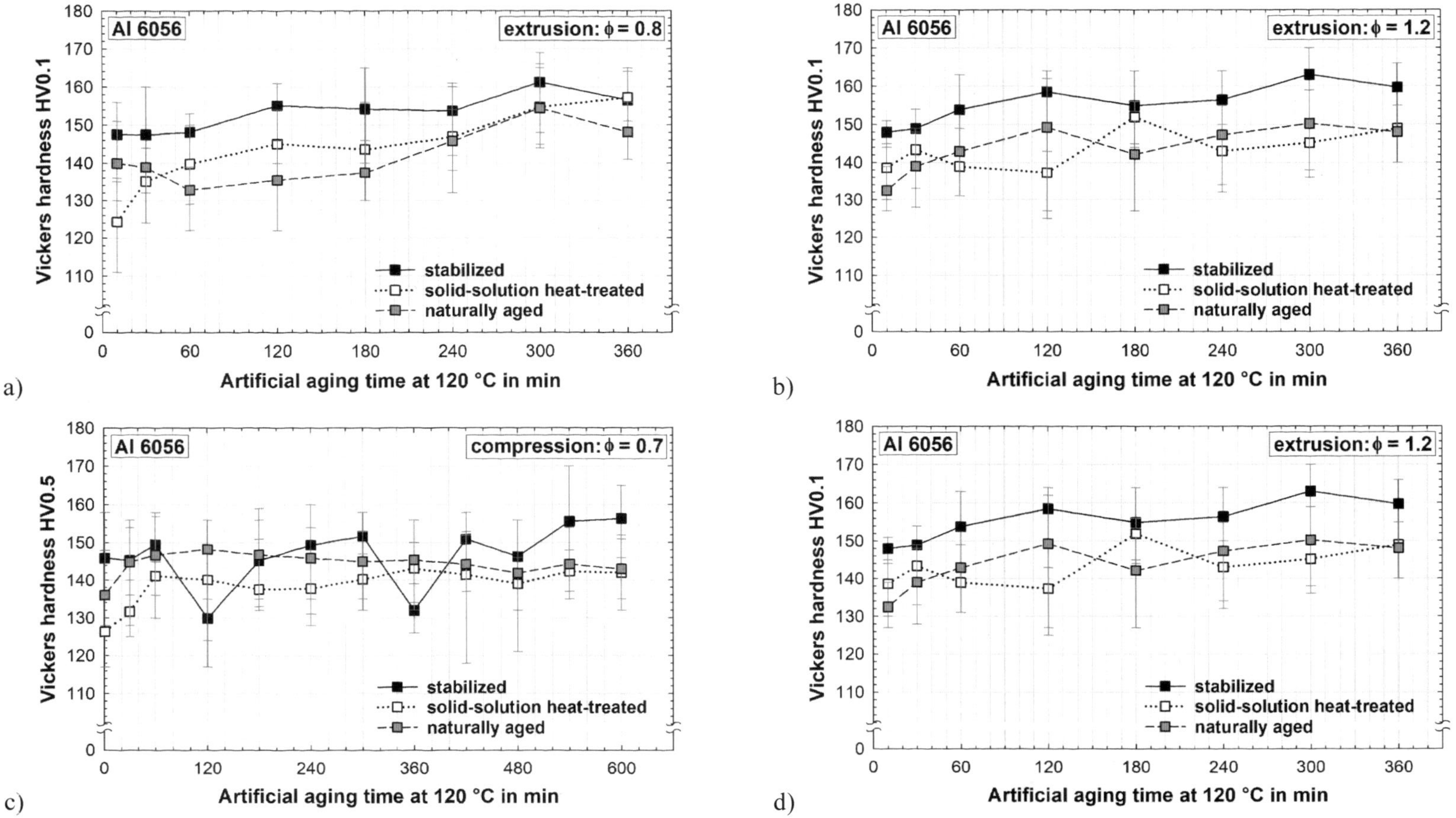


**Figure 2.** Vickers hardness of the aluminum alloy 6056 as a function of the artificial aging time at 120 °C and dependent on the initial heat-treatment condition: (a, b) after extrusion with an introduced strain of (a) $\varphi_E = 0.8$ and (b) $\varphi_E = 1.2$, (c, d) after compression with an introduced strain of (c) $\varphi_C = 0.55$ and (d) $\varphi_C = 0.7$. Scatter bars depict the maximum and minimum values. Stabilization annealing of the aluminum alloy prior to plastic deformation results in the highest hardness values for both deformation processes and both introduced strains.

## 4. Discussion

The precipitation sequence for Al-Si-Mg-alloys containing copper is described as follows [10–12]:

$$\begin{matrix} \alpha/\text{ super-saturated} \\ \text{solid-solution} \end{matrix} \rightarrow \begin{matrix} \text{cluster/} \\ \text{GP(I)-zones} \end{matrix} \rightarrow \begin{matrix} \text{metastable } \beta''/ \\ \text{GP(II)-zones} \end{matrix} \rightarrow \begin{matrix} \text{metastable} \\ \beta' + Q' \end{matrix} \rightarrow \begin{matrix} \text{stable} \\ \beta + Q + Si \end{matrix}$$

Guinier-Preston (GP)-zones are nanoscaled precipitates with an ordered arrangement of Mg-, Si- and Cu-atoms which structurally evolve during aging [9,13]. The needle-shaped β'' and the rod-like β' are metastable ternary precursor phases to the stable phase β ($Mg_2Si$). Whereas, for Al-Mg-Si-alloys with Cu, the lath-shaped, metastable quaternary Q'-phase occurs in addition and is a precursor for the stable Q-phase ($Al_5Cu_2Mg_8Si_6$) [14]. The significant strengthening of these aluminum alloys during artificial aging is attributed in majority to the hardening phase β'' ($Mg_5Si_6$ [15]) [14,16,17]. The phase Q' [10,14,17] gives an additional contribution. A very high density of these semi-coherent phases in combination with very fine fully coherent GP(I)-zones leads to major strengthening [15].

The significant increase in hardness during natural aging after solid-solution heat-treatment is a result of aggregated Mg- and Si-clusters, which form stable Mg-Si co-clusters, which are also described as GP(I)-zones [11,18–20]. The formation of these co-clusters is mainly controlled by the concentration of mobile vacancies [9,11]. Stabilization annealing after solid-solution treatment and quenching annihilates these vacancies and the diffusion controlled formation of the co-clusters is inhibited [9], which explains the increased room temperature stability (see figure 1). Instead, GP(II)-zones are formed during stabilization annealing [9,10,21–23].

The improved artificial aging response of the stabilized conditions, if compared to the naturally aged ones (see figure 2), is in good accordance to the literature [24]. The reason for this effect are the GP(II)-zones, which form during stabilization annealing and their transformability into β'' during subsequent artificial aging due to the similarity in composition, structure and size [10,18,19,22,25]. Further, the high number of GP(II)-zones initiated during stabilization treatment results in an increase in the number of the hardening phase β'' in the following artificial aging treatment, as numerous nucleation sites are provided [18,19]. The pronounced drops in the hardness profile during artificial aging indicate the formation of metastable strengthening phases, which lead to a following peak in hardness after the precipitation transformation [12].

In contrast, the hardening response of the naturally aged conditions during subsequent artificial aging is limited. The precipitation kinetics is retarded by the previous natural aging and the formation of GP(II)-zones is suppressed by the formation of Mg-Si co-clusters [11,19,22,25,26].

Of all three heat-treated and subsequently plastically deformed ($\varphi = 0.7$) conditions, the solid-solution heat-treated condition attains peak-hardness most rapidly during artificial aging. The accelerated precipitation kinetics of this condition is a result of the supported diffusion by the quenched-in vacancies after solid-solution heat-treatment [11] and the high dislocation density due to the plastic deformation [1,4,5,12]. Both effects lead to numerous nucleation sites and an increased nucleation rate of the β''-phase and therefore faster hardening of the aluminum alloy. The precipitation sequence remains unaltered by the plastic deformation [27]. The overall minor hardness of the solid-solution heat-treated condition (if compared to the stabilized condition) is supposedly due to the precipitation formation and the introduced strain. Through artificial aging, the hardening response of the initial GP(II)-zones formed by stabilization annealing predominates the strengthening effect of the combined solid-solution treatment and plastic deformation. The latter strategy is strongly dependent on the dislocation density and it is assumed that the introduced strain was too low to take advantage of the full potential of this method.

## 5. Conclusions

The influence of the initial heat-treatment in combination with a following plastic deformation process on the artificial aging behavior and the achieved hardness of the aluminum alloy 6056 was investigated. Three initial heat-treatment conditions were compared: a solid-solution heat-treated, a naturally aged and a stabilized condition. Extrusion, respectively compression, were chosen as plastic deformation processes with two different introduced strains and were performed on each of the three initial heat-treatment conditions followed by artificial aging. Conclusions can be drawn as follows:

(1) Stabilization annealing at low temperatures for several hours immediately after the solid-solution heat-treatment inhibits natural aging of the aluminum alloy. This increase in room temperature stability is attributed to the inhibited formation of stable Mg-Si co-clusters/ GP(I)-zones during the stabilization treatment and causes the formation of GP(II)-zones instead.

(2) The artificial aging behavior of the three tested initial heat-treatment conditions after plastic deformation was not influenced by the type of the deformation process. The extruded, respectively compressed, conditions show a similar hardness evolution during artificial aging.

(3) Stabilization treatment before plastic deformation leads to major hardness, if compared to the naturally aged respectively solid-solution heat-treated and plastically deformed conditions. This improved artificial aging response is a result of the formed GP(II)-zones during stabilization annealing, which promote a pronounced nucleation of the hardening phase β'' due to their structural and compositional similarity.

(4) The combination of solid-solution heat-treatment and subsequent plastic deformation results in a more rapid hardening of the conditions with higher introduced strain, if compared to the two other initial heat-treatments. The introduced strain hardening and the high dislocation density lead – in combination with enhanced diffusion due to quenched-in vacancies – to an accelerated precipitation kinetics of the solid-solution treated and plastically deformed condition.

In summary, stabilization annealing of the aluminum alloy 6056 is a promising strategy for a resource and cost efficient manufacturing process, as the room temperature stability during interim storage periods is improved and, additionally, the artificial aging response and therefore the achieved hardness is increased.

## Acknowledgments

The authors gratefully acknowledge funding of the Collaborative Research Center SFB 692 "High-strength aluminum-based lightweight materials for safety components" (project number: SFB692T7) received from the German Research Foundation (Deutsche Forschungsgemeinschaft DFG).

## References

[1] Kim J K, Jeong H G, Hong S I, Kim Y S and Kim W J 2001 Effect of aging treatment on heavily deformed microstructure of a 6061 aluminum alloy after equal channel angular pressing *Scr. Mater.* **45** 901–7

[2] Kim W J, Chung C S, Ma D S, Hong S I and Kim H K 2003 Optimization of strength and ductility of 2024 Al by equal channel angular pressing (ECAP) and post-ECAP aging *Scr. Mater.* **49** 333–8

[3] Kim W J, Kim J K, Kim H K, Park J W and Jeong Y H 2008 Effect of post equal-channel-angular-pressing aging on the modified 7075 Al alloy containing Sc *J. Alloys Compd.* **450** 222–8

[4] Hockauf M, Meyer L W, Zillmann B, Hietschold M, Schulze S and Krüger L 2009 Simultaneous improvement of strength and ductility of Al-Mg-Si alloys by combining equal-channel angular extrusion with subsequent high-temperature short-time aging *Mater. Sci. Eng. A* **503** 167–71

[5] Hockauf K, Meyer L W, Hockauf M and Halle T 2010 Improvement of strength and ductility for a 6056 aluminum alloy achieved by a combination of equal-channel angular pressing and aging treatment *J. Mater. Sci.* **45** 4754–60
[6] Kim W J, Kim J K, Park T Y, Hong S I, Kim D I, Kim Y S and Lee J D 2002 Enhancement of strength and superplasticity in a 6061 Al alloy processed by equal-channel-angular-pressing *Metall. Mater. Trans. A* **33** 3155–64
[7] Zhao Y H, Liao X Z, Cheng S, Ma E and Zhu Y T 2006 Simultaneously Increasing the Ductility and Strength of Nanostructured Alloys *Adv. Mater.* **18** 2280–3
[8] Cheng S, Zhao Y H, Zhu Y T and Ma E 2007 Optimizing the strength and ductility of fine structured 2024 Al alloy by nano-precipitation *Acta Mater.* **55** 5822–32
[9] Ostermann F 2014 *Anwendungstechnologie Aluminium* (Springer Berlin Heidelberg)
[10] Miao W F and Laughlin D E 2000 Effects of Cu content and preaging on precipitation characteristics in aluminum alloy 6022 *Metall. Mater. Trans. A* **31** 361–71
[11] Pogatscher S, Antrekowitsch H, Leitner H, Ebner T and Uggowitzer P J 2011 Mechanisms controlling the artificial aging of Al–Mg–Si Alloys *Acta Mater.* **59** 3352–63
[12] Roven H J, Liu M and Werenskiold J C 2008 Dynamic precipitation during severe plastic deformation of an Al–Mg–Si aluminium alloy *Mater. Sci. Eng. A* **483–484** 54–8
[13] Liu S, Li K, Lu J, Sha G, Wang J, Yang M, Ji G, Song M, Wang J and Du Y 2018 On the atomic model of Guinier-Preston zones in Al-Mg-Si-Cu alloys *J. Alloys Compd.* **745** 644–50
[14] Chakrabarti D J and Laughlin D E 2004 Phase relations and precipitation in Al–Mg–Si alloys with Cu additions *Prog. Mater. Sci.* **49** 389–410
[15] Andersen S J, Zandbergen H W, Jansen J, Træholt C, Tundal U and Reiso O 1998 The crystal structure of the β″ phase in Al–Mg–Si alloys *Acta Mater.* **46** 3283–98
[16] Takeda M, Ohkubo F, Shirai T and Takeda M 1998 Stability of metastable phases and microstructures in the ageing process of Al–Mg–Si ternary alloys *J. Mater. Sci.* **33** 2385–90
[17] Delmas F, Casanove M J, Lours P, Couret A and Coujou A 2004 Quantitative TEM study of the precipitation microstructure in aluminium alloy Al(MgSiCu) 6056 T6 *Mater. Sci. Eng. A* **373** 80–9
[18] Dutta I and Allen S M 1991 A calorimetric study of precipitation in commercial aluminium alloy 6061 *J. Mater. Sci. Lett.* **10** 323–6
[19] Murayama M and Hono K 1999 Pre-precipitate clusters and precipitation processes in Al-Mg-Si alloys *Acta Mater.* **47** 1537–48
[20] Edwards G A, Stiller K, Dunlop G L and Couper M J 1998 The precipitation sequence in Al-Mg-Si alloys *Acta Mater.* **46** 3893–904
[21] Österreicher J A, Kirov G, Gerstl S S A, Mukeli E, Grabner F and Kumar M 2018 Stabilization of 7xxx aluminium alloys *J. Alloys Compd.* **740** 167–73
[22] Saga M, Sasaki Y, Kikuchi M, Yan Z and Matsuo M 1996 Effect of Pre-Aging Temperature on the Behavior in the Early Stage of Aging at High Temperature for Al-Mg-Si Alloy *Mater. Sci. Forum* **217–222** 821–6
[23] Takata K, Takahashi J, Saga M, Ushioda K, Hibino A and Kikuchi M 2014 Effect of Two-Step Aging on Cluster Formation in Al-Mg-Si Alloys *Mater. Trans.* **55** 885–91
[24] Suzuki H, Kanno M and Itoh G 1981 A consideration of the two-step aging process in an Al-Mg-Si alloy *Aluminium* **57** 628–9
[25] Serizawa A, Hirosawa S and Sato T 2008 Three-Dimensional Atom Probe Characterization of Nanoclusters Responsible for Multistep Aging Behavior of an Al-Mg-Si Alloy *Metall. Mater. Trans. A* **39** 243–51
[26] Pogatscher S, Werinos M, Antrekowitsch H and Uggowitzer P J 2014 The Role of Vacancies in the Aging of Al-Mg-Si Alloys *Mater. Sci. Forum* **794–796** 1008–13
[27] Zhao Y H, Liao X Z, Jin Z, Valiev R Z and Zhu Y T 2004 Microstructures and mechanical properties of ultrafine grained 7075 Al alloy processed by ECAP and their evolutions during annealing *Acta Mater.* **52** 4589–99

IOP Conf. Series: Materials Science and Engineering **480** (2019) 012032 doi:10.1088/1757-899X/480/1/012032

# Mean stress sensitivity of the fatigue strength after equal-channel angular pressing of the aluminum alloys 6082 and 6060

**L Winter*, K Hockauf and T Lampke**

Materials and Surface Engineering Group, Institute of Materials Science and Engineering, Chemnitz University of Technology, D-09125, Germany

* e-mail: lisa.winter@mb.tu-chemnitz.de

**Abstract.** The present study is focused on the influence of equal-channel angular pressing (ECAP) on the mean stress sensitivity of the fatigue strength. Systematic work has been done comparing the two precipitation-hardening aluminum alloys 6082 and 6060 in peak-aged and ECAP-processed condition, respectively. The fatigue strength was determined at alternating compression-tension and tension stresses with load ratios of R = -1, 0 and 0.1. Both aluminum alloys in peak-aged condition exhibit a strong dependence of the fatigue strength on the load ratio; i.e. and therefore on the applied mean stress. An increase in load ratio, which implies an increase in mean stress, results in a decrease of the fatigue strength amplitude. In contrast, ECAP-processing neutralizes the mean stress sensitivity of the fatigue strength for both aluminum alloys. The fatigue strength for the ECAP-processed conditions is not influenced by the load ratio: the fatigue strength amplitudes at tension mean stresses (R = 0 and 0.1) are just as high as for zero mean stress at alternating compression-tension stresses (R = -1). The neutralization of the mean stress sensitivity of two aluminum alloys due to ECAP-processing is shown for the first time in this study.

## 1. Introduction

High-strength aluminum alloys are of major importance for lightweight structural applications and to meet the requirements, a further severe plastic deformation (SPD) can be beneficial. As one SPD-method for bulk material in industrial scale [1], equal-channel angular pressing (ECAP) leads to improved mechanical properties and in particular a major tensile strength and an enhanced high-cycle fatigue strength [2–6]. Outstanding mechanical properties are enabled by the introduced strain hardening and the refined microstructure with grain sizes in sub-micrometer range [7–9].

In-service fatigue loadings often exhibit changing mean stresses over operation time. Therefore, the mean stress sensitivity of materials for mechanical components is of particular interest. For most metals and their alloys, the high cycle fatigue behavior usually depends on the load ratio R and therefore the mean stress [10]. An increase in mean stress results in a reduction of the fatigue strength.

HAIGH or SMITH diagrams illustrate the influence of the mean stress. The mean stress sensitivity M can be calculated according to Schütz [11] as:

$$M = \frac{\sigma_{ad}(R=-1)}{\sigma_{ad}(R=0)} - 1 \quad (1)$$

M is calculated from the fatigue strength amplitudes under alternating compression-tension (R = -1, zero mean stress) and under tension loading (R = 0, zero minimum stress) and ranges from 0 (fatigue strength is not sensitive to mean stress) to 1 (major influence of mean stress on fatigue strength).

Aluminum alloys exhibit a significant dependence of the fatigue strength on the load ratio and therefore on the applied mean stress [12–16]. The mean stress sensitivity, calculated according to equation 1, is listed with M = approx. 0.2, depended on the alloy composition [10,17]. To the best of our knowledge, the mean stress sensitivity of the fatigue strength of ECAP-processed aluminum alloys has not been investigated, yet.

Therefore, the focus of the present study is the effect of ECAP-processing on the mean stress sensitivity of the fatigue strength. As bulk material, the aluminum alloys 6082 and 6060 were chosen, due to their wide usage in structural applications and their good workability and suitability for ECAP-processing.

## 2. Experimental and materials

For this study, the precipitation-hardening aluminum alloys 6082 and 6060 were used. For both alloys ECAP-processing was performed at room temperature in a friction-optimized ECAP tool with an internal angle of 90° and a cross-section of 15×15mm², equipped with movable walls and a bottom slider.

The mechanical properties were determined by quasi-static tensile tests at room temperature at a strain rate of $10^{-3}\cdot s^{-1}$ in a Zwick Roell servohydraulic testing machine (Zwick Roell, Germany). For each alloy and each condition, three samples were tested.

The fatigue testing was performed with axial fatigue specimens in a RUMUL Testronic resonant testing machine (Russenberger Prüfmaschinen AG, Switzerland) under alternating compression-tension and tension-tension loading with load ratios of R = -1, R = 0 and R = 0.1. The fatigue tests were carried out until the endurance limit of $N_E = 10^7$ cycles for the 6082 and $N_E = 5\cdot 10^7$ cycles for the 6060 aluminum alloy was reached or until a crack occurred, which was detected by a drop in the resonant frequency of 2 Hz or more.

### *2.1. Aluminum alloy 6082*

The chemical composition of the aluminum alloy 6082 (AlSiMgMn) is given in table 1. For this alloy a peak-aged and an ECAP-processed condition were compared, hereafter referred to as condition "6082 T6" and "6082 E1". To achieve peak strength, the material was solid-solution treated at 530 °C for 1 h, quenched in water and subsequently aged at 170 °C for 65 h. To produce the condition "6082 E1" the material was solid-solution treated at 530 °C for 1 h, quenched in water and immediately afterwards ECAP-processed at room temperature with a processing speed of 50 mm/min for one pass and peak-aged at 170 °C for 40 min. The mechanical properties for the peak-aged and the ECAP-processed 6082 aluminum alloy are given in table 2.

**Table 1.** Chemical composition of the aluminum alloy 6082 (AlSi1MgMn).

| Element | Si | Mg | Mn | Fe | Cu | Zn | Al |
|---|---|---|---|---|---|---|---|
| wt.-% | 1.00 | 0.90 | 0.76 | 0.21 | 0.05 | 0.02 | balance |

**Table 2.** Mechanical properties determined by tensile testing of the aluminum alloy 6082 in peak-aged condition (T6) and after one ECAP-pass (E1). The standard deviation is given in absolute values.

| Condition | Yield strength in MPa | Ultimate tensile strength in MPa | Uniform elongation in % | Elongation to failure in % |
|---|---|---|---|---|
| 6082 T6 | 297 ± 1 | 307 ± 4 | 5.2 ± 0.8 | 23.7 ± 0.9 |
| 6082 E1 | 332 ± 12 | 345 ± 12 | 5.4 ± 0.4 | 18.9 ± 0.5 |

### *2.2. Aluminum alloy 6060*

The chemical composition of the aluminum alloy 6060 (AlMgSi0.5) is given in table 3. For the aluminum alloy 6082, a peak-aged and an ECAP-processed condition were compared, as well, hereafter referred to as condition "6060 T6" and "6060 E2". The peak-aged condition was solid-solution treated at 525 °C for 1 h, quenched in water and subsequently aged at 170 °C for 16 h. ECAP-processing was performed on the peak-aged condition at room temperature with 25 mm/min processing speed for two passes. The mechanical properties for this alloy in peak-aged and ECAP-processed condition are given in table 4.

**Table 3.** Chemical composition of the aluminum alloy 6060 (AlMgSi0.5).

| Element | Mg | Si | Fe | Mn | Zn | Cu | Al |
|---|---|---|---|---|---|---|---|
| wt.-% | 0.52 | 0.43 | 0.16 | < 0.01 | < 0.01 | < 0.01 | balance |

**Table 4.** Mechanical properties determined by tensile testing of the aluminum alloy 6060 in peak-aged condition (T6) and after two ECAP-passes (E2). The standard deviation is given in absolute values.

| Condition | Yield strength in MPa | Ultimate tensile strength in MPa | Uniform elongation in % | Elongation to failure in % |
|---|---|---|---|---|
| 6060 T6 | 187 ± 2 | 214 ± 2 | 6.6 ± 0.2 | 24.3 ± 0.8 |
| 6060 E2 | 301 ± 4 | 305 ± 3 | 1.0 ± 0.1 | 16.9 ± 0.5 |

## 3. Results and discussion

The load-ratio dependent high-cycle fatigue behavior in peak-aged and in ECAP-processed condition are given in figure 1 (for Al 6082) and in figure 2 (for Al 6060). The fatigue strength at the three tested load ratios are listed in table 5 (for Al 6082) and table 6 (for Al 6060).

In general, the aluminum alloy 6082 exhibits a higher fatigue strength, if compared to the alloy 6060 due to its higher tensile strength (see figures 1a) and 2a). The peak-aged condition of both tested aluminum alloys shows a significant influence of the applied load ratio R on the fatigue strength amplitude. At alternating tension-compression stresses ($R = -1$), the fatigue strength amplitude is the highest for the respective alloys. With an increasing load ratio to $R = 0$ and $R = 0.1$, and therefore an increasing mean stress, the fatigue strength amplitude decreases significantly for both alloys. This effect is confirmed by the literature [12–15].

ECAP-processing results in a higher fatigue strength for both tested alloys due to the enhanced tensile strength as a consequence of the severe plastic deformation [2–5] (see figures 1b) and 2b). Moreover, the fatigue strength of both ECAP-processed alloys is not influenced by the load ratio. For the respective alloy, the fatigue strength amplitudes of the ECAP-processed condition at alternating compression-tension stresses ($R = -1$) are just as high as for tension and tension-tension stresses ($R = 0$ and $R = 0.1$).

IOP Conf. Series: Materials Science and Engineering **480** (2019) 012032 doi:10.1088/1757-899X/480/1/012032

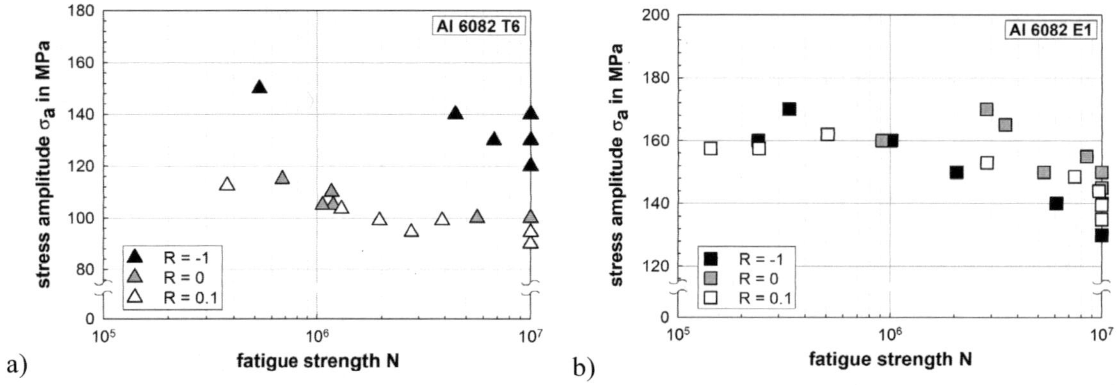


**Figure 1.** Fatigue strength depending on the load ratio of the aluminum alloy 6082 in a) peak-aged condition and b) after 1 ECAP-pass. ECAP-processing leads to an enhanced fatigue strength and a diminished influence of the applied load ratio on the fatigue strength amplitudes.

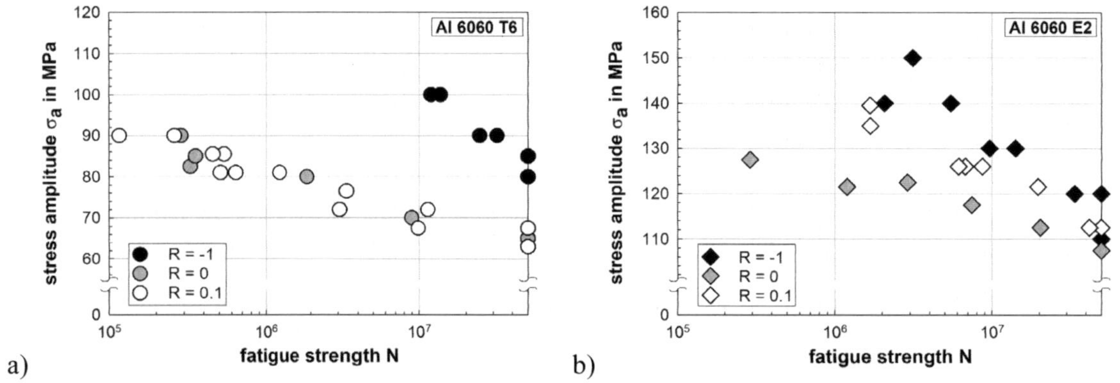


**Figure 2.** Fatigue strength depending on the load ratio of the aluminum alloy 6060 in a) peak-aged condition and b) after 2 ECAP-passes. ECAP-processing leads to an enhanced fatigue strength and a diminished influence of the applied load ratio on the fatigue strength amplitudes.

**Table 5.** Fatigue strength of the aluminum alloy 6082 in peak-aged condition and after one ECAP-pass at $N_E = 10^7$ cycles at the tested load ratios.

| Condition | Load ratio R | Maximum stress $\sigma_o$ in MPa | Stress amplitude $\sigma_a$ in MPa | Mean stress $\sigma_m$ in MPa |
|---|---|---|---|---|
| 6082 T6 | -1 | 120 | 120 | 0 |
| | 0 | 180 | 90 | 90 |
| | 0.1 | 200 | 90 | 110 |
| 6082 E1 | -1 | 135 | 135 | 0 |
| | 0 | 290 | 145 | 145 |
| | 0.1 | 310 | 139.5 | 170.5 |

**Table 6.** Fatigue strength of the aluminum alloy 6060 in peak-aged condition and after two ECAP-passes at $N_E = 5 \cdot 10^7$ cycles at the tested load ratios.

| Condition | Load ratio R | Maximum stress $\sigma_o$ in MPa | Stress amplitude $\sigma_a$ in MPa | Mean stress $\sigma_m$ in MPa |
|---|---|---|---|---|
| 6060 T6 | -1 | 85 | 85 | 0 |
| | 0 | 130 | 65 | 65 |
| | 0.1 | 140 | 63 | 77 |
| 6060 E2 | -1 | 110 | 110 | 0 |
| | 0 | 220 | 110 | 110 |
| | 0.1 | 240 | 108 | 132 |

The mean stress sensitivity of the fatigue strength for the aluminum alloys 6082 and 6060 is represented in the form of a HAIGH diagram in figure 3. The slope of the secant of the data points for the fatigue strength indicates the mean stress sensitivity M, which was calculated according to equation 1.

The secant of the fatigue strength shows a considerable slope for the peak-aged condition of the tested alloys. The fatigue strength amplitude decreases with an increased load ratio R and therefore an increased mean stress. For both aluminum alloys, the calculated mean stress sensitivity is approx. M = 0.2, which is in good accordance with the literature [10,17].

In contrast, for the ECAP-processed conditions, the means stress sensitivity is diminished to M = 0, which shows in a horizontal course of the secant in the Haigh diagram. After ECAP-processing, the fatigue strength amplitudes of the aluminum alloys 6082 and 6060 are not influenced by the applied load ratio and therefore are independent from the mean stress. This neutralization of the mean stress influence of the fatigue strength after ECAP-processing is shown for the first time here. However, this effect contradicts with the accepted theory, that an increase in strength results in an increase in mean stress sensitivity [10].

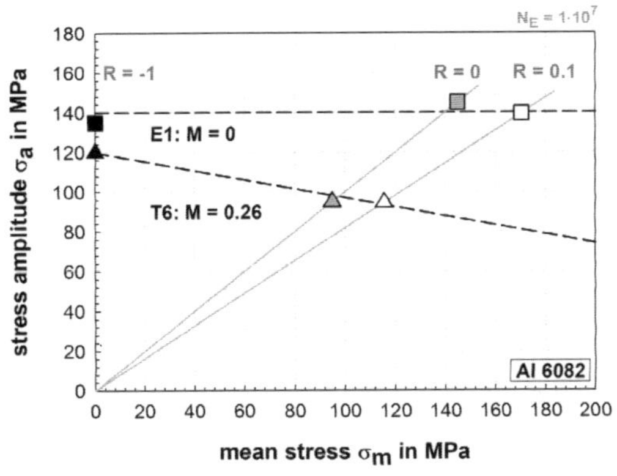


a)

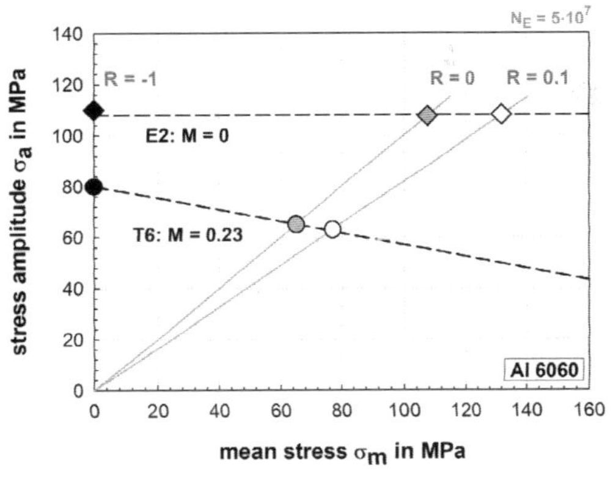


b)

**Figure 3.** Haigh diagram representing the fatigue strength amplitudes depending on the load ratio for the a) 6082 aluminum alloy and for the b) 6060 aluminum alloy, both in peak-aged condition and after ECAP-processing. The slope of the secant indicates the mean stress sensitivity M, which is calculated according to Schütz [11]. For both aluminum alloys, ECAP-processing leads to a neutralized influence of the applied load ratio and therefore to a diminished mean stress influence on the fatigue strength.

## 4. Summary and conclusions

The influence of equal-channel angular pressing (ECAP) on the mean stress sensitivity of the fatigue strength was investigated. For the aluminum alloys 6082 and 6060 in peak-aged as well as in ECAP-processed condition, the fatigue strength at three different load ratios was compared. Conclusions can be drawn as follows:

(1) Equal-channel angular pressing generally results in a significant enhancement in fatigue strength. This enhanced high cycle fatigue behavior is a consequence of the grain refinement and the increased tensile strength of the ECAP-processed material.

(2) The fatigue strength of both aluminum alloys, 6082 and 6060, in unprocessed, peak-aged condition, is dependent on the applied mean stress. With an increase in load ratio and therefore an increase in mean stress, the fatigue strength is reduced. The mean stress sensitivity is approx. $M = 0.2$ for both alloys.

(3) ECAP-processing results in a neutralized influence of the mean stress on the fatigue strength. Both ECAP-processed aluminum alloys exhibit a mean stress sensitivity of $M = 0$. This contradicts to the accepted theory, that an increase in strength results in an increase in mean stress sensitivity. The influence of ECAP on the mean stress sensitivity of the fatigue strength of aluminum alloys is investigated for the first time in this study and further research is required to clarify this point.

## Acknowledgments

The authors gratefully acknowledge funding of the Collaborative Research Center SFB 692 "High-strength aluminum-based lightweight materials for safety components" received from the German Research Foundation (Deutsche Forschungsgemeinschaft DFG).

## References

[1] Frint S, Hockauf M, Frint P and Wagner M F-X 2016 Scaling up Segal's principle of Equal-Channel Angular Pressing *Mater. Des.* **97** 502–11
[2] Hockauf K, Meyer L W, Halle T and Hockauf M 2009 Crack growth and high cycle fatigue behaviour of an AA6060 aluminium alloy after ECAP combined with a subsequent heat treatment *Materwiss. Werksttech.* **40** 493–9
[3] Chung C S, Kim J K, Kim H K and Kim W J 2002 Improvement of high-cycle fatigue life in a 6061 Al alloy produced by equal channel angular pressing *Mater. Sci. Eng. A* **337** 39–44
[4] Vinogradov A, Nagasaki S, Patlan V, Kitagawa K and Kawazoe M 1999 Fatigue properties of 5056 Al-Mg alloy produced by equal-channel angular pressing *Nanostructured Mater.* **11** 925–34
[5] Mughrabi H and Höppel H W 2010 Cyclic deformation and fatigue properties of very fine-grained metals and alloys *Int. J. Fatigue* **32** 1413–27
[6] Winter L, Hockauf K and Lampke T 2018 High cycle fatigue behavior of the severely plastically deformed 6082 aluminum alloy with an anodic and plasma electrolytic oxide coating *Surf. Coatings Technol.* **349** 576–83
[7] Furukawa M, Horita Z and Langdon T G 2001 Developing ultrafine grain sizes using severe plastic deformation *Adv. Eng. Mater.* **3** 121–5
[8] Valiev R Z and Langdon T G 2006 Principles of equal-channel angular pressing as a processing tool for grain refinement *Prog. Mater. Sci.* **51** 881–981
[9] Iwahashi Y, Wang J, Horita Z, Nemoto M and Langdon T G 1996 Principle of equal-channel angular pressing for the processing of ultra-fine grained materials *Scr. Mater.* **35** 143–6
[10] Haibach E 2006 *Betriebsfestigkeit* (Berlin Heidelberg: Springer Verlag)
[11] Schütz W 1967 Über eine Beziehung zwischen der Lebensdauer bei konstanter und bei veränderlicher Beanspruchungsamplitude und ihre Anwendbarkeit auf die Bemes sung von Flugzeugbauteilen *Zeitschrift für Flugwissenschaften* **15** 407–19

[12] Zhao T and Jiang Y 2008 Fatigue of 7075-T651 aluminum alloy *Int. J. Fatigue* **30** 834–49
[13] Mayer H, Schuller R and Fitzka M 2013 Fatigue of 2024-T351 aluminium alloy at different load ratios up to 1010 cycles *Int. J. Fatigue* **57** 113–9
[14] Benedetti M, Fontanari V and Monelli B D 2010 Plain fatigue resistance of shot peened high strength aluminium alloys: Effect of loading ratio *Procedia Eng.* **2** 397–406
[15] El-Shabasy A B, Hassan H A and Lewandowski J J 2012 Effects of load ratio, R, and test temperature on high cycle fatigue behavior of nano-structured Al–4Y–4Ni–X alloy composites *Mater. Sci. Eng. A* **558** 211–6
[16] Winter L, Morgenstern R, Hockauf K and Lampke T 2016 The effect of plasma electrolytic oxidation on the mean stress sensitivity of the fatigue life of the 6082 aluminum alloy *IOP Conf. Ser. Mater. Sci. Eng.* **118** 012033
[17] Ostermann F 2014 *Anwendungstechnologie Aluminium* (Berlin Heidelberg: Springer)

IOP Conf. Series: Materials Science and Engineering **480** (2019) 012033 doi:10.1088/1757-899X/480/1/012033

# Experimental study on the magnetic pulse welding process of large aluminum tubes on steel rods

**J Bellmann[1,2,*], S Schettler[2,3], S Dittrich[4], J Lueg-Althoff[5], S Schulze[2], M Hahn[5], E Beyer[1,2] and A E Tekkaya[5]**

[1] Institute of Manufacturing Science and Engineering, Technische Universität Dresden, George-Baehr-Str. 3c, 01062 Dresden, Germany
[2] Fraunhofer IWS Dresden, Winterbergstr. 28, 01277 Dresden, Germany
[3] Chair of Mechanics of Materials and Failure Analysis, Institute of Material Science, Technische Universität Dresden, Helmholtzstr. 7, 01069 Dresden, Germany
[4] Dresden High Magnetic Field Laboratory (HLD-EMFL), Helmholtz-Zentrum Dresden-Rossendorf, Bautzner Landstr. 400, 01328 Dresden, Germany
[5] Institute of Forming Technology and Lightweight Components, TU Dortmund University, Baroper Str. 303, 44227 Dortmund, Germany

* e-mail: joerg.bellmann@tu-dresden.de

**Abstract.** Solid state welding technologies enable dissimilar metal welding without critical intermetallic phase formation. Magnetic Pulse Welding (MPW) is a promising joining method for hybrid sheet connections in car body production or for manufacturing of dissimilar tube connections. Given a suitable MPW process design, the shear testing of MPW joints usually leads to failure in the weaker base material. This finding emphasizes the high strength level of the joining zone itself. Consequently, the transmission of higher forces or torques, respectively, requires stronger materials or adapted geometries. In the present experimental study, the diameter of an exemplary driveshaft was doubled to 80 mm at constant tube wall thickness to increase the load bearing capability. The characteristic impact flash was recorded at different positions around the tube's circumference and it was used to adjust the most relevant process parameters, i.e. working length and acceleration gap, at the lower process boundary. In metallographic analysis, the final shapes of both joining partners were compared with the original driveshaft dummies on macroscopic and microscopic scale. The typical wavy interface between aluminum and steel was analyzed in detail. Doubling the tube diameter lead to four times higher torque levels of failure during quasistatic and cyclic torsion tests.

## 1. Introduction

Lightweight designs are of great importance in the transportation sector in order to reduce the energy consumption of vehicles. At the same time, reducing the mass leads to less energy that has to be absorbed in case of a crash. Multi material concepts are essential to reach this goal and require suitable methods for joining dissimilar metals. In most cases, conventional fusion welding processes are not applicable due to the differences in the thermophysical material properties and the formation of brittle intermetallic phases in the weld seam. Purely mechanical connections based on form or force fits do not show these disadvantages but often require additional elements like fasteners or elaborate preparation steps to achieve close fits. There is a wide variety of joining by forming processes that reduce or overcome these drawbacks [1]. For example, high speed forming processes can be utilized to connect dissimilar metals by form fit, force fit or even by welding. Magnetic pulse welding (MPW) is a representative of the last mentioned methods and is applicable for the production of hybrid drive shafts [2–6]. The torsional load bearing capabilities of these connections have already been proved in former studies [7, 8]. It was found that sound welds can be produced even with non-contoured inner parts by fine-tuning the energetic and geometric parameters and, thus, reducing the preparation effort compared to [4, 6]. Failure occurred in the outer tube's base material next to the weld seam. Consequently, this tube has to be toughened in order to increase the overall torsional load bearing capability. Since the material is often a preset boundary condition due to cost or process related factors, the outer tube's geometry has to be adapted. Table 1 highlights the advantage of an increased tube diameter (setup 3) compared to an increased tube wall thickness (setup 2). On the one hand, the cross-sectional areas of the tubes and their weights, respectively, are comparable. On the other hand, the polar section modulus calculated according to equation 1 is more than doubled for the larger tube diameter and increased by a factor of four compared to the reference setup 1. Aiming for higher load bearing capabilities, setup 3 is favorable with respect to weight saving aspects. Furthermore, the high speed forming process of the thicker tube in setup 2 would require a comparatively high magnetic pressure which would have negative influence on the lifetime of the tool coils.

**Table 1**. Characteristic values for different outer tube geometries.

| | Outer tube diameter $D_e$ (mm) | Tube wall thickness (mm) | Inner tube diameter $D_i$ (mm) | Polar section modulus $Z_p$ [a] ($mm^3$) | Cross-sectional area ($mm^2$) |
|---|---|---|---|---|---|
| Setup 1[8] | 40 | 2 | 36 | 4,322 | 239 |
| Setup 2 | 40 | 4 | 32 | 7,419 | 452 |
| Setup 3 | 80 | 2 | 76 | 18,648 | 490 |

$$^{a}\ Z_p = \frac{\pi(D_e{}^4 - D_i{}^4)}{16 D_e} \qquad (1)$$

In this paper, the process adjustment for MPW of aluminum tubes with an outer diameter of 80 mm on steel parent parts is described. Previously developed methods [9] are applied to adjust the diameter of the non-contoured parent parts and axial position of the flyer tube in the working coil in order to ensure suitable collision conditions within the welding window. Finally, the welding interface is analyzed in detail.

## 2. Experimental setups and materials

### *2.1. Joining setup*

The principal joining setup and the geometrical definitions are depicted in figure 1. The chemical compositions of the materials are given in table 2, while table 3 lists the specific geometries and characteristics of the deployed pulse generators for both setups. The quasistatic yield strength of the aluminum alloy was determined in tube tensile tests with approximately 222 MPa. During the joining experiments the joining partners were positioned coaxially inside a single turn tool coil made of CuCr1Zr. When the capacitor of the pulse generator is discharged via the tool coil in a damped, sinusoidal current (see exemplary course in figure 2), the magnetic pressure accelerates the flyer radially. After the joining standoff $g$ the flyer collides with the polished parent surface ($R_a < 1$ µm). A weld seam is generated, if the impact velocity and the collision angle are in a proper range, called "welding window". The joining standoff $g$, the charging energy and the working length $l_w$ are key parameters to ensure these favorable collision conditions and have to be adjusted carefully. [9]

IOP Conf. Series: Materials Science and Engineering **480** (2019) 012033 doi:10.1088/1757-899X/480/1/012033

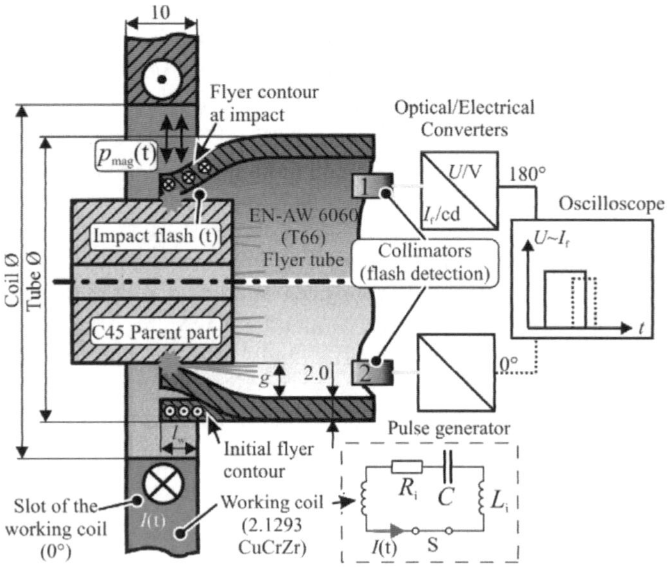


**Figure 1.** Geometrical definitions of the joining setup and measurement system for flash detection (not true to scale, all values in mm).

**Table 2.** Aluminum EN AW-6060 alloy composition [10] and steel C45 (1.0503) alloy composition [11].

| Flyer part EN AW-6060, T66 | | Parent part C45 (1.0503), normalized | |
|---|---|---|---|
| Element | Weight % | Element | Weight % |
| Mg | 0.35-0.6 | C | 0.42-0.5 |
| Mn | ≤0.1 | Mn | 0.5-0.8 |
| Fe | 0.1-0.3 | P | <0.045 |
| Si | 0.3-0.6 | S | <0.045 |
| Cu | ≤0.1 | Si | <0.4 |
| Zn | ≤0.15 | Ni | <0.4 |
| Cr | ≤0.05 | Cr | <0.4 |
| Ti | ≤0.1 | Mo | <0.1 |

**Table 3.** Characteristics of the deployed pulse generators for the complete RLC-circuit with working coil and workpieces.

| | Setup 1 [8] | Setup 3 |
|---|---|---|
| Machine | Bmax MPW50/25 | EmGen5 |
| Coil diameter (mm) | 41 | 81 |
| Outer tube diameter (mm) | 40 | 80 |
| Tube wall thickness (mm) | 2 | 2 |
| Max. discharge energy $E_{max}$ (kJ) | 32 | 160 |
| Max. discharge voltage $U_{max}$ (kV) | 20 | 24 |
| Capacitance $C_i$ (µF) | 160 | 560 |
| Inductance $L$ (nH) | 372 | 111 |
| Discharge frequency $f$ (kHz) | approx. 21 | approx. 25 |
| Damping coefficient $\gamma$ ($s^{-1}$) | 16,500 | 8,500 |

### *2.2. Measurement devices*

High speed collisions between metals are accompanied by so called impact flashes, as reported in [12, 13]. The evaluation of the time depended light intensity can be utilized to study and adjust collision welding processes like MPW [9, 14–16]. Therefore, collimators are positioned at different azimuthal locations as depicted in figure 1. The light intensities at these positions are recorded with the device described in [15] and can be plotted over time together with the tool coil current, exemplary shown in figure 2. The maximum intensity and the flash appearance time $t_{f,start}$ are key values for process analysis and optimization. Since a circumferential weld is required, special attention was paid to the position of the coil's slot (0° in figure 1), where the magnetic field intensity is reduced. The coil current $I(t)$ was measured by Rogowski current probes CWR 3000 B from Power Electronic Measurements Ltd.

IOP Conf. Series: Materials Science and Engineering **480** (2019) 012033 doi:10.1088/1757-899X/480/1/012033

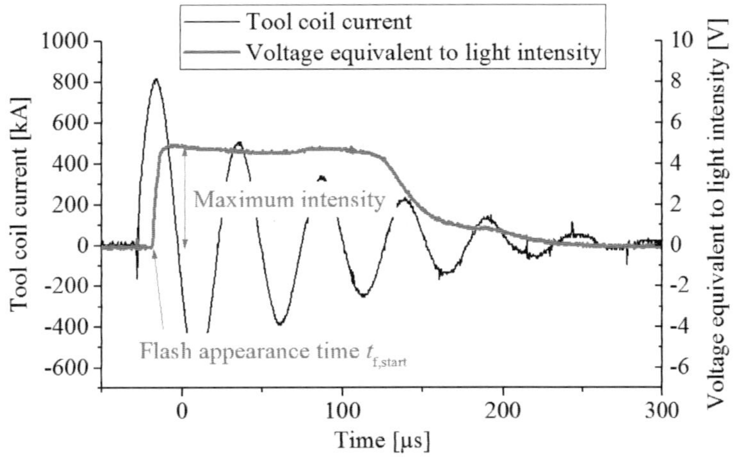


**Figure 2.** Tool coil current and light intensity over time for MPW experiment with 18 kJ charging energy.

### *2.3. Process adjustment and peel testing*

The joining standoff $g$, the charging energy and the working length $l_w$ in the MPW setup were adjusted using the method presented in [9] while the intensity of the impact flash served as a welding criterion. During this process adjustment, the circumferential weld quality was checked by a manual peel test, as described in [14].

### *2.4. Metallographic and metallurgical analysis*

Weld seams formed during high-speed collisions are of special interest from the metallurgical point of view since the prevalent pressures, temperatures and process durations differ significantly from conventional fusion welding processes. In order to correlate the influence of certain collision conditions with the morphology of the interface, metallographic investigations were performed on cross-sections polishes. The weld seam length and the shape of the interfacial waves are schematically depicted in figure 3 and were compared for both welding setups. Furthermore, the deformations of both joining partners were determined. Scanning electron microscopy (SEM) and energy dispersive X-ray spectroscopy (EDS) were applied to identify the element distribution perpendicular to the weld interface.

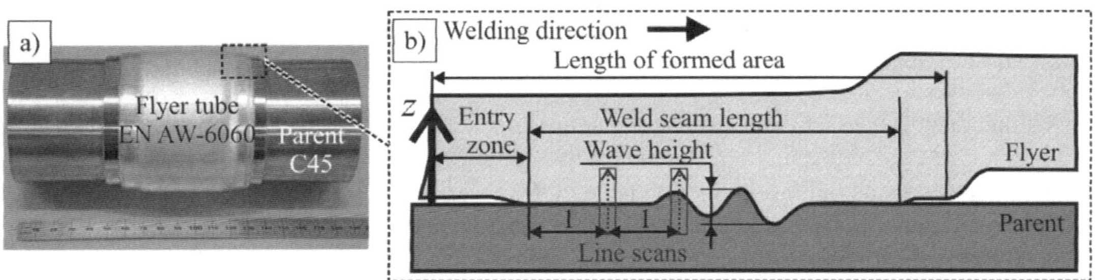


**Figure 3.** a) Joined sample with location of the cross section, b) schematic cross section with characteristic values of the weld seam.

### *2.5. Torsion testing*

Samples for torsion testing were manufactured with the identified welding parameters, see figure 3a). The two weld seams between the aluminum tube and the steel rods were manufactured successively. Compared to real driveshafts, the length of the outer tube was reduced to 70 mm in order to avoid buckling and to assure the ability to test the weld seam itself. A servo-hydraulic axial-torsional testing machine from Inova was used for the torsion tests. The axial force in the samples was controlled to zero during all tests. For quasistatic testing, a constant angular velocity of 10° / minute was applied while the torque was measured. The cyclic testing was performed according to DIN 50100:2016-12 by using an alternating torque (stress ratio R = -1), a constant load amplitude, 20 Hz test frequency and a maximum number of two million load cycles as truncation criterion [17]. The so called staircase test method was applied to evaluate the torque amplitude with 50 % failure probability. The tests were aborted and the samples were classified as failed, if the angle amplitude overstepped the initial value by more than 10 %. Additionally, the aluminum's outer surface in the joining zone of each sample was analyzed in detail with optical microscopy and dye penetrant testing in order to identify any cracks.

## 3. Results and discussion

### *3.1. MPW experiments*

The MPW experiments were divided into two parts. First, simplified parts were joined in order to identify suitable combinations of the joining standoff *g*, the charging energy and the working length $l_w$ in the MPW setup depicted in figure 1. The maximum light intensities detected by the optical sensors at the positions 0° (slot of the working coil) and 180° served as necessary criterions for the tuning of the charging energy and the joining standoff *g*. Then, the collision angle needed for a proper weld formation was adjusted by the working length $l_w$. The details of the procedure for process parameter identification are described [9] and the results are compared to the initial setup 1 in table 4.

**Table 4.** Experimental data for both joining setups.

| | Setup 1 | Setup 3 |
|---|---|---|
| Machine | Bmax MPW50/25 | EmGen5 |
| Outer tube diameter (mm) | 40 | 80 |
| Tube wall thickness (mm) | 2 | 2 |
| Joining standoff *g* (mm) | 1.5 | 1 |
| Working length $l_w$ (mm) | 6 | 6 |
| Charging energy (kJ) | 8 | 18 |
| Maximum tool coil current (kA) | 499±5 (12 tests) | 507±1 (4 tests) |
| Flash appearance time (µs) | 10.49±0.57 (12 tests) | 8.88±0.20 (12 tests) |
| Calculated impact velocity[a] (m/s) | 300 | 239 |

[a] Based on a steady acceleration and 0.5 µs delay between collision and mean flash appearance time [18].

In the second step, generic hybrid driveshafts were produced as demo parts and were partly cut for metallographic analysis, see figure 4. Samples for torsion testing were joined with the same MPW parameters, but consisted of 70 mm long aluminum tubes to avoid buckling during the static tests.

IOP Conf. Series: Materials Science and Engineering **480** (2019) 012033 doi:10.1088/1757-899X/480/1/012033

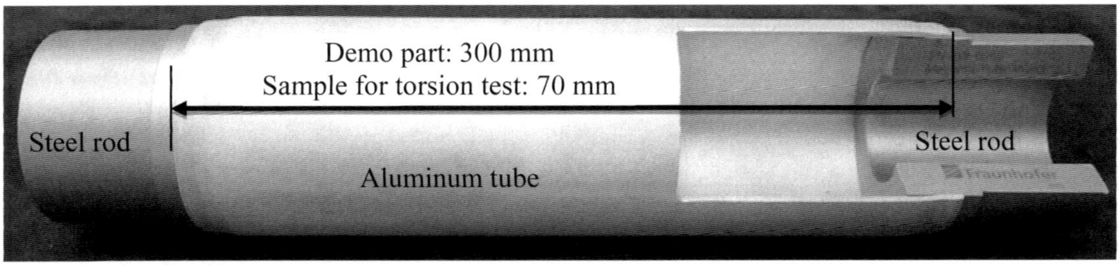


**Figure 4.** Driveshaft dummy, partly cut to reveal the joining zone.

*3.2. Metallographic analysis*

Cross-section polishes of the welding interface were prepared at different locations and optical microscopy was performed to study the morphology of the weld seam. Figure 5 compares the global and local deformation of both joining partners for the reference setup 1 with the actual setup 3.

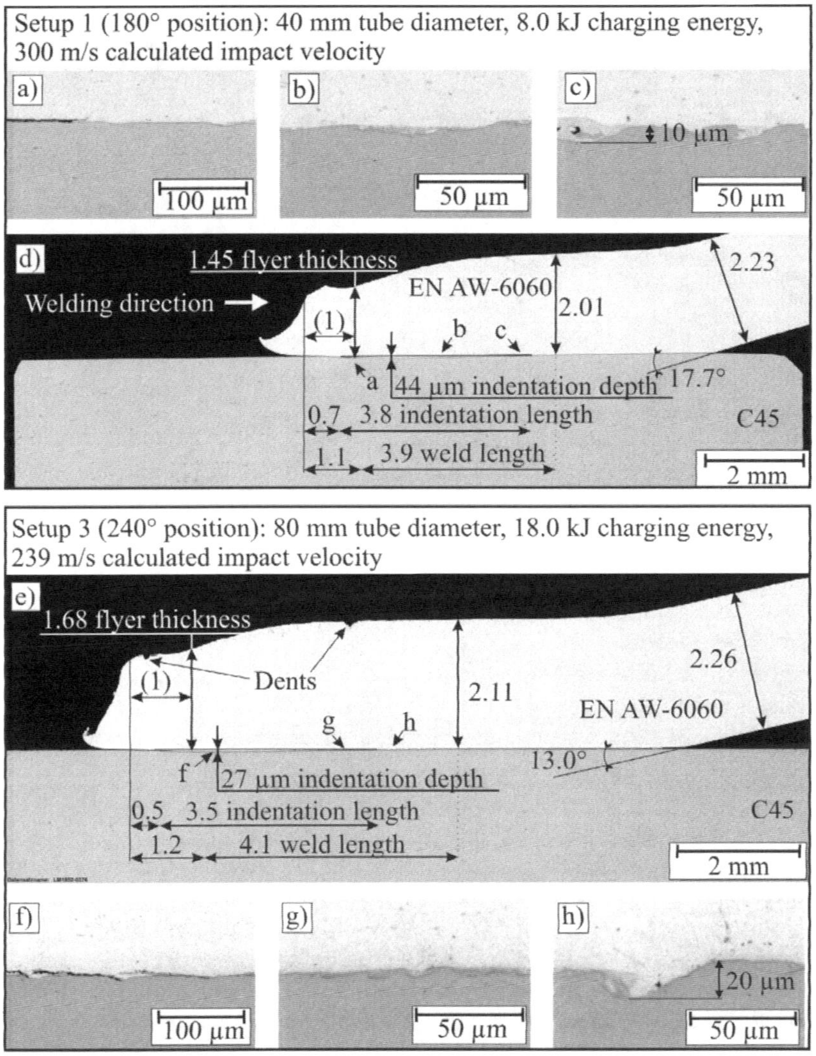


**Figure 5.** Polished cross sections showing the joining zones of selected positions of setup 1 (a-d) [8] and setup 3 (e-h) (all values in millimeter, if not indicated).

The charging energy was significantly increased in setup 3 since the tube diameter and the accelerated material volume, respectively, were doubled. Assuming uniform flyer acceleration, the flash appearance times and corresponding joining standoffs *g* allows for the calculation of the impact velocities to be 300 m/s for setup 1 and 239 m/s for setup 3. The reduction of the kinetic flyer energy at the time of impact did not affect the weld seam length. It corresponds to the decreased indentation depth of the flyer into the parent material and a higher flyer wall thickness very well, especially at the indicated position 1 mm behind the free flyer edge. The final angle between parent and flyer surface was measured and it decreased significantly from 17.7° (figure 5 d) to 13.0° (figure 5 e). It is assumed that the reduced joining standoff *g* leads to a smaller collision angle and a higher collision front velocity in welding direction, respectively. From explosive welding experiments it is known that this leads to higher temperatures in the joining gap [19] and wave heights [20], especially at the end of the joining zone. Dents are visible at the outer flyer surface at setup 3. A structure, which is typical for melted and solidified material is indicated in figure 6. A possible explanation for the extensive heating during the high speed forming process might be the energy input to the outer flyer surface subjected to joule heating. It is increased for the MPW setup 3, since its damping factor is smaller compared to setup 1, see table 3.

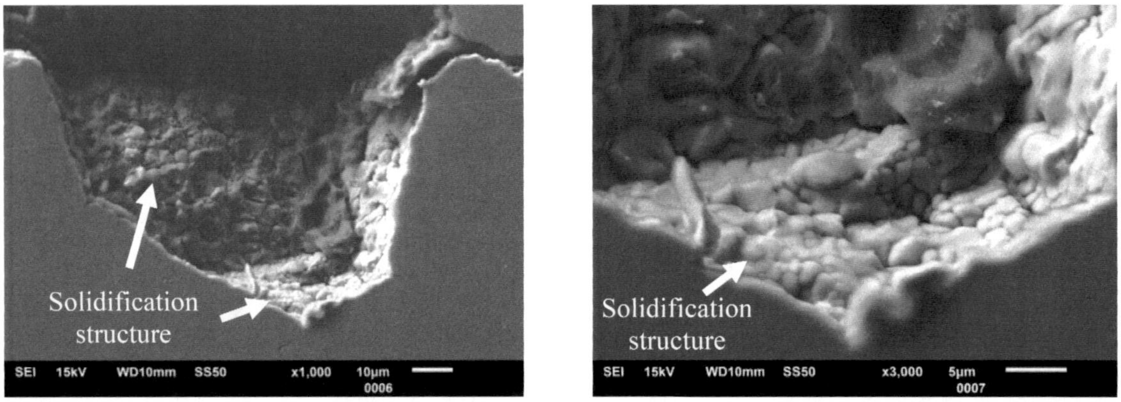


**Figure 6.** Secondary electron images of the solidified material in the dent at the outer flyer surface in figure 5 e) at two different magnifications.

The morphology of the wavy interface changes in welding direction as shown in figure 5. The height of the waves increases and, simultaneously, the mixed zones at the transition from steel to aluminum are extended. Figure 7 indicates the locations of ten parallel line scans for EDS-analysis one millimeter behind the weld seam beginning. The obtained average is plotted in figure 8 and exhibits a zone with a roughly linear transition between the two elements for $0\ \mu m < z < 1\ \mu m$. The linear transition might be attributed to the activation volume during EDS measurements and needs not inevitably reflect the real element distribution within this narrow zone. The element distribution for $z > 1\ \mu m$ is assumed to be a real effect, since it also occurs distinctly at other locations along the weld front propagation. Figure 9 was taken two millimeters behind the weld seam beginning and includes ten line scans, averaged in figure 10. Here, the major part of the roughly 5 µm thick transition zone shows an almost homogenous distribution and consists of approx. 90 atomic percent aluminum and 10 atomic percent iron. A similar ratio was found in a previous study for MPW of aluminum and steel, where the effect of reactive interlayers was investigated [21]. At faraway positions from the initial impact point, the jetting effect acts on the surfaces over a longer time before they get in contact. Thus, it can be assumed that the introduced energy and the resulting surfaces temperatures are higher. Consequently, more material is involved in exothermic reactions that occur at the interface and the transition zone is enlarged compared to figure 7.

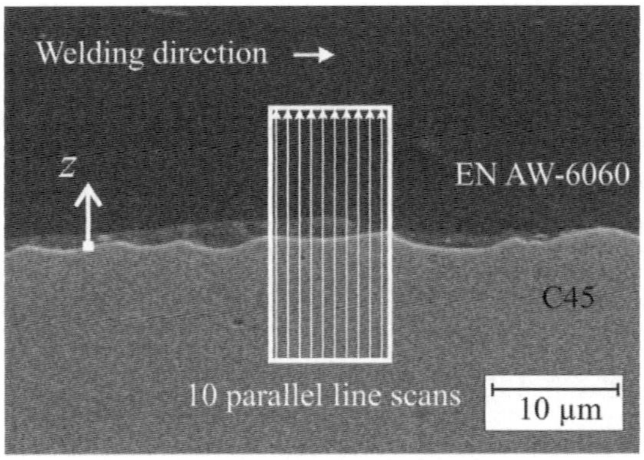


**Figure 7.** Secondary electron image of the interface 1 mm behind weld seam beginning.

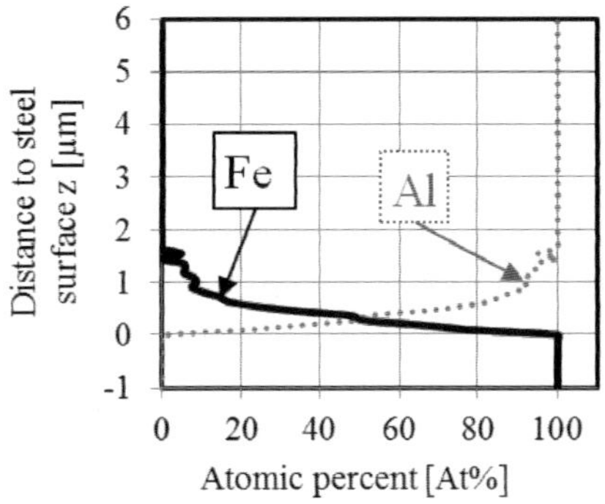


**Figure 8.** Average element distribution from 10 line scans in figure 7 perpendicular to the steel surface.

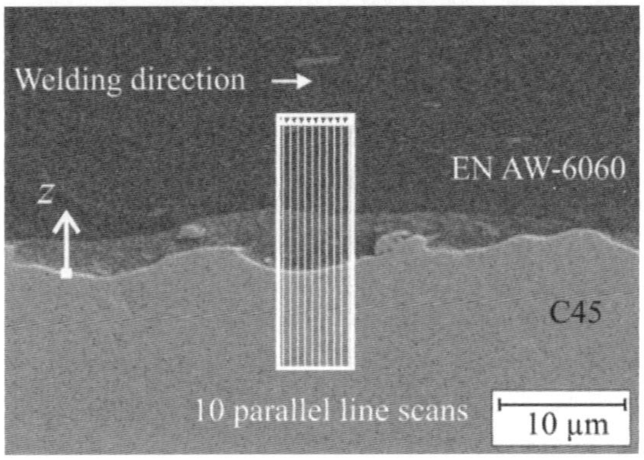


**Figure 9.** Secondary electron image of the interface 2 mm behind weld seam beginning.

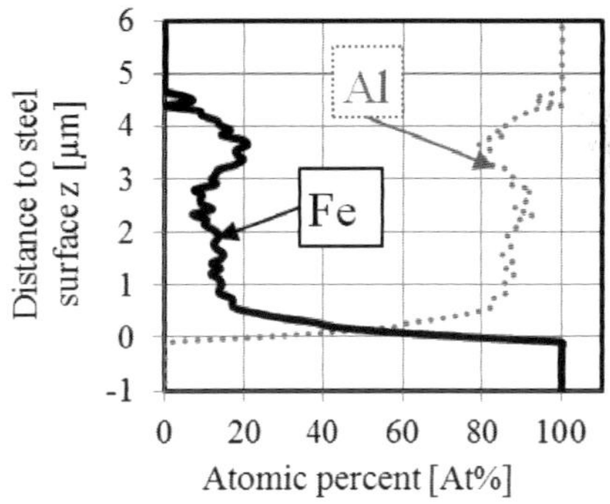


**Figure 10.** Average element distribution from 10 line scans in figure 9 perpendicular to the steel surface.

*3.3. Torsion testing*

Table 5 lists the results of the static and cyclic torsion tests with the parameters described in chapter 2.5. Although statistical evidence in support of the results is not provided by the small sample number, clear tendencies can be derived. For both loading conditions, the maximum torque or torque amplitude until failure, respectively, was four times higher for the setup with 80 millimeter tube diameter. This result was expected since the same tube material was used in both setups, but the polar section modulus $Z_p$ of setup 3 was four times higher, see table 1. During the quasistatic tests, no buckling occurred and all samples failed by a circumferential rupture in the tube's base material next to the weld. Cyclic loading lead to crack initiation in the aluminum tube at the tapered area next to the weld seam and crack growing was detected in both axial orientations. Once the crack reached the welded zone, the propagation direction became radial, leading to a circumferential rupture next to the weld seam.

Finally, the maximum static torque is divided by the cross-sectional area of the tube, which is proportional to its weight, see table 5. This specific value is doubled for setup 3 compared to setup 1, which is a clear improvement in the design from a lightweight point of view.

**Table 5.** Comparison of the static and cyclic torsional load bearing capabilities of magnetic pulse welded drive shaft dummies in two different joining setups.

| | Setup 1 | Setup 3 |
|---|---|---|
| Outer tube diameter $D_e$ (mm) | 40 | 80 |
| Tube wall thickness (mm) | 2 | 2 |
| Polar section modulus $Z_p$ (mm$^3$) | 4322 | 18,648 |
| Cross-sectional area (mm$^2$) | 239 | 490 |
| Static tests: average maximum static torque (Nm) (3 samples) | 755±2 | 3,181±12 |
| Cyclic tests: torque amplitude for 50 % failure probability (Nm) (6 samples) | 124±8 | 510±14 |
| Maximum static torque per cross-sectional area (Nm/mm$^2$) | 3.16 | 6.49 |

## 4. Conclusion

The aim of the study was to increase the torsional load bearing capabilities of hybrid drive shafts consisting of aluminum tubes and steel rods, which are joined by MPW. To reach this goal, the tube diameter was doubled, while the material and tube wall thickness were kept equal. Adjusting the joining standoff, charging energy and working length was realized by using the impact flash intensity as a necessary welding criterion according to [9]. Aluminum tubes (EN AW-6060 T66) with an outer diameter of 80 mm were successfully welded with non-contoured steel parent parts (C45) at the lower process boundary. An impact velocity of approx. 239 m/s was sufficient to establish a four millimeter long circumferential weld seam. The impact velocity of the flyer was decreased compared to previous joining setups, which is reflected in the reduced global plastic deformation of both joining partners. Interestingly, the height of the characteristic interfacial waves increased. It is assumed that the reduction of the joining standoff and the collision angle, respectively, lead to an increased collision front velocity and thus, higher temperature in the joining gap [19], which allows for increased local deformations. EDS line scans perpendicular to the weld interface revealed a transition zone with an almost homogenous element distribution of 90 atomic percent aluminum and 10 atomic percent iron. Depending on the distance to the start of the weld seam, they differ with respect to their width. The mechanical performance of the generic drive shafts was proved in quasistatic and cyclic torsion tests and found to be on a four times higher level compared to the reference setup.

## Acknowledgments

This work is based on the results of subproject A1 of the priority program 1640 ("joining by plastic deformation"); the authors would like to thank the German Research Foundation (DFG) for its financial support. We also acknowledge the support of the HLD-HZDR, member of the European Magnetic Field Laboratory (EMFL).

## References

[1] Groche P, Wohletz S, Brenneis M, Pabst C and Resch F 2014 Joining by forming - A review on joint mechanisms, applications and future trends *Journal of Materials Processing Technology* **214** 1972–1994

[2] Cai W, Daehn G, Li J, Mishra R, Vivek A, Khan H and Komarasamy M 2018 A state-of-the-art review on solid-state metal joining *J. Manuf. Sci. Eng.* doi: 10.1115/1.4041182

[3] Bmax 2016 Magnetic Pulse Welding - The ultimate solution for driveshaft manufacturers https://www.bmax.com/wp-ontent/uploads/2018/03/wp2015_mpw_driveshaft3.pdf accessed 02 Mar 2016

[4] Shanthala K, Sreenivasa T N, Choudhury H, Dond S and Sharma A 2018 Analytical, numerical and experimental study on joining of aluminium tube to dissimilar steel rods by electromagnetic pulse force *J Mech Sci Technol* **32** 1725–1732 doi: 10.1007/s12206-018-0328-0

[5] Shribman V, Gafri O and Livshitz Y 2001 Magnetic pulse welding for automotive materials *Proceedings of the 2001 Global Powertrain Congress*, ed D Roessler pp 148–156

IOP Conf. Series: Materials Science and Engineering **480** (2019) 012033 doi:10.1088/1757-899X/480/1/012033

[6] Yablochnikov B A 2006 Method of magnetic pulse welding an end fitting to a driveshaft tube of a vehicular driveshaft (US 7,015,435) http://pdfstore.patentorder.com/pdf/us/435/us7015435.pdf. accessed 19 Apr 2017

[7] Bellmann J, Lueg-Althoff J, Kirchhoff G, Schulze S, Gies S, Beyer E and Tekkaya A E 2016 Magnetic pulse welding: Joining within microseconds –High strength forever *Proceedings of the 10th International Conference on Trends in Welding Research and 9th International Welding Symposium of Japan Welding Society*, ed American Welding Society and Japan Welding Society pp 91–94

[8] Bellmann J, Beyer E, Lueg-Althoff J, Gies S, Tekkaya A E, Schettler S and Schulze S 2017 Targeted weld seam formation and energy reduction at magnetic pulse welding (MPW) *eBIS* **2017** 91–102 doi: 10.17729/ebis.2017.5/10

[9] Bellmann J, Lueg-Althoff J, Schulze S, Gies S, Beyer E and Tekkaya A E 2018 Parameter identification for magnetic pulse welding applications *KEM* **767** 431–438 doi: 10.4028/www.scientific.net/KEM.767.431

[10] Seeberger 2016 Datasheet AlMgSi (EN AW-6060) http://www.seeberger.net/_assets/ pdf/werkstoffe/aluminium/de/AlMgSi.pdf accessed 08 Sep 2016

[11] Schramm G 2016 Datasheet C45 (1.0503) http://www.guenther-schramm-stahl.de/files/datasheets/C45(1.0503).pdf accessed 08 Sep 2016

[12] Eichhorn G 1976 Analysis of the hypervelocity impact process from impact flash measurements *Planetary and Space Science* **24** 771–781 doi: 10.1016/0032-0633(76)90114-8

[13] Appelgren P, Bjarnholt G, Brenning N, Elfsberg M, Hurtig T, Larsson A, Novac B M and Nyholm S E 2008 Small helical magnetic flux-compression generators: Experiments and analysis *IEEE Transactions on Plasma Science* **36** 2673–2683 doi: 10.1109/TPS.2008.2003966

[14] Bellmann J, Lueg-Althoff J, Schulze S, Gies S, Beyer E and Tekkaya A E 2016 Measurement and analysis technologies for magnetic pulse welding: Established methods and new strategies *Adv. Manuf.* **4** 322–339 doi: 10.1007/s40436-016-0162-5

[15] Bellmann J, Beyer E, Lueg-Althoff J, Gies S, Tekkaya A E and Schulze 2017 Measurement of collision conditions in magnetic pulse welding processes *Journal of Physical Science and Application* **7** 1–10 doi: 10.17265/2159-5348/2017.04.001

[16] Bellmann J 2016 Verfahren und Vorrichtung zur Prozessüberwachung bei einer mittels Kollisionsschweißen gebildeten Schweißnaht DE 10 2016 217 758 B3

[17] DIN Deutsches Institut für Normung e.V. 2016 Load controlled fatigue testing –Execution and evaluation of cyclic tests at constant load amplitudes on metallic specimens and components (DIN 50100)

[18] Lueg-Althoff J, Bellmann J, Gies S, Schulze S, Tekkaya A E and Beyer E 2018 Influence of the flyer kinetics on magnetic pulse welding of tubes *Journal of Materials Processing Technology* **262** 189–203 doi: 10.1016/j.jmatprotec.2018.06.005

[19] Ishutkin S N, Kirko V I and Simonov V A 1979 Thermal action of shock-compressed gas on the surface of colliding plates *Fizika Goreniya / Vzryva* **16** 69–73

[20] Bergmann O R 1984 The Scientific Basis of Metal Bonding with Explosives *High Energy Rate Fabrication - 1984 The 8th Int. ASME Conference* pp 197–202

[21] Bellmann J, Lueg-Althoff J, Schulze S, Gies S, Beyer A and Tekkaya A E 2018 Effects of reactive interlayers in magnetic pulse welding *Proceedings of the 8th International Conference on High Speed Forming*, ed G S Daehn, A E Tekkaya

IOP Conf. Series: Materials Science and Engineering **480** (2019) 012034 doi:10.1088/1757-899X/480/1/012034

# Influence of nitrogen in brazing atmospheres on the hardness of the microstructural constituents of brazed stainless steel joints

**V Fedorov[1], T Uhlig[1], G Wagner[1], A Langohr[2] and U Holländer[2]**

[1] Chemnitz University of Technology, Institute of Materials Science and Engineering
[2] Leibniz University Hannover, Institute of Materials Science

**Abstract.** Stainless steel components, such as heat exchangers for energy and air-conditioning technologies are commonly manufactured using nickel-based brazing fillers in continuous furnaces or vacuum furnaces. In the continuous furnace, the brazing process is often supported by a protective gas. As protective gas is commonly used nitrogen or mixtures of nitrogen and hydrogen. In the vacuum furnace, nitrogen is often used as cooling gas. The arising nitrogen enrichment of the braze metal and the base material influences the mechanical properties of the microstructural constituents of the brazed joints, especially the hardness. In this work, the influence of the nitrogen enrichment on the hardness of the microstructural constituents of the joints was investigated with regard to the process conditions. The amount of nitrogen in the braze metal as well as in the base material was determined using a carrier gas hot extraction technique. The hardness of the microstructural constituents of the brazed joints was determined using nanoindentation due to their small size (few microns). The results of samples, brazed with and without the influence of nitrogen, were compared.

## 1. Introduction

Brazing of stainless steel components is commonly carried out in continuous furnaces or vacuum furnaces. A significant part of these components is brazed with nickel-based fillers, which provide the highest corrosion and oxidation resistance of the resulting joints. Nickel-based brazing fillers offer a great potential for joining of stainless steels [1-3]. By alloying with the elements silicon and phosphorus, the melting temperature of the nickel or the nickel-chromium alloy is reduced and thus brazing temperatures between 1000°C and 1200°C are possible. The processing, structure and properties of the resulting brazed joints have been extensively reported [4-6]. The alloying elements provide excellent flowing and wetting properties, because they are able to eliminate the oxide film on the surfaces of the base materials. However, the alloying elements adversely affect the microstructure and the mechanical properties of the joints when the gap size exceeds 50 µm [7]. In case of continuous furnace brazing, the process is often supported by a process gas. Nitrogen ($N_2$) or mixtures of nitrogen and hydrogen ($N_2$+$H_2$) are commonly used as process gases. In previous work, the influence of the process parameters holding time and $N_2$ content in process gas on the $N_2$ contents of the base material and the braze metal of the joints, brazed using different brazing fillers, was investigated [8]. The brazing process in the continuous furnace was carried out using different holding times of 5 min and 10 min and different process gases, 100% $N_2$ and 50% $N_2$/50% $H_2$. As reference, a brazing process in the vacuum furnace was carried out using a holding time of 10 min and a free cooling. During continuous furnace brazing, a nitrogen enrichment takes place in the braze metal and in the near-surface area of the base material. It was found out that the process parameters holding time and $N_2$

content in process gas do not have a significant influence on the $N_2$ contents of base material and braze metal. The results of corrosion tests show that the arising nitrogen enrichment of the braze metal and the base material have a significant influence on the corrosion behavior of the brazed joints [8]. Due to the nitrogen enrichment, the mechanical properties of the microstructural constituents in the brazed joints may also be affected. These local properties, especially the hardness, influence the mechanical properties of the joints globally. The knowledge about the properties of the microstructural constituents allows the determination of potential points of crack initiation inside the joints. In this work, the influence of the nitrogen enrichment on the hardness of the microstructural constituents in brazed joints is investigated with regard to the process conditions. The nitrogen content in the braze metal as well as in the base material is determined using the carrier gas hot extraction technique (CGHE). The hardness of the microstructural constituents in the brazed joints is determined using nanoindentation. The results of samples, brazed with and without the influence of nitrogen, are compared.

## 2. Experimental procedures

Austenitic stainless steel (AISI 304) was used as base material. The components to be brazed (circular and semi-circular blanks with a diameter of 30 mm) were made of finely rolled sheets by laser cutting, which leads to burr-free and distortion-free samples. The thickness of the produced brazed joint was adjusted at 50 μm by means of inserted spacer foils. The quantity of applied filler was the same for all samples and was dimensioned, such that a joint thickness of at least 50 μm was ensured on average. The brazing filler Ni60CrPSi was applied on the base material as powder. The use of binder was avoided to eliminate carbon residues and their consequences, for example, the formation of chromium carbide in the braze metal. The brazed single lap samples, used for the analysis of the nitrogen enrichment were produced at the temperature of 1125°C in a conveyor belt furnace applying a process gas. The brazing temperature was recorded as sample temperature by a trailing thermocouple attached just below the sample surface. A content of 100% $N_2$ and a holding time of 10 min were used, which should lead to noticeable effects according to [8]. As reference, a brazing process in the vacuum furnace was carried out using a holding time of 10 min and a free cooling. The brazing temperature was measured by a thermocouple, attached in the base material below the brazed surface. The heating rate and cooling rate was adjusted at 50 K/min in both furnaces. The quantitative determination of the nitrogen enrichment in the braze metal as well as in the base material is possible by means of CGHE. For this purpose, sample masses of about 100 mg are required. The process principle is described by Bagel [9]. Specific areas of the overlap samples are extracted by milling using a cutter with an associated cemented carbide indexable insert. The sample preparation for the analysis of the nitrogen enrichment is schematically presented in figure 1. The surface of the base material is cut off on the upper part of the sample in a depth of 100 μm. The braze metal is cut out of the braze fillet and the wetted surface without contact to the base material. The emerging material chips are used for CGHE.

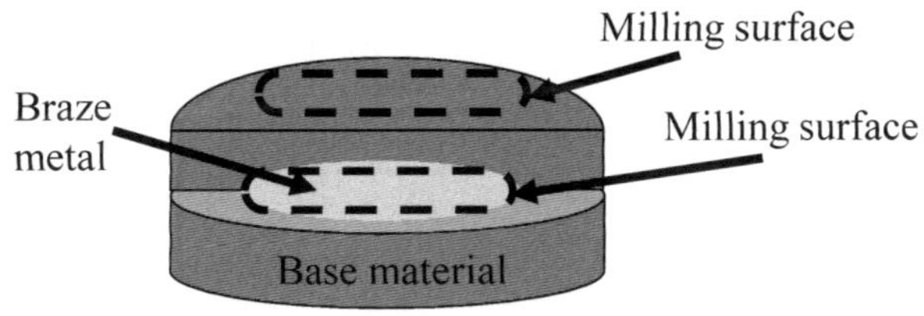


**Figure 1.** Scheme of sample preparation for the analysis of nitrogen enrichment

The nanoindentation experiments were carried out using a fully calibrated nanoindentator UNAT of the company Asmec GmbH equipped with a standard Berkovich indenter. The process principle is described in DIN EN ISO 14577-1 [10]. The indentation test consisted typically of three steps: loading

IOP Conf. Series: Materials Science and Engineering **480** (2019) 012034 doi:10.1088/1757-899X/480/1/012034

(10 s), holding at a maximum force of 5 mN (5 s) and unloading (4 s). The calibration for the maximum force took place on sapphire and quartz. The evaluation of the measurement results was performed using a Poisson's ratio of $\nu = 0.31$ according to the Oliver and Pharr method [11].

## 3. Results and discussion

### *3.1. Microstructure of the brazed joint*

In figure 2, the microstructure of the vacuum brazed joint inside the gap is presented. The microstructure of joints, brazed in process gas, is similar. Due to the diffusion of Fe into the braze metal, Fe-enriched Ni-Cr-Si solid solutions (1) are formed at the interface to the stainless steel and inside the braze metal. Additionally, the resulting braze metal consists of Cr-enriched Ni-P intermetallics (2) and chromium phosphides (3).

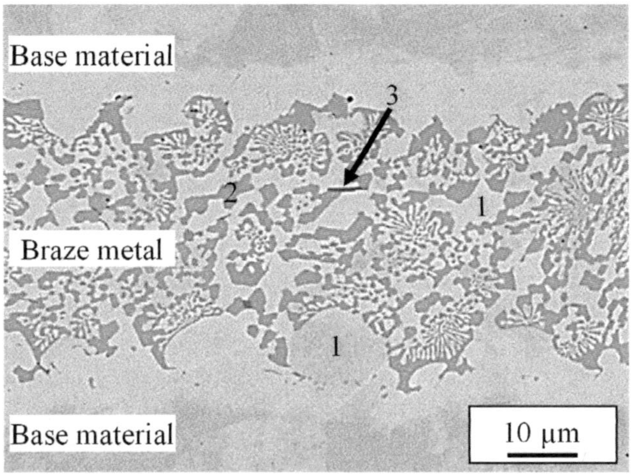


**Figure 2.** Microstructure of the vacuum brazed joint:
1 – Fe-enriched Ni-Cr-Si solid solution,
2 – Cr-enriched Ni-P intermetallic,
3 – chromium phosphide

### *3.2. Analysis of the nitrogen enrichment*

In the as-delivered condition, the nitrogen contents are 0.04 wt.% in the brazing filler Ni60CrPSi and 0.07 wt.% in the base material AISI 304. As expected, these values are relatively low. The vacuum brazing process does not influence the $N_2$ contents of the base material and the braze metal in comparison to the as-delivered condition. After vacuum brazing, $N_2$ contents of 0.04 wt.% in the braze metal and 0.07 wt.% in the base material are determined. In contrast to that, samples, brazed in the continuous furnace, are strongly affected by the process gas. The $N_2$ content in the braze metal increases to 1.14 wt.% while the base material is enriched by $N_2$ to a value of 0.4 wt.%. In comparison to the as-delivered condition, the $N_2$ contents of the base material and the braze metal increase significantly. The enhanced diffusion of nitrogen atoms into the braze metal and the base material is caused by the high brazing temperature as it is known from nitriding processes of steels [12]. Additionally, it can be seen that the braze metal is more intensively enriched with $N_2$ in comparison to the base material. The difference can be explained by the fact that liquid metals can absorb a higher amount of $N_2$ in comparison to solid metals, which is dissolved in the braze metal after solidification.

### *3.3. Hardness of the microstructural constituents in the brazed joints*

To investigate the influence of the arising nitrogen enrichment on the mechanical properties of the microstructural constituents in the brazed joints, nanoindentation experiments were carried out on the braze metal and the base material. In order to determine a potential influence of the measuring position within the joints, measurements were carried out on the braze gap in the center of the overlap as well as on the braze fillet. At the adjusted maximum force of 5 mN, indents of about 2 µm in size arise. Due to the small size of the chromium phosphides (less than 2 µm), the determination of their hardness is not possible with this measuring technique. For other microstructural constituents less than 5 µm in size, edge effects of the surrounding phases can occur during the nanoindentation experiments. To avoid these effects, all indents were investigated using a scanning electron microscopy (REM). Various indents in the Fe-enriched Ni-Cr-Si solid solution at different positions are shown in figure 3a. In figure 3b, an indent in the Fe-enriched Ni-Cr-Si solid solution, used for the evaluation of the measured results, is presented. Indents with a distance of less than 2 µm to the surrounding phases were ignored. This procedure was also carried out for the Cr-enriched Ni-P intermetallics. After that, at least 10 values per phase could be used for the evaluation of the measured results.

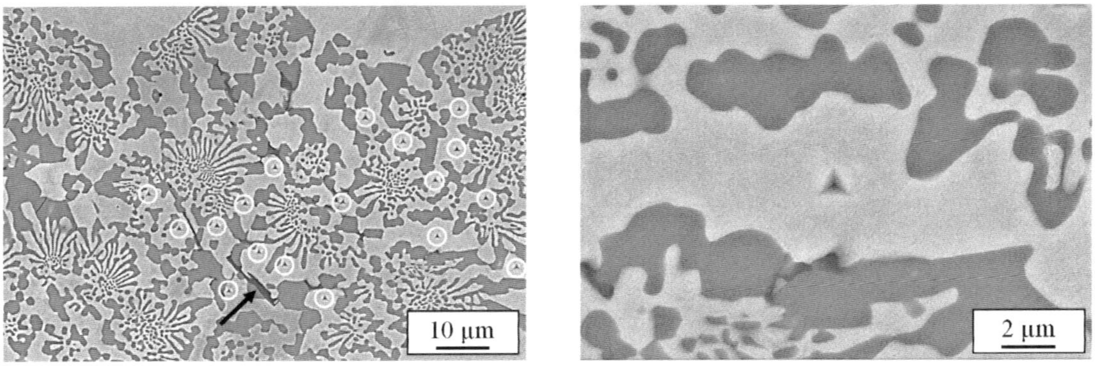


a) b)

**Figure 3.** REM images: a) various indents in the Fe-enriched Ni-Cr-Si solid solution; b) example of an indent, used for the evaluation of the measured results

In figure 4, the results of the nanoindentation experiments are summarized. The nitrogen enrichment does not influence the hardness of the base material significantly. It can be seen that the base materials of vacuum brazed as well as process gas brazed joints show similar hardness values of 3000 MPa on average. The hardness of the Fe-enriched Ni-Cr-Si solid solutions does not differ with regard to the measuring position in the braze gap and in the braze fillet. There is also no significant influence of the brazing process on the hardness of the Ni-Cr-Si solid solutions. The values of 4000 MPa on average are almost the same for joints brazed in vacuum and in process gas. Consequently, the nitrogen enrichment does not have an influence on the hardness of the Fe-enriched Ni-Cr-Si solid solutions. In the braze gap, the hardness values of the Cr-enriched Ni-P intermetallics are not affected by the brazing process. In all cases 14000 MPa are measured at this position. In contrast to that, the hardness of the Cr-enriched Ni-P intermetallics, measured in the braze fillet, strongly depends on the brazing process. In vacuum brazed joints an average value of 13000 MPa was determined. Their hardness in joints, brazed in process gas, increases to 18000 MPa, which corresponds to an increase of 22%. Consequently, the difference in hardness of the Cr-enriched Ni-P intermetallics, formed in the braze fillet, is caused by the nitrogen enrichment. The small differences in the hardness values of the microstructural constituents of the brazed joints can be explained by the scatter, shown as standard deviation in the diagram. Additionally, the hardness of the Cr-enriched Ni-P intermetallics is significantly higher than that of the base material and the Fe-enriched Ni-Cr-Si solid solutions.

Therefore, the crack initiation and the damage of the brazed joints during mechanical testing will preferably take place in these intermetallics.

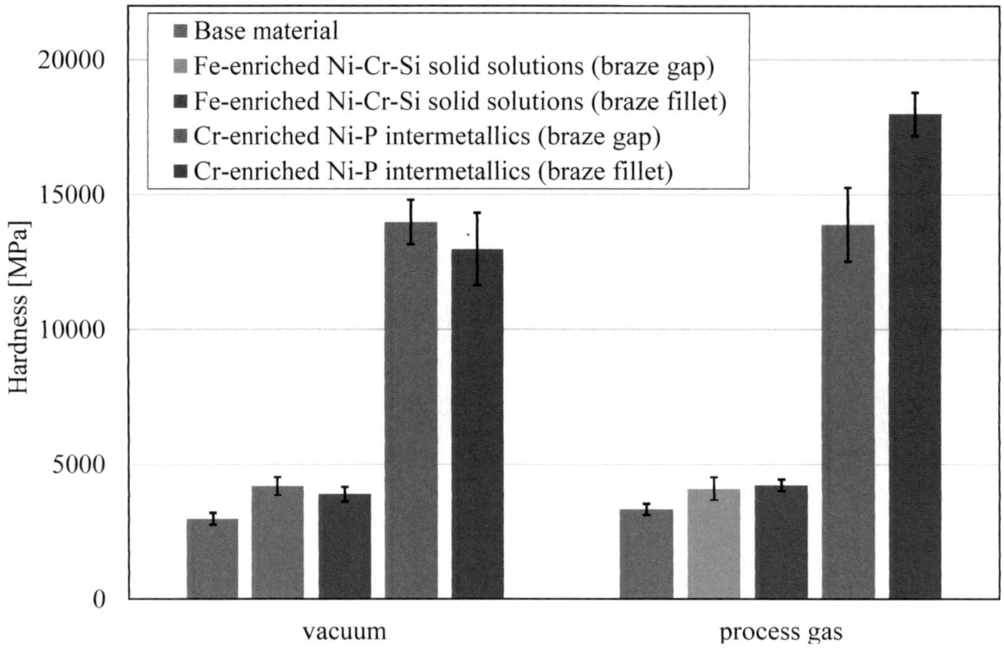


**Figure 4.** Hardness values of the microstructural constituents in the brazed joints

## 4. Conclusions

The investigated joints were produced using the brazing filler Ni60CrPSi at a temperature of 1125°C in a vacuum furnace and in a conveyor belt furnace applying nitrogen as process gas. The $N_2$ content in the braze metal as well as in the base material was determined by CGHE. It was determined that the braze metal is more intensively enriched with nitrogen in comparison to the base material. The influence of the nitrogen enrichment on the hardness of the microstructural constituents of the brazed joints was investigated using nanoindentation. It was found out, that the hardness of the Cr-enriched Ni-P intermetallics, measured in the braze gap as well as in the braze fillet, is significantly higher than that of all other microstructural constituents in the brazed joints. In the braze gap, the hardness of the intermetallics is 14000 MPa, while the base material and the Fe-enriched Ni-Cr-Si solid solutions show hardness values of 3000 and 4000 MPa. In the braze fillet, the hardness of the Cr-enriched Ni-P intermetallics increases to 18000 MPa, while the hardness values of the base material and the Fe-enriched Ni-Cr-Si solid solutions do not change. The high hardness of the Cr-enriched Ni-P intermetallics is caused by the nitrogen enrichment. In further investigations, these results will be correlated with the results of investigations on the fatigue behavior of the brazed joints in order to identify preferred the crack initiation areas in the brazed joints.

**Acknowledgments**
Part of the research was carried out within the scope of the project # 19056 BR funded by the Bundesministerium für Wirtschaft und Energie (BMWi) via the Arbeitsgemeinschaft industrieller Forschungsvereinigungen "Otto von Guericke" e.V. The authors thank the BMWi for financing this project.

IOP Conf. Series: Materials Science and Engineering **480** (2019) 012034 doi:10.1088/1757-899X/480/1/012034

**References**

[1] Hoffmann E E, Patriarca P and Leitten Jr. C F 1956 An Evaluation Of The Corrosion And Oxidation Resistance Of High-Temperature Brazing Alloys *Welding Journal*
[2] Achete M, Lugscheider E and Pelster H 1987 Korrosionsbeständigkeit von Hochtemperaturlötverbindungen in organischen und wässrigen Lösungen *Schweißen & Schneiden* **49** 280-284
[3] Brandl W, Minarski P and Podieschny R 1990 Korrosionsverhalten von gelöteten CrNi-Stahl-verbindungen in Trinkwässern *Hart- und Hochtemperaturlöten* vol 132 (Düsseldorf: DVS)
[4] Rabinkin A 2004 Brazing with (NiCoCr)–B–Si amorphous brazing filler metals: alloys, processing, joint structure, properties *Science and Technology of Welding and Joining* **9** 181-199
[5] Hartmann T and Nuetzel D 2010 Nickel-chromium-based amorphous foils for continuous brazing of stainless steel *Hart- und Hochtemperaturlöten und Diffusionsschweißen* (Düsseldorf: DVS)
[6] Persson U and Bornegård N 2016 New nickel based brazing alloy suitable for very corrosive environments *DVS-Berichte* vol 325 p 224-226
[7] Lugscheider E, Schittny T and Halmoy E 1989 Metallurgical Aspects of Additive-Aided Wide-Clearance Brazing with Nickel-Based Filler Metals *Welding Research Supplement* **9-13**
[8] Fedorov V, Uhlig T, Wagner G, Langohr A and Holländer U 2018 Investigation on the influence of nitrogen in process atmospheres on the corrosion behavior of brazed stainless steel joints *In IOP Conference Series: Materials Science and Engineering* **373**
[9] Bagel A 1985 *Handbuch für das Eisenhüttenlaboratorium* vol 2A (Düsseldorf: Stahleisen)
[10] DIN EN ISO 14577-1 2015 Instrumentierte Eindringprüfung zur Bestimmung der Härte und anderer Werkstoffparameter
[11] Oliver W C and Pharr G M 1992 An improved technique for determining hardness and elastic modulus using load and displacement sensing indentation Experiments *Journal of Materials Research* **7** 1564-1583
[12] Liedtke D 2006 *Wärmebehandlung von Eisenwerkstoffen, II: Nitrieren und Nitrocarburieren* (Renningen-Malmsheim: Expert)

IOP Conf. Series: Materials Science and Engineering **480** (2019) 012035 doi:10.1088/1757-899X/480/1/012035

# Automated surface inspection of friction stir welds by means of structured light projection

**R Hartl*, A Bachmann, S Liebl, A Zens and M F Zaeh**

Institute for Machine Tools and Industrial Management, Technical University of Munich, Boltzmannstrasse 15, 85748 Garching, Germany

* e-mail: roman.hartl@iwb.mw.tum.de

**Abstract.** Friction stir welding is an innovative joining technology that is particularly suitable for aluminium alloys. Various studies have shown a significant dependence of the weld seam quality on the welding speed and the rotational speed of the tool. Frequently, an unsuitable setting of these parameters can be detected by visual examination of resulting surface defects, such as increased flash formation, surface galling, or irregular formation of the surface arc texture. The visual inspection for these defects is often conducted manually and is therefore associated with increased costs and personnel effort. In this work, a method to automatically detect irregularities and features on the weld surface is introduced. It is based on a three-dimensional shape detection of the surface features using structured light projection. For this purpose, the topography of EN AW-5754-H111 sheets, welded in a butt joint configuration, was measured. The data were evaluated and characteristic key structures of the weld seam surface were derived. It was shown that welding irregularities can be detected automatically by an evaluation of the weld seam topography. The results are the basis for the future development of an inline quality monitoring and parameter control method for friction stir welding.

## 1. Introduction

The annual primary aluminium production has increased from 19.1 million tonnes in 1994 to 60 million tonnes in 2017 [1]. The main reasons for this trend are the versatile applications and the properties of aluminium alloys, such as the high specific strength compared to other metals [2]. The increasing demand of aluminium alloys in the industry also increases the need for suitable processes to join aluminium components. Friction stir welding (FSW) is a solid state welding process, which makes it very suitable for materials that are difficult or impossible to weld by fusion welding processes due to problems with the formation of brittle phases, gas pores and cracking [3]. Hence, friction stir welding has already found applications in various sectors of industry. A popular example are the friction-stir-welded tailored blanks, which are subsequently deep-formed to make the center closing panel of the Audi R8 [4]. KAHNERT ET AL. [5] describe the use of FSW for launcher systems at MT Aerospace.

The weld quality in friction stir welding is affected by various characteristic defects produced during the welding process [6]. The DIN EN ISO 25239-5 standard [7] provides an overview of the different types of defects that can occur at FSW of aluminium and its alloys. The standard enumerates seven surface irregularities and two internal irregularities. The mentioned surface irregularities are: incomplete penetration (also known as lack of penetration, LOP), excessive root elevation, excessive flash formation, edge offset, seam underfill, irregular seam width, and irregular seam surface. The standard also specifies admissible acceptance criteria for some surface defects. For example, the calculation rule for the permissible seam underfill $h$ for sheets with a minimum thickness $t$ of 2 mm is:

$$h \leq 2\ \mathrm{mm} + 0.1 \cdot t\,. \tag{1}$$

In accordance with DIN EN ISO 25239-5 [7], the surface defects listed therein can be detected by visual inspection, with the exception of insufficient penetration welding. As per DIN EN ISO 25239-4 [8], a 100 % visual inspection of finished friction stir welds is mandatory in order to meet the requirements for the qualification of a welding process for friction stir welding of aluminium. The tunnel defect as well as internal deformations are regarded as internal irregularities according to the standard [7].

ZETTLER ET AL. [9] divided the causes of seam defects at friction stir welding into defects from welding at excessively high temperatures, defects from welding at too low temperatures, and defects caused by geometrical mistakes. Flaws that are generated under too high processing temperatures are visually identifiable through the surface appearance of the welded joint. In extreme cases, the surface appears to contain blisters or surface galling. In addition, distinct heat generation can lead to excessive surface flash formation. Visually discernable flaw types as a result of welding at too low temperatures are a surface lack of fill (surface LOF) and tunnel defects. Tunnel defects can easily be confirmed by inspecting the exit hole left after the retraction of the tool probe from the workpiece. Thus, most flaw types that develop due to welding at too cold conditions produce coarse and visually identifiable flaws [10]. Defects from geometrical mistakes include a gap between the workpieces, a thickness mismatch, or a plate thickness variation.

SINHA ET AL. [11] report the possibility to automatically detect surface defects of friction stir welds by means of image data processing with the aim of gathering information about the seam quality. The images of the FSW seams were captured by a camera tracking the weld behind the tool and then processed offline, for example by histogram equalization, contrast stretching, and various filtering techniques. For sound welds, the processed images showed a low variation in the pixel intensities along a selected line in the contouring data. However, the contour plot as well as the grey values along an evaluated line showed larger variations when a probe failure occurred. The authors also noted that an online monitoring system can improve the productivity and quality of the friction stir welding process. They propose a system that uses a camera to record images of the seam, which are then processed using algorithms to detect any defects on the surface. In addition, the suggested online system can estimate the appropriate process parameters for new materials. RANJAN ET AL. [12] describe the automated detection and classification of various surface imperfections resulting from FSW using digital image processing. The localization and expression of the severity of the defects are evaluated individually for each defect type by intensity plots along the weld seam. In order to recognize the defects in the pictures more easily, grey value filters, median filters, Hough transform, image pyramid based processing techniques, and other methods were used. RANJAN ET AL. [12] also express the use of their gained knowledge for online monitoring. The calculation time to run the MATLAB code is given as approximately 3.9 seconds for one frame. This enables an online adaption of the welding parameters. The computer architecture used is not specified.

DENG ET AL. [13] have investigated the influence of the welding speed $v_s$ and the rotational speed $n$ of the tool on the surface topography of the friction stir weld. On the one hand, the authors found that there is a linear relationship between the input parameters and the distance between the individual arc textures on the surface. If, on the other hand, the ratio between the rotational speed $n$ and the welding speed $v_s$ remains constant, the distance between the arc lines on the surface is the same. ZUO ET AL. [14] also investigated the influence of process parameters on the surface topography of friction stir welds. According to the authors, the surface topography plays an important role in the quality of the joint. Another finding was that the distance between the arc lines on the surface decreases with increasing rotational speed $n$ or decreasing welding speed $v_s$. However, the depth of the surface arc texture was independent of the input variable magnitudes.

The cited works on the automated analysis of the seam surface at FSW are based either on the detection of defects through image data or on the influence of process parameters on the topography of the weld seam. For the detection of defects based on image data, a series of image processing steps is necessary. The detection of defects based on the weld topography has not yet been investigated.

In the context of this work, it was studied how surface defects of friction stir welds can be detected and quantified by evaluating the three-dimensional point cloud obtained by structured light projection.

## 2. Hypothesis and methodology

### *2.1. Hypothesis*

Previous work has shown that the welding speed $v_s$ and the rotational speed $n$ of the tool have a significant influence on the surface and thus on the topography of the friction stir weld. An unsuitable setting of these input parameters often leads to weld seam surface defects.

The hypothesis is that it is possible to derive one-dimensional characteristic structures from the three-dimensional point cloud representing the surface topography, which allow an automated identification and quantification of surface features. This would be an extension of the existing approaches and a step towards a fully automated surface inspection during friction stir welding.

### *2.2. Approach*

First, a sufficient parameter window was defined for the welding tests. Both the welding speed $v_s$ and the rotational speed $n$ were varied in a large parameter window in order to produce welds with different surface defects as well as welds without surface irregularities. Two studies, labelled study 1 and study 2, were conducted in order to prove the hypothesis posed in section 2.1. In both studies, the aluminium alloy EN AW-5754-H111 was used. The welds in study 1 (experiments 1 to 16) were produced with a concave shoulder geometry, whereas in study 2 (experiment 17 to 32) a flat shoulder with a spiral contour was employed. After welding, the generated seams were first inspected visually and the obtained surface defects were determined. The weld surface was then recorded by means of structured light projection. The produced three-dimensional point cloud was further processed in order to generate one-dimensional key indicators, which were intended to quantify the investigated quality of the surface.

### *2.3. Experimental set-up*

*2.3.1. Welding experiments.* The welding machine used during the studies was a 4-axis horizontal milling machine MCH 250 from Gebr. Heller Maschinenfabrik GmbH, which had been adapted to perform friction stir welding. The maximum axial force of the system was 30 kN. Two-piece tools consisting of a shoulder and a conical welding probe with an M6 thread and three flats were used for the two studies. The tool geometry for study 1 is given in figure 1. The tool for study 2 was identical with the exception of the shoulder contour. Table 1 lists the most important dimensions for both tools.

Welds with a length of 150 mm were made on 4 mm thick plates of EN AW 5754 H111. The plates were fixed on a backing made of unalloyed tool steel. In both studies, the welding speed vs and the rotational speed n were systematically varied.

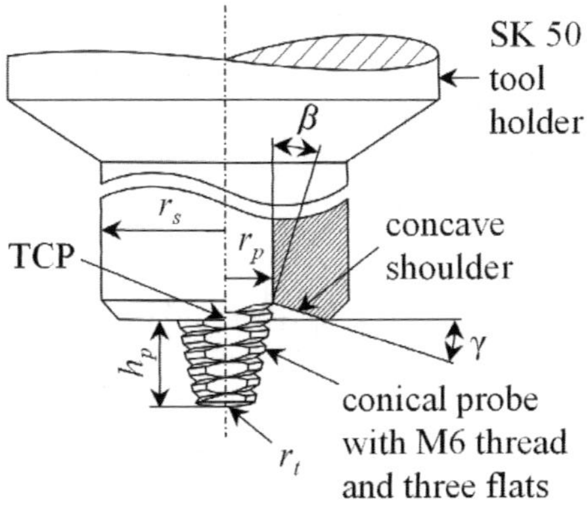


**Figure 1.** Tool geometry for study 1 with the most important dimensions (adapted from BACHMANN ET AL. [15]).

**Table 1.** Dimensions for the used tools.

| | |
|---|---|
| **Tool type 1 and 2** | |
| Probe radius $r_p$ | 3 mm |
| Shoulder radius $r_s$ | 7 mm |
| Conical probe angle $\beta$ | 10° |
| Probe length $h_p$ | 3.85 mm |
| Probe tip radius $r_t$ | 10 mm |
| **Tool type 1** | |
| Concave shoulder angle $\gamma$ | 10 ° |
| **Tool type 2** | |
| Width of the spirals | 0.7 mm |
| Depth of the spirals | 0.7 mm |
| Spiral angle | 360 ° |
| Spiral mean start radius | 3 mm |
| Spiral mean end radius | 6 mm |

The tilt angle of the tool was kept constant at 2° and the dwell time at start of weld was 1 second in all experiments. The plunge depth of the Tool Center Point (TCP) was 0.1 mm throughout the experiments. The geometry of the samples is shown in figure 2. The specimens were dimensioned with a length of 342 mm and a width of 88 mm. A rigid clamping device avoided gaps between the joining partners. To reduce the experimental effort, two 150 mm long welds were made on each sheet configuration.

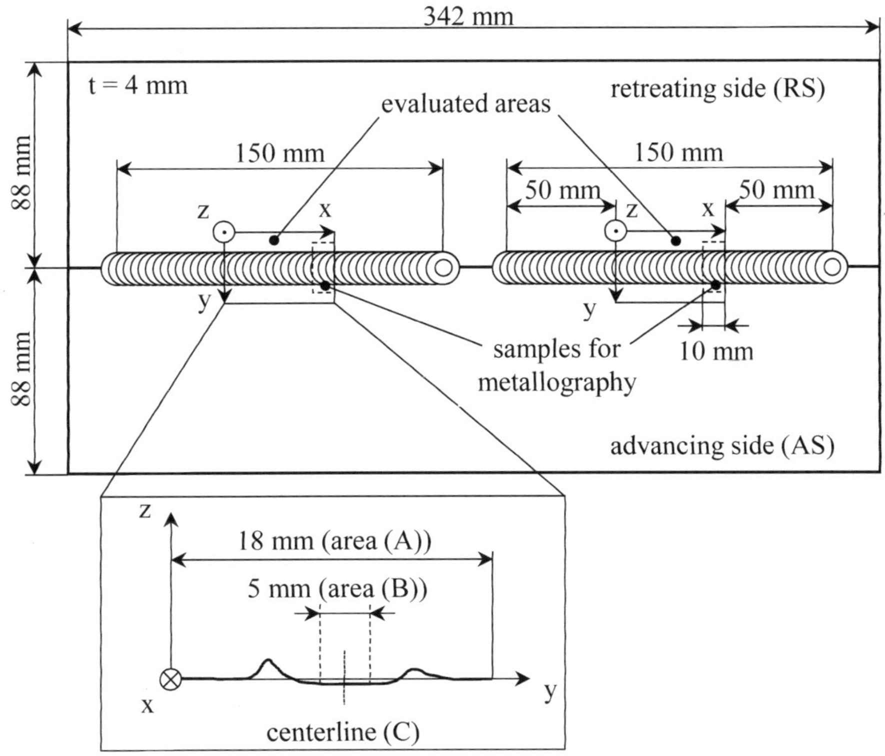


**Figure 2.** Dimensions of the welded plates and positions of the evaluated surface areas.

The investigated parameter window is depicted in figure 3 and was adapted from ZAEH ET AL. [16]. In figure 3, the tested parameter sets are marked as circles and labelled with the experiment numbers for study 1 and study 2 as well as the ratio $n/v_s$ between the rotational speed $n$ and the welding speed $v_s$, respectively. In order to protect the welding equipment, the minimum $n/v_s$ ratio was limited to 1 $mm^{-1}$. The welding speed $v_s$ and the rotational speed $n$ were varied in four equidistant steps, respectively. This resulted in a full factorial experimental plan with 16 welds each for study 1 and study 2.

*2.3.2. Visual inspection.* After welding, the seams were inspected visually. Particular attention was paid to surface defects such as flash formation, surface galling and irregular formation of the surface texture. Additionally, the plunge site and the exit hole were inspected. Often, welding irregularities can already be noticed at the plunge point. Tunnel defects can be detected via the exit hole.

*2.3.3. Topography analysis.* After the visual inspection, the topography of the welds was examined using a three-dimensional profilometer VR-3100 from KEYENCE DEUTSCHLAND GmbH based on phase-coded structured light projection. The white LEDs projected light from two places onto the welds with an angle of incidence of 35°. The reflected light was measured by a CMOS sensor. The resolution of the height measurement (z axis) was 0.1 µm. The smallest measurable difference in the z-direction was 1 µm. The distance between the individual topography points in the width (x-y plane) was 23.549 µm. The measurement accuracy was 3 µm in the height and 5 µm in the width. The repeatability was 0.5 µm in the height and 1 µm in the width. A 12-fold magnification was used, resulting in a viewing measurement range of 24 mm x 18 mm for each frame. The first 50 millimeters as well as the last 50 millimeters of the welds were not used for evaluation, as the welding conditions may not have been stationary there. This resulted in a 50 mm long seam section that was used for the evaluation of the surface. To enable an evaluation of the 50 mm long weld piece, at least two single frames had to be stitched. Depending on whether the topography of the entire seam, a section of the inner seam surface, or along a single line should be evaluated, either the area (A) or (B), or the centerline (C) was investigated (see figure 2). The zero height was always referenced to the sheet surface. Additionally, existing inclinations of the sheet metal were averaged out. These can occur due to sheet unevenness or distortion during welding. As the distance between the individual topography points in the x-y plane was 23.549 µm, this led to 2123 height information points along the welding direction. The weld seam width on the surface was approximately 14 mm as the shoulder diameter was 14 mm. This resulted in 594 height information points over the weld width. Thus, approximately 1.26 million topography points were evaluated in total per weld seam.

*2.3.4. Metallography.* Metallographic examinations were conducted on individual welds in order to investigate the weld surface more closely. Both, cross sections and longitudinal sections, were produced for this purpose. The samples were taken from the sheets with the aid of a wet cut-off grinder. The positions at which the metallographic samples were taken are labelled in figure 2. The samples were embedded in epoxy resin to stabilize the surface characteristics created during welding and to prevent damage during grinding and polishing. The embedded samples were first surface ground with P320 silicon carbide wet abrasive paper according to the FEPA P grit standard [17] and then finish-ground with P1000 silicon carbide wet abrasive paper, also according to the FEPA P grit standard. Finally, the cuts were micro-etched according to Kroll [18]. The cross sections were recorded with the three-dimensional profile measuring device VR-3100 from KEYENCE DEUTSCHLAND GmbH. Thereby the image mode of the profilometer was used. The longitudinal sections were analyzed with the Nikon MM-40 Measuring Microscope.

IOP Conf. Series: Materials Science and Engineering **480** (2019) 012035 doi:10.1088/1757-899X/480/1/012035

## 3. Results and discussion

### *3.1. Visual surface inspection*

The most important surface defects detected during the visual inspection are indicated in figure 3 by differently coloured rings.

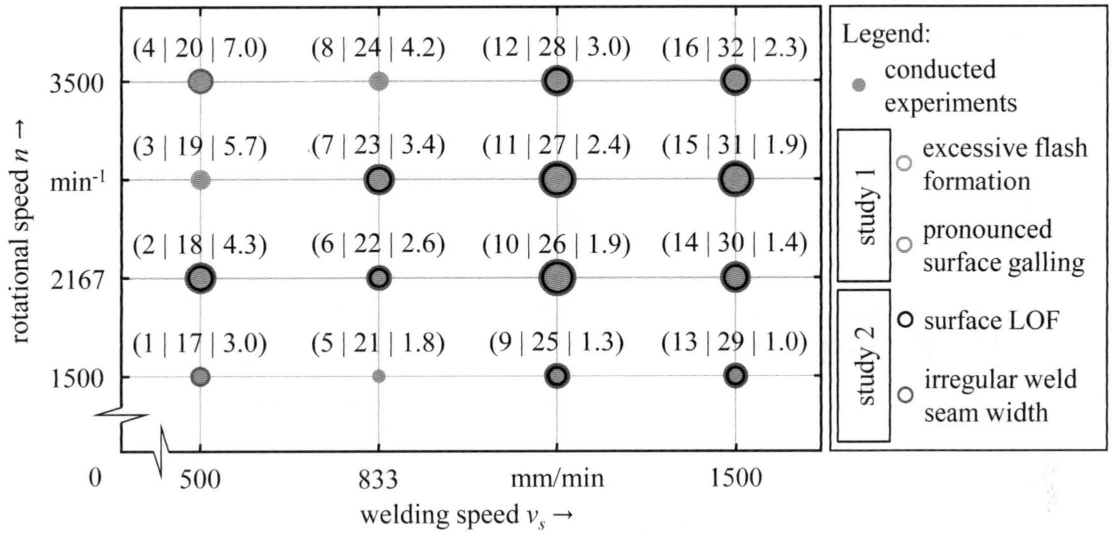


**Figure 3.** Experimental plan with numbering (experiment no. for study 1 | experiment no. for study 2 | $n/v_s$ ratio in $mm^{-1}$) and surface defects (evaluated by visual inspection) noted for each point within the process window.

The visual examination of the experiments from study 1 (tool type 1) led to the following results:

- The most notable surface defects were the occurrence of increased flash formation and surface galling.
- The best welding result by means of the visual inspection was achieved in experiment no. 1 with $v_s$ 500 mm/min and $n$ 1500 mm/min. In this experiment, the flash formation on both the advancing side and the retreating side was very low and uniform. Excessive surface galling or an irregular surface arc texture could not be detected. Furthermore, the weld seam width was very uniform.
- The flash formation was more pronounced on the retreating side than on the advancing side in all experiments conducted in study 1.
- At $n/v_s$ ratios smaller than 3 $mm^{-1}$, the flash formation was more irregular and was chip- or fringe-shaped. If the $n/v_s$ ratio was at least 3 $mm^{-1}$, the flash formation was more continuous and wave-like.
- Surface galling was noticeably detectable in the experiments no. 10, 11, 14, and 15 (see figure 3). Within the examined parameter window, these tests were performed with a high welding speed $v_s$ and a medium rotational speed $n$.
- During the inspection of the exit hole a tunnel defect was not detected in any of the welds.

The visual examination of the experiments from study 2 (tool type 2) revealed the following results:

- The most notable surface defects were the surface lack of fill (surface LOF) as well as a very irregular weld seam width.
- Based on the visual inspection, the best welding results in study 2 were obtained in experiments no. 19 and 20.
- The surface lack of fill appeared as a crack or seam collapse on the surface, so that the seam surface was not closed (all experiments except no. 17, 19, 20, 21, and 24).

IOP Conf. Series: Materials Science and Engineering **480** (2019) 012035 doi:10.1088/1757-899X/480/1/012035

- An irregular weld seam width occurred in almost all experiments with tool type 2. Only in the experiments no. 19, 21, and 24 the weld width was evaluated as regular on the basis of the visual inspection.

Thus, when using the concave tool shoulder (study 1), excessive flash formation and pronounced surface galling occurred and when using the spiral tool shoulder (study 2), the surface lack of fill and an irregular weld seam width were the primary surface defects. The annotated surface defects were therefore specific to the used tool shoulder geometry.

*3.2. Topography analysis*

The topography of the weld seams was investigated with the aim of deriving one-dimensional characteristic structures from the three-dimensional point cloud. The key structures are intended to enable an automated detection of surface defects and features, and thus mirror the visual inspection of the seam surface. Based on the results of the visual inspection, in study 1 the flash formation, the occurrence of surface galling and the regularity of the surface texture were examined through the structured light projection. In contrast, the weld seams from study 2 were investigated with regard to the surface lack of fill and an irregular weld seam width.

*3.2.1. Flash formation.* Since the flash formation was more pronounced on the retreating side than on the advancing side in all experiments from study 1, only the retreating side was used to evaluate the flash formation. Over a weld seam length of 50 mm (area (A) in figure 2), the complete flash volume $f_{Vol}$, the maximum flash height $f_{max}$, and the standard deviation of the flash height $S_f$ were determined. As described in section 2.3.3, a height of zero corresponded to the position of the sheet surface.

**Figure 4.** Cross section of the weld from experiment no. 1 with clearly visible cavities under the formed flash.

When determining the flash volume $f_{Vol}$ by means of structured light projection, it was disadvantageous that in some evaluated experiments cavities below the flash were included in the measurement (see figure 4). This led to measurement errors that could not be avoided. The flash volume $f_{Vol}$ was therefore not suitable to evaluate the flash formation. The cross section in figure 4 was prepared as described in section 2.3.4. Figure 5 displays the maximum flash height $f_{max}$ as well as the standard deviation of the flash height $S_f$ on the retreating side for the 16 experiments conducted in study 1. In experiment no. 1 the maximum flash height $f_{max}$ was 0.708 mm and the standard deviation of the flash height $S_f$ was 0.016 mm. This were the lowest values out of all experiments conducted in study 1, which was consistent with the finding that experiment no. 1 was the best welding result in study 1 by means of the visual examination (see section 3.1). The lower the flash height and the more even the flash formation, the better the result of the visual inspection of the weld surface was in terms of the flash formation. The maximum flash height $f_{max}$ and the standard deviation of the flash height $S_f$ therefore appear to be the most suitable key-indicators to evaluate the flash formation of friction stir welds by means of the surface topography.

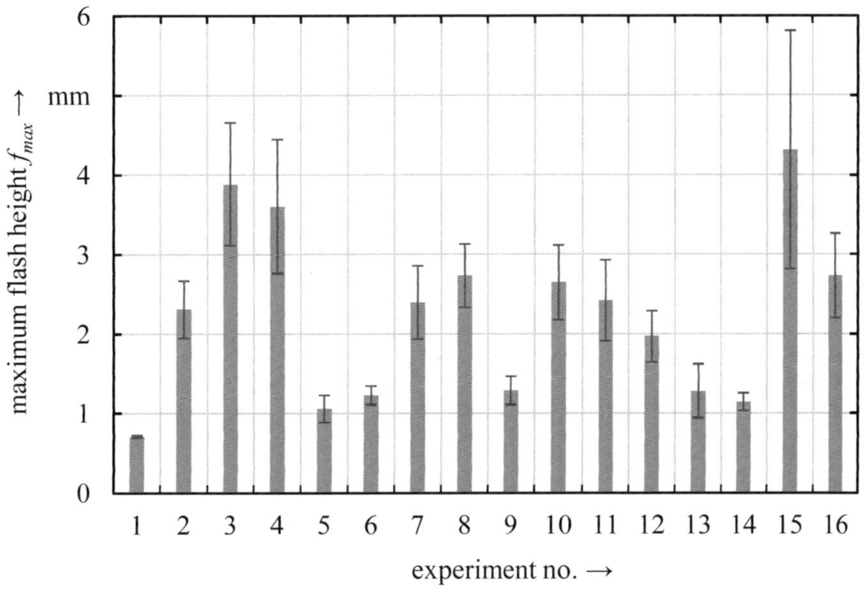


**Figure 5.** Maximum flash height ($f_{max}$) and the standard deviation of the flash height ($S_f$).

*3.2.2. Surface galling.* Another surface defect, which was found during the visual inspection, is surface galling. Surface galling on friction stir welds is embodied by material particles that either deposit loosely on the surface of the weld seam or protrude from the weld surface. To the best of the authors' knowledge, no procedure to quantitatively evaluate the occurrence of surface galling has been described in literature to date. According to the visual examination, the experiment with the most pronounced surface galling in study 1 was experiment no. 10. Figure 6 depicts a comparison of longitudinal sections of experiment no. 1, in which no pronounced surface galling was visible in the visual examination, with experiment no. 10.. In the longitudinal section of experiment 10, the protruding material artifacts are clearly visible, whereas the surface in experiment 1 is very regular and without surface galling. The longitudinal sections in figure 6 were prepared and recorded as described in section 2.3.4.

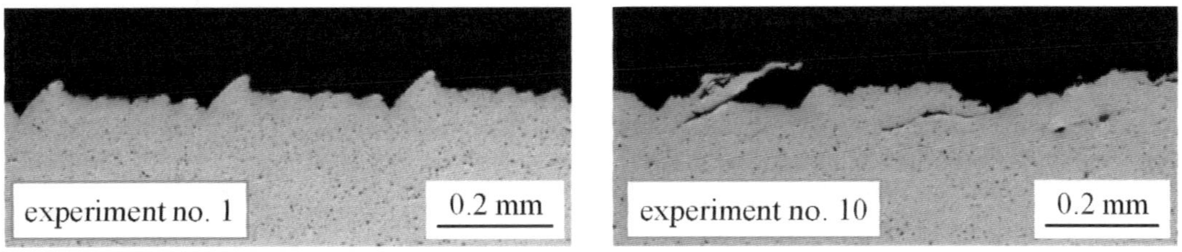


**Figure 6.** Longitudinal sections of the experiments no. 1 and 10.

Figure 7 (a) depicts an image and the elevation picture of the area (B) in top view for the experiments no. 1 and 10. The occurring surface galling in experiment no. 10 gets especially evident in the elevation picture. In particular the red areas with a height of approximately 0.2 mm above the sheet surface display strong surface galling.

IOP Conf. Series: Materials Science and Engineering **480** (2019) 012035 doi:10.1088/1757-899X/480/1/012035

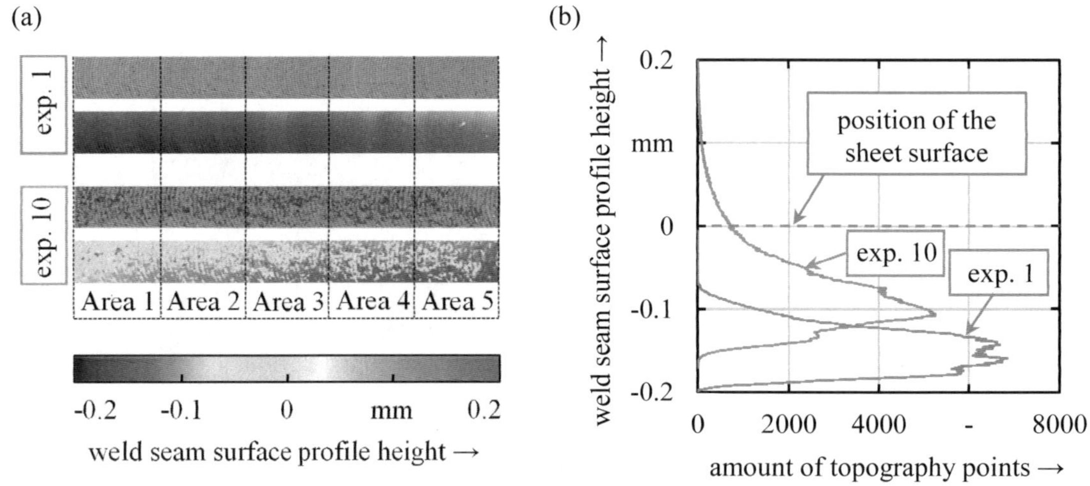


**Figure 7.** (a) Weld surface image and elevation picture of the experiments no. 1 and 10; (b) Distribution of the evaluated surface topography for experiment no. 1 and 10.

Since the surface galling in experiment no. 10 was easy to see from the elevation image in figure 7 (a), an attempt was made to use the height distribution of the topography points to find a way of quantifying surface galling. For that reason, the distributions of the topography points were compiled for the experiments no. 1 and 10 (see figure 7 (b)). In both experiments there were 450076 elevation points in area (B) (see figure 2), which was measuring 50 mm x 5 mm. Since the smallest measurable height difference, as specified in section 2.3.3, was 1 µm, the point cloud was discretized to 1 µm in the direction of the weld profile height (z-axis). Due to the seam underfill, most of the points were located below the sheet surface, which was adjusted at zero height (see section 2.3.3). Figure 7 (b) illustrates that the distribution of topography points in experiment no. 10 had a higher variance than in experiment no. 1. Additionally, the positive skewness of the distribution in experiment no. 10 was more pronounced than in experiment no. 1. This is visible through the longer tail of the distribution in the positive direction of the weld seam surface profile height in experiment no. 10 in figure 7 (b).

By accumulation and normalization of the height distribution of the surface profile, the material ratio curve (also known as Abbott-Firestone curve) results [19]. Figure 8 (a) shows a schematic representation of the material ratio curve. The figure also illustrates the peak material volume $V_{mp}$, the core material volume $V_{mc}$, the core void volume $V_{vc}$, and the dale void volume $V_{vv}$, according to the DIN EN ISO 25178-2 standard [20]. The peak material volume $V_{mp}$ is defined as the area-related volume at a given material content $p$. The value for $p$ can be taken from the ISO 25178-3 standard [21] and is 10 %. Due to the greater positive skewness of the height distribution of the topography points (see figure 7 (b)), it was assumed that the peak material volume $V_{mp}$ gets increased by pronounced surface galling. This hypothesis is reinforced by figure 8 (b). There the material ratio curves are displayed for the experiments no. 1 and 10 for area (B). The colored filled areas below the material ratio curve with a material ratio that is smaller than $p$ (10 %) mark the peak material volume $V_{mp}$. This area is larger in experiment no. 10, which reveals the larger $V_{mp}$ compared to experiment no. 1.

The mean peak material volume as well as the standard deviation of the peak material volume for the areas 1 to 5 (see figure 7 (a)), each measuring 10 mm x 5 mm, were determined. The result is charted in figure 9.

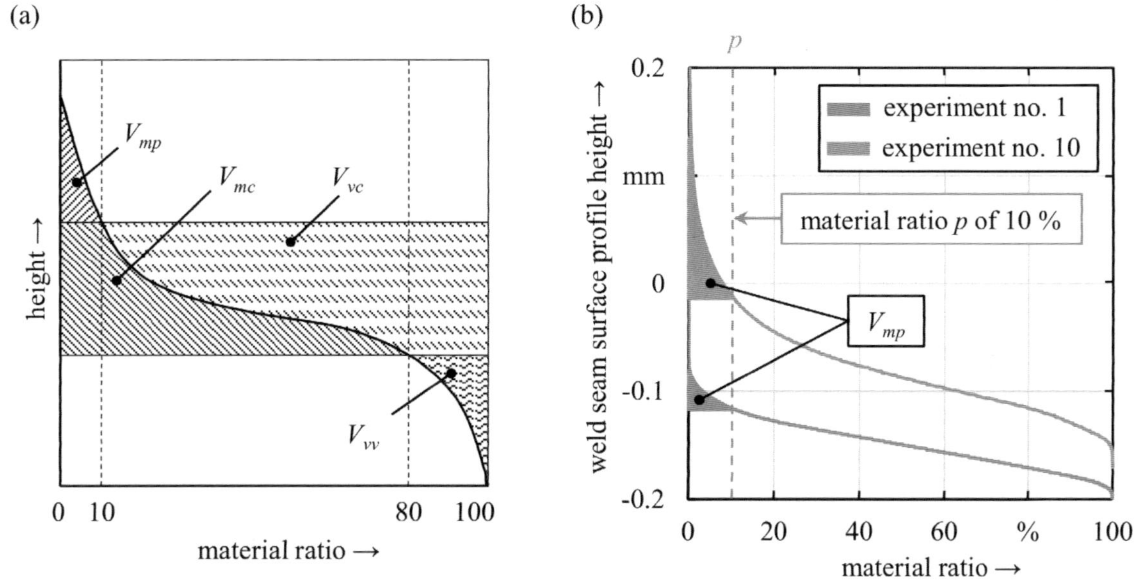


**Figure 8.** (a) Schematic representation of the parameters for the void volume ($V_{vc}$, $V_{vv}$) and the material volume ($V_{mc}$, $V_{mp}$); (b) Material ratio curves for the experiments no. 1 and 10.

Experiment no. 10, which revealed the largest amount of surface galling in the visual examination, also had the highest peak material volume $V_{mp}$ out of all experiments in study 1. Experiment no. 1, which showed the best welding result in study 1 in terms of the visual examination, showed a low area-related peak material volume $V_{mp}$ of 1.02 ml/m². In total, the evaluation of the peak material volume $V_{mp}$ resulted in a good agreement with the detected surface galling by means of the visual inspection. When more surface galling occurred, higher amounts of peak material volume $V_{mp}$ were measured. The peak material volume $V_{mp}$ can therefore be used to detect and quantify occurring surface galling.

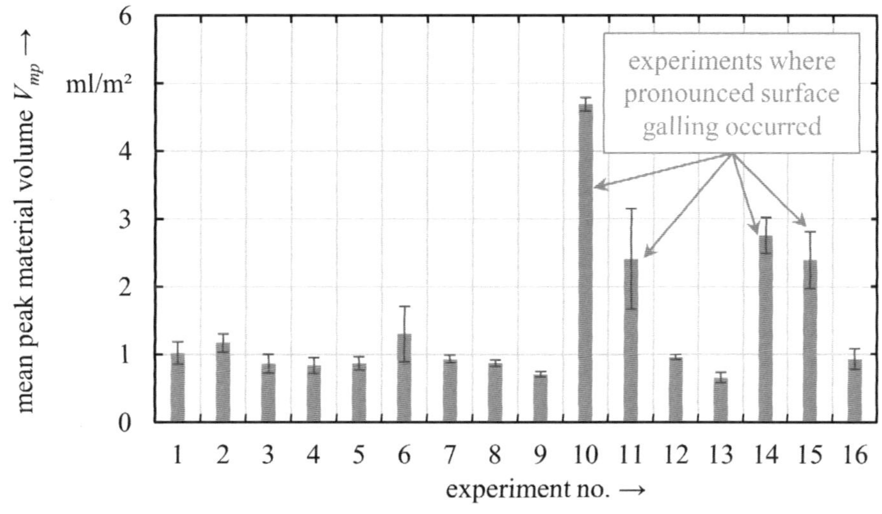


**Figure 9.** Mean peak material volume and standard deviation of $V_{mp}$ for area (B).

*3.2.3. Irregular arc texture formation.* A further quality feature in the visual inspection of FSW seams is the regular formation of the arc texture on the weld surface. The periodicity of the surface arc texture depends on the welding speed $v_s$ and the rotational speed $n$ of the tool. In contrast, the height difference in the texture is independent of both welding parameters [14]. The results of this study suggest that this is an idealization. Depending on the setting of $v_s$ and $n$, irregularities occur on the weld surface, which influence the measured depth of the arc texture. For the evaluation of the surface, the arc texture along the centerline (C) (see figure 2) was examined. Figure 10 illustrates an excerpt of the evaluated arc texture of experiment no 1.

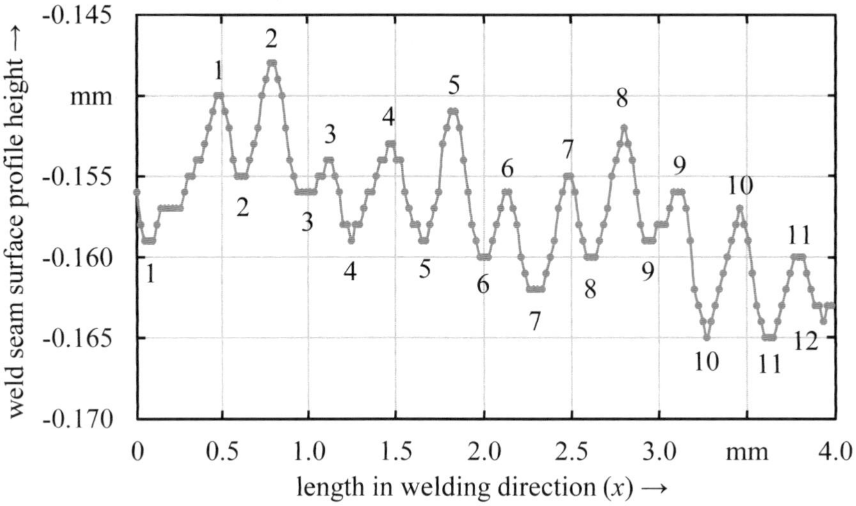


**Figure 10.** Weld seam arc texture for the first 4 mm of the evaluated centerline (C) (see figure 2) of experiment no. 1.

The calibration of 23.549 µm in the x-y plane mentioned in section 2.3.3 resulted in 170 measured points for the topography on the chosen 4 mm long section. The measured points of the topography are connected with linearly interpolated lines in figure 10.

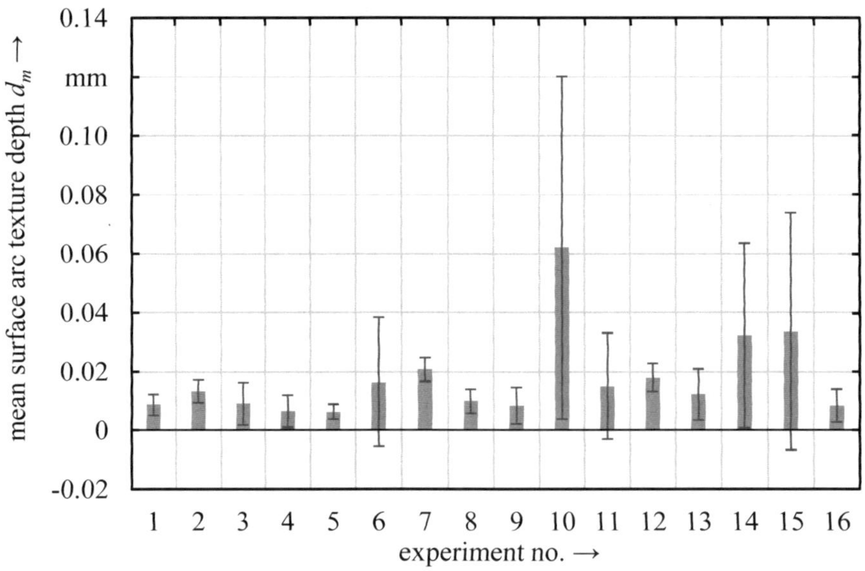


**Figure 11.** Mean surface arc texture depth ($d_m$) and its standard deviation ($S_d$).

By using a MATLAB code, the local minima and maxima along the weld seam surface were identified and the height difference of every local minimum and the following local maximum was calculated. Then, for the evaluation of the regularity of the arc formation, the standard deviation of the arc texture depth was calculated for the 16 experiments conducted in study 1. The results can be seen in figure 11. In experiment no. 5, the standard deviation of the arc depth was the lowest with 0.0026 mm.

Additionally, the numbers of local minima and maxima were counted in the evaluation of the weld seam arc texture (see the numeration in figure 10). The evaluation of the arc texture clarified that the measured number $n_{meas}$ of local minima and maxima derived from the point cloud deviated from the theoretical number $n_{theoret}$ of arcs for some conducted experiments. The theoretical number $n_{theoret}$ of arcs results from the ratio of the rotational speed $n$ to the welding speed $v_s$, multiplied by the length $l$ of the evaluated distance of the weld seam, which was 50 mm:

$$n_{theoret} = \frac{n}{v_s} \cdot l \,. \tag{2}$$

It should be noted that $n_{meas}$ depends on the accuracy of the measurement. Information about the profilometer used in this paper is given in section 2.3.3. By evaluating the ratio $r_{arc}$ between the amount of measured arc numbers $n_{meas}$ to the amount of theoretical arc numbers $n_{theoret}$, a statement can be made about the correct setting of the parameters rotational speed $n$ and welding speed $v_s$:

$$r_{arc} = \frac{n_{meas}}{n_{theoret}} \,. \tag{3}$$

If the ratio $r_{arc}$ was significantly below 1, the $n/v_s$ ratio was too high and vice versa. This is also shown by the trendline depicted in figure 12. Remarkably, the experiment with the best welding result after the visual inspection (no. 1) is almost exactly at the intersection between the trendline and the ideal ratio, which is 1. The closer the experiments in figure 12 were to the intersection of the two lines, the better the surface quality of the weld seam was.

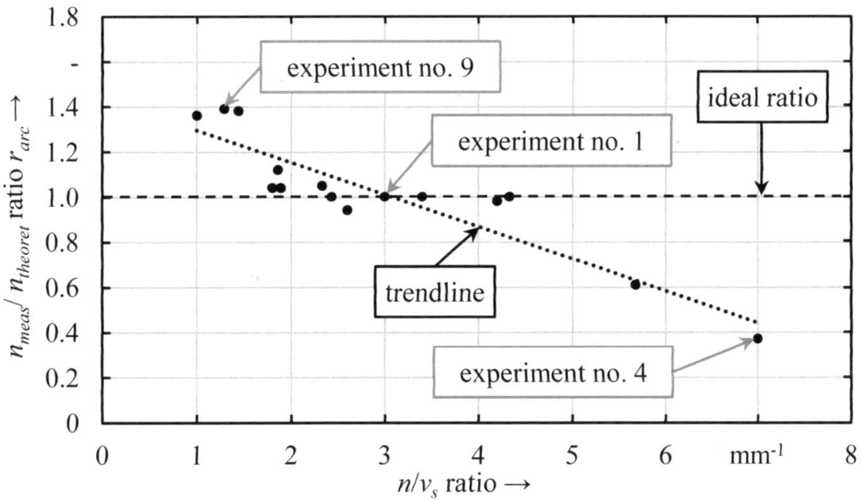


**Figure 12.** Ratio of the measured and the theoretical number of surface arcs $r_{arc}$ as a function of the $n/v_s$ ratio of the experiments conducted in study 1.

Figure 13 compares the surface profiles of longitudinal sections of experiments no. 1 (ideal $r_{arc}$), 4 (lowest $r_{arc}$), and 9 (highest $r_{arc}$). Evidently, experiment no. 1 had a very regular surface arc texture. Both the periodicity of the arc texture and the depth of the single arc textures are more regular than in the other two tests illustrated in figure 13.

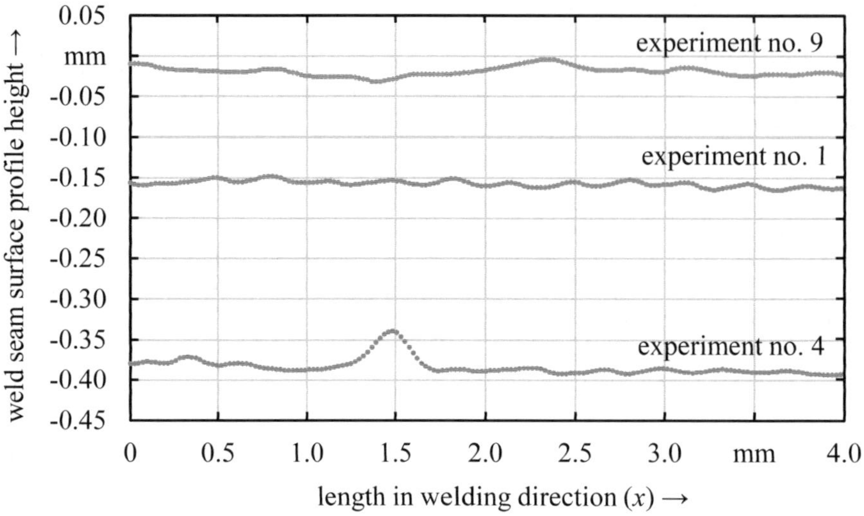


**Figure 13.** Comparison of the surface profile heights along the centerline (C) between experiments no. 1, 4 and 9.

Furthermore, the topography signals along the centerline (C) were subjected to a Fast Fourier Transformation (FFT). The ideal welding result would show a perfect sinusoidal surface profile and thus only the frequency $f_1$ in the frequency spectrum. Thereby the frequency $f_1$ corresponds to the $n/v_s$ ratio.

The results of the FFT for experiments no. 1, 4, and 9 are given in figure 14. In all experiments from study 1, with the exception of experiment no. 4, pronounced spectral components occurred for the frequency $f_1$. The $n/v_s$ ratio was 3 in experiment no. 1, 7 in experiment no. 4, and 1.3 in experiment no. 9 (see figure 3). In some experiments the surface profile also contained remarkable spectral components with the double frequency $f_2$ and in experiment no. 1 even with the triple frequency $f_3$ of the $n/v_s$ ratio. Additionally, pronounced amplitudes outside the frequency $f_1$ and their integer multiples occurred for some experiments.

Experiment no. 1, which showed the highest quality after the visual inspection, evinced the greatest similarity with a perfect sinusoidal oscillation. The frequency spectrum in figure 14 (a) is dominated by the frequency $f_1$. The higher the $n/v_s$ ratios of the experiments were, the lower were the profile components of the frequencies $f_1$, $f_2$, and $f_3$. In experiment no. 4, $f_1$ even showed no pronounced amplitude at all for $f_1$, $f_2$, and $f_3$. This is due to the fact that the surface arc texture characteristic decreases with higher $n/v_s$ ratios. At experiments with lower $n/v_s$ ratios, the noise between the frequencies $f_1$ and its integer multiples increased. Stochastically distributed irregularities occurred over the entire frequency spectrum. This can also be seen in the frequency spectrum of experiment no. 9 in figure 14 (c). Consequently, the frequency spectrum of the surface arc texture can also be suitable for the evaluation of the surface regularity of friction stir welds. The closer the surface arc texture of the friction stir weld comes to the ideal sinusoidal profile, the better the surface quality.

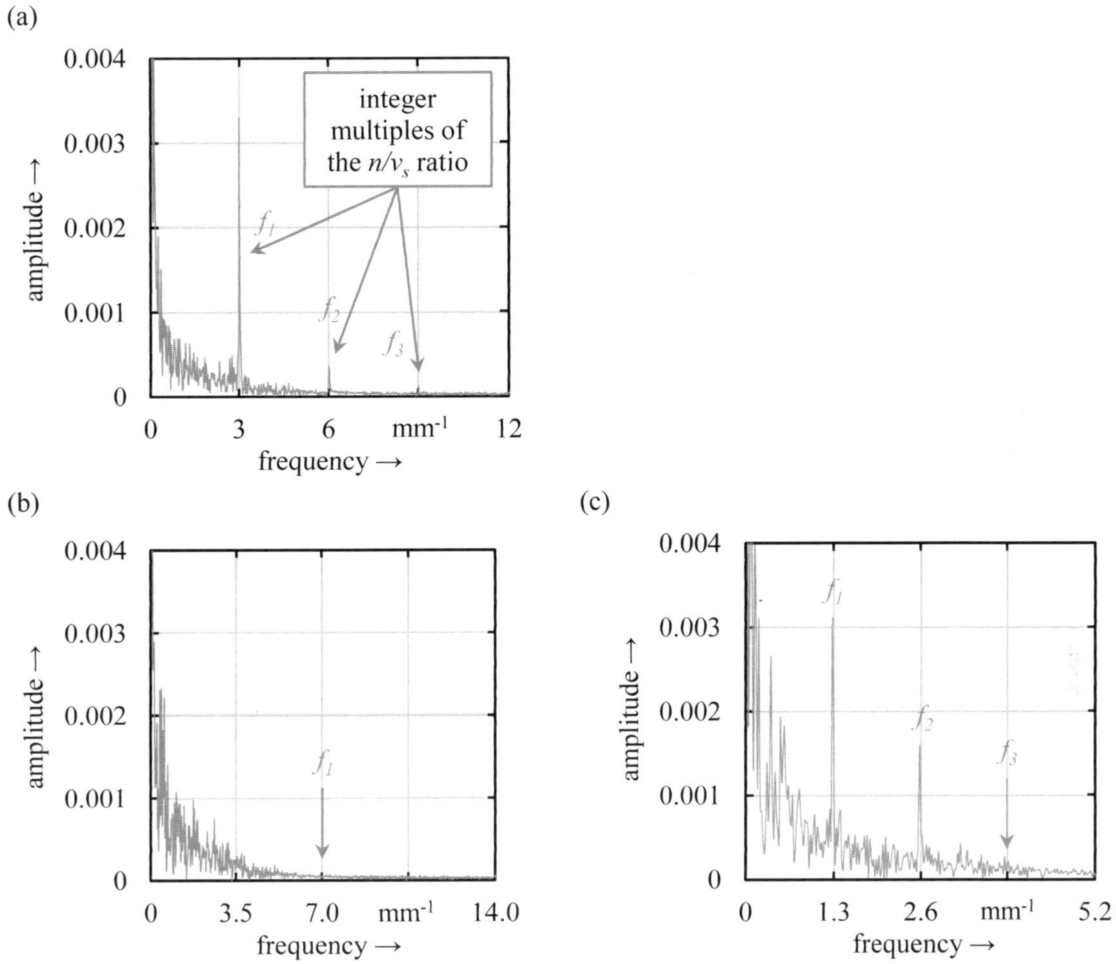


**Figure 14.** Frequency spectra of the surface topographies along the centerline (C) for experiment no. 1 (a), experiment no. 4 (b), and experiment no. 9 (c).

*3.2.4. Surface lack of fill.* The surface lack of fill (surface LOF) occurred exclusively in study 2 using the welding tool with spiral shoulder contour. Figure 15 illustrates the surface height profile of a cross section from experiment no. 22. Thereby, the crack on the surface is clearly visible.

The surface lack of fill can be identified by evaluating the standard deviation of the minima of the cross section profiles because of the very irregular depth of the crack. If the standard deviation of the profile minima Sp,min along the welding direction is small, this indicates a weld seam without surface lack of fill. A high standard deviation of the profile's minima indicates a crack or seam collapse in the seam surface. Figure 16 shows the standard deviations of the profile minima Sp,min along the evaluated 50 mm section (area (A) in figure 2) for each experiment.

Figure 17 demonstrates sections of the experiments no. 19, 25 and 32. In experiment no. 19 no surface LOF has occurred. In experiments no. 25 and 32 the surface lack of fill is clearly visible.

IOP Conf. Series: Materials Science and Engineering **480** (2019) 012035 doi:10.1088/1757-899X/480/1/012035

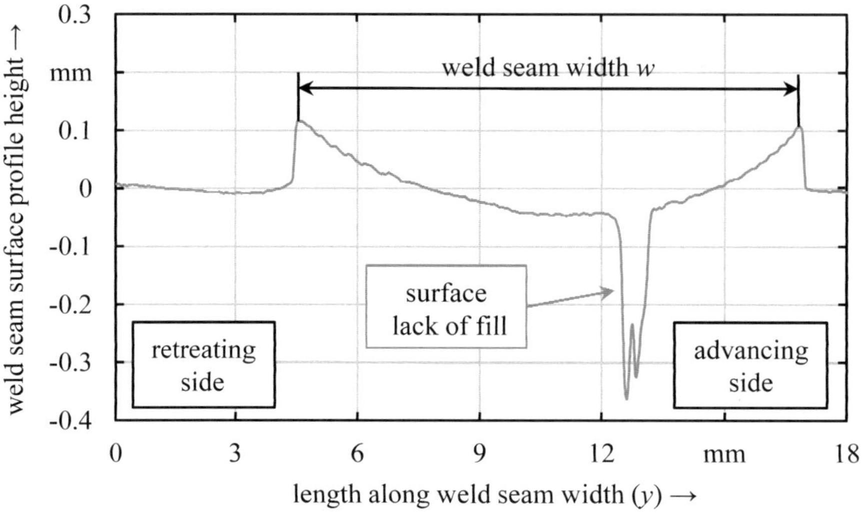


**Figure 15.** Surface profile height from experiment no. 22 with occurring surface lack of fill.

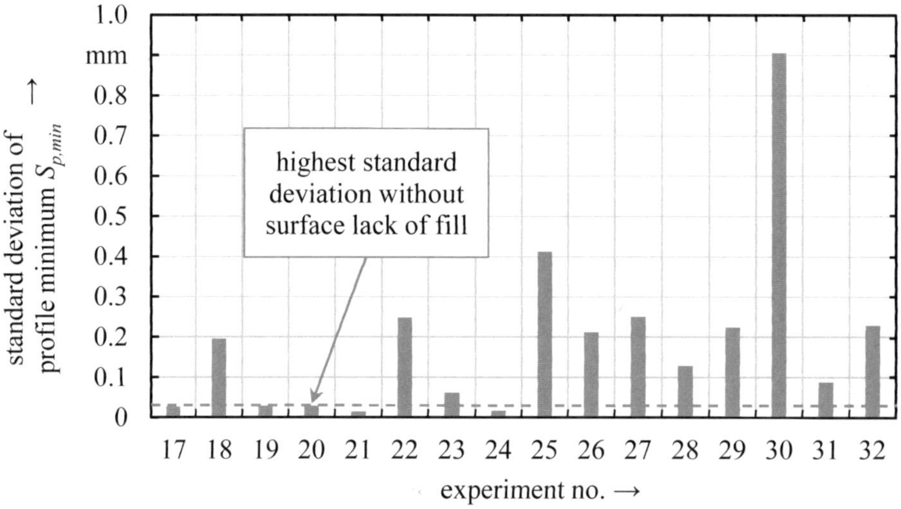


**Figure 16.** Standard deviations of the profile minima ($S_{p,min}$) for study 2.

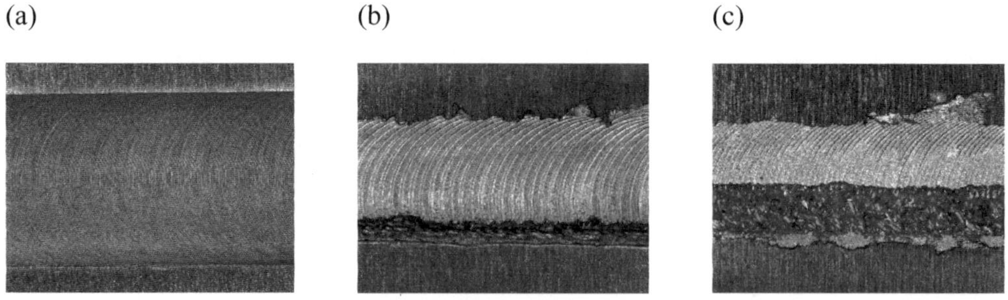


**Figure 17.** (a) Exp. no. 19; (b) Exp. no. 25; (c) Exp. no. 32.

*3.2.5. Irregular weld seam width.* The width of some welds produced in study 2 was very irregular. To evaluate the weld seam width, the y-coordinates of the profile height maxima on the retreating side $max_{RS}$ and on the advancing side $max_{AS}$ were determined for each cross section of the weld in question. The difference of the two determined y-coordinates was defined as the weld seam width $w$ (see figure 15):

$$w = max_{AS} - max_{RS} . \qquad (4)$$

In order to evaluate the uniformity of the weld seam width $w$, the standard deviation along the evaluated 50 mm section of the weld seam was used. Figure 18 shows the mean weld seam width $w_m$ and the standard deviations of the weld seam width $S_w$ from study 2.

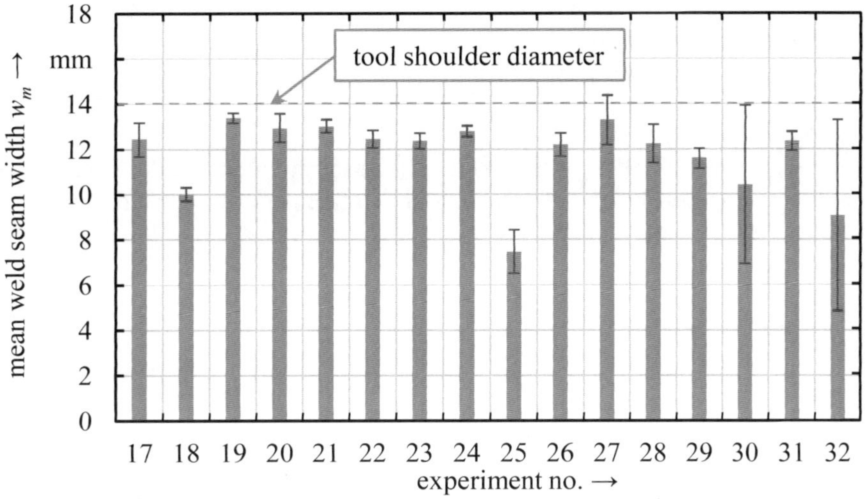


**Figure 18.** Mean weld seam width ($w_m$) and its standard deviation ($S_w$).

Figure 17 illustrates pictures of the experiments no. 19, 25 and 32. In experiment no. 19, the standard deviation $S_w$ of the weld width $w$ was the smallest and thus the weld width the most regular. Experiment no. 25 showed the smallest mean weld seam width $w_m$ and experiment no. 32 the largest $S_w$ and thus the largest irregularity in the weld seam width.

## 4. Conclusions and future work

The aim of this work was to derive one-dimensional indicators from the surface topography of friction stir welds, which enable the detection and quantification of surface defects and features. For this purpose, two test series with two different shoulder geometries were performed. The resulting welds were inspected visually and the surface topography was recorded using structured light projection. Key indicators were derived from the resulting three-dimensional point cloud to describe the surface defects and characteristics that occurred. The key figures determined are summarized in table 1. With the help of the key indicators listed in table 2, frequently occurring surface defects in friction stir welds can be detected and quantified.

The subject of future investigations will be the transfer of the knowledge gained in this work to inline applications. The goal is to control the process parameters in order to optimize the weld surface and thus to increase the weld quality.

**Table 2.** Summary of the determined key indicators to describe the surface defects and properties at FSW by means of structured light projection.

| Surface defect | Key indicator 1 | Key indicator 2 |
|---|---|---|
| Flash formation | Maximum flash height $f_{max}$ | Standard deviation of the flash height $S_f$ |
| Surface galling | Peak material volume $V_{mp}$ | n/a |
| Irregular arc texture formation | Standard deviation of the arc texture depth $S_d$ | Ratio of the measured and the theoretical number of surface arcs $r_{arc}$ |
| Surface LOF | Standard deviation of the profile minima $S_{p,min}$ | n/a |
| Irregular weld seam width | Standard deviation of the weld seam width $S_w$ | n/a |

**Acknowledgments**

The IGF-research project no. 19389 N of the "Research Association on Welding and Allied Processes of the DVS" has been funded by the AiF within the framework for the promotion of industrial community research (IGF) of the Federal Ministry for Economic Affairs and Energy because of a decision of the German Bundestag.

Supported by:

Federal Ministry for Economic Affairs and Energy

on the basis of a decision by the German Bundestag

**References**

[1] U.S. Geological Survey 1995 - 2018 Mineral Commodity Summaries *U.S. Department of the Interior, U.S. Geological Survey (USGS)*

[2] Askeland D R 1996 *Materialwissenschaften* ('Material sciences') (Heidelberg: Springer Spektrum)

[3] Colligan K J 2010 The friction stir welding process: an overview Lohwasser D and Chen Z Friction stir welding - From basics to applications 15–41

[4] Kallee S W 2010 Industrial applications of friction stir welding Lohwasser D and Chen Z Friction stir welding - From basics to applications 118–163

[5] Kahnert M, Mestek M, Windisch M, Tessier I and Okualla M 2012 Preparation of Friction Stir Welding of the Aluminum Alloy AA 2219 for Launcher Applications *9th International Friction Stir Welding Symposium* (Huntsville, Alabama) May 15th – 17th 2012

[6] Soni N, Chandrashekhar S, Kumar A and Chary V R 2017 Defects Formation during Friction Stir Welding: A Review *International Journal of Engineering and Management Research* **Volume-7, Issue-3** 121–125

[7] DIN Deutsches Institut für Normung e. V. DIN EN ISO **25239-5:2011** Rührreibschweißen – Aluminium – part 5: Qualitäts- und Prüfungsanforderungen (Berlin: Beuth Verlag GmbH)

[8] DIN Deutsches Institut für Normung e. V. DIN EN ISO **25239-4:2011** Rührreibschweißen – Aluminium – part 4: Anforderung und Qualifizierung von Schweißverfahren (Berlin: Beuth Verlag GmbH)

[9] Zettler R, Vugrin T and Schmuecker M 2010 Effects and defects of friction stir welds Lohwasser D and Chen Z Friction stir welding - From basics to applications 245–276

[10] Bachmann A, Gamper J, Krutzlinger M, Zens A and Zaeh M F 2017 Adaptive model-based temperature control in friction stir welding *Int J Adv Manuf Technol* **93** 1157–1171

[11] Sinha P, Muthukumaran S, Sivakumar R and Mukherjee S K 2008 Condition monitoring of first mode of metal transfer in friction stir welding by image processing techniques *Int J Adv Manuf Technol* **36** 484–489

[12] Ranjan R, Khan A R, Parikh C, Jain R, Mahto R P, Pal S, Pal S K and Chakravarty D 2016 Classification and identification of surface defects in friction stir welding: An image processing approach *Journal of Manufacturing Processes* **22** 237–253

[13] Deng Y, Zuo D and Song B 2014 Analysis of the main factors affecting the surface morphology of FSJ joint *Advanced Materials Research* **1027** 183–186

[14] Zuo L, Zuo D, Zhu Y and Wang H 2018 Effect of process parameters on surface topography of friction stir welding *Int J Adv Manuf Technol*

[15] Bachmann A, Krutzlinger M and Zaeh M F 2018 Influence of the welding temperature and the welding speed on the mechanical properties of friction stir welds in EN AW-2219-T87 *IOP Conf. Ser.: Mater. Sci. Eng.* **373** 1–14

[16] Zaeh M F, Gebhard P, Roos E, Noveva R, Kraetschmer D and Kleih L 2010 Eignung von Anlagen für das Rührreibschweißen und Übertragbarkeit von Schweißparametern ('Suitability of equipment for friction stir welding and transferability of welding parameters') *Düsseldorf* vol. 267 ed Deutscher Verband für Schweißen und verwandte Verfahren e. V. DVS-Berichte (Düsseldorf: DVS Media GmbH 410–416

[17] Federation of European Producers of Abrasives (FEPA) **43-2:2017** Grains of fused aluminium oxide, silicon carbide and other abrasive materials for coated abrasives Microgrits P 240 to P 5000

[18] Petzow G 2015 *Metallographisches, keramographisches, plastographisches Ätzen* ('Metallographic, ceramographic, plastographic etching') (Stuttgart: Gebrüder Borntraeger)

[19] Stachowiak G and Batchelor A 2005 *Engineering Tribology* (Oxford: Butterworth-Heinemann)

[20] DIN Deutsches Institut für Normung e. V. DIN EN ISO **25178-2:2012** Geometrische Produktspezifikation (GPS) - Oberflächenbeschaffenheit: Flächenhaft – part 2: Begriffe und Oberflächen-Kenngrößen (Berlin: Beuth Verlag GmbH)

[21] DIN Deutsches Institut für Normung e. V. DIN EN ISO **25178-3:2012** Geometrische Produktspezifikation (GPS) - Oberflächenbeschaffenheit: Flächenhaft – part 3: Spezifikationsoperatoren (Berlin: Beuth Verlag GmbH)

**Institute of Physics**
Dirac House, Temple Back
Bristol BS1 6BE UK

ISSN: 1757-8981
ISBN 978-1-5108-8272-0